Methods in Plant Biochemistry

Volume 4
Lipids, Membranes and
Aspects of Photobiology

METHODS IN PLANT BIOCHEMISTRY

Series Editors

P. M. DEY
Department of Biochemistry, Royal Holloway and Bedford New College, UK

J. B. HARBORNE
Plant Science Laboratories, University of Reading, UK

1 Plant Phenolics: J. B. HARBORNE

2 Carbohydrates: P. M. DEY

3 Enzymes of Primary Metabolism: P. J. LEA

4 Lipids, Membranes and Aspects of Photobiology: J. L. HARWOOD and J. R. BOWYER

5 Amino Acids, Proteins and Nucleic Acids: L. J. ROGERS

Methods in Plant Biochemistry

Series editors
P. M. DEY and J. B. HARBORNE

Volume 4
Lipids, Membranes and Aspects of Photobiology

Edited by

J. L. HARWOOD

Department of Biochemistry
University of Wales, Cardiff, UK

and

J. R. BOWYER

Biochemistry Department
Royal Holloway and Bedford New College, University of London, UK

ACADEMIC PRESS
Harcourt Brace Jovanovich, Publishers
London San Diego New York
Boston Sydney Tokyo Toronto

ACADEMIC PRESS LIMITED
24–28 Oval Road
London NW1 7DX

US edition published by
ACADEMIC PRESS INC
San Diego, CA 92101

Copyright © 1990, by
ACADEMIC PRESS LIMITED

All Rights Reserved
No part of this book may be reproduced in any form, by photostat, microfilm or any other means, without written permission from the publishers

This book is printed on acid-free paper.

British Library Cataloguing in Publication Data is available

ISBN 0-12-461014-5

Filmset by Bath Typesetting Limited, Bath, Avon
Printed by Galliard (Printers) Ltd, Great Yarmouth, Norfolk

Contents

Contributors	vii
Preface to the Series	ix
Preface	xi
1 Fatty Acids—Structural Identification F. D. Gunstone	1
2 Triacylglycerol Biosynthesis A. K. Stobart and S. Stymne	19
3 Phospholipids T. S. Moore	47
4 Glycolipid Analyses and Synthesis in Plants R. Douce, J. Joyard, M. A. Block and A.-J. Dorne	71
5 Waxes, Cutin and Suberin T. J. Walton	105

6 Polyacetylenes and Related Compounds:
   Analytical and Chemical Methods
   J. Lam and L. Hansen
	159

7 Phytochrome and Other Receptors L. H. Pratt, H. Senger and P. Galland	185
8 Absorption Techniques in the Study of Photosynthesis P. Mathis	231
9 Chlorophyll Fluorescence Transients P. Horton and J. R. Bowyer	259

10 Structure and Dynamics of Plant Membranes 297
   P. J. Quinn and W. P. Williams

Index 341

Contributors

M. A. Block, Laboratoire de Physiologie Cellulaire Végétale, UA CNRS no. 576, Departement de Biologie Moléculaire et Structurale, Centre d'Etudes Nucléaires de Grenoble, et Université Joseph Fourier, 85X, F-38041, Grenoble-cedex, France.
J. R. Bowyer, Biochemistry Department, Royal Holloway and Bedford New College, University of London, Egham Hill, Egham, Surrey TW20 0EX, UK.
A.-J. Dorne, Laboratoire de Physiologie Cellulaire Végétale, UA CNRS no. 576, Departement de Biologie Moléculaire et Structurale, Centre d'Etudes Nucléaires de Grenoble, et Université Joseph Fourier, 85X, F-38041, Grenoble-cedex, France.
R. Douce, Laboratoire de Physiologie Cellulaire Végétale, UA CNRS no. 576, Departement de Biologie Moléculaire et Structurale, Centre d'Etudes Nucléaires de Grenoble, et Université Joseph Fourier, 85X, F-38041, Grenoble-cedex, France.
P. Galland, Fachbereich Biologie-Botanik, Philipps-Universität, 3550 Marburg, FRG.
F. D. Gunstone, Department of Chemistry, The Purdie Building, University of St Andrews, St. Andrews, Fife, Scotland KY16 9ST, UK.
L. Hansen, BST-Center, 7000 Fredericia, Denmark.
P. Horton, Robert Hill Institute, Department of Molecular Biology and Biotechnology, University of Sheffield, Western Bank, Sheffield S10 2TN, UK.
J. Joyard, Laboratoire de Physiologie Cellulaire Végétale, UA CNRS no. 576, Departement de Biologie Moléculaire et Structurale, Centre d'Etudes Nucléaires de Grenoble, et Université Joseph Fourier, 85X, F-38041, Grenoble-cedex, France.
J. Lam, University of Aarhus, 8000 Aarhus C, Denmark.
P. Mathis, Service de Biophysique, Département de Biologie, Centre d'Etudes Nucléaires de Saclay, 91191 Gif-sur-Yvette Cedex, France.
T. S. Moore, Jr, Department of Botany, Louisiana State University, Baton Rouge, LA 70803-1705, USA.
L. H. Pratt, Department of Botany, University of Georgia, Athens, GA 30602, USA.
P. J. Quinn, Department of Biochemistry, King's College London, Campden Hill, London W8 7AH, UK.
H. Senger, Fachbereich Biologie-Botanik, Philipps-Universität, 3550 Marburg, FRG.
A. K. Stobart, Department of Botany, The University, Bristol BS8 1UG, UK
S. Stymne, Department of Plant Physiology, Swedish University of Agricultural Sciences, S-750 07, Uppsala, Sweden.

T. J. Walton, Biochemistry Research Group, School of Biological Sciences, University College, Singleton Park, Swansea SA2 8PP, Wales, UK.

W. P. Williams, Department of Biochemistry, King's College London, Campden Hill, London W8 7AH, UK.

Preface to the Series

Scientific progress hinges on the continual discovery and extension of new laboratory methods and nowhere is this more evident than in the subject of biochemistry. The application in recent decades of novel techniques for fractionating cellular constituents, for isolating enzymes, for electrophoretically separating nucleic acids and proteins and for chromatographically identifying the intermediates and products of cellular metabolism has revolutionised our knowledge of the biochemical processes of life.

While there are many books and series of books on biochemical methods, volumes specifically catering for the plant biochemist have been few and far between. This is particularly unfortunate in that the isolation of DNA, enzymes or metabolites from plant tissues can often pose special problems not encountered by the animal biochemist. For a long time, the Springer series *Modern Methods in Plant Analysis*, which first appeared in the 1950s, provided the only comprehensive guide to experimental techniques for the investigation of plant metabolism and plant enzymology. This series, however, has never been completely updated; a second series has recently appeared but this is organised on a techniques basis and thus does not provide the comprehensive coverage of the first series. One of us (JBH) wrote a short guide to modern techniques of plant analysis *Phytochemical Methods* in 1976 (second edition, 1984) which showed the need for an expanded comprehensive treatment, but which by its very nature could only provide an outline of available methodology.

The time therefore seemed ripe to us to produce an entirely new multi-volume series on methods of plant biochemical analysis, which would be both thoroughly up-to-date and comprehensive. The success of *The Biochemistry of Plants*, edited by P. K. Stumpf and E. E. Conn and published by Academic Press, was an added stimulus to produce a complementary series on the methodology of the subject. With these thoughts in mind, we planned individual volumes covering: phenolics, carbohydrates, amino acids, proteins and nucleic acids, terpenoids, nitrogen and sulphur compounds, lipids, membranes and light receptors, enzymes of primary and secondary metabolism, plant molecular biology and biological techniques in plant biochemistry. Thus we have tried to cover all the major areas of current endeavour in phytochemistry and plant biochemistry.

The main aim of the series is to introduce to the scientist current knowledge of techniques in various fields of biochemically-related topics in plant research. It is also intended to present the historical background to each topic, to give experimental details of methods and analyses and appraisal of them, pointing out those methods that are most suitable for immediate application. Wherever possible illustrations and structures have been used and one or more case treatments presented. The compilation of known data and properties, where appropriate, is included in many chapters. In addition, the reader is directed to relevant references for further details. However, for the sake of clarity and completeness of individual reviews, some overlap between chapters of volumes has been allowed.

Finally, we extend our warmest thanks to our volume editors for undertaking the important task of organising each volume and cooperating in preparing the contents lists. Our special thanks go to the staff of Academic Press and to the many colleagues who have made this project a success.

P. M. DEY
J. B. HARBORNE

Preface to Volume 4

In recent years there has been an explosion of interest in the structure and function of membranes. So far as plants are concerned this interest has shown particular emphasis in the area of photosynthesis. Because lipids are key components of membranes we have chosen to group chapters in this volume so that they begin with discussions about individual types of lipids and then progress on to chapters concerned with membrane function and the different physical techniques used in the examination of particular processes.

Of course, although each chapter highlights the specific experimental methods very well, they cannot hope to cover every aspect of lipid, membrane and photosynthetic biochemistry. Indeed, one chapter on electron paramagnetic resonance was not submitted and, therefore, this area is not covered although it was originally intended to do so.

We would like to thank the individual authors very much indeed for the professional way in which they complied with our original instructions and were also, clearly, at pains to take great care in making their chapters both readable and very informative. Some authors had to wait quite a time after submitting their chapters before the final volume was produced and for their forbearance we are grateful. In particular, also, we would like to thank Terry Walton for producing his chapter at very short notice after the original author had failed to deliver!

To produce a volume such as this, a great deal of work has been involved. For their original idea of the series and for their encouragement throughout, we are indebted to Prakash Dey and Jeffrey Harborne. On the publishing side, several people have made valuable contributions but, in particular, we should mention Andrew Richford and Sarah Robertson who have dealt with the volume from its infancy to its final mature state. Ian Ashton took on the arduous job of compiling an index with enthusiasm and skill. Finally, we have to thank our families who yet again coped admirably with two disgruntled and weary editors and saw even less than usual of us during this project!

JOHN HARWOOD
JOHN BOWYER

1 Fatty Acids—Structural Identification

F. D. GUNSTONE

Department of Chemistry, University of St. Andrews. St Andrews, Fife, Scotland KY16 9ST, UK

I.	Introduction	1
II.	Separation procedures	3
	A. Thin layer chromatography	3
	B. High performance liquid chromatography	3
	C. Gas chromatography	4
III.	Identification procedures	6
	A. Ultraviolet spectroscopy	6
	B. Infrared and Raman spectroscopy	7
	C. ^{1}H NMR spectroscopy	7
	D. ^{13}C NMR spectroscopy	8
	E. Mass spectrometry	10
	F. Chemical procedures	15
IV.	References	16

I. INTRODUCTION

Most lipid chemists and biochemists only deal with about 20 fatty acids during their whole research career. These compounds are generally recognised by comparison with authentic samples, usually in terms of their gas chromatographic behaviour on one or more columns. However, it is pertinent to note that the number of natural fatty acids exceeds 1000 and that this number continues to grow with the study of less common lipid sources such as marine lipids and with the investigation of fatty acid metabolites.

Most natural acids have structures in accord with the following generalisations, to all of which, however, there are exceptions

(a) Natural fatty acids, whether saturated or unsaturated, are usually straight-chain compounds with an even number of carbon atoms in each molecule. Despite this, branched-chain, cyclic systems, and acids with an odd number of carbon atoms per molecule are known.

(b) Most fatty acids contain 16, 18, 20 or 22 carbon atoms per molecule but structures ranging from two to >80 carbon atoms per molecule also occur.

(c) Unsaturated acids may contain up to six double bonds. These are usually of *cis* (*Z*) configuration and, in the most important examples, double bonds are separated from each other by a single methylene function. These points are illustrated in the structure of arachidonic acid (**1**). In addition, however, *trans* (*E*) olefinic, allenic, and acetylenic patterns of unsaturation are known. Also polyene acids may show conjugated unsaturation or have their double bonds separated by more than one methylene group.

arachidonic acid (20 : 4 5*Z*, 8*Z*, 11*Z*, 14*Z*)

(**1**)

(d) Unsaturation usually occurs in certain preferred positions along the carbon chain. This is sometimes correlated with the carboxyl group (e.g. arachidonic acid with $\Delta 5,8,11,14$, unsaturation) or with the end methyl group (e.g. arachidonic acid with *n*-6 tetraene unsaturation).

(e) Functional groups other than the carboxylic acid and the unsaturated centre(s) are less common but hydroxy, epoxy, methoxy, oxo, and halogenated acids are known.

To define the structure of a fatty acid it is necessary to know the following:

(a) the chain length and whether the molecule contains any branched or cyclic systems or any less common functional groups;
(b) the number, nature, configuration, and position of all unsaturated centres.

If the acid is already known then its structure can usually be defined by comparison with an authentic sample or with compounds of closely related structure. This comparison will usually be made on the basis of chromatographic behaviour but may include some spectroscopic comparisons. If the acid has a completely novel structure then a more intensive investigation will be needed with spectroscopy providing most, if not all, of the necessary information.

Evidence of unusual structure may appear in unexpected behaviour during thin layer chromatography, gas–liquid chromatography, or systematic spectroscopic examination. When this is suspected it is desirable to isolate or concentrate the unusual acid so that it can be identified in an appropriate way. Separation procedures—generally chromatographic—and analytical procedures—usually spectroscopic—are now frequently combined in a single operation of which gas chromatography–mass spectrometry (GC-MS) is the most widely practised.

II. SEPARATION PROCEDURES

Only those separation processes which are used in conjunction with identification procedures are considered here.

The classical separation procedures of distillation and crystallisation are only of limited value to lipid chemists. With polyunsaturated acids it is desirable to avoid the high temperatures normally associated with fractional distillation since these may promote stereo-mutation, double bond migration, cyclisation and dimerisation. Distillation will allow separation by chain length but hardly any through differences confined to the number of double bonds. Despite this, Ackmann (1988) has described the distillation of fish fatty acids using a Pope wiped-wall molecular still claiming that the short passage time of about one minute limits thermal abuse.

Many of the fatty acids to be separated differ so little in solubility that crystallisation—a single unit separation—does not effect a useful fractionation process.

Most useful separations are now effected by chromatographic procedure.

A. Thin Layer Chromatography

Thin layer chromatography is still a useful isolation or concentration process. Simple adsorption on silica will not usefully separate acids or esters differing only in chain length and/or degree of unsaturation. Nevertheless there is some subfractionation according to these properties. Long-chain compounds are slightly less polar than their medium-chain analogues and polyunsaturated compounds are slightly more polar than their more saturated homologues. Acids or esters containing additional polar functional groups such as hydroxy or epoxy groups can be easily separated from acids or esters without these extra groups and this provides a valuable way of isolating or separating such compounds.

Other separations are possible if the normal silica layer is modified in some way. The best known example of this is silver ion chromatography. When silver nitrate is incorporated into a silica layer (5–20%) then, because of interaction between silver ions and double bonds, *cis* and *trans* esters can be separated from each other, as can esters differing in the number of unsaturated centres they contain. If a sample contains only esters with *cis* unsaturation then it is possible to obtain concentrates of monoenes, dienes, trienes, etc.

Peers and Coxon (1986) have described a partition process for separating esters with three or more double bonds from those with less unsaturation. The esters (up to 1 g) are partitioned between 2,2,4-trimethylpentane (10 ml) and an equal volume of a 25% solution of silver nitrate in 1:1 aqueous ethanol. This is a useful way of concentrating polyene esters originally present at low levels.

Borates and arsenites have also been incorporated into silica layers to assist the separation of polyhydroxy esters differing in the number of hydroxyl groups and also in their stereochemistry. This is illustrated in a recent account of work by Hamberg (1987) in which he separated diastereoisomeric methyl 9,12,13(S)-trihydroxyoctadec-10-enoates.

B. High Performance Liquid Chromatography

Separations effected by thin layer chromatography can usually be achieved more efficiently in columns using high performance liquid chromatography (HPLC) systems.

Christie (1987) has described a silver ion HPLC system which he has used in conjunction with his structural studies by GC-MS of appropriate derivatives (Section III.E). Bascetta *et al.* (1984) used reversed phase HPLC systems with methanol as eluting solvent to isolate pure esters from appropriate sources such as methyl oleate from olive oil, methyl linoleate from maize or sunflower seed oil, methyl α-linolenate from linseed oil, and methyl γ-linolenate from evening primrose oil. These separations furnished material of 98–99% purity on a gram scale in a few minutes.

C. Gas Chromatography

Gas chromatography (GC) is the technique most commonly employed to separate methyl esters of fatty acids using either packed or capillary columns. The latter give higher resolution than the former. Though used for quantitative analysis of ester mixtures for the most part, this procedure can be used in a preparative mode or combined with mass spectrometry for structural identification of the component in each GC peak.

TABLE 1.1. Equivalent chain lengths and relative retention times of some unsaturated esters on packed columns with EGSS-X and EGSS-Y as stationary phases. Reproduced from Christie (1982), with permission.

Methyl ester	ECLs		Relative retention times[a]	
	EGSS-X	EGSS-Y	EGSS-X	EGSS-Y
16:0	16.00	16.00	0.58	0.57
16:1 (*n*-9)	16.57	16.62	0.69	0.68
16:2 (*n*-6)	17.65	17.45	0.90	0.85
18:0	18.00	18.00	1.00	1.00
18:1 (*n*-9)	18.53	18.52	1.19	1.16
18:2 (*n*-6)	19.42	19.20	1.52	1.41
18:3 (*n*-6)	20.00	19.67	1.80	1.60
18:3 (*n*-3)	20.40	20.02	2.02	1.78
18:4 (*n*-3)	21.05	20.52	2.40	2.04
20:0	20.00	20.00	1.77	1.76
20:1 (*n*-9)	20.50	20.45	2.08	2.01
20:2 (*n*-6)	21.40	21.15	2.67	2.44
20:3 (*n*-9)	21.63	21.33	2.87	2.57
20:3 (*n*-6)	21.77	21.53	2.99	2.72
20:3 (*n*-3)	21.95	21.60	3.13	2.78
20:4 (*n*-6)	22.43	22.00	3.59	3.10
20:4 (*n*-3)	23.00	22.47	4.18	3.54
20:5 (*n*-3)	23.50	22.80	4.85	3.91
22:0	22.00	22.00	3.14	3.12
22:1 (*n*-9)	22.43	22.35	3.59	3.44
22:4 (*n*-6)	24.45	24.00	6.34	5.48
22:5 (*n*-6)	24.57	23.85	6.55	5.27
22:5 (*n*-3)	25.53	24.80	8.60	6.92
22:6 (*n*-3)	26.18	25.20	9.10	7.74
24:0	24.00	24.00	5.59	5.50

[a] Relative to 18:0.
Data obtained with 7 ft × ¼ inch o.d. glass columns packed with 15% (w/w) stationary phase on Chromosorb W (100–120 mesh, acid-washed and silanised). Carrier gas—nitrogen at 50 ml min^{-1}. Column temperatures, 194°C (EGSS-Y) and 178°C (EGSS-X).

TABLE 1.2. ECL of a range of polyene esters related to linolenate. Reproduced from Gunstone (1983), with permission (adapted from Jamieson, 1975).

ω-6					ω-3					
18:3	**20:3**	**20:4**	**22:4**	**22:5**	**18:3**	**18:4**	**20:4**	**20:5**	**22:5**	**22:6**
19.17	21.03	21.27	23.23	23.47	19.50	19.88	21.70	21.99	23.93	24.18
19.27	21.12	21.40	23.35	23.62	19.60	20.01	21.84	22.14	24.08	24.36
19.36	21.21	21.53	23.48	23.77	19.70	20.14	21.98	22.29	24.23	24.54
19.45	21.29	21.66	23.61	23.92	19.80	20.27	22.11	22.45	24.38	24.72
19.54	21.38	21.79	23.73	24.07	19.90	20.40	22.25	22.60	24.54	24.91
19.63	21.47	21.91	23.86	24.22	20.00	20.53	22.39	22.75	24.69	25.09
19.73	21.55	22.04	23.99	24.37	20.10	20.66	22.53	22.91	24.84	25.27
19.82	21.64	22.17	24.11	24.52	20.20	20.79	22.66	23.06	25.00	25.45
19.91	21.73	22.30	24.24	24.67	20.30	20.92	22.80	23.21	25.15	25.64
20.00	21.81	22.43	24.37	24.82	20.40	21.05	22.94	23.37	25.30	25.82
20.09	21.90	22.56	24.49	24.97	20.50	21.18	23.08	23.52	25.45	26.00
20.18	21.99	22.69	24.62	25.12	20.60	21.31	23.21	23.67	25.61	26.18
20.28	22.07	22.82	24.75	25.27	20.70	21.44	23.35	23.83	25.76	26.36
20.37	22.16	22.95	24.87	25.42	20.80	21.57	23.49	23.98	25.91	26.55
20.46	22.25	23.07	25.00	25.57	20.90	21.70	23.63	24.13	26.06	26.73
20.55	22.33	23.20	25.13	25.72	21.00	21.83	23.76	24.29	26.22	26.91
18:3³	**16:2⁶**	**16:2⁴**	**16:3⁶**	**16:3⁴**	**16:3³**	**16:4³**	**16:4¹**	**20:2⁹**	**20:2⁶**	**20:3⁹**	**20:3³**
19.50	16.80	17.02	17.20	17.39	17.53	17.81	18.02	20.50	20.73	20.65	21.40
19.60	16.87	17.08	17.28	17.50	17.64	17.96	18.17	20.57	20.80	20.77	21.50
19.70	16.94	17.15	17.36	17.61	17.75	18.10	18.31	20.64	20.87	20.88	21.60
19.80	17.01	17.22	17.45	17.71	17.86	18.25	18.45	20.71	20.94	20.99	21.70
19.90	17.08	17.28	17.53	17.82	17.97	18.39	18.60	20.79	21.00	21.10	21.80
20.00	17.15	17.35	17.61	17.92	18.07	18.54	18.74	20.86	21.07	21.21	21.90
20.10	17.22	17.42	17.70	18.03	18.18	18.68	18.88	20.93	21.14	21.33	22.00
20.20	17.29	17.48	17.78	18.13	18.29	18.83	19.03	21.00	21.20	21.44	22.10
20.30	17.37	17.55	17.87	18.24	18.40	18.97	19.17	21.08	21.27	21.55	22.20
20.40	17.44	17.62	17.95	18.34	18.51	19.12	19.31	21.15	21.34	21.66	22.30
20.50	17.51	17.68	18.03	18.45	18.62	19.26	19.45	21.22	21.41	21.77	22.40
20.60	17.58	17.75	18.12	18.55	18.73	19.41	19.59	21.30	21.47	21.89	22.50
20.70	17.65	17.82	18.20	18.66	18.84	19.55	19.74	21.37	21.54	22.00	22.60
20.80	17.72	17.88	18.28	18.76	18.95	19.70	19.88	21.44	21.61	22.11	22.70
20.90	17.79	17.95	18.37	18.87	19.06	19.84	20.02	21.51	21.68	22.22	22.80
21.00	17.86	18.02	18.45	18.98	19.16	19.99	20.17	21.59	21.74	22.34	22.90

The ECL of polyene esters vary with the polarity of the stationary phase. This table gives values for the common polyene esters related to those observed for methyl α-linolenate in the range 19.5–21.0. For a column on which methyl α-linolenate has an ECL of 20.0, for example, the ECL of other polyene esters will be close to the values quoted in the horizontal line which includes 20.00 for 18:3 (ω-3).

By a careful study of the elution behaviour of fatty acid methyl esters it is possible to learn a lot about the structure of the ester. In favourable circumstances (i.e. when there are not too many unusual structural features) it may be possible to indicate the chain length, degree of unsaturation, and (in part) the position of each unsaturated centre. For this purpose retention behaviour is expressed in relation to that of one component in the mixture (e.g. stearate) or in terms of the equivalent chain length (ECL) (Miwa *et* al., 1960). This is the notional number of carbon atoms in a saturated fatty acid whose methyl ester would be co-eluted with the ester in question. Significant collections of such information have been made by Jamieson (1970) and by Ackman (1984) and some examples are given in Tables 1.1 and 1.2. Such figures show, for example, that among 18:3 isomers on a polar column the order of elution is given by Δ5,9,12 before Δ6,9,12 before Δ9,12,15.

III. IDENTIFICATION PROCEDURES

Reference has been made in Section II.C to the use of GLC retention behaviour in the identification of fatty acid structure. Such a procedure may verify the structure of a known ester or may provide useful pointers toward structure: it is unlikely to permit the full identification of a very rare or unknown compound. For this purpose one or more of the spectroscopic procedures must be employed.

A. Ultraviolet Spectroscopy

Ultraviolet spectroscopy provides little information about saturated and monoene acids and for those polyene acids in which unsaturated centres are separated from each other by one or more methylene groups. There are some acids—mainly of plant origin—with 9,11,13 or 8,10,12 patterns of triene configuration (Hopkins, 1972; Gunstone, 1986) but these are not very common. Parinaric acid with a conjugated tetraene system (18:4 9,11,13,15) has attracted some interest recently as a label in membrane lipids. Many of the metabolites of arachidonic and other C_{20} polyene acids contain conjugated systems as part of their total unsaturation (e.g. leukotriene A_4, **2**).

(2)

Some examples of absorption maxima are collected together in Table 1.3.

TABLE 1.3. Absorption maxima (nm) of some ultraviolet chromophores present in natural fatty acids. Reproduced from Galliard and Mercer (1975), with permission.

Chromophore	Fatty acid[1]	$\lambda_{max}(\log \varepsilon_{max})^2$
Diene	18:2(9t,11t)[a]	227 (infl), 231 (4·52), 239 (infl)
Enyne	18:2(9a,11t)[a]	229 (4·22), 240 (infl)
Diyne	16:2(8a,10a)[b]	215 (2·55), 226 (2·62), 239 (2·62), 254 (2·38)
Dienoic acid	10:2(2t,4c)[a]	263 (4·58)
Triene	18:3(9c,11t,13t)[a]	261 (4·56), 271 (4·67), 281 (4·58)
Dienyne	18:3(9a,11t,13t)[e]	267 (4·59), 277 (4·48)
Enediyne	18:3(9a,11a,13c)[c]	214 (?), 229 (3·58), 240 (3·84), 253 (4·19), 268 (4·35), 284 (4·26)
Tetraene	18:4(9t,11t,13t,15t)[d]	286 (4·77), 299 (4·97), 313 (4·94)
Enetriyne	18:4(9a,11a,13a,15e)[e]	212, 223, 232, 244, 258, 274, 291

[1] Solvent: a ethanol, b cyclohexane, c hexane, d methanol, e unknown.
[2] infl = inflexion.

Lopez and Gerwick (1987) recently identified two novel icosapentaenoic acids in red seaweed. The two conjugated triene systems present in the acids (**3** and **4**) were identified from their ultraviolet spectra.

(**3**) (20:5 5Z, 7E, 9E, 14Z, 17Z)

(**4**) (20:5 5E, 7E, 9E, 11Z, 14Z)

B. Infrared and Raman Spectroscopy

Infrared spectroscopy assists in the recognition of some of the less common functional groups present in fatty acids. These include hydroxyl (absorption at 3450 cm^{-1}), oxo (1725 cm^{-1}), cyclopropene (1850 and 1010 cm^{-1}), epoxide (850 and 825 cm^{-1}), allene (2220 and 1960 cm^{-1}), vinyl (990 and 910 cm^{-1}), and conjugated enyne (950 cm^{-1}).

The most common use of infrared spectra in this field, however, is the recognition of *trans* (E) unsaturation. Compounds with one *trans* double bond have characteristic absorption at 968 cm^{-1}. If there is more than one *trans* double bond and these are not conjugated then there is an enhancement in the intensity of absorption but no change in the position. Conjugated polyene systems with one or more *trans* double bonds give more complex absorption patterns which may be of value in identification. Some useful figures are given in Table 1.4.

Infrared spectroscopy is of little help in detecting unsaturation which is *cis* olefinic (Z) or acetylenic. But such compounds have characteristic absorption bands in their Raman spectra at 1656 ± 1 cm^{-1} for *cis* alkenes, 1670 ± 1 cm^{-1} for *trans* alkenes, and 2232 ± 2 cm^{-1} for alkynoic acids (Davies *et al.*, 1972).

TABLE 1.4. Infrared absorption maxima in compounds containing *trans* unsaturation in conjugated polyenoic systems. Reproduced from Galliard and Mercer (1975), with permission.

Chromophore	Configuration	Absorption maxima (cm^{-1})
Monoene	*t*	968
Diene	*tt*	988
	tc	985, 950
Triene	*ttt*	994
	ttc	993, 965
	ctc	988, 937
Tetraene	*tttt*	997
	cttc	993, 952

C. ^{1}H NMR Spectroscopy

The ^{1}H NMR spectrum of a fatty acid or ester shows characteristic signals for the end methyl group, methylene groups not influenced by any other functional group, methylene groups α and β to the carboxylic acid or ester unit, allylic and doubly allylic methylene groups, and olefinic centres. As an example methyl linoleate with thirty four hydrogen atoms shows the following eight signals:

$$CH_3(CH_2)_3CH_2CH=CHCH_2CH=CHCH_2(CH_2)_4CH_2CH_2COOCH_3$$
$$abdhhfhhdbceg$$

δ_H(300 MHz spectrum p.p.m.): (a) 0.89, (b) 1.31, (c) 1.62 (d) 2.05, (e) 2.30, (f) 2.77, (g) 3.67, and (h) 5.35.

This spectrum indicates the presence of a pentadiene unit in the mid-chain area. If the diene (or other methylene-interrupted polyene unit) gets closer to the carboxyl or methyl end then it may be possible to locate it. Thus methyl α-linolenate (like other n-3 acids) has a different and characteristic δ_H value for the end methyl group.

$$CH_3CH_2CH=CHCH_2CH=CHCH_2CH=CHCH_2(CH_2)_4CH_2CH_2COOCH_3$$
$$adhhfhhfhhdbceg$$

δ_H(300 MHz spectrum p.p.m.): (a) 0.97, (b) 1.30, (c) 1.60, (d) 2.05, (e) 2.28, (f) 2.78, (g) 3.65, and (h) 5.35.

Some characteristic signals for n-3 and n-6 (and other families) acids/esters are clearly apparent even in mixtures and natural oils. These signals and the integrals associated with them give much structural identification about individual compounds. For example, it is easy to distinguish between the Δ5,9,12, Δ6,9,12 and Δ9,12,15 C_{18} trienes merely on the basis of their ^{1}H spectra.

The ^{1}H NMR spectra of natural oils will also indicate the presence of hydrogen atoms in the glycerol unit and will distinguish between the four α and the single β hydrogen atom. The latter overlaps with olefinic hydrogen atoms and may complicate the interpretation of the ^{1}H NMR spectra.

Some useful data are collated in Table 1.5 and more information is given by Frost and Gunstone (1975).

D. ^{13}C NMR Spectroscopy

^{13}C NMR spectra of long-chain acids/esters usually provide more signals for individual carbon atoms than are found in ^{1}H spectra. Thus methyl linoleate with 8 ^{1}H signals will have at least 18 ^{13}C signals. Some useful figures are given in Table 1.5 and valuable data have been given by several authors (see Table 1.8).

The methyl ester of a saturated acid shows at least eight signals as in the following figures for methyl palmitate.

$$\underset{14.1422.8432.11*25.1234.18}{CH_3CH_2CH_2(CH_2)_{10}CH_2CH_2C}\overset{\overset{O}{\parallel}}{\underset{\underset{51.27}{OCH_3}}{}}{}^{174.08}$$

* 29.37–29.85—some resolution of this complex signal is sometimes possible (see references cited in Table 1.8).

TABLE 1.5. ^{1}H and ^{13}C chemical shift values for fatty acids and lipids. Adapted with permission from Pollard (1986, pp. 410–411).

Structural unit	Chemical shift (ppm) ^{1}H	Chemical shift (ppm) ^{13}C
CH$_3$	0.89	14.1
CH$_2$ ω-1		22.6
CH$_2$ ω-2		32.1
CH$_2$ ω-3		29.5
CH$_2$ (bulk methylenes)	1.25	29.8
Allylic CH$_2$ (*cis*)	2.0	27.3
Allylic CH$_2$ (*trans*)	1.95	32.7
Diallylic CH$_2$ (*cis,cis*)	2.72	25.6
CH$_2$ (α COOH)	2.29	34.3
CH$_2$ (β COOH)	1.6	24.8
CH$_2$ (α COOCH$_3$)	2.2	34.2
CH$_2$ (β COOCH$_3$)	1.6	25.1
Glycerides α CH$_2$	4.12, 4.28	62.1
Glycerides β CH	5.25	68.9
Olefinic (*cis*)	5.28	129.9
Olefinic (*trans*)	5.32	130.4
COOH	>10	180.6
COOCH$_3$	—, 3.65	174.1, 51.5

TABLE 1.6. ^{13}C chemical shift values for methyl linoleate and methyl linolenate.

Linoleate			α-Linolenate		
Carbon atom	δ$_C$	Intensity	Carbon atom	δ$_C$	Intensity
1	174.03	3.2	1	174.16	3.2
13	130.16	11.3	16	131.92	10.7
9	130.00	11.1	9	130.24	15.3
10	128.15	12.0	12, 13	128.29	18.4
12	128.01	11.2	10	127.80	16.1
			15	127.18	11.1
MeO	51.31	8.2	MeO	51.36	10.0
2	34.11	11.7	2	34.11	15.5
16	31.65	11.3			
7 (?)	29.71	13.0	7 (?)	29.63	17.5
15 (?)	29.47	11.2	4–6	29.21	21.2
4–6	29.28	15.6		29.18	24.6
	29.24	16.8			
	29.21	16.3			
8, 14	27.29	17.3	8	27.25	16.9
11	25.73	12.2	11	25.68	15.9
			14	25.58	14.7
3	25.04	13.6	3	24.99	16.9
17	22.68	10.5	17	20.60	9.7
18	14.10	9.0	18	14.29	9.4

With unsaturated esters there are additional peaks for all the olefinic carbon atoms and for allylic and doubly allylic carbon atoms. These have different values depending on whether the double bond is E or Z. Thus methyl linoleate with 19 carbon atoms has 18 distinct signals in its ^{13}C NMR spectrum and methyl α-linolenate has 17 signals (Table 1.6).

Most monoene esters show separate signals for the two olefinic carbon atoms. The difference between these two values changes with the distance of the double bond from the acyl group and permits assignment of the double bond position in monoenes up to about Δ10 (Table 1.7).

TABLE 1.7. Chemical shift induced in *cis* olefinic carbon atoms (129.90 ppm when undisturbed) in monoenoic acids and esters by COOH, COOCH$_3$ and CH$_3$ groups.

	Acids[a]			Esters[a]		
	n	$n+1$	Diff.	n	$n+1$	Diff.
Δ2	—	—	—	−10.49	+20.79	31.3
3	−9.85	+4.27	14.1	−8.90	+3.50	12.4
4	−2.84	+2.05	4.9	−2.37	+1.64	4.0
5	−1.70	+1.53	3.2	−1.46	+1.31	2.8
6	−0.88	+0.72	1.6	−0.72	+0.58	1.3
7	−0.44	+0.41	0.9	−0.41	+0.32	0.7
8	−0.25	+0.25	0.5	−0.22	+0.23	0.5
9	−0.11	+0.15	0.3	−0.12	+0.12	0.2
10	−0.07	+0.10	0.2	—	—	—
11	−0.01	+0.06	0.1	0	0	0
	m	$m+1$				
ω-4	+0.23	−0.25	0.5	—	—	—
ω-3	−0.53	+1.63	2.2	—	—	—
ω-2[b]	+1.01	−6.33	7.3	—	—	—

[a] Shift refers to carbon atoms n and $n+1$ for the Δn series and $m+1$ and m for the ω-m series.
[b] ω-1 values (acid) are 139.12 and 114.15 for ω-1 and ω-2 carbon atoms, respectively.

The Table is to be read as follows. In methyl oleate the two olefinic carbon atoms have δ_C of 129.79 and 130.05 for C(9) and C(10), respectively. These values differ from 129.90 by −0.11 and +0.15 and from one another by 0.3. This final value is characteristic of a double bond in the Δ9 position.

Relating δ_C values with a particular carbon atom has been achieved in a number of ways. These involve the use of model compounds, empirical relationships and literature tabulations, the use of specifically labelled compounds, of shift reagents, and of a variety of special NMR experiments including ^{1}H decoupling, off-resonance experiments, spectra editing techniques, and two-dimensional NMR spectroscopy. Some of these are described at greater length by Pollard (1986). Williams and Fleming (1987) describe methods of predicting δ_C values based on substituent constants, a 'steric' correction, and a conformational increment for γ-substituents. These are useful in interpreting ^{13}C NMR spectra. Other useful sources of information are listed in Table 1.8.

E. Mass Spectrometry

Mass spectrometry is perhaps the most important method of determining fatty acid structure. Since it is best applied to individual compounds rather than to mixtures it has to be preceded by or combined with appropriate separation procedures. Of these GC-MS is the most widely employed, though mass spectrometry can also be linked to liquid chromatography.

TABLE 1.8. References to ^{13}C NMR spectroscopic data.

Tallent et al. (1966)	Conjugated polyenes
Suzuki et al. (1970)	Conjugated polyenes
Gunstone et al. (1976)	Saturated and acetylenic
Bus et al. (1976, 1977)	Olefinic and acetylenic
Tulloch and Mazurek (1976), Tulloch (1977)	Saturated, unsaturated and oxygenated
Gunstone et al. (1977)	Olefinic
Miller et al. (1977)	Acetylenic
Khan and Scheinmann (1978)	General
Tulloch and Bergter (1979), Tulloch (1982)	Conjugated polyenes
Rakoff et al. (1979)	Hydroxy
Bergter and Seidl (1984)	Conjugated polyenes
Gaydou et al. (1987)	Conjugated polyenes

Recognition of a molecular ion is a useful start for structural identification. In most electron impact systems the molecular ions for long-chain acids/esters are either very small or cannot be detected. The problem can sometimes be overcome by carrying out electron impact MS at 20 eV rather than at the usual level of 70 eV. However, a more certain way of solving this problem is to use chemical ionisation mass spectrometry (CIMS) with butane or ammonia (Games, 1978).

The identification of oxygen-containing functional groups (hydroxy, oxo, epoxy, methoxy) in a long-chain acid and their position in the chain is often a tedious task by classical degradation procedures, though the recognition of the degradation products is now more easily achieved by gas chromatography. These oxygenated acids/esters can usually be identified directly mass spectrometrically by their fragment ions, though better results are sometimes obtained after appropriate derivatisation. In the following formulations symbols such as ⌊ and ⌋ are to be interpreted as cleavage as shown by the vertical line with the charged fragment being that to the right or left, respectively.

$$R\lfloor CH(OH)\rfloor R' \qquad R\lfloor CH(OSiMe_3)\rfloor R'$$
$$(5) \qquad\qquad\qquad (6)$$

$$R\left[C\overset{O}{\underset{}{\diagdown\diagup}}CH\right]R' \qquad \begin{array}{c}RCH(OH)\;\lceil\;CH(OMe)R'\\ RCH(OMe)\;\lfloor\;CH(OH)R'\end{array}$$
$$(7) \qquad\qquad\qquad (8)$$

$$\begin{array}{c}RCH(OSiMe_3)\;\lceil\;CH(OMe)R'\\ RCH(OMe)\;\lfloor\;CH(OSiMe_3)R'\end{array}$$
$$(9)$$

$$R = CH_3(CH_2)_m \qquad R' = (CH_2)_nCOOCH_3$$

Hydroxy esters give two fragments as shown (5) though the results are often clearer if the hydroxyl group is first converted to its trimethylsilyl ester (6). Epoxy esters (7) can be identified directly though their fragmentation can be complex (Gunstone and Jacobsberg, 1972; Gunstone *et al.*, 1975). The problem is simplified if the epoxide is treated with methanolic boron trifluoride to give (**8**) and silylated thereafter (**9**). The epoxide (**7**) gives two products as (**8**) or as (**9**), but this does not greatly complicate the interpretation of the mass spectrum.

Branched methyl groups may also be identified and positioned through fragmentation on either side of the tertiary methyl group (**10**). Cyclopropane units (like the double bonds to be discussed later) are unstable under electron bombardment in such a way that the cyclic unit cannot be located. They have therefore to be 'fixed' by a chemical reaction leading to products which can more easily be identified by gas chromatography. Hydrogenolysis, oxidation, and ring cleavage with methanolic boron trifluoride have been recommended.

$$R[CH(Me)]R'$$

(**10**)

$$\underset{\text{RCOCHCHCH}_2\text{R}'}{\overset{\text{CH}_2}{\triangle}} + \underset{\text{RCH}_2\text{CHCHCOR}'}{\overset{\text{CH}_2}{\triangle}}$$

$\uparrow$ CrO$_3$

$$\underset{\text{RCH}_2\text{CHCHCH}_2\text{R}'}{\overset{\text{CH}_2}{\triangle}} \xrightarrow{H_2} \underset{\text{RCH}_2\overset{|}{\text{CH}}(\text{CH}_2)_2\text{R}'}{\overset{\text{CH}_3}{|}} + \underset{\text{R}(\text{CH}_2)_2\overset{|}{\text{CH}}\text{CH}_2\text{R}'}{\overset{\text{CH}_3}{|}}$$

$$+ \text{R}(\text{CH}_2)_5\text{R}'$$

$\downarrow$ MeOH, BF$_3$

$$\underset{\text{RCH}_2\overset{|}{\text{CH}}(\text{CH}_2)_3\text{R}'}{\overset{\text{OCH}_3}{|}} \quad \underset{\text{R}(\text{CH}_2)_2\overset{|}{\text{CH}}\text{CH}_2\text{R}'}{\overset{\text{CH}_2\text{OCH}_3}{|}} \quad \underset{\text{RCH}_2\overset{|}{\text{CH}}(\text{CH}_2)_2\text{R}'}{\overset{\text{CH}_2\text{OCH}_3}{|}}$$

$$\underset{\text{R}(\text{CH}_2)_3\overset{|}{\text{CH}}\text{CH}_2\text{R}'}{\overset{\text{OCH}_3}{|}} + \underset{\underset{\text{CH}_3}{|}}{\overset{\text{OCH}_3}{\underset{|}{\text{RCH}_2\text{CHCHCH}_2\text{R}'}}} + \underset{\underset{\text{OCH}_3}{|}}{\overset{\text{CH}_3}{\underset{|}{\text{RCH}_2\text{CHCHCH}_2\text{R}'}}}$$

However, the structural problems to be solved most often involve the position of one or more unsaturated centres in a carbon chain. Double bonds are mobile during electron bombardment so that many isomeric species are formed prior to fragmentation. Thus the mass spectra of methyl oleate ($\Delta 9$) and its isomer methyl petroselinate ($\Delta 6$) are virtually identical and neither allows the position of the double bond to be determined. The solution to this problem has been along two lines. Either the double bonds are 'fixed' in one of several possible ways, or the mass spectrum is obtained not with the acid or its methyl ester but on some amide or other ester in which the double bond is less labile. Most of the procedures which have been described work well with monoene compounds but only a few are satisfactory with polyenes having two to six double bonds. This topic has been reviewed by Minnikin (1978) and by Schmitz and Klein (1986).

A range of the more important procedures which have been applied to monoenes (and occasionally to polyenes) are set out in Table 1.9. Most of these proceed through the epoxide or the corresponding diol.

TABLE 1.9. Derivatives used for mass spectrometric investigation of olefinic esters.

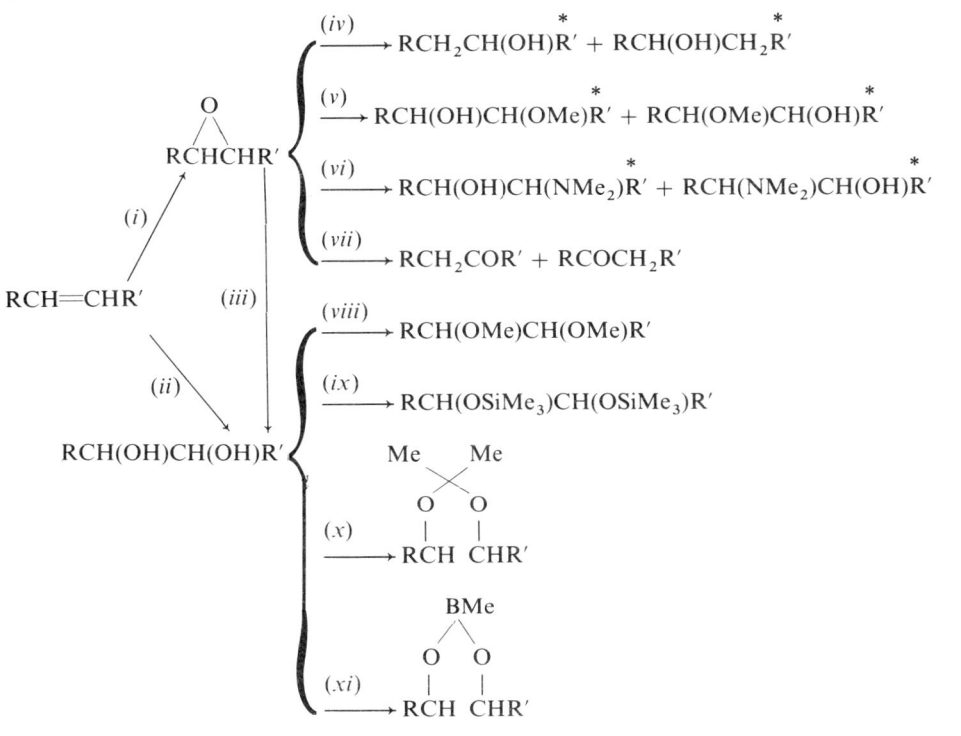

Reagents: (*i*), peracid; (*ii*), OsO$_4$. (*iii*), H$_3$O$^+$; (*iv*), LiAlH$_4$; (*v*), BF$_3$, MeOH; (*vi*), Me$_2$NH; (*vii*), NaI; (*viii*), MeI; (*ix*), (Me$_3$Si)$_2$NH; (*x*), COMe$_2$, H$^+$; (*xi*), MeB(OH)$_2$.
* These compounds containing a free hydroxyl group may give clearer mass spectrometric results after conversion to their trimethylsilyl ethers.

If trimethylsilyl ethers are replaced by tertbutyl dimethylsilyl ethers then, in addition to valuable fragment ions, peaks at M-57 (Bu^t) and M-89 (Bu^t and MeOH) are useful indicators of molecular size.

When acids contain an oxygenated group and one or more olefinic centres the latter are somewhat less labile and it is sometimes possible to identify these compounds by mass spectrometry of their methyl esters (Kleiman and Spencer, 1973; Gunstone and Schuler, 1975).

In an interesting paper Fellenberg *et al.* (1987) claim to distinguish *n*-3, *n*-6 and *n*-9 methylene-interrupted polyene acids by mass spectrometry of their methyl esters. These three classes of esters produce fragment ions of *m/e* 108, 150 and 192, respectively, probably arising from the ions CH$_3$(CH$_2$)$_n$(CH═CH)$_3$H where n = 1, 4 or 7. Comparison of the intensities of these fragment ions (Table 1.10) shows that there are significantly different patterns for these three families of polyene esters, though it should be noted that the conclusion for *n*-9 esters is based on a single example.

A proposal to convert double bonds to bisalkylthiols is only appropriate for monoenes (Francis, 1981).

TABLE 1.10. Peak heights relative to height of an m/e 150 fragment ion ($= 1.00$).

Acid type	Number of examples	m/e		
		108	150	192
n-3	6	4–80	1.0	0.1–1.0
n-6	7	0.3–0.8	1.0	0.01–0.4
n-9	1	1.6	1.0	1.7

$$RCH\!=\!CHR' \xrightarrow[\text{24 h, room temp.}]{I_2,\ Me_2S_2,} RCH(SMe)\!\!\mid\!\!CH(SMe)R'$$

The best procedure for examining polyunsaturated acids and esters involves the preparation of a polyol by reaction with osmium tetroxide in presence of pyridine and its conversion to a poly(trimethylsilyl)ether. Fragmentation of this product occurs between vic-CH(OSiMe$_3$) groups with the fragments further stabilised by loss of 1–6 trimethylsilyl groups each of 90 amu (Schmitz and Egge, 1979). Molecular ions are not detected. This procedure has also been applied to polyenes containing a conjugated diene unit (Janssen et al., 1988).

An alternative to fixing double bonds is to examine acyl derivatives other than simple alkyl esters. In particular acyl pyrrolidides (**11**), picolinyl esters (**12**), and some other heterocyclic compounds have been recommended and used.

RCON⟨pyrrolidine⟩   RCOOCH$_2$–⟨pyridine⟩

(**11**)   (**12**)

The mass spectrum of a monoacyl pyrrolidide consists of a series of clusters, each containing a dominant peak resulting from direct fragmentation of the alkyl chain. In a saturated pyrrolidide these dominant peaks differ by 14 amu. In a monoene one pair of peaks differ by only 12 amu and this occurs in a manner related to the position of the double bond. The values obtained with oleylpyrrolidide are set out below.

```
        320     292     264     236     208   182     154     126    98
    CH₃CH₂CH₂CH₂CH₂CH₂CH₂CH₂CH=CHCH₂CH₂CH₂CH₂CH₂CH₂CH₂CON⟩
            335     306     278     250     222   196     168     140    113
```

Most of the numbers differ from adjacent numbers by 14 except for the pair 196 (C$_8$ side chain) and 208 (C$_9$ side chain). This is a general observation for all Δ9 unsaturated pyrrolidides.

Significant fragments in linoleylpyrrolidide occur at 262 (C$_{13}$ side chain), 248, 236, 222, 208, 196 and 182. Differences of 12 amu are between 196 and 208 (C$_8$ and C$_9$ side chains) and between 236 and 248 (C$_{11}$ and C$_{12}$ side chains). The method is less satisfactory for amides with more than two double bonds. It has been reviewed by Andersson (1978).

Christie *et al.* (1986, 1987a,b) have developed the proposal made earlier by Harvey (1982, 1984a,b) that picolinyl esters (**12**) are useful for the GC-MS studies of fatty acids. The acids are converted first to mixed anhydrides with trifluoroacetic anhydride and then reacted with 3-hydroxymethylpyridine in tetrahydrofuran. Mixtures, such as those obtained from complex materials like pig testis lipids and cod liver oil, are separated by capillary gas chromatography and identified by mass spectrometry. In this way 19 acids were fully identified in the pig testis lipid and 22 in the cod liver oil.

The mass spectra show useful molecular ions along with fragment ions of m/e 92 (usually the base peak), 108, 151 (or 150) and 164. Monoene esters also contain two fragment ions 14 amu apart which stand out prominently from other ions. These are usually 30–40% of the base peak height and are directly related to double bond position thus: $\Delta 7$, 246 and 260; $\Delta 8$, 260 and 274; $\Delta 9$, 274 and 288; $\Delta 11$, 302 and 316; and $\Delta 13$, 330 and 344. Polyene esters are identified by groups of peaks with gaps of 26, 14 and 26 amu. For example, the arachidonic picolinyl ester has fragment ions at m/e 324, 298, 284 and 258. These contain 14, 12, 11 and 9 carbon atoms in the acyl chain and indicate double bonds at position 11 and 14. These studies have been extended by Christie *et al.* (1987a,b). The same author (Christie *et al.*, 1988) has demonstrated the power of this method when combined with silver ion liquid chromatography. Fatty acids from two Black Sea invertebrates (*Mytilus galloprovincialis* and *Rapana thomasiana*) were separated into 41 and 46 defined components (58 different acids in total).

It is likely that acyl derivatives of other amines and alcohols and of other systems will also provide useful structural information by mass spectrometry. A recent paper recommends the 2-alkenyl-4,4-dimethyloxazolines (**13**) (Zhang *et al.*, 1988) among other possibilities. For mono-alkyl derivatives peaks are normally separated by 14 amu but one pair differs by 12 amu and this is related to double bond position, e.g. $\Delta 9$, 196 and 208 (C_8 and C_9 alkyl derivatives); $\Delta 11$, 224 and 236 (C_{10} and C_{11}); $\Delta 13$, 252 and 264 (C_{12} and C_{13}); and $\Delta 15$, 280 and 292 (C_{14} and C_{15}).

(**13**)

F. Chemical Procedures

With the increasing use of spectroscopic procedures to determine fatty acid structure, less use is made of chemical procedures. Two, however, are still deserving of mention: reduction (hydrogenation) and oxidative cleavage.

Complete catalytic hydrogenation of an unsaturated acid (ester) using a nickel, palladium or platinum catalyst followed by gas chromatographic examination of the products will show whether the carbon chain is straight or branched or contains a cyclic unit and also its total length.

Partial reduction is effected by incomplete reaction with hydrazine. This non-catalytic procedure is chosen because it occurs without double bond movement and without stereo-mutation. Double bonds remaining in a partially reduced product therefore have the same position and the same configuration as in the starting material. This procedure is most often used to determine the position of *cis* and *trans* double bonds in a polyene containing both types of unsaturation. For example, the identification of 18:3 9t, 12t, 15c requires partial reduction with hydrazine to give a mixture of saturated, monoene, diene and triene esters (Perkins and Smick, 1987). This is followed by isolation of *cis* and *trans* monoenes as separate fractions by silver ion chromatography, and finally the identification of each of these by spectroscopic procedures or by oxidative cleavage.

Oxidation of unsaturated acids (esters, alcohols) is usually effected by ozonation followed by splitting of the ozonide. This can be achieved in different ways to produce alcohols (lithium aluminium hydride or hydrogen–palladium charcoal), aldehydes (triphenylphosphine, dimethylsulphide, or hydrogen—Lindlar's catalyst), or acids (silver oxide, potassium permanganate). The oxidation products are usually identified by gas chromatography. This procedure can also be applied in a (semi) quantitative manner to determine the composition of mixtures such as those produced by partial catalytic hydrogenation of polyene acids (Johnston et al., 1978).

$$RCH{=}CHR' \xrightarrow{O_3} \text{ozonide} \begin{cases} \xrightarrow{*} RCH_2OH + R'CH_2OH \\ \xrightarrow{*} RCHO + R'CHO \\ \xrightarrow{*} RCOOH + R'COOH \end{cases}$$

* For reagents, see text.

REFERENCES

Ackman, R. G. (1984). *In* "CRC Handbook of Chromatography" (H. K. Mangold, ed.), Lipids Vol. 1, pp. 95–240. CRC Press, Boca Raton, FL.
Ackman, R. G. (1988). *Chem. and Ind.* 139–145.
Andersson, B. A. (1978). *Progr. Chem. Fats Other Lipids* **16**, 279–308.
Bascetta, E., Gunstone, F. D. and Scrimgeour, C. M. (1984). *Lipids* **19**, 801–803.
Bergter, L. and Seidl, P. (1984). *Lipids* **19**, 44–47.
Bus, J., Sies, I. and Lie Ken Jie, M. S. F. (1976). *Chem. Phys. Lipids* **17**, 501–518.
Bus, J., Sies, I. and Lie Ken Jie, M. S. F. (1977). *Chem. Phys. Lipids* **18**, 130–144.
Christie, W. W. (1982). "Lipid Analysis", 2nd edn, p. 67. Pergamon Press, Oxford.
Christie, W. W. (1987). *J. High Res. Chromatogr. Chromatogr. Commun.* **10**, 148–150.
Christie, W. W., Brechany, E. Y., Johnson, S. B. and Holman, R. T. (1986). *Lipids* **21**, 657–661.
Christie, W. W., Brechany, E. Y. and Holman, R. T. (1987a). *Lipids* **22**, 224–228.
Christie, W. W., Brechany, E. Y., Gunstone, F. D., Lie Ken Jie, M. S. F. and Holman, R. T. (1987b). *Lipids* **22**, 664–666.
Christie, W. W., Brechany, E. Y. and Stefanov, K. (1988). *Chem. Phys. Lipids* **46**, 127–135.
Davies, J. E. D., Hodge, P., Barve, J. A. and Gunstone, F. D. (1972). *J. Chem. Soc. (Perkin Trans. II)*, 1557–1561.
Fellenberg, A. J., Johnson, D. W., Poulos, A. and Sharp, P. (1987). *Biomed. Mass Spectrom.* **14**, 127–129.
Francis, G. W. (1981). *Chem. Phys. Lipids* **29**, 369–374.
Frost, D. J. and Gunstone, F. D. (1975). *Chem. Phys. Lipids* **15**, 53–85.
Galliard, T. and Mercer, E. I. (eds) (1975). "Recent Advances in the Chemistry and Biochemistry of Plant Lipids", pp. 24–25. Academic Press, London.

Games, D. E. (1978). *Chem. Phys. Lipids* **21**, 389–402.
Gaydou, E. M., Miralles, J. and Rasoazanakolona, V. (1987). *J. Am. Oil Chem. Soc.* **64**, 997–1000.
Gunstone, F. D. (1983). *Brit. J. Clin. Practice* **38**, (Symp. Suppl. 31), 17.
Gunstone, F. D. (1986). *In* "The Lipid Handbook" (F. D. Gunstone, F. B. Padley and J. L. Harwood, eds), pp 11–12. Chapman and Hall, London.
Gunstone, F. D. and Jacobsberg, F. R. (1972). *Chem. Phys. Lipids* **9**, 26–34.
Gunstone, F. D. and Schuler, H. R. (1975). *Chem. Phys. Lipids* **15**, 198–208.
Gunstone, F. D., Pollard, M. R. and Scrimgeour, C. M. (1976). *Chem. Phys. Lipids* **17**, 1–13.
Gunstone, F. D., Pollard, M. R., Scrimgeour, C. M. and Vedanayagam, H. S. (1977). *Chem. Phys. Lipids* **18**, 115–129.
Hamberg, M. (1987). *Chem. Phys. Lipids* **43**, 55–67.
Harvey, D. J. (1982). *Biomed. Mass Spectrom.* **9**, 33–38.
Harvey, D. J. (1984a). *Biomed. Mass Spectrom.* **11**, 187–192.
Harvey, D. J. (1984b). *Biomed. Mass Spectrom.* **11**, 340–347.
Hopkins, C. Y. (1972). *In* "Topics in Lipid Chemistry" (F. D. Gunstone, ed.), Vol. 3, 37–87. Logos Press, London.
Jamieson, G. R. (1970). *In* "Topics in Lipid Chemistry" (F. D. Gunstone, ed.), Vol. 1, pp. 107–159. Logos Press, London.
Jamieson, G. R. (1975). *J. Chromatogr. Sci.* **13**, 491–497.
Janssen, G., Verhulst, A. and Parmentier, G. (1988). *Biomed. Mass Spectrom.* **15**, 1–6.
Johnston, A. E., Dutton, H. J., Scholfield, C. R. and Butterfield, R. O. (1978). *J. Am. Oil. Chem. Soc.* **55**, 486–490.
Khan, G. R. and Scheinmann, F. (1978). *Progr. Chem. Fats Other Lipids* **15**, 343–367.
Kleiman, R. and Spencer, G. F. (1973). *J. Am. Oil Chem. Soc.* **50**, 31–38.
Lopez, A. and Gerwick, W. H. (1987). *Lipids* **22**, 190–194.
Miller, R. W., Weisleder, D., Plattner, R. D. and Smith, C. R. (1977). *Lipids* **12**, 669–675.
Minnikin, D. E. (1978). *Chem. Phys. Lipids* **21**, 313–347.
Miwa, T. K., Mikolajczak, K. L., Earle, F. R. and Wolff, I. A. (1960). *Anal. Chem.* **32**, 1739–1742.
Peers, K. E. and Coxon, D. T. (1986). *J. Food Technol.* **21**, 463–469.
Perkins, E. G. and Smick, C. (1987). *J. Am. Oil Chem. Soc.* **64**, 1150–1155.
Pollard, M. (1986). *In* "Analysis of Oils and Fats" (R. J. Hamilton and J. B. Rossell, eds), pp. 401–434. Elsevier Applied Science, London.
Rakoff, H., Weisleder, D. and Emken, E. A. (1979). *Lipids* **14**, 81–83.
Schmitz, B. and Egge, H. (1979). *Chem. Phys. Lipids* **25**, 287–298.
Schmitz, B. and Klein, R. A. (1986). *Chem. Phys. Lipids* **39**, 285–311.
Suzuki, O., Hashimoto, T., Hayamizu, K. and Yamamoto, O. (1970). *Lipids* **5**, 457–462.
Tallent, W. H., Harris, J. and Wolff, I. A. (1966). *Tetrahedron Lett.*, 4329–4334.
Tulloch, A. P. (1977). *Can. J. Chem.* **55**, 1135–1142.
Tulloch, A. P. (1982). *Lipids* **17**, 544–550.
Tulloch, A. P. and Bergter, L. (1979). *Lipids* **14**, 996–1002.
Tulloch, A. P. and Mazurek, M. (1976). *Lipids* **11**, 228–234.
Williams, D. H. and Fleming, I. (1987). "Spectroscopic Methods in Organic Chemistry", 4th edn, pp. 126–128. McGraw-Hill, London.
Zhang, J. Y., Yu, Q. T., Liu, B. N. and Huang, Z. H. (1988). *Biomed. Mass Spectrom.* **15**, 33–44.

2 Triacylglycerol Biosynthesis

ALLAN KEITH STOBART[1] and STEN STYMNE[2]

[1] Department of Botany, The University, Bristol BS8 1UG, UK
[2] Department of Plant Physiology, Swedish University of Agricultural Sciences, S-750 07, Uppsala, Sweden

I.	Introduction	20
II.	Triacylglycerol structure and biosynthesis	20
	A. Structure	20
	B. Biosynthesis	22
	C. Role of phosphatidylcholine in the synthesis of the C18 polyunsaturated fatty acids	23
III.	Experimental material	25
	A. Developing seeds	25
	B. Microsomal membrane preparations	26
	C. Oil body preparations	27
IV.	Substrates	27
	A. Acyl-CoA	27
	B. Acyl-CoAs as substrates for lipid synthesis	28
	C. Ammonium salts of fatty acids	29
	D. [^{32}P]Glycerol 3-phosphate	29
V.	Lipid extraction, purification and quantitative analysis	29
	A. Extraction	29
	B. Purification of complex lipids	31
	C. Fatty acids and acyl constituents of complex lipids	31
	D. Silver nitrate thin layer chromatography of fatty acid methyl esters	32
	E. Purification of the acyl-CoA fraction	33
VI.	Stereospecific analysis of fatty acids in complex lipids	34
	A. Phosphatidylcholine and phosphatidic acid	34
	B. Triacylglycerols	34
VII.	Determination of radioactivity in lipid samples	35
VIII.	Protocol for a typical experiment	37

IX.	Assays of the enzymic steps in triacylglycerol biosynthesis	37
	A. Acyl-CoA: lysophosphatidylcholine acyltransferase	37
	B. Acyl-CoA: lysophosphatidylcholine acyltransferase and acyl exchange	38
	C. Desaturases and C18 polyunsaturated fatty acid synthesis	39
	D. Glycerol 3-phosphate acylating enzymes	41
	E. Phosphatidic acid phosphohydrolase (phosphatidase; PAP)	42
	F. Interconversion of diacylglycerol with phosphatidylcholine	42
	G. Diacylglycerol acyltransferase	44
X.	Concluding remarks	44
	Acknowledgements	45
	References	45

I. INTRODUCTION

The plant triacylglycerols (triglycerides) are of agricultural and economic importance and are the 'fats' and 'oils' with which most people are familiar. They are generally abundant in specialised plant tissues and constitute important storage reserves, particularly in oleaceous seeds. The natural diversity in the composition of plant triacylglycerols gives them a wide range of application in the food and chemical industries. It is considered that the storage triacylglycerols, the so-called vegetable oils, have potential for the manipulation and tailoring of their fatty acid constituents in order to suit particular requirements. The interest in the plant triacylglycerols has increased considerably in recent years, particularly with the expectation that the current methodologies of genetic engineering could play a major role in realising the above objectives. The production of useful varieties of oil-crops, through sustained programmes of conventional plant breeding, perhaps coupled to the chemo-generation of fatty acid mutants, will also gain more impetus if it is strongly interactive with studies in lipid biochemistry. Although expectations are high, it is essential for success that a thorough understanding of the biosynthesis of triacylglycerols in plants and its regulation is forthcoming and that the attempts at acyl manipulation, through the current technologies, are solidly underpinned with fundamental research (see Hatch, 1986). This chapter is, therefore, devoted to the methodologies that the authors have used in studies on triacylglycerol biosynthesis in oil-seeds, in the hope that it will benefit not only the established lipid biochemist, but, perhaps more importantly, form a technical base for those who anticipate entering the field.

II. TRIACYLGLYCEROL STRUCTURE AND BIOSYNTHESIS

In order to appreciate more fully the techniques required in studies on the biochemistry of the triacylglycerols, it is necessary for the reader to have some understanding of triacylglycerol structure and to be familiar, at least to some extent, with our present knowledge regarding their biosynthesis in plants (Gurr, 1980; Roughan and Slack, 1982; Stymne and Stobart, 1987; Griffiths et al., 1988a).

A. Structure

The basic structure of the triacylglycerol molecule consists of a glycerol backbone to which three fatty acid residues (acyl groups) are esterified, one to each of the three hydroxy groups (Fig. 2.1). The glycerol molecule does not possess rotational symmetry and hence all the carbon atoms can be distinguished from each other. By convention, a stereospecific numbering system (-*sn*) is favoured, in which, in a Fischer projection of the natural L-glycerol derivative, the secondary hydroxyl group is located to the left of the *sn*-2 carbon (Brockerhoff, 1972). The carbon atoms directly above and below the central *sn*-2 carbon become the *sn*-1 and *sn*-3 atoms, respectively. On occasion, the groups attached to the primary and the secondary hydroxyl groups are designated as in the α- and β-position, respectively. The intramolecular structure of the glycerophospholipids is designated in a similar manner to the triacylglycerol with, in this case, the *sn*-3 position possessing a characteristic headgroup. Those phospholipids which are intimately involved in the biosynthesis of the triacylglycerols are illustrated in Fig. 2.1.

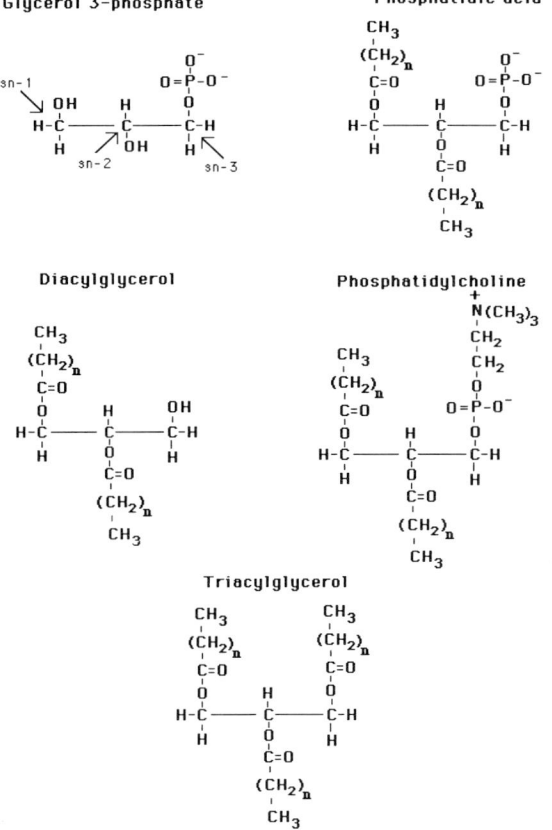

FIG. 2.1. Structures of the major complex lipids involved in triacylglycerol biosynthesis.

The most abundant fatty acids in the plant oils and fats are unbranched, even-numbered, monocarboxylic acids (Table 2.1) in which all the carbon atoms are either saturated (e.g. capric, C10:0; lauric, C12:0; myristic, C14:0; palmitic, C16:0; stearic, C18:0) or unsaturated to varying degrees (e.g. oleic, C18:1; linoleic, C18:2; linolenic,

C18:3; erucic, C22:1). Various nomenclatures have been developed to distinguish the number and position of the double bonds in the acyl chain. The shorthand C18, for example, gives the number of carbon atoms in the acyl chain, and a colon followed by a value gives the number of double bonds in the molecule. Generally, biochemists tend to locate the double bonds from the carboxyl carbon (the Δ carbon) and the carbon atoms are numbered from this end of the molecule. Thus C18:3 Δ9,12,15, would show that the double bonds were at carbon atoms 9 to 10, 12 to 13 and 15 to 16, and the fatty acid would be the common polyunsaturated acid, α-linolenic acid. The uncommon, but nevertheless important, linolenic acid isomer, γ-linolenic acid, is C18:3 Δ6,9,12. Lipid chemists, on the other hand, tend to operate with a different system of fatty acid nomenclature based upon the form $(n-x)$, where n is the chain length of the acid and x the number of carbon atoms from the last double bond to the terminal methyl group. The double bonds in each family of non-conjugated fatty acids (polyunsaturated fatty acids) are separated by a single methylene group, and all the individuals would have the same terminal structure. Thus, α-linolenic acid would be C18:3 $(n-3)$ and γ-linolenic acid, C18:3 $(n-6)$. Since fatty acids are elongated from the carboxylic end of the molecule, the $(n-x)$ terminology has many advantages and hence the designations for the unsaturated fatty acids, which are relevant to this chapter, are also given in Table 2.1.

TABLE 2.1. Fatty acids commonly found in plant triacylglycerols.

Trivial name	Symbol	Systemic name	$(n-x)$	Source
Capric	10:0	decanoic		Cuphea
Lauric	12:0	dodecanoic		Palm kernel
Myristic	14:0	tetradecanoic		Coconut
Palmitic	16:0	hexadecanoic		Palm mesocarp
Stearic	18:0	octadecanoic		Cocoa, shea
Oleic	18:1 Δ9	octadec-9-enoic	n-9	Peanut, olive
Linoleic	18:2 Δ9,12	octadec-9,12-dienoic	n-6	Sunflower, safflower
α-Linolenic	18:3 Δ9,12,15	octadec-9,12,15-trienoic	n-3	Linseed
γ-Linolenic	18:3 Δ6,9,12	octadec-6,9,12-trienoic	n-6	Evening primrose, borage
Erucic	22:1 Δ13	docos-13-enoic	n-9	Mustard, rape

A combination of fatty acid quality and the stereospecific acyl distribution in the triacylglycerol molecule determines the chemical and physical properties of the oil, and thus its commercial use and value (Gunstone and Norris, 1983). Although hundreds of fatty acids and their derivatives are found in the plant kingdom (Hilditch and Williams, 1964), many of which could be of significant industrial value, the fatty acids that are the most widespread in oleaceous tissues of the major agricultural oil crops are largely restricted to those given in Table 2.1.

B. Biosynthesis

Plant tissues, and in particular developing oil-seeds, appear to synthesise the bulk of their triacylglycerols in the membranes of the endoplasmic reticulum (ER) and various schemes for the ontogeny of the oil-body have been presented (Wanner *et al.*, 1981; Stobart *et al.*, 1986; Herman, 1987). The synthesis of fatty acids up to the C16:0, C18:0

and C18:1 level occurs in the plastids and, where present, the chloroplasts, of the maturing cotyledons or endosperm of the developing seed (see Stumpf, 1987, and Jaworski, 1987, for further references; Browse and Slack, 1985). These fatty acids are made available to the endoplasmic reticulum, where they are utilised in triacylglycerol biosynthesis and the production of the C18-polyunsaturated fatty acids, C18:2 (n-6), C18:3 (n-3) and C18:3 (n-6). The developing oil-seeds of many species yield microsomal membrane preparations capable of high rates of triacylglycerol formation (Stymne and Stobart, 1987). With such preparations, it has been possible to establish the mode of triacylglycerol biosynthesis and its relationship to the formation of the C18-polyunsaturated fatty acids (Fig. 2.2). Microsomal membranes from the many species studied catalyse triacylglycerol formation, essentially through the reactions of the, so-called, Kennedy pathway (Kennedy, 1961). Initially, the sn-glycerol 3-phosphate is acylated to give phosphatidic acid (sn-1,2 diacylglycerol 3-phosphate). The formation of phosphatidate is a two-step event in which the sn-glycerol 3-phosphate is first acylated at position sn-1, through the action of a glycerophosphate acyltransferase (EC 2.3.1.15), to yield a $lyso$-phosphatidic acid. The $lyso$-phosphatidate so formed is the substrate for a second enzyme, sn-1-acyl-glycerol 3-phosphate acyltransferase (EC 2.3.1.51), which completes the acylation at position sn-2. The acyl donor species for these two reactions are CoA esters rather than the acyl carrier protein (ACP) derivatives (unpubl. obs.). It is the acyl selectivity properties of these enzymes that govern the non-random distribution of fatty acid species often observed at positions sn-1 and -2 of the phosphatidate, and hence of the other phospholipids, diacylglycerols and triacylglycerols that are subsequently derived from it. The generated phosphatidate serves as a substrate for the enzyme, phosphatidate phosphohydrolase (EC 3.1.3.4) (PAP), and it is through the action of this enzyme that the diacylglycerols are formed. Although the phosphohydrolase may have regulatory properties in mammalian tissue (Brindley, 1984), there is as yet little evidence for a similar function in plants. The diacylglycerol can follow one of two possible routes and be either channelled into phosphatidylcholine or undergo acylation at position sn-3 to yield triacylglycerol. In the first of these reactions an equilibration between the diacylglycerol and the phosphatidylcholine can occur, and this may be catalysed by a cholinephosphotransferase (EC 2.7.8.2) (Roughan and Slack, 1982; Slack et al., 1985). The interconversion of these two complex lipid pools allows oleate, which is present in the diacylglycerol molecule, to enter the phosphatidylcholine and there be desaturated further on the phospholipid substrate (see below). In the final acylation step, a diacylglycerol acyltransferase (EC 2.3.1.20), the only enzyme which can be considered dedicated to the triacylglycerol biosynthetic pathway, catalyses the transfer of fatty acid from the acyl-CoA to position sn-3 and so is important in regulating the quality of the fatty acids in the final triacylglycerol.

C. Role of Phosphatidylcholine in the Synthesis of the C18 Polyunsaturated Fatty Acids

It is now widely accepted that the phosphatidylcholine in the membranes of the endoplasmic reticulum is the complex lipid substrate for the desaturation of oleate to linoleate and, where applicable, further to linolenate (Jaworski, 1987). It appears that these C18-polyunsaturated fatty acids, which are being formed in the phosphatidylcholine, can be made available for triacylglycerol assembly in two ways. One involves the interconversion of the diacylglycerol and phosphatidylcholine pools during the flow of glycerol backbone towards triacylglycerol through the Kennedy pathway (Stobart and Stymne, 1985a). The other involves the oleate, from oleoyl-CoA, which can be transferred to position *sn*-2 of the *sn*-phosphatidylcholine, by the action of an acyl-CoA: lysophosphatidylcholine acyltransferase (LPCAT). The LPCAT can also operate in a reversible mode and this activity can account for the entry of linoleate and the linolenates, which are being formed at position *sn*-2 of the *sn*-phosphatidylcholine, back into the acyl-CoA pool from where they can be further utilised in the acylation of the glycerol backbone. The LPCAT and the diacylglycerol–phosphatidylcholine interconversion are, therefore, key reactions which, in intimate association with the desaturase enzymes, regulate the quality of the fatty acids for triacylglycerol assembly.

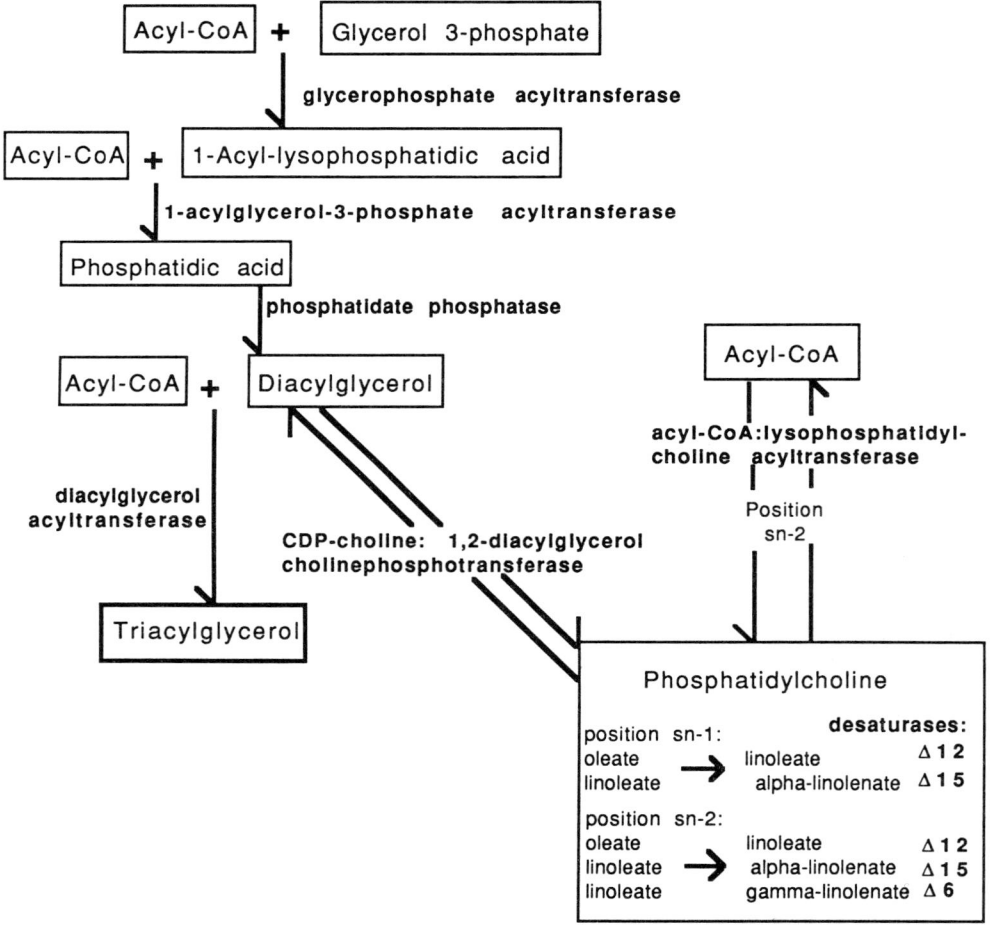

FIG. 2.2. Biosynthetic pathway for C18-polyunsaturated fatty acid formation and triacylglycerol assembly in oil seeds (see text for details).

Our knowledge of triacylglycerol synthesis in plants and its regulation is far from complete. Much basic research is still required in order to understand the details of *de novo* fatty acid synthesis in the plastid and its relationship to the production of the C18-polyunsaturated fatty acids and the triacylglycerols in the endoplasmic reticulum. The assembly of triacylglycerols with the more uncommon of the fatty acids, such as erucate and the medium-chain fatty acids, still requires intensive study. The enzymes in the membranes of the endoplasmic reticulum are strongly hydrophobic and tend to lose their activity rapidly during solubilisation and attempts at purification. This is particularly the case with the desaturase enzymes which, to be active, also require the participation of electron transport components. The electron transport chain, involved in the synthesis of the C18-polyunsaturated fatty acids in the endoplasmic reticulum of plant cells is still ill-defined, and awaits proper characterisation (Gennity and Stumpf, 1985; Jaworski, 1987). It is, therefore, necessary to establish the precise details of triacylglycerol biosynthesis and the regulatory reactions which govern quantity and fatty acid quality. Here, we detail the techniques of lipid analyses required in studies on triacylglycerol biosynthesis, the methods used by the authors to assess aspects of the biosynthetic pathway (Fig. 2.2) and possible assays for the enzymes involved.

III. EXPERIMENTAL MATERIAL

A. Developing Seeds

It should go without saying that biochemical studies on triacylglycerol biosynthesis in plants will only be as good as the 'state' of the plant material used in the experimentation and will depend critically on the use made of the techniques available. It is most important to utilise tissues that are actively synthesising triacylglycerols and are in such a stage of development that they will yield good active membrane preparations for *in vitro* studies. Unfortunately, quantity of material is not necessarily the same as quality. Because of their agricultural importance, work has largely centred on experimentation with the oil-seeds of the various crop species. The many species studied accumulate triacylglycerols in a precise and synchronised fashion and the kinetics of this must be assessed before embarking on experimentation. The developing seeds of safflower, for example, provide a good 'model' system for study. Here, oil deposition begins at about 12 days after flowering and well over 70% of the oil is generally laid down in a window of about four days, commencing 16 to 18 days after flowering (Ichihara and Noda, 1980; Slack *et al.*, 1985). Seed cotyledonary tissue, at this stage of development, yields excellent microsomal membrane preparations capable of sustained C18-polyunsaturated fatty acid synthesis and triacylglycerol assembly. Some other species, such as oilseed rape, accumulate their storage seed-oil over a much longer period (Ching *et al.*, 1974), but even here active membrane preparations, suitable for biosynthetic studies, are better obtained from developing cotyledons in the early phases of expansion (Stymne *et al.*, 1987; Griffiths *et al.*, 1988a). Developing seed material that is reaching maturity and undergoing dehydration is usually extremely poor for triacylglycerol studies and can give rise to many artifactual observations. At the moment, little of the biosynthetic work on oil deposition has taken into consideration the homogeneity of the tissue involved. It is usually assumed that the first phase in cotyledonary development is the rapid division of the cells which, after the requisite number in the cell population has been reached, will enter an expansion phase accompanied by the deposition of the storage reserves (Griffiths *et al.*, 1989). How much cell division still accompanies cell expansion, however, may have a critical bearing on future attempts at genetic engineering and needs to be fully established. Although most of the oil species that have received attention have seed cotyledons as their storage organs, there is great interest in the developing mesocarps of certain fruits and, in particular, the oil-palms. The mesocarp of the oil-palm starts to accumulate lipid at about 18 weeks after anthesis and, again, the bulk of the storage oil is deposited over a relatively short period of about three weeks (Bafor and Osagi, 1986), so the earlier comments on the state of seed material will also apply to these situations.

Although good, developing seed material is somewhat laborious to harvest, the time spent is certainly well invested, especially when rewarding results are there to be obtained. The delicate cotyledons, after removal of the seed coat, should be stored on ice, in phosphate buffer, pH 7.2, until sufficient quantities have been harvested. The cotyledons may be used directly in *in vivo* studies or immediately utilised in the preparation of microsomal membranes (see below). In some cases, certain species will yield microsomal preparations which can be stored at $-80°C$ almost indefinitely and still retain good activity. In the authors' laboratories, sunflower preparations that had been stored for some four years were found to be as active as the original preparation. It is, however, important that workers assess and compare both the quantitative and qualitative activity of the stored preparations since many species (e.g. oil-seed rape, turnip rape, linseed, borage) lose their activity extremely rapidly and have to be used as freshly prepared material. The storage of the developing whole-seed, at low temperature, before use in metabolic studies, is to be strongly advised against as this most certainly gives rise to artifacts and poor membrane preparations which yield erroneous results.

B. Microsomal Membrane Preparations

In the many oil-seed species under study, it has proved possible, providing that the plant material is at the 'optimum' state of development, to obtain microsomal preparations with excellent activities and with the minimum addition of cofactors, etc., to the homogenisation medium. It is essential, however, that as little of the seed-coat as possible is included with the isolated cotyledons. In nearly all cases, the cotyledons (one part) are ground in a mortar at $2°C$ with five parts 0.1 M potassium phosphate buffer, pH 7.2, containing 0.1% bovine serum albumin and 0.33 M sucrose. In some instances, depending on the species, the desaturase enzymes may be subject to deactivation and the addition of catalase to the extraction medium (1000 units of catalase per 1 ml of the extraction buffer) may have some value (Browse and Slack, 1981). After homogenisation, which with young cotyledons takes only 2 min, the homogenate is diluted 20-fold with fresh grinding medium and then filtered through a double layer of Miracloth™ before centrifugation at $20\,000 \times g$ for 10 min. The supernatant is again passed through Miracloth™ and then centrifuged at $100\,000 \times g$ for 90 min. The resulting microsomal pellet is suspended in a small volume of homogenisation medium and either used immediately or, if the species permits, stored at $-80°C$ until required.

C. Oil-body Preparations

In some instances it may be necessary to purify oil-bodies for biochemical study. This is easily achieved and a number of methods have been employed. The procedure of Slack and co-workers (Slack et al., 1980) is particularly successful and yields oil-bodies which differ little in size to those which are observed in vivo. Cotyledons (1–2 g) are homogenised in 8 ml 50 mM Tris-HCl buffer, pH 7.5, which contains 500 mM sucrose, 0.5 mM EDTA and 0.5 mM EGTA, in a polytron at half speed for 2 s. After passing through three layers of Miracloth™, the filtrate (c. 7 ml) is transferred to a 15 ml centrifuge tube and overlaid with two 3 ml layers of the above buffer mixture containing 250 mM and 125 mM sucrose, respectively. Centrifugation at $18\,000 \times g$, 20 min, yields a 'pellicle' of oil-bodies at the surface. The oil-body 'pellicle' is readily removed with a spatula, and after washing in buffer, dispersed in 0.5 ml 500 mM sucrose buffer by squeezing against the side of the centrifuge tube with a glass rod and, after the addition of a further 1 ml of buffer, rapidly squirting through a Pasteur pipette. The suspension can be adjusted to 7 ml with 500 mM sucrose buffer and overlayed with further volumes of more dilute sucrose solution and re-centrifuged. The procedure may be repeated depending on the purity of the preparation required.

IV. SUBSTRATES

A. Acyl-CoA

Of major importance, for in vitro work on the biosynthesis of the triacylglycerols and the formation of the C18-polyunsaturated fatty acids, is the availability of adequate supplies of acyl-CoA substrates, both radioactive and non-radioactive. Although some of these are available from commercial sources, they are usually costly and occasionally impure. Moreover, the more 'uncommon' fatty acids are usually not available as their CoA derivatives. The authors find it more convenient, and the product more reliable, to synthesise acyl-CoAs 'in-house' since one can accurately manipulate their specific radioactivity and be sure of their purity. The rapid and convenient method for the preparation of acyl-CoA from CoA and the free fatty acid, based on that of Sanchez et al. (1973), is most reliable. The synthesis has an average overall yield of some 60% acyl-CoA when 8–40 µmol of fatty acid are used as the initial starting material. The method has proved most successful for nearly all fatty acids varying in chain length from 10 to 22 carbon atoms. The following protocol is for the preparation of about 20 µmol of acyl-CoA, but this is easily scaled down to one-fifth by reducing all volumes and concentrations proportionately.

Dissolve 36 µmol of the free fatty acid in 500 µl water-free methylene chloride. Add 5 µl triethylamine (which has been stored over sodium hydroxide) to the solution and allow it to stand at room temperature for 30 min. Cool to 0°C in an ice bath and add 3.6 µl ethylchloroformate and leave the mixture for a further 2 h at 0°C. Remove the methylenechloride under a stream of nitrogen at room temperature and add 2 ml peroxide-free tetrahydrofuran to the residue. Place the mixture in a sonication bath and fully disperse the precipitate before centrifugation in a bench centrifuge. Remove the supernatant and evaporate to dryness at room temperature under a stream of nitrogen.

Add 300 µl peroxide-free tetrahydrofuran to the residue and transfer the solution to a small test tube which can be thoroughly sealed. Add 1 ml 1 M KHCO$_3$, in which is dissolved 36 µmol CoA as either the lithium or sodium salt. Purge the solution with nitrogen and rapidly seal the tube. Vigorously mix the solution in a flask-shaker or vortex apparatus for 90 min. It is essential, at this stage, that the mixing is thorough and continuous for the production of good yields. Acidify the solution by the careful dropwise addition of 2 M HCl until no further precipitation occurs and recover the acyl-CoA by centrifugation. Remove the supernatant and wash the pellet three times with 2 ml quantities of diethylether followed twice with 2 ml quantities of 0.1 M HCl to remove the unreacted anhydrides, free fatty acids and CoA. It is preferable to disperse the acyl-CoA in the washing medium by ultrasonication followed by recovery through centrifugation. Dissolve the acyl-CoA in distilled water and adjust the pH to 4.5 with 1 M KHCO$_3$. If it is necessary, for economic reasons, the unreacted radioactive fatty acid can be recovered by bulking the supernatant and diethylether washings. The ether phase can be removed, evaporated to dryness under nitrogen, and the fatty acid residue re-used for the preparation of the triethylamine salt and the anhydride as described above. Excellent purity of even the highly polyunsaturated fatty acids, such as arachidonic acid, has been achieved by this method.

B. Acyl-CoAs as Substrates for Lipid Synthesis

In aqueous solutions long-chain acyl-CoAs generally form micelles in the µM concentration range; this can give rise to difficulties in studies on the kinetics of the acyl-CoA utilising enzymes. The concentration of the acyl-CoA substrate and that of the enzyme is critical and must be kept as low as possible for the estimation of K_m values. It is essential to keep the acyl-CoA substrate in its monomeric form for the reaction to proceed linearly. Moreover, since long-chain acyl-CoA tends to bind strongly to biological membrane fractions, as well as to hydrophobic proteins, the actual concentration of the acyl-CoA in the assay mixture will be unknown. If the concentration of the acyl-CoA in the assay is raised above the critical micelle concentration many of the enzymes involved in the biosynthesis of the triacylglycerols, such as the glycerol acylating enzymes, can be strongly inhibited whereas others, such as the acyl-CoA thiohydrolases, are stimulated. It can be assumed that the acyl-CoA concentration *in vivo* in the plant cell is always low and is never allowed to increase above the critical micelle concentration. In the cell, the monomeric acyl-CoA concentration is probably kept low through its reversible binding to membrane proteins coupled to the high turnover of these metabolic intermediates. We find that a useful approach in acyl specificity and selectivity studies of the enzymes involved in triacylglycerol biosynthesis is to bind, reversibly, the acyl-CoA to bovine serum albumin (BSA). The use of bovine serum albumin in the reaction mixture prohibits micelle formation, even at high concentrations of acyl-CoA, and thus permits the use of high enzyme levels and prolonged incubation times without the risk of the depletion of the acyl-CoA substrate. The use of bovine serum albumin, in this manner, probably closely mimics the situation *in vivo* in the cells and allows the enzyme-protein to have a constant, and yet low, supply of the fatty acid substrate, through the equilibration between the protein-bound and monomeric forms of the acyl-CoA.

It is possible that the use of bovine serum albumin in experiments designed to study acyl specificity and selectivity of the acylating enzymes could lead to erroneous observations, especially if the dissociation constants for the acyl-CoA–protein complexes are different for each species of acyl-CoA. No such data, on the dissociation of such complexes, are, as yet, available. It is noteworthy, however, that the acyl selectivity properties of the glycerol acylating enzymes observed in *in vitro* studies are in close agreement to those predicted from the intramolecular distribution of fatty acids in the complex lipids which have been purified from whole tissues and from *in vivo* experiments (Griffiths *et al.*, 1988b,c). In the assay of 'acyl exchange', between acyl-CoA and position sn-2 of phosphatidylcholine, the inclusion of the bovine serum albumin in the incubations is essential, since the process is strongly inhibited by high concentrations of free acyl-CoA.

C. The Ammonium Salt of the Free Fatty Acid

In vivo experiments, with whole tissues or tissue slices, which require acyl substrates, are best achieved using the ammonium salt of the free fatty acid. The ammonium-salt derivative of a radioactive fatty acid is easily synthesised by reacting the fatty acid (up to 2 μmol) with 2 M NH_4OH (0.2 ml) for 30 min at 60°C. The ammonia is removed under nitrogen and the salt dissolved in distilled water to give the required concentration and activity.

D. [^{32}P]-sn-Glycerol 3-Phosphate

This substrate appears to be unavailable commercially and yet is useful in the assay of the enzyme, phosphatidate phosphohydrolase, which is found in the microsomal membrane preparations of oil synthesising tissues (see below). It is easily produced in the laboratory by incubation of 1.5 μmol [γ-^{32}P]ATP (100 μCi) with 10 μmol glycerol and 10 units of glycerol kinase in 0.25 ml, 0.1 M glycine-HCl buffer, pH 9.8, containing 10 μmol $MgCl_2$, at 37°C for 60 min. The labelled sn-glycerol 3-phosphate in the incubation mixture is purified by TLC on pre-coated plates of cellulose (Merck; 0.1 mm thick) with ethanol-ammonium acetate (14:9, by volume), pH 7.8. Reference sn-glycerol 3-phosphate can be visualised with molybdate reagent and the corresponding sample area removed. Elute the radioactive product from the gel in distilled water and adjust the specific radioactivity by the addition of unlabelled sn-glycerol 3-phosphate.

V. LIPID EXTRACTION, PURIFICATION AND QUANTITATIVE ANALYSIS

A. Extraction

It is possible that lipid derivatives may arise as artifacts during the storage of plant material or during the extraction of tissue, or membrane preparations, with organic solvents. For example, phosphatidic acid is generally present in only trace amounts in fresh developing seeds but can be generated upon the freezing and thawing of tissue prior to analysis (Wilson and Rinne, 1976). Certainly, developing seeds of most species, if stored at −20°C prior to use, may contain unusual amounts of this lipid. Most microsomal membrane preparations, on the other hand, which have been stored at −80°C, show no detrimental effect as evidenced by lipid analysis. Lipid artifacts, perhaps largely phosphatidylmethanol, may be formed by the breakdown of phosphatidylcholine and phosphatidylethanolamine during the extraction of fresh material with methanol (Roughan et al., 1978). The development of these artifacts appears to be due to a number of interacting factors such as slight differences in the homogenising procedures, variation in the time the tissue is exposed to the organic solvents and the relative volume of solvent to tissue. The steam killing of the tissue, usually for 15 min prior to extraction, has been found a convenient means of ensuring that little breakdown of the lipids occurs during extraction (Roughan et al., 1978). In the hands of the authors, however, very little breakdown of the complex lipids has been observed during the rapid extraction of the seed tissue and microsomal membrane preparations in modified Bligh and Dyer extraction media (Bligh and Dyer, 1959) and we favour the following extraction method.

1. Whole tissue

Tissue slices (up to 0.3 g) can be conveniently extracted in a glass boiling tube in 37.5 ml chloroform–methanol–0.15 M acetic acid (10:20:7.5, v/v) by homogenation with a polytron-type homogeniser at full speed for 30 s. Transfer the extract to a 50 ml glass-stoppered measuring cylinder and rinse the polytron probe and boiling tube with a further 10 ml chloroform. Bulk the chloroform washing with the extract. After the addition of 10 ml distilled water and thorough shaking, the chloroform and aqueous phases are allowed to phase separate for 30 min. The lower chloroform phase (20 ml), which contains the complex lipids and free fatty acids, is easily removed with a Pasteur pipette and transferred to a clean glass tube or flask and reduced to dryness under a stream of nitrogen at 50°C. Dissolve the lipid residue, immediately, in a small volume of chloroform and store at −20°C if further analyses are not to be carried out forthwith.

2. Microsomal membranes

Microsomal membranes, which have been utilised in *in vitro* experiments, can be rapidly extracted in a scaled down version of the Bligh and Dyer extraction. Normally (see below), incubation of the membranes is carried out in 1 ml reaction medium which contains all the desired substrates and cofactors. The reaction is terminated by the addition of 1 ml 0.15 M acetic acid and the lipids extracted by the addition of 3.75 ml chloroform–methanol (1:2; v/v) followed by 1.25 ml chloroform. The incubation and the subsequent extraction of the lipid, is conveniently carried out in a ground glass-stoppered tube (10 cm × 1.5 cm) which can be directly centrifuged (30 s) in a suitable bench centrifuge (Wifug) to aid the separation of the chloroform and aqueous phases. The lower chloroform phase is readily removed with a Pasteur pipette, and again this can be stored at low temperature as required or used immediately for further analyses. It is most convenient to transfer the chloroform extracts, which are to be stored, or otherwise, to tapered ground-glass stoppered tubes, 8.5 cm × 1 cm (Labora Ab, Sweden), as these enable the handling of small volumes and facilitate accurate analytical procedure. It should be noted, however, that if a quantitative recovery of *lyso*-phosphatidic acid is required then the membranes should be extracted in acidified butan-1-ol (Bjerve et al., 1974; see Section VI.A).

B. Purification of the Complex Lipids

The separation of the neutral and polar complex-lipids is best achieved by thin layer chromatography (TLC) on silica gel. Although such plates can be prepared in the laboratory, these do not generally give the degree of separation that is available with the commercial product. Particularly good resolution is obtained with Merck silica gel 60 pre-coated plates and these are used routinely in the authors' laboratories. A reactivation of the plate, at 110°C for 15 min, often helps to increase further the resolution of the chromatographic separation. Typically, the chloroform, which contains the lipid in the tapered tube, is reduced to dryness at 50°C under a stream of nitrogen and the lipids redissolved in a known volume of fresh chloroform (50–100 µl, depending on the concentration of lipid). It is usual to keep the tubes, containing the lipid in chloroform, on ice to reduce evaporation and breakdown. Aliquots of the chloroform can be applied to the thin-layer plate using a suitable syringe. We find that rapid and accurate application, of the sample in chloroform, to the gel is readily achieved with a small-volume automatic pipette of the Oxford or Finn type. The known volume in the pipette tip rapidly enters the gel by capillary action if the tip is placed on the origin and the body of the pipette raised vertically to the plate.

Although a number of solvent mixtures have been described for the purification of complex lipids, for most applications concerned with the lipids associated with the biosynthesis of the triacylglycerols, adequate separation of the polar lipid components is obtained with chloroform–methanol–acetic acid–water (170:30:20:6; v/v) and the neutral lipids with hexane–diethylether–acetic acid (70:30:1; v/v), respectively. After development and the evaporation of the solvent, the lipid areas on the plate can be located by the 'non-destructive' exposure to iodine vapour, either in a glass tank or applied by expelling air over iodine crystals loaded in a Pasteur pipette. Thus, by the relatively rapid and simple procedure of thin layer chromatography the major lipids, which are associated with nearly all the aspects of triacylglycerol biosynthesis (i.e. phosphatidic acid, phosphatidylcholine, diacylglycerol, triacylglycerol and unesterified fatty acid), are easily and rapidly resolved and purified for further analysis.

The analysis of the molecular species of complex lipid in a mixture has also been achieved by capillary gas–liquid chromatography (Geeraert and Sandra, 1987) and high-performance liquid chromatography (Kesselmeier and Heinz, 1985; Norman and St. John, 1986; Shukla, 1988). The application of these techniques, however, to biochemical studies has been somewhat limited and will not be dealt with here.

C. Fatty Acids and Acyl Constituents of Complex Lipids

The methyl esters of the unesterified fatty acids and the fatty acid constituents in the acyl lipids (FAMES) are routinely measured by gas–liquid chromatography (GLC) and numerous protocols for this are available in the literature. Whilst capillary GLC can offer some advantages (notably speed of separation) we prefer the standard GLC procedure since good base-line separations up to the C18:3 level can still be obtained in a few minutes. A glass column (2 m × 2 mm) containing 10% BDS on Chromosorb W (HP, 80–100 mesh) will give excellent separation of all the appropriate fatty acids.

Numerous methods exist for the preparation of the methyl ester derivative of the fatty acids, and again the laboratory concerned may have legitimate preferences. The only criteria which should be applied to the method are that it is relatively rapid, can adequately deal with the small amounts of lipid generally encountered in biochemical studies and does not give rise to background contamination. In these respects the methanolic-HCl method of Kates (1964) is dependable and can deal with unesterified fatty acids as well as the transmethylation *in situ* of the complex lipid fatty acid constituents. The gel, containing the 'lightly' iodine-stained lipid area, is dampened with distilled water from a chromatography spray, removed from the plate into a glass-stoppered tube (15 cm × 1.5 cm; screw-capped tubes with Teflon inserts are also adequate for the methylation procedure) and a few drops of dry methanol added. The methanol–water mixture is removed from the gel under a stream of nitrogen at 50°C and the dried gel covered with 2 ml 2.5% HCl in methanol (by weight). (The HCl–methanol mixture is best prepared by bubbling pure gaseous HCl through anhydrous methanol until the required concentration is reached. Providing the methylation mixture is well stopped and stored in the cold, it will remain potent for many months). The stoppered tube, which contains the gel and methanolic-HCl, should be refluxed in a heating block at 80°C for 20 min and allowed to cool. Extract the methyl esters of the fatty acids by the addition of 3 ml hexane followed by 2 ml distilled water and shake thoroughly. The hexane phase is removed to a tapered tube and reduced to 'just' dryness with nitrogen. Care must be taken not to overheat since the more unsaturated fatty acid derivatives are volatile and easily lost. Cool the tube on ice and dissolve the residue in hexane (about 20 µl), aliquots of which can be used for analysis by GLC and radio-GLC.

If a quantitative analysis of the methyl esters is required, then they can be measured relative to an internal standard such as methylheptadecanoic acid, which should be added in hexane just prior to the addition of the volumes of hexane and water for phase separation. The method outlined is also appropriate for the quantification of the acyl-CoA preparations and the determination of the concentration of particular complex lipids in whole tissue and in the microsomal membranes.

D. Silver Nitrate Thin Layer Chromatography of Fatty Acid Methyl Esters

Argentation thin layer chromatography utilises the property of silver salts for forming reversible polar complexes with the double bonds in the aliphatic moieties of lipids (Morris, 1966). The method is useful for the separation of fatty acids based on their degree of unsaturation, and is most convenient for the purification of the mono-, di-, and tri-enoic acid fractions which precedes the determination of radioactivity. If radio-GLC techniques are unavailable then argentation-TLC of the fatty acid methyl esters, from the complex lipids, can give adequate information on the synthesis of the C18-polyunsaturated fatty acids and the assembly of complex lipids from labelled precursors. Argentation chromatography has also proved useful in the separation of molecular species of triacylglycerol in plant oils (Shewry *et al.*, 1972; Gurr, 1980).

Silver nitrate impregnated thin-layer plates are best prepared from commercial silica gel pre-coated plates (Merck silica gel 60, 0.25 mm thick), since these yield excellent resolution of the fatty acid species. Dissolve silver nitrate (0.5 g) in 2 ml distilled water and add acetone (about 12 ml), taking care not to precipitate the silver salt. Pour the solution evenly over one 20 × 20 cm silica gel plate giving a 15% (w/w) of silver nitrate in the gel. After the acetone has evaporated, re-activate the plate at 100°C for 20 min. Exposure to light can cause a rapid blackening of the gel and for this reason plates should be prepared just prior to use and the chromatographic separation carried out in dim light. Apply the methylated fatty acid samples in hexane to the plate and develop in hexane–diethyl ether–acetic acid (85 : 15 : 1; v/v). Depending on the relative concentration of the fatty acids in the sample, further resolution can sometimes be achieved by a double development in the same solvent. The methyl esters can be visualised under UV light after spraying the plate with 2,7-dichlorofluorescein solution (0.02% dichlorofluorescein in ethanol, w/v). The gel, which contains the fatty acid methyl esters of interest, can be removed from the plate and assayed for radioactivity taking care to use a liquid scintillation cocktail (see below) that does not react with the silver nitrate and/or acetic acid. If necessary, the methyl esters can be eluted from the gel and the individual fatty acid components in each fraction analysed by GLC.

E. Purification of the Acyl-CoA Fraction

A number of methods have been published for the purification of acyl-CoA in crude enzyme-assay systems or from whole tissue (Baker and Schooley, 1981; Juguelin and Cassagne, 1984; Woldegiorgis et al., 1985). Acyl-CoA differs sufficiently in polarity from most of the other complex lipids present in biological preparations that a chloroform–methanol–water extraction and phase separation results in the CoA derivatives largely entering the aqueous phase, with some 95% of the complex lipid in the chloroform phase. The recovery of the acyl-CoA in the aqueous phase is quantitative, providing bovine serum albumin is included in the reaction mixture (Stymne and Glad, 1981), and is largely present associated with denatured protein. If the analysis only requires the determination of radioactivity of the acyl component in the acyl-CoA, an alkaline hydrolysis of the aqueous phase, followed by acidification and extraction of the unesterified fatty acid in diethylether, is often adequate. If, on the other hand, the acyl composition of the acyl-CoA fraction is to be determined by GLC then the acyl-CoA has to be purified further by TLC or HPLC. Since most of the acyl-CoA is tightly bound to protein (even in the absence of bovine serum albumin in the reaction mixtures), any such chromatographic method must be preceded by the removal of the protein from the mixture. This can be achieved by passing the sonically dispersed proteins in the methanol–water phase through a C_{18} silica gel cartridge (SEPAK™), onto which essentially all the acyl-CoA becomes bound. The acyl-CoA fraction can be eluted from the gel in 0.4 M NH_4OH in methanol–water and no interfering precipitate of salt is obtained. Although some of the acyl-CoA undergoes hydrolysis in the alkaline ammonium solution, the rapid evaporation of the eluting solvent in a vacuum rotary evaporator, at 30°C, minimises the loss to some 10–20%. We carry out subsequent purification by TLC on silica gel with butan-1-ol–acetic acid–water (5 : 2 : 3; v/v) as the developing solvent (Stymne and Stobart, 1985). The acyl-CoA, on the thin layer plate, is visualised as an intense white area after gently spraying the plate with water. The methyl esters of the fatty acids in the acyl-CoA can be preapred using the methanolic-HCl outlined above. Since there may be substantial losses in the acyl-CoA during such a laborious purification it is necessary, for quantification, to add an internal standard, usually heptadecanoyl-CoA, as early as possible in the extraction procedure.

VI. STEREOSPECIFIC ANALYSIS OF FATTY ACIDS IN COMPLEX LIPIDS

A. Phosphatidylcholine and Phosphatidic Acid

Phospholipase A_2 will hydrolyse the ester bond in position *sn*-2 of glycerophosphatides and release the free fatty acid (Christie, 1982). The products of the enzymic digestion, free fatty acid and the lysophosphatide, can be purified and the fatty acids present at both the intramolecular positions determined.

The 'dampened' gel, which contains the separated phospholipid, is removed from the thin-layer plate (see above) and transferred to a glass column for elution. A Pasteur pipette is most suitable for this. The lipids elute relatively rapidly in 3 ml methanol–chloroform–acetic acid (2 : 1 : 0.1; v/v) followed by 3 ml methanol. The bulked eluates should be reduced to dryness in a glass tube. Redissolve the lipid residue in 1 ml diethylether and disperse by sonication, for 1 min. Add 1 ml 0.1 M buffer, pH 8.9, containing 5 mM $CaCl_2$ (Tris-HCl for the phosphatidylcholine and borate-buffer with the phosphatidic acid) and 25 units of commercial phospholipase A_2. Phospholipase A_2 preparations are available from a number of sources; however, the enzyme from Indian cobra venom (*Naja naja*) is particularly potent. After incubation for 60 min at 25°C, with constant sonication, the ether is removed under nitrogen and the products extracted in 1 ml acidified butan-1-ol saturated with water (butan-1-ol–water–acetic acid; 1 : 9.5 : 0.05; v/v) followed by a further 1 ml of butan-1-ol (Bjerve *et al.*, 1974). The butan-1-ol extracts are bulked and evaporated to dryness under nitrogen. The lipid residue is dissolved in a small volume of chloroform and the *lyso*-phosphatides and free fatty acids separated from each other by TLC in the solvent system given above for the phospholipids. The *lyso*-derivatives and the fatty acids are particularly well visualised after spraying with distilled water and the gel can be removed either for direct radioassay or methylation and further analysis by GLC or radio-GLC, as required.

B. Triacylglycerols

No efficient method is available for the determination of the intramolecular distribution of fatty acids in micro-quantities of triacylglycerol. Providing, however, that reasonable amounts of material, usually mg quantities, are available, then suitable procedures can be utilised (Brockerhoff, 1972). A stereospecific analysis of triacylglycerol generally requires the preparation of partial glycerides which can be readily converted to a phospholipid (Myher and Kuksis, 1979). The resulting phospholipid, in its turn, can be hydrolysed with stereospecific phospholipase. In the most efficient approach, the triacylglycerol is partially hydrolysed with a Grignard reagent to yield equimolar mixtures of the 1,2- and the 2,3-diacylglycerol isomers from which the phosphatidylcholines can be prepared. A typical procedure (see Christie, 1982), which has worked well in the authors' laboratories, is outlined. The triacylglycerol, up to 40 mg, is dissolved in 2 ml dry diethylether and freshly prepared ethyl MgBr in ether (1 ml, 0.5 M) is added. After sonication for 1 min, 0.05 ml acetic acid and 2 ml distilled water are added. The products are extracted in 3 × 10 ml portions of ether and washed with 5 ml aqueous potassium bicarbonate (2%, w/v) followed by 5 ml distilled water. After drying the ether extract over anhydrous sodium sulphate the ether is removed under nitrogen and the residue dissolved in a small volume of chloroform. The isomers of the diacylglycerol can be separated by thin layer chromatography on silica gel, which contains 5% (by weight) boric acid, with hexane–diethylether (1:1; v/v) as the developing solvent. After visualising with iodine vapour the gel, which contains the diacylglycerols, is removed and the lipids eluted in 100 ml diethylether. Evaporate the ether under nitrogen, dissolve the residue in 0.65 ml chloroform–pyridine–phosphorus oxychloride (47:47:5; v/v) and keep at 4°C for 1 h followed by a further 60 min at room temperature. Add 200 mg dry, powdered, choline chloride and stir at 30°C for 12 h and then for 30 min after the addition of 20 µl distilled water. Remove the bulk of the solvent under nitrogen and extract the products in 12 ml chloroform–methanol–water–acetic acid (50:39:10:1; v/v) and phase separate against 4 ml 4 M ammonia solution. Re-extract the aqueous phase with 2 ml of fresh lower phase. Bulk the organic phases and evaporate to dryness under nitrogen. Redissolve the residue in chloroform and apply to a CM-Cellulose column (Whatman CM-52, Na$^+$ form, packed into a Pasteur pipette in chloroform) and pass through 10 ml chloroform to elute the unreacted diacylglycerols. The phosphatidylcholines are eluted in 10 ml chloroform–methanol (9.25:0.75; v/v). The recovered phosphatidylcholines can then be hydrolysed with phospholipase A_2 as outlined above and the resulting fatty acids and *lyso*-derivatives purified, methylated and analysed. The distribution of fatty acid at the *sn*-3 position is deduced from the acyl composition of the total triacylglycerol and those present at positions *sn*-1 and -2 of the derived phosphatidylcholine.

$$\text{Position } sn\text{-}3 = 3[\text{total triacylglycerol}] - [sn\text{-}1] - [sn\text{-}2]$$
$$\text{(mol \%)} \qquad \qquad \text{(mol \%)}$$

In some instances (Christie, 1982) the phospholipase A_2 hydrolysis of the phosphatidylcholine is good for the fatty acids present in the *lyso*-derivative, but may not be too accurate for the liberated fatty acids from the *sn*-2 position. With the phospholipase A_2 from cobra venom, however, we invariably find that the mol ratio of fatty acids from positions *sn*-1 and -2 of the phosphatidylcholine is unity. This, of course, should be checked during the development of the method.

VII. DETERMINATION OF RADIOACTIVITY IN LIPID SAMPLES

The radioassay of the individual lipid components, separated by thin layer chromatography, is best achieved by liquid scintillation techniques. Whilst equipment is available for the radioscanning of plates it is generally difficult, with these methods, to obtain accurate quantitative measurements which can be relied upon. The 'water-dampened' gel, which contains the lipid sample, is easily removed from the plate and placed directly into a liquid scintillation counting vial for radioassay. Although many suitable commercial liquid scintillant cocktails are available, the xylene-based formulation, Phase Combined System (PCS™; Amersham/Searle) is particularly good and gives low background and quenching even for samples which contain large quantities of the gel material. PCS™, however, is unsuitable for the radioassay of lipids and the methyl esters of fatty acids which have been purified by argentation thin layer chromatography with solvents which contain acetic acid, since a rapid quenching is often experienced. Other scintillant fluids, such as the PPO-POPOP combinations, are more desirable.

The radioactivity in fatty acid methyl esters, obtained either from the unesterified fatty acids or the transmethylated complex-lipids, can be determined after argentation-TLC or, more conveniently, directly by radio-GLC.

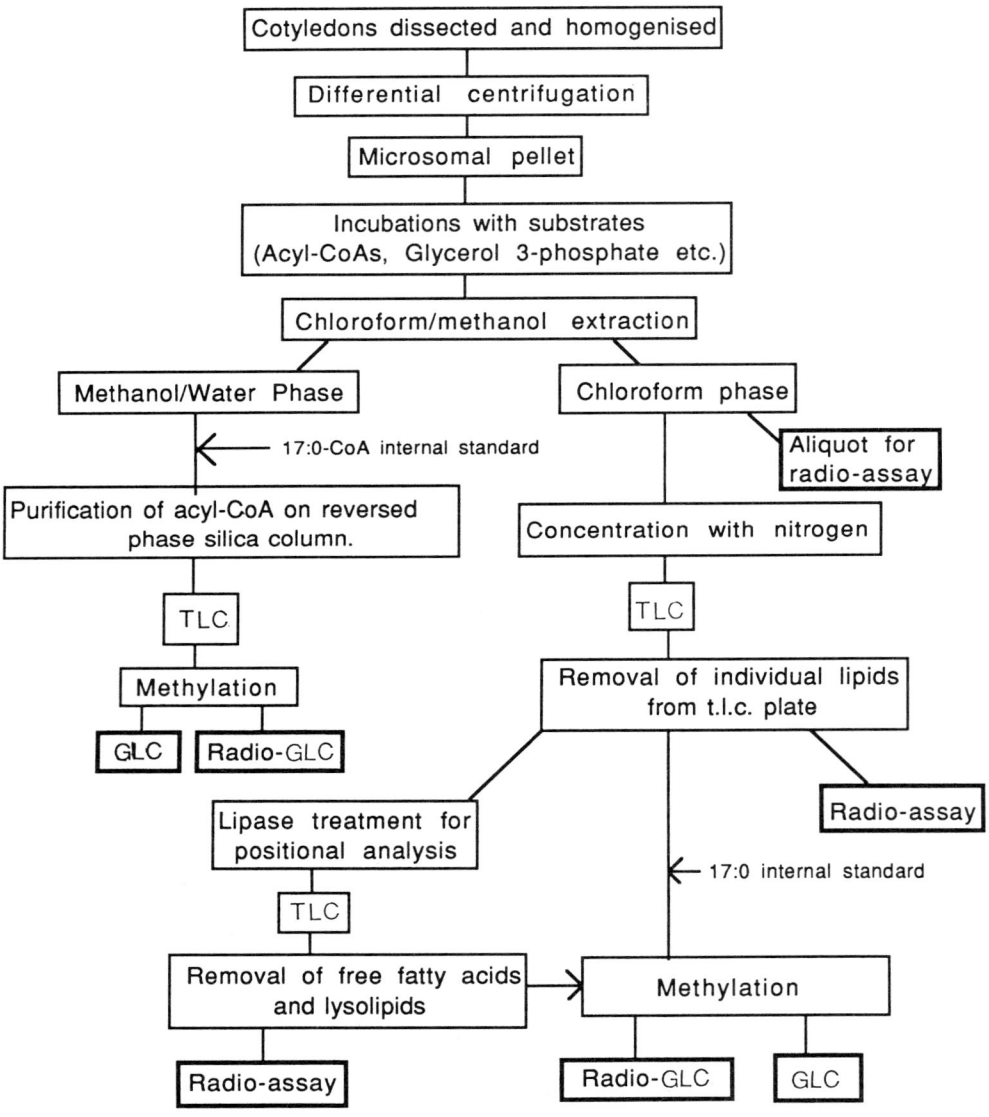

FIG. 2.3 Protocol for a typical experiment on triacylglycerol biosynthesis (see text for details).

VIII. PROTOCOL FOR A TYPICAL EXPERIMENT

The protocol for a typical experiment, which utilises radioactive substrates and a microsomal membrane preparation from developing seed cotyledons, is depicted in Fig. 2.3. For nearly all the seed species studied, 1–2 g fresh weight of developing cotyledonary tissue is ample to provide a suitable microsomal preparation which can be incubated with radioactive substrate in, for example, a time course study of triacylglycerol assembly and/or C18-polyunsaturated fatty acid synthesis. Aliquots of the microsomal membrane preparation, usually containing 50–150 nmol phosphatidylcholine (0.2–0.7 mg microsomal protein), depending on the experiment, are incubated, in glass tubes at 30°C with constant shaking, with radioactive acyl-CoA (usually 200 nmol), *sn*-glycerol 3-phosphate (400 nmol), bovine serum albumin (10 mg) and $MgCl_2$ (20 mM), in a total volume of 1 ml with 0.1 M-phosphate buffer, pH 7.2. After the desired time of incubation, the reaction is stopped by the addition of the modified Bligh and Dyer medium and the two phases separated by the direct centrifugation of the incubation tube. The chloroform phase is removed into a small tapered glass tube and an aliquot, usually one hundredth, removed for radioassay by liquid scintillation. The remaining chloroform is reduced to dryness under nitrogen and the lipid redissolved in 50–100 µl chloroform. Five-microlitre aliquots can then be taken and applied to TLC plates for the purification of the polar and neutral lipids, and after visualisation in iodine vapour the lipid areas of the gel are removed and each individual lipid assayed for radioactivity. It is also necessary to remove the remaining gel, other than that which contains the lipids of interest, and to assay it for radioactivity. The results can then be expressed as a percentage of the total radioactivity recovered from the plate and calculated in absolute terms based upon the total radioactivity in the chloroform phase and the known specific radioactivity of the substrates used in the experiment.

The remaining chloroform extract (after the removal of the two 5 µl aliquots for the lipid analysis) can be applied to further TLC plates and the lipids separated and transmethylated for GLC and radio-GLC analysis. If required at this stage, the purified phospholipids could be analysed for the intramolecular distribution of their acyl components and the radioactivity which they contain. The aqueous methanol phase, on the other hand, can be used for the determination of the radioactivity and the fatty acid content of the acyl-CoA fraction.

IX. ASSAY OF THE ENZYMIC STEPS IN TRIACYLGLYCEROL BIOSYNTHESIS

A. Acyl-CoA: Lysophosphatidylcholine Acyltransferase (LPCAT; EC 2.3.1.23)

$$\text{Acyl-CoA} + \textit{lyso}\text{-phosphatidylcholine} \rightarrow \text{phosphatidylcholine} + \text{CoA} \qquad (1)$$

The enzyme catalyses what could be considered a pivotal reaction in the regulation of the fatty acid quality of the triacylglycerols. The enzyme directs the oleate, from oleoyl-CoA, towards position *sn*-2 of *sn*-phosphatidylcholine, and so competes, to some extent, with those enzymes which are involved in the acylation of the glycerol backbone. The

LPCAT can be monitored both by a discontinuous method, which utilises labelled acyl-CoA substrates, or by a continuous spectrophotometric assay.

1. Discontinuous assay

Microsomal membrane preparations (equivalent to 0.1 mg protein) are incubated with palmitoyl-*lyso* phosphatidylcholine (100 nmol) and [^{14}C]oleoyl-CoA (160 nmol) for 2 min, after which the lipids are extracted from the membranes, the phosphatidylcholine purified by TLC, and the incorporation of the [^{14}C]oleate determined. Microsomal membranes from the developing cotyledons of safflower are particularly active and specific activity values of 300 nmol oleate incorporated min^{-1} (mg protein)$^{-1}$ are often observed.

2. Continuous assay

The continuous formation of the coloured complex, which forms between free CoA and 5,5-dithiobis(2-nitrobenzoic acid) (DTNB), is monitored spectrophotometrically at 405 nm. Theoretically, the reaction of the DTNB with liberated CoA can be used in the assay of all the acyltransferase enzymes involved in the biosynthesis of the triacylglycerols. In crude enzyme preparations (microsomal membrane, homogenate) from most oil seeds, however, the spectrophotometric assay is only reliable with the LPCAT. The activities of the glycerol 3-phosphate acylating enzymes, for example, are usually too low, relative to the background hydrolysis of the acyl-CoA and other non-specific reactions which occur between the DTNB and some proteins. During attempts at the purification of oil seed acyltransferases, the spectrophotometric assay may provide a rapid screen for the appropriate enzyme.

To assay the LPCAT, the reaction mixture contains DTNB (50 nmol), oleoyl-CoA (20 nmol), *lyso*-phosphatidylcholine (30 nmol) and the enzyme–membrane preparation, in a total volume of 1 ml with 0.1 M phosphate buffer, pH 7.2. The reference cuvette should contain all the reactants but with the omission of the *lyso*-phosphatidylcholine. The reaction is started by the addition of the enzyme preparation. The increase in the absorbance at 405 nm is generally linear for 5 min. An increase of 0.01 A is equivalent to the release of 0.73 nmol free CoA in a volume of 1 ml (mol extinction coefficient = 13 600).

B. Acyl-CoA: Lysophosphatidylcholine Acyltransferase and Acyl-exchange

The combined forward and reverse reactions of the LPCAT results in the exchange of fatty acids between position *sn*-2 of *sn*-phosphatidylcholine and the acyl-CoA (Reaction 2):

$$\text{X-phosphatidylcholine} + \text{Y-CoA} \rightarrow \text{X-CoA} + \text{Y-phosphatidylcholine} \quad (2)$$

The initial step in the exchange is the trans-esterification of the fatty acids, from the *sn*-2 position of the *sn*-phosphatidylcholine, to the CoA by the reverse reaction of the LPCAT (Reaction 3):

X-phosphatidylcholine + CoA → *lyso*-phosphatidylcholine + X-CoA   (3)

The next stage in the exchange is the reacylation of the generated *lyso*-phosphatidylcholine by the forward reaction of the LPCAT (see Reaction 1).

The exchange of fatty acids between the acyl-CoA and phosphatidylcholine is readily measured by following the incorporation of [^{14}C]acyl groups, from labelled acyl-CoA, in the presence of a high concentration of bovine serum albumin and free CoA (Stymne and Stobart, 1984). The LPCAT catalyses the equilibration of fatty acids between the phosphatidylcholine and the acyl-CoA pool and the incorporation of radioactive fatty acids rapidly ceases as the acyl-CoA becomes progressively diluted with acyl groups derived from the phosphatidylcholine. In microsomes of safflower some 30% of the fatty acids in position *sn*-2 of the *sn*-phosphatidylcholine are exchanged within 5 min. Prolonged incubation times, however, with a large pool of acyl-CoA will yield a maximum exchange of 60% of the fatty acids in position *sn*-2 of the *sn*-phosphatidylcholine. Confirmation of the acyl exchange can be achieved by an analysis of the fatty acid constituents in the acyl-CoA fraction (Stymne and Stobart, 1987). A typical reaction mixture consists of [^{14}C]oleoyl-CoA (200 nmol), bovine serum albumin (10 mg), free CoA (400 nmol) and microsomal membranes (0.4 mg microsomal protein) in 0.1 M phosphate buffer, pH 7.2, in a final volume of 1 ml. After 5 min incubation at 30°C, the reaction is stopped and the incorporation of the radioactivity in the phosphatidylcholine determined.

C. Desaturase Enzymes and C18-polyunsaturated Fatty Acid Synthesis

The C18-polyunsaturated fatty acids, linoleic acid (C18:2, *n*-6), α-linolenic acid (C18:3, *n*-3) and γ-linolenic acid (C18:3, *n*-6) in oil seeds, are synthesised by the action of specific desaturase enzymes acting on the acyl substrates which are associated with the phospholipids. Although it seems that phospholipids other than phosphatidylcholine can undergo desaturation (Stymne *et al.*, 1987), it is only this complex lipid which is actively turned over and is involved in the supply of C18-polyunsaturated fatty acids for triacylglycerol assembly in oil seeds. Microsomal membrane preparations from the developing seeds of many oil seed species (safflower, soy, sunflower, borage) all possess active Δ12 desaturase enzymes and will catalyse the synthesis of linoleate from oleate in the phosphatidylcholine (Stymne and Stobart, 1987). Similar preparations from the cotyledons of borage also have Δ6 desaturase activity and convert linoleate into γ-linolenate (Stymne and Stobart, 1986a; Griffiths *et al.*, 1988c). The formation of α-linolenate (i.e. the activity of the Δ15 desaturase) has, however, proved difficult to demonstrate with any consistency in *in vitro* membrane preparations. Homogenates of linseed cotyledons, however, can yield reproducible Δ12 and Δ15 desaturase activity, under optimum conditions (Stymne *et al.*, 1989).

Since complex lipids are the substrates for the desaturase enzymes, problems are encountered in the introduction of these hydrophobic compounds to the active site of the enzyme in *in vitro* assays. Radioactive oleate in oleoyl-phosphatidylcholine, presented as an emulsion in detergent (Pugh and Kates, 1975) or dissolved in ethanol (Slack *et al.*, 1979), is converted to linoleoyl-phosphatidylcholine in microsomal preparations of yeast and safflower, respectively. The rates of desaturation, however, are an order of magnitude less than observed with *in situ* labelled substrates. The acyl exchange between acyl-CoA and the phosphatidylcholine (see above) in the microsomal preparations of most oil seed species, allows the more natural introduction of the acyl substrate into the complex lipid for the study of the desaturase enzyme activity. Thus, microsomal phosphatidylcholine can be readily loaded with [^{14}C]acyl substrate under non-desaturating conditions, i.e. in the absence of NADH, and desaturase activity initiated by the addition of reductant to the incubation mixture (Stymne and Stobart, 1987). In a typical assay the microsomal membrane preparation (equivalent to some 100 nmol phosphatidylcholine) is incubated with [^{14}C]oleoyl-CoA (100 nmol), bovine serum albumin (10 mg) and CoA (400 nmol) in a total volume of 1 ml with 0.1 M phosphate buffer, pH 7.2, at 30°C, for 20 min. The incubation mixtures are then diluted with buffer and the membranes recovered by a further centrifugation. The resulting pellet is virtually free of the acyl-CoA and, after resuspending, yields membranes containing sn-phosphatidylcholine which is enriched with [^{14}C]oleate at the sn-2 position. Upon the addition of NADH (3 µmol), the Δ12 desaturase will catalyse the conversion of the [^{14}C]oleate to [^{14}C]linoleate. At intervals, the radioactivity in the acyl substrate/product in the phosphatidylcholine can be determined, either by radio-GLC or argentation TLC of the fatty acid methyl esters. In microsomal preparations from high linoleate varieties of safflower, rates of Δ12 desaturase activity of some 12 nmol [^{14}C]oleate desaturated min^{-1} (mg microsomal protein)$^{-1}$ are found. When rates of desaturation are measured, it is also essential to know accurately the specific radio-activity of the acyl substrate and the extent of its dilution by the endogenous fatty acids in the phosphatidylcholine. It should be noted, however, that standard kinetics of the desaturase enzymes, which utilise complex lipid substrates, are difficult to study. The desaturase enzyme-protein is, in effect, moving in a bilayer of its own complex-lipid substrate and thus the actual concentration of the substrate, which is associated with the enzyme, is impossible to determine or to manipulate easily. Acyl exchange, however, offers the opportunity to study the kinetics of desaturation in more detailed manner since it allows the amount of the acyl substrate, at position sn-2 of the sn-phosphatidylcholine, to be altered in a more precise fashion. In microsomal membrane preparations from safflower, the oleate in the phosphatidylcholine has been increased from 2 to 30% in this way and under these conditions the rate of desaturase activity is directly related to the concentration of the oleate in the complex lipid (Stymne and Stobart, 1986b).

The desaturation of oleate to yield linoleate (Δ12 desaturase activity) will occur with the acyl substrate at both positions sn-1 and sn-2 of the microsomal sn-phosphatidylcholine (Slack et al., 1979; Stobart and Stymne, 1985b; Griffiths et al., 1988b). Whether this represents the presence of different isoenzymes of the same desaturase remains to be determined. Desaturase activity at both the acyl-containing positions of the phosphatidylcholine is readily demonstrated. Microsomal membranes (5 mg microsomal protein), which have good triacylglycerol synthesising capacity, can be incubated with glycerol 3-phosphate (2 µmol), bovine serum albumin (40 mg) and MgCl$_2$ (10 µM) in a total volume of 4 ml with phosphate buffer. At regular 5 min intervals, [^{14}C]oleoyl-CoA (60 nmol) is added in a total incubation time of 45 min (i.e. a total of 480 nmol substrate added throughout the incubation period). Under these conditions the [^{14}C]oleate enters both the sn-1 and sn-2 positions of the sn-phosphatidylcholine (see Fig. 2.2). After washing and the recovery of the membranes, aliquots can be taken, desaturation initiated by the addition of NADH and the conversion of radioactive substrate into product determined at both positions in the purified phosphatidylcholine.

In some situations a further approach can be used to measure desaturase activities. It is possible to monitor, by GLC, the mass of endogenous fatty acids in the microsomal phosphatidylcholine under desaturating and non-desaturating conditions (Stymne et al., 1987; Griffiths et al., 1988c). In microsomal preparations from developing oil seeds, incubated at 25°C, there is little turnover of the fatty acids in the phosphatidylcholine and this is due to the lack of essential substrates, such as acyl-CoA and glycerol 3-phosphate. The addition of NADH to such membranes results in the desaturation of the acyl substrate in the complex lipid. The rate of desaturation is relatively easily determined by comparing the fatty acid composition, as determined by GLC, of the purified phospholipids. If desaturation rates, at the different positions of the phospholipid, are required, then the purified lipid can be partially hydrolysed with phospholipase A_2 and the methyl esters of the free fatty acids (position 2) and the resulting *lyso*-derivative (position 1) determined by GLC. In microsomal preparations from the developing seeds of borage, the desaturation of the endogenous oleate at both positions of the phosphatidylcholine remained linear for 20 min and the desaturation of the linoleate to γ-linolenate (Δ6 desaturase activity and which occurs essentially at position *sn*-2) for 60 min (Griffiths et al., 1988c). The same approach has been used successfully to study both the Δ12 and the Δ15 desaturase enzymes in dilute homogenate preparations from the developing cotyledons of linseed (Stymne et al., 1989). The method, however, is less suitable for oil seeds, such as the high linoleate varieties of safflower, with very high desaturase activities. Here, the activity of the desaturase enzymes appears to be non-rate-limiting and there is little endogenous acyl substrate actually present in the phosphatidylcholine. Desaturase activity, under these circumstances, which is measured on the basis of mass changes in the endogenous fatty acids of the complex lipid, tends to be non-linear and well below maximal rates.

It may also be possible to study, in some cases, desaturase activity *in vivo* in whole or sliced cotyledons by manipulating the oxygen tension which surrounds the experimental material. Whole cotyledons of linseed, suspended in buffer which has been purged with nitrogen gas for 5 min, and then incubated, under nitrogen, with radioactive acyl substrates (as their ammonium salts) rapidly take up and incorporate the radioactive substrates into the complex lipids. Desaturase activity is prevented by lack of oxygen and the subsequent release of air into the system results in the rapid formation of the C18-polyunsaturated products in the phospholipids. In the case of chlorophyllous cotyledonary tissue, it is necessary to carry out the incubations in the dark in order to prevent the photosynthetic production of oxygen.

D. Glycerol 3-Phosphate Acylating Enzymes

Glycerol 3-phosphate + acyl-CoA → *sn*-1 acyl-*lyso*-phosphatidic acid
sn-acyl-*lyso*-phosphatidic acid + acyl-CoA → phosphatidic acid

The formation of phosphatidic acid, from glycerol phosphate, is a two-step event and requires the participation of two distinct enzymes. The first enzyme, a glycerophosphate acyltransferase (EC 2.3.1.15), brings about the acylation of *sn*-glycerol 3-phosphate at the *sn*-1 position to form a *lyso*-phosphatidic acid. The *lyso*-phosphatidic acid forms the substrate for the second enzyme, 1-acylglycerol-3-phosphate acyltransferase (EC 2.3.1.51), which completes the acylation at the *sn*-2 position with the formation of phosphatidic acid. In most oil seed preparations, the limiting step is at the first acylation and the *lyso*-phosphatidic acid intermediate is usually not observed. In microsomal preparations, the total glycerol phosphate acylating activity is easily determined. The membranes are incubated with [^{14}C]glycerol 3-phosphate (400 nmol), the appropriate acyl-CoA (100 nmol) and bovine serum albumin (10 mg) in a total volume of 1 ml with 0.1 M phosphate buffer, pH 7.2. After incubation for the desired time interval, the incorporation of label in the total lipid is determined and is usually linear for 30 min. Because of the differing acyl specificity properties of the glycerol phosphate acylating enzymes it is more usual to incubate the microsomal preparations with a mixed acyl-CoA substrate (100 nmol each of palmitate and linoleate) in order to promote optimal activities (Griffiths *et al.*, 1985). In safflower preparations, rates of incorporation can be as high as 10 nmol glycerol 3-phosphate incorporated $\min^{-1}$ (mg protein)$^{-1}$.

It may, at times, be more appropriate to determine the incorporation of glycerol into phosphatidic acid and under conditions which limit its further metabolism. In safflower preparations this is possible by incubation with the usual substrates and cofactors but in the presence of EDTA (10 mM), which inhibits the action of the phosphatidate phosphohydrolase and under these conditions the phosphatidate accumulates (Stobart and Stymne, 1985a). The phosphatidate is easily purified by TLC and, if necessary, the acyl complement at positions *sn*-1 and *sn*-2 determined after treatment with phospholipase A_2.

A 20 000 × g particulate fraction from maturing cotyledons of safflower was also reported to have glycerol 3-phosphate acyltransferase activity (Ichihara, 1984) and the membranes, in the presence of glycerol phosphate and acyl-CoA (palmitoyl- or linoleoyl-CoA), accumulated *lyso*-phosphatidate which was exclusively esterified at the *sn*-1 position. We find, however, that microsomal preparations from safflower will only accumulate *lyso*-phosphatidate, in any quantity, with saturated species of acyl-CoA.

The 1-acyl-glycerol 3-phosphate acyltransferase is relatively easy to determine in microsomal preparations from oil seeds. In this case the reaction mixture contains *lyso*-phosphatidate (1-acyl-glycerol 3-phosphate; 100 nmol), [^{14}C]linoleoyl-CoA (100 nmol), bovine serum albumin (10 mg), EDTA (10 mM) and membranes, in a total volume of 1 ml with phosphate buffer. After the desired time of incubation the incorporation of radioactivity in the phosphatidate can be determined. Recently, Ichihara *et al.* (1987) have examined the enzyme in preparations from safflower, using a spectrophotometric assay based on the reduction of NbS_2 by the sulphhydryl group of the liberated free CoA. The reaction mixture included sorbitol (0.4 M), EDTA (0.25 M), spermidine (2 mM), NbS_2 (0.2 mM), linoleoyl-CoA (15 µM) and 35 µg protein in a final volume of 3 ml with 0.15 M Tris-HCl buffer, pH 8.2. The coloured complex was monitored at 405 nm for 1 min at 30°C and the liberated CoA determined using a molar extinction coefficient of 13 600. The addition of the polyamine is interesting and was found to stimulate the activity of the enzyme by 'alleviating the inhibitory effects of divalent cations' such as Mn^{2+}, and, to a lesser extent, Mg^{2+}.

E. Phosphatidic Acid Phosphohydrolase (Phosphatidase; PAP)

$$\text{Phosphatidic acid} + H_2O \rightarrow \text{diacylglycerol} + P_i$$

The dephosphorylation of phosphatidate is an important step in the biosynthesis of glycerolipids and the products of the reaction, *sn*-1, 2 diacylglycerols, serve as direct precursors of phosphatidylethanolamine, phosphatidylcholine and the triacylglycerols. The activity of the PAP (EC 3.1.3.4) in microsomal preparations of most oil seed species tends to be limiting to some extent. In incubations, with glycerol phosphate and acyl-CoA substrates, phosphatidate is usually an obvious product on TLC of the polar lipid fraction (Stobart and Stymne, 1985b). In incubations *in vivo* with cotyledon slices and the appropriate substrates, however, little phosphatidate is found (Griffiths *et al.*, 1988b). The enzyme appears to be weakly bound to the microsomal membrane and may be lost, in some instances, to the supernatant during preparation.

Assays for PAP activity have been based upon the release of inorganic phosphate, which can be measured spectrophotometrically as the molybdate complex (Moore *et al.*, 1973) or as radioactivity if a ^{32}P-substrate is used (Hosaka *et al.*, 1975). Phosphatidate, which is labelled in its fatty acid moiety, is, however, commercially available and this should be most useful in the assay of PAP in oil seed preparations. The animal enzyme has been assayed with such a radioactive phosphatidate substrate, presented in the reaction mixtures as a mixed emulsion with phosphatidylcholine (Butterworth *et al.*, 1984). The emulsion is considered to be a better substrate than pure phosphatidate and it probably resembles the natural form of the substrate in the microsomal membrane. After the appropriate time of incubation, the radioactivity incorporated into the diacylglycerol can be determined. In the case of oil seed preparations, it is also necessary to analyse the phosphatidylcholine since most preparations will interconvert, to some extent, this phospholipid with the formed diacylglycerol.

In some instances it is possible to generate radioactive phosphatidate in the microsomal membrane and then follow its consumption in diacylglycerol formation (Stobart and Stymne, 1985a), and again this can be used as a measure of PAP activity. Microsomal preparations (2 mg protein) are incubated with [^{14}C]glycerol 3-phosphate (2 µmol) and EDTA (10 mM). At 5 min intervals, aliquots of acyl-CoA (160 nmol) are added to the incubation for up to 80 min, after which, the mixture is diluted with fresh phosphate buffer and the membranes recovered by centrifugation at $100\,000 \times g$, 60 min. The procedure effectively removes residual substrate and will yield membranes which contain enhanced levels of radioactive phosphatidate. The PAP can then be reactivated in the resuspended membranes by the addition of $MgCl_2$ (20 mM) and the radioactivity which enters the diacylglycerol determined. With adequate controls the activity of the PAP is easily established.

F. Interconversion of Diacylglycerol with Phosphatidylcholine

During the flow of glycerol backbone through the so-called Kennedy pathway, towards triacylglycerol, there is an interconversion of the diacylglycerol and phosphatidylcholine pools. Such a reaction will make available more polyunsaturated species of diacylglycerol for the final assembly of the triacylglycerol and so may have some importance in determining the acyl quality of the final oil. The enzyme, CDP-choline: 1,2-diacylglycerol cholinephosphotransferase (cholinephosphotransferase; EC 2.7.8.2), is a possible candidate for catalysing this reaction (Roughan and Slack, 1982). Our understanding of this enzyme in oil seed systems is still somewhat scanty. The activity of the cholinephosphotransferase has been measured in microsomal preparations of safflower (Slack *et al.*, 1985) using the following assay system. Microsomes (0.1 mg protein) were incubated at 30°C with [^{14}C]-CDPcholine (0.5 mM), MgCl$_2$ (15 mM), EDTA (1 mM), EGTA (0.5 mM), DTT (5 mM) in 60 mM Tricine-KOH buffer, pH 8.0, in a total volume of 0.1 ml. The incorporation of the radioactive choline into the phosphatidylcholine was assayed against time and the rate of activity was 6–10 nmol min^{-1} (mg protein)$^{-1}$. The rate of the back reaction from phosphatidylcholine to diacylglycerol was also measured and was some 1 nmol min^{-1} (mg protein)$^{-1}$.

Studies on the interconversion of diacylglycerol with the phosphatidylcholine in safflower, however, would indicate that the cholinephosphotransferase is almost freely reversible (Stobart and Stymne, 1985a). Measurements of this can be achieved by generating large quantities of non-radioactive phosphatidate in the microsomal membrane by the methods outlined above and then incubating the resuspended and washed microsomes (0.2 mg protein) with 1-palmitoyl-sn-lyso-phosphatidylcholine (46 nmol) and [^{14}C]oleoyl-CoA (20 nmol) for 5 min. At this stage, the membranes contain non-radioactive phosphatidate and labelled phosphatidylcholine. On the addition of MgCl$_2$ (20 mM) to the incubation mixtures the phosphatidate is rapidly converted to diacylglycerol and the appearance of label in the diacylglycerol is a measure of the reverse reaction of the cholinephosphotransferase.

G. Diacylglycerol Acyltransferase

$$\text{Diacylglycerol} + \text{acyl-CoA} \rightarrow \text{triacylglycerol} + \text{CoA}$$

The ultimate step in the biosynthesis of the triacylglycerols is the acylation of the 1,2-diacylglycerol at position sn-3 and this is catalysed by the activity of a diacylglycerol acyltransferase (EC 2.3.1.20). The final acylation of the glycerol backbone at position 3 is, in fact, the only reaction which can be considered unique to the triacylglycerol biosynthetic pathway. The enzyme has been studied in a number of oil seed species (Ichihara and Noda, 1982; Cao and Huang, 1986; Ichihara et al., 1988). Difficulties arise in the presentation of the diacylglycerol substrate to the enzyme/membrane preparation. Cao and Huang (1986) have used the detergent, Tween 20, and Ichihara and co-workers (1982, 1988), gelatin. A typical assay procedure is as follows (Cao and Huang, 1986). The 1,2-dioleoyl-glycerol (2 mM) is emulsified with Tween 20 (10 mg) in 5 ml H$_2$O and 50 µl of the preparation added to the reaction mixtures. The reaction mixture contains sucrose (12 mM), MgCl$_2$ (6 mM), DTT (4 mM), [^{14}C]oleoyl-CoA (20 µM) and the diolein (0.4 mM) in Tween 20, in a total volume of 250 µl with 40 mM Tris-HCl buffer, pH 7.0. The reaction is initiated by the addition of the membrane preparation and terminated after 2 min. If necessary, carrier triolein may be added and the radioactivity in the purified triacylglycerol determined. Of the plant species studied, preparations from maize scutella and soybean were the most active with specific activity values of over 3 nmol min^{-1} (mg microsomal protein)$^{-1}$. Safflower preparations also gave activities in the same range (Ichihara et al., 1988).

X. CONCLUDING REMARKS

The understanding of triacylglycerol biosynthesis and its regulation in plants is still only in its infancy. A great deal of work is required in order to elucidate the intracellular compartments involved in the provision of substrates and in the assembly of the triacylglycerol molecule and their relationships to each other. At present, any advance at the molecular genetic level is limited by the unavailability of purified proteins. Unfortunately, many of the enzymes involved in triacylglycerol biosynthesis are membrane bound and highly hydrophobic. These properties make such proteins difficult to solubilise and purify. As soon as protocols become available for the successful identification of the enzyme peptides, which catalyse the synthesis of the C18-polyunsaturated fatty acids and the assembly of the triacylglycerol, then a new and exciting dimension of research will become available to the lipid scientist. Interspecies gene transfer will produce plants with modified oils and this will provide biochemists with novel experimental material for further research. Eventually, this will lead to a more complete understanding of triacylglycerol biosynthesis. Armed with such knowledge, it may then become possible to manipulate the acyl quality of the oil in a precise fashion and produce new varieties which will yield more useful and valuable plant products.

ACKNOWLEDGEMENTS

The authors are grateful for financial support (past and present) to the Swedish Natural Science Research Council, Swedish Council for Forestry and Agricultural Research, National Swedish Board for Technical Development, Stiftelsen Karlshamn Forskning (Karlshamn Ab, Sweden), Nuffield Research Foundation (UK), The Royal Society (UK), Science and Engineering Research Council (UK) and Scotia Pharmaceuticals (Efamol Ltd, UK).

REFERENCES

Bafor, M. E. and Osagi, A. U. (1986). *J. Sci. Food Agric.* **37**, 825–832.
Baker, F. C. and Schooley, D. A. (1981). *In* "Methods in Enzymology" (L. Lowenstein, ed.), Vol. 72, pp. 41–52. Academic Press, New York.
Bligh, E. G. and Dyer, W. J. (1959). *Can. J. Biochem. Physiol.* **37**, 911–917.
Bjerve, K. S., Daae, L. N. W. and Bremer, J. (1974). *Anal. Biochem.* **58**, 238–245.
Brindley, D. N. (1984). *Progr. Lipid Res.* **23**, 115–133.
Brockerhoff, H. (1972). *Lipids* **6**, 942–956.
Browse, J. A. and Slack, C. R. (1981). *FEBS Lett.* **131**, 111–114.
Browse, J. A. and Slack, C. R. (1985). *Planta* **166**, 74–80.
Butterworth, S. C., Hopewell, R. and Brindley, D. N. (1984). *Biochem. J.* **220**, 825–833.
Cao, Y.-Z. and Huang, A. H. C. (1986). *Plant Physiol.* **82**, 813–820.
Ching, T. M., Crane, J. M. and Stamp, D. L. (1974). *Plant Physiol.* **54**, 748–751.
Christie, W. W. (1982). "Lipid Analysis", 2nd edn. Pergamon Press, London.
Geeraert, E. and Sandra, P. (1987). *J. Am. Oil Chem. Soc.* **64**, 100–106.
Gennity, J. M. and Stumpf, P. K. (1985). *Arch. Biochem. Biophys.* **239**, 444–454.
Griffiths, G., Stobart, A. K. and Stymne, S. (1985). *Biochem. J.* **230**, 379–388.
Griffiths, G., Stymne, S. and Stobart, A. K. (1988a). *In* "Proceedings of the World Conference on Biotechnology for the Fats and Oils Industry" (T. H. Applewhite, ed.), pp. 23–29. American Oil Chemists Society, Champaign, Ill., USA.
Griffiths, G., Stymne, S. and Stobart, A. K. (1988b). *Planta* **173**, 309–316.
Griffiths, G., Stobart, A. K. and Stymne, S. (1988c). *Biochem. J.* **252**, 641–647.
Griffiths, G., Hakman, S., Tillberg, E., Hellman, M., Stymne, S. and Stobart, A. K. (1989). *In* "Lipids: Targets for Manipulation" (N. J. Pinfield, ed.). Monograph No. 17. The British Plant Growth Regulator Group (N. J. Pinfield and A. K. Stobart, eds), pp. 11–34. Bristol.

Gunstone, F. D. and Norris, F. A. (1983). "Lipids in Foods: Chemistry, Biochemistry and Technology." Pergamon Press, Oxford.
Gurr, M. I. (1980). *In* "The Biochemistry of Plants" (P. K. Stumpf and E. E. Conn, eds), Vol. 4, pp. 205–248. Academic Press, New York.
Hatch, M. D. (1986). *Trends Biochem. Sci.* **11**, 9–11.
Herman, E. M. (1987). *Planta* **172**, 336–345.
Hilditch, T. P. and Williams, P. N. (1964). "The Chemical Constitution of Natural Fats." Chapman and Hall, London.
Hosaka, K., Yamashita, S. and Numa, S. (1975). *J. Biochem.* **77**, 501–509.
Ichihara, K. (1984). *Arch. Biochem. Biophys.* **232**, 685–688.
Ichihara, K. and Noda, M. (1980). *Phytochemistry* **19**, 49–54.
Ichihara, K. and Noda, M. (1982). *Phytochemistry* **21**, 1895–1901.
Ichihara, K., Asahi, T. and Fuzii, S. (1987). *Eur. J. Biochem.* **167**, 339–345.
Ichihara, K., Takahashi, T. and Fuzii, S. (1988). *Biochim. Biophys. Acta* **958**, 125–129.
Jaworski, J. G. (1987). *In* "The Biochemistry of Plants" (P. K. Stumpf, ed.), Vol. 9, pp. 159–174. Academic Press, New York.
Juguelin, H. and Cassagne, C. (1984). *Anal. Biochem.* **142**, 329–335.
Kates, M. (1964). *J. Lipid Res.* **5**, 132–135.
Kennedy, E. P. (1961). *Fed. Proc. Fed. Am. Soc. Exp. Biol.* **20**, 934–940.
Kesselmeier, J. and Heinz, E. (1985). *Anal. Biochem.* **144**, 319–328.
Moore, T. S., Lord, J. M., Kagawa, T. and Beevers, H. (1973). *Plant Physiol.* **52**, 50–53.
Morris, L. J. (1966). *J. Lipid Res.* **7**, 717–732.
Myher, J. J. and Kuksis, A. (1979). *Can. J. Biochem.* **57**, 117–124.
Norman, H. A. and St. John, J. B. (1986). *J. Lipid Res.* **27**, 1104–1107.
Pugh, E. L. and Kates, M. (1975). *Biochem. Biophys. Acta* **380**, 442–453.
Roughan, P. G. and Slack, C. R. (1982). *Ann. Rev. Plant Physiol.* **33**, 97–132.
Roughan, P. G., Slack, C. R. and Holland, R. (1978). *Lipids* **13**, 497–503.
Sanchez, M., Nicholls, D. G. and Brindley, D. N. (1973). *Biochem. J.* **132**, 697–706.
Shewry, P. R., Pinfield, N. J. and Stobart, A. K. (1972). *Phytochemistry* **11**, 2149–2154.
Shukla, V. K. S. (1988). *Progr. Lipid Res.* **27**, 5–38.
Slack, C. R., Roughan, P. G. and Browse, J. (1979). *Biochem. J.* **179**, 649–656.
Slack, C. R., Bertaud, W. S., Shaw, B. P., Holland, R., Browse, J. A. and Wright, H. (1980). *Biochem. J.* **190**, 551–561.
Slack, C. R., Roughan, P. G., Browse, J. A. and Gardiner, S. E. (1985). *Biochem. Biophys. Acta* **833**, 438–448.
Stobart, A. K. and Stymne, S. (1985a). *Biochem. J.* **232**, 217–221.
Stobart, A. K. and Stymne, S. (1985b). *Planta* **163**, 119–125.
Stobart, A. K., Stymne, S. and Hoglund, S. (1986). *Planta* **169**, 33–37.
Stumpf, P. K. (1987). *In* "The Biochemistry of Plants" (P. K. Stumpf, ed.), Vol. 9, pp. 121–136. Academic Press, New York.
Stymne, S. and Glad, G. (1981). *Lipids* **16**, 298–305.
Stymne, S. and Stobart, A. K. (1984). *Biochem. J.* **223**, 305–314.
Stymne, S. and Stobart, A. K. (1985). *Biochem. Biophys. Acta* **837**, 239–250.
Stymne, S. and Stobart, A. K. (1986a). *Biochem. J.* **240**, 385–393.
Stymne, S. and Stobart, A. K. (1986b). *Physiol. Veg.* **24**, 45–51.
Stymne, S. and Stobart, A. K. (1987). *In* "The Biochemistry of Plants" (P. K. Stumpf, ed.), Vol. 9, pp. 175–214. Academic Press, New York.
Stymne, S., Griffiths, G. and Stobart, A. K. (1987). *In* "The Metabolism, Structure, and Function of Plant Lipids" (P. K. Stumpf, J. B. Mudd and W. D. Nes, eds), pp. 405–412. Plenum Press, New York.
Stymne, S., Green, A. G. and Tonnet, M. L. (1989). *In* "Proceedings of the 8th International Symposium on the Biological Role of Plant Lipids," Budapest (P. A. Biacs, K. Gruiz and T. Kremmer, eds), pp. 147–150. Plenum, New York and London.
Wanner, G., Formanek, H. and Theimer, R. R. (1981). *Planta* **151**, 109–123.
Wilson, R. F. and Rinne, R. W. (1976). *Plant Physiol.* **57**, 270–273.
Woldegiorgis, G., Spennetta, T., Corkey, B. E., Williamson, J. R. and Shrago, E. (1985). *Anal. Biochem.* **150**, 8–12.

3 Phospholipids

THOMAS S. MOORE, JR

Department of Botany, Louisiana State University, Baton Rouge, LA 70803-1705 USA

I.	Introduction	47
II.	Selection of tissue	49
III.	Phospholipid extraction	49
	A. Intact tissue	49
IV.	Lipid class separation	50
V.	Chromatography	51
	A. Thin layer chromatography	52
	B. High performance liquid chromatography (HPLC)	58
VI.	Sidedness of phospholipids in membranes	63
	A. Chemical probes	64
	B. Enzymatic alterations	64
	C. Transfer proteins	64
	D. Summary	66
VII.	Phospholipid precursors	66
	A. Lipid precursors	66
	B. Headgroup precursors	66
VIII.	Overview	69
	Acknowledgement	69
	References	69

I. INTRODUCTION

The phospholipid composition of plants and their membranes has been the subject of investigation for many years. As a result, techniques have been developed which allow for efficient extraction and sensitive detection of those phospholipids which occur. On the other hand, we also have learned of a number of problems which may exist in such studies, and the modern investigator must be aware of these difficulties in order to obtain accurate results.

The major classes of phospholipids according to headgroup (Fig. 3.1) are phosphatidic acid (PtdOH), phosphatidylcholine (PtdCho), phosphatidylethanolamine (PtdEtn), phosphatidylinositol (PtdIns), phosphatidylglycerol (PtdGro), phosphatidylserine (PtdSer) and bis-phosphatidylglycerol (cardiolipin, diphosphatidylglycerol, Bis-PtdGro). Within these major headgroup species, however, a broad spectrum of subspecies exists depending on the acyl constituents found on positions 1 and 2, and monoacyl derivatives are formed by removal of one of the acyl units. Thus, identification can be aimed at several levels, depending on the intent of the research programme.

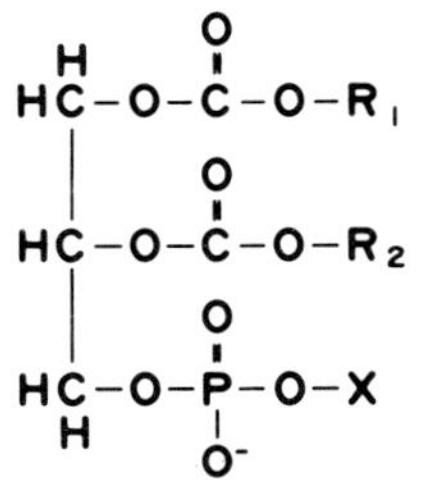

$$\begin{array}{c} \text{H} \quad \text{O} \\ \text{HC}-\text{O}-\overset{\parallel}{\text{C}}-\text{O}-\text{R}_1 \\ | \quad \text{O} \\ \text{HC}-\text{O}-\overset{\parallel}{\text{C}}-\text{O}-\text{R}_2 \\ | \quad \text{O} \\ \text{HC}-\text{O}-\overset{\parallel}{\text{P}}-\text{O}-\text{X} \\ \text{H} \quad | \\ \quad \text{O}^- \end{array}$$

Moiety Identity	Polarity	Phospholipid
R_1, R_2 = Acyl	Nonpolar	
X = Hydrogen	Acidic	PtdOH
Serine	Acidic	PtdSer
Choline	Basic	PtdCho
Ethanolamine	Basic	PtdEtn
Glycerol	Neutral	PtdGro
Inositol	Neutral	PtdIns
PtdGro	Nonpolar/Neutral	Bis-PtdGro

FIG. 3.1. General structures of phospholipids commonly found in plants. All headgroups are polar, with the most abundant phospholipids (PtdCho and PtdEtn) being basic, the two commonly least abundant (PtdOH and PtdSer) acidic, and the remainder neutral. The more non-polar nature of Bis-PtdGro derives from the fact that a second phosphatidyl moiety is linked through the headgroup (glycerol) to the first. Abbreviations are: Bis-PtdGro, cardiolipin or bis-phosphatidylglycerol; PtdCho, phosphatidylcholine; PtdEtn, phosphatidylethanolamine; PtdGro, phosphatidylglycerol; PtdIns, phosphatidylinositol; PtdOH, phosphatidic acid; PtdSer, phosphatidylserine.

The individual needs of a large number of researchers, and specific requirements of particular plant tissues, has led to a variety of methods for extraction, separation and characterisation of the phospholipids. The descriptions below are not exhaustive, and a particular method may not be suitable for all tissues or organelles. For that reason, more than one technique is presented where possible, and where specialised options exist references are given which will provide adequate details of these methods.

II. SELECTION OF TISSUE

Tissues chosen for investigations of phospholipids often are selected for reasons other than suitability for phospholipid isolation. As a result, difficulties are encountered commonly. Plant tissues frequently present problems not found with animal cells, in particular the occurrence of high concentrations of phospholipases in many cells, and the investigator must be aware of the characteristics of the system chosen. For example, spinach and bean may be high in phospholipase activity, while pea and castor bean endosperm are low. However, the latter tissue clearly is unsuitable for studies on photosynthesis. In summary, the tissue chosen may well be difficult to work for phospholipid studies; the investigator must take care to determine the suitability and difficulties, and ascertain the method best suited to overcoming these problems. Several options are presented below.

III. PHOSPHOLIPID EXTRACTION

Phospholipid purifications may start from intact tissue or, more commonly now, from isolated organelles. Various techniques have been described which are aimed at alleviating the artifacts arising from high phospholipase activities occurring in many plant tissues. The artifacts which might occur, and therefore for which one should be on the lookout, are 'unusually high' concentrations of phosphatidic acid or lysophospholipids. The former indicates the presence of phospholipase D, while the latter reflects phospholipase A activities. Phospholipase C also may be present, leading to the production of diacylglycerol, a result which is not as obvious in casual observations. When working with a new tissue, two or three different extraction methods should initially be compared for elimination of artifacts as well as efficiency of extraction.

A. Intact Tissue

Methods for extraction of phospholipids from intact plant tissues are several, and some are designed specifically for eliminating phospholipase activities. The various methods can be grouped into: (1) traditional, with no special precautions; (2) use of non-substrate solvents; (3) use of heat to inactive enzymes; (4) use of cold to retard enzyme activity; and (5) a combination of two or more of these approaches. Normally, (1) or (5) best describes the methods used, with the latter being a combination of utilising non-substrate solvents (generally 2-propanol) and inactivation by heat.

1. Traditional extractions

The simplest methods for phospholipid extraction are based on the procedures described by Folch *et al.* (1957) or on the Bligh and Dyer (1959) modifications of the Folch *et al.* method. These procedures rely on chloroform–methanol–water solutions to extract the lipids. These extracts are washed with several aqueous washes, usually containing high salt concentrations, and sometimes a buffer to help avoid acid or base hydrolysis (e.g. Garbus *et al.*, 1963). Bligh and Dyer (1959) point out the importance of having the correct proportions of chloroform, methanol and water (about 1:2:0.8; v/v) for the initial extraction. This provides a single phase and allows for both non-polar and polar solvent molecules to intercalate and disrupt the membranes. A procedure based on the Bligh and Dyer (1959) method that we commonly use with isolated endoplasmic reticulum is:

(1) Add 3.3 ml of chloroform–methanol–water (1 : 2 : 0.3; v/v) to 0.5 ml of ER suspension. Mix well and leave at room temperature for 30 min.
(2) Add 1 ml of chloroform, mix, then add 3 ml of 1 M KCl and mix again.
(3) Centrifuge (3–5 min at about 5000 rpm in a benchtop centrifuge) to break the emulsion.
(4) Aspirate off the upper aqueous layer, then add 2–3 ml of 1 M KCl, mix and repeat Step 3.
(5) Repeat Step 4; finally, decant the lower chloroform layer into a scintillation or storage vial.

2. Phospholipase inhibition

While the above procedure is straightforward and easy to do, it is subject to considerable artifacts if phospholipases are present. Extreme cases of this are readily recognisable if high amounts of phosphatidate or lysophospholipids are identified in the extract. Even if they are not present at obviously elevated levels, however, the careful researcher will compare the results of this method with others designed to greatly reduce or eliminate lipase activity. Commonly, this is accomplished by first treating the tissue or organelle fraction with hot (70°C) 2-propanol (isopropanol) to inactivate the enzymes. Isopropanol is preferable to *n*-butanol, since butanol can be utilised by phospholipase D to exchange with a headgroup (Christie, 1982). Hot 2-propanol can be simply substituted for methanol in the procedure described above. A mixture of alcohol and the organelle (or tissue) fraction is heated for 30 min at 70°C, following which it is cooled, appropriate amounts of chloroform and water added, and the extraction Steps 2–5 carried out as usual.

3. Non-chloroform methods

As a result of improved, highly sensitive analytical techniques, concerns have been expressed about reactive impurities found in chloroform (HCl and phosgene), and other problems associated with monitoring elution pattern of lipids following chromatography. This has led to the recent introduction of hexane–2-propanol extraction methods (Kolarovic and Traitler, 1985; Kolarovic and Fournier, 1986). The procedure described by those researchers leads to an improved extraction, but requires both sonication and a vacuum evaporator. The technique appears to hold promise and should be considered, but has not been tested with plant tissue as yet.

IV. LIPID CLASS SEPARATION

Once the lipids have been extracted, it may be possible to chromatograph the material directly. On the other hand, with tissue containing high concentrations of other lipids (galactolipids, triacylglycerols, etc.) this could severely restrict the quantity of extract that can be applied to a chromatogram without overlapping occurring among some of the fractions of interest. A solution to this is to pass the sample through a column under conditions which would separate the fractions based on polarity. A number of procedures have been described for such separations (Christie, 1982), and these can be scaled up or down according to need. A column of silica gel can be eluted with chloroform or diethyl ether to obtain neutral or simple lipids, then acetone to obtain the glycolipids, and finally methanol to obtain phospholipids (Christie, 1982; Rouser et al., 1976).

Norman and St. John (1986) recently applied this principle to preparation of samples for HPLC by passing extracted samples through a Sep-Pak cartridge (Waters Assoc., Milford, MA) and eluting the lipid classes with chloroform, acetone and methanol as above. We have found similar cartridges from other manufacturers also work well. Unfortunately, there is at least one case of a manufacturer changing the properties of their cartridges without notice, which led to problems with some newly developed techniques for lipid separations. It is suggested that these separations be regularly tested for efficiency and recovery, particulary when a new batch of cartridges is received.

V. CHROMATOGRAPHY

A great many chromatography systems have been developed for separation of phospholipids (Radwan, 1984). The advent of silica gel thin layer chromatography (TLC) resulted in major improvements of separation methods for all lipids. Most laboratories working in the area of phospholipid metabolism use TLC routinely, and have adapted methods to suit their particular conditions. More recently, high performance liquid chromatography (HPLC) has begun to receive attention as a method allowing rapid separations which avoid excessive oxidation through contact with the air. There are some problems with this method, however, and these are discussed below.

Phospholipid separations by chromatography are affected by the properties imposed by both the polar headgroups and the non-polar acyl units. Indeed, some smearing or tailing of phospholipid spots can result from a partial separation of the lipids based on acyl composition within the spot obtained on chromatographic separation designed for headgroup separations. It generally is necessary to confirm identifications of phospholipids after chromatography by colorimetric assays or by identification of hydrolysis products since R_fs will not necessarily be the same as those published. A recent example of this is contained in a paper concerning development of a TLC system. These researchers found that phosphatidylinositols from animals and plants chromatographed at distinctly different R_fs in the same system (Heape et al., 1985).

Most phospholipid research has involved separations based on the headgroups, but more recently attention has been given to following such separations with additional purifications to separate subspecies based on acyl composition. Most recently, separations based on headgroup properties utilising TLC, followed by subclass separation based on acyl composition by HPLC, have been reported for plant phospholipids. These are discussed below.

A. Thin Layer Chromatography (TLC)

Thin layer chromatography has become the method of choice for separation of the primary phospholipid classes. It is rapid, straightforward, and readily modified for special cases. Depending on the requirements, one- or two-dimensional, preparative, or the new high performance thin layer chromatography (HPTLC) may be utilised. These are discussed below.

1. Standard TLC

Silica gel plates rely on the negative charges of the support for interaction with the sample. Plates which have been stored under humid conditions absorb water, and thus the number of active spots is reduced. As a result, it is commonly accepted practice to 'activate' the plates by removing some of this water. The normal method for activating the silica is to heat the plates in an oven at 110°C for 1 h, followed by storage in a desiccator until use. These extremes may not be necessary, however, since for most cases an 11–12% (w/w) water content allows for appropriate separations. Such concentrations are obtained on plates stored at 50% relative humidity, conditions near those which exist in many environmentally controlled laboratories today. Thus, activation of the plates at 70–80°C followed by equilibration in the laboratory may be adequate for most purposes. Storage over a saturated solution of $Mg(NO_3)_2.6H_2O$ maintains the plates at about 52% relative humidity and would reduce the rapid changes in water content occurring upon transfer of plates stored over a desiccant into the laboratory environment. The plates transferred into the room would begin absorbing water as soon as they are removed from the low humidity of the desiccator and will do so over a period of time. It may be difficult to obtain precisely the same conditions with each run.

There is a temptation to heat TLC plates at a very high temperature to assure that almost all water is lost. This should be resisted, however, as such treatment can lead to revealing highly reactive spots which can react with, and degrade, the sample. Also, with plates containing $CaSO_4$ as a binder, the binder's water of hydration is lost at 130°C or higher, thus reducing binding action. As a general rule, silica plates should not be heated above 110°C.

(a) *One-dimensional separation.* Single dimension chromatography has the advantage of simplicity and speed, often being completed in under an hour. In many instances the separations may be adequate. This holds for cases in which the general lipid composition is known and it is certain that no interference with the lipid of interest will occur from other phospholipids, glycolipids, diacylglycerol, etc. Sometimes problems arising from the presence of overlapping or masked lipids may be negated by a judicious choice of detection or quantification methods (see below). Most often the use of single dimension TLC is best for isolated membrane functions in which a limited number of phospholipids are being studied and large numbers of samples are to be chromatographed.

The chromatogram in Fig. 3.2(a) is an example of the separations which can be obtained in one dimension. A silica gel G, 250 μm thin layer on a glass plate was developed with chloroform–methanol–7 N ammonium hydroxide (65:30:4; v/v) and separations of many of the phospholipids were obtained. The results in Fig. 3.2(b) are those obtained with an acidic solvent system (chloroform–methanol–acetic acid–water, 170:25:25:6; v/v). Other systems have been used for similar separations (Christie, 1982, 1987).

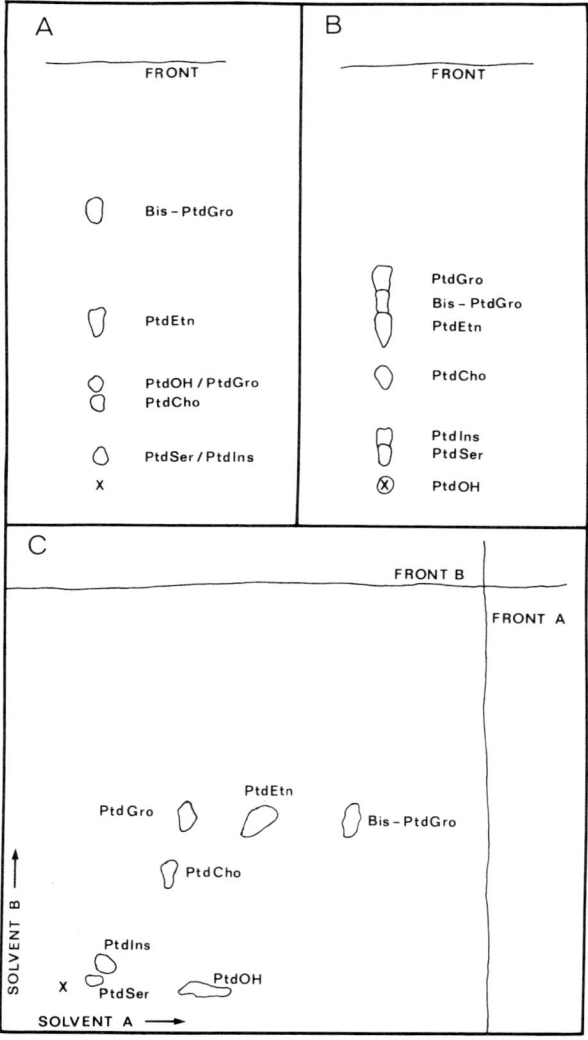

FIG. 3.2. Thin layer chromatographic separation of phospholipids. One-dimensional (a,b) and two-dimensional (c) are represented. The first (a) was developed with a basic solvent system of chloroform–methanol–7N NH$_4$OH (65:30:4; v/v), the second (b) in an acidic solvent of chloroform–methanol–acetic acid–water (170:25:25:6; v/v). The two-dimensional chromatogram (c) was developed in the first dimension with the basic solvent and in the second dimension with the acidic solvent. These techniques were based on the work of Nichols (1964). The chromatograms all were silica gel G (from Analtech, Newark, DE, USA) and were activated at 100°C followed by storing in a laboratory drawer. Note that using Nichols' solvent systems without change did not utilise the full area of the plates. This could be adjusted by modifying the water content. Abbreviations: see Fig. 3.1.

A number of specialised one-dimensional chromatography systems have been described. These include some used for identification of specific metabolic intermediates or for clear separation of a single phospholipid from others. Examples include systems for PtdGro (Hostetler et al., 1971), PtdEtn and intermediates to PtdCho (Horrocks, 1963; Glaser et al., 1974), as well as PtdIns and its phosphorylated derivatives (Crosby and Dale, 1985). In cases where chromatography is used for identification, all conclusions should be confirmed by testing migrations in several solvent systems, using appropriate spray reagents, and often it is best to determine the hydrolysis products (described below).

(b) *Two-dimensional separation.* Complete lipid extracts from plant tissues include glycolipids, neutral lipids and/or lysophospholipids which do not separate well from the phospholipids in one dimension. In addition, the acyl composition in one phospholipid could result in co-chromatography with a second. In such cases two-dimensional chromatography generally may be utilised to achieve the desired separations. Such chromatography can be designed to accomplish a variety of tasks, two common ones being (1) to separate the more neutral lipids in one dimension by chromatographing them to near the top followed by a second dimension to separate the phospholipids, or (2) use of two phospholipid-separating mobile phases with different properties to resolve particularly recalcitrant phospholipids.

An example of two-dimensional separations based on the procedure of Nichols (1964) is presented in Fig. 3.2(c). The first separation utilises a basic, ammonium hydroxide-containing mobile phase, and is followed by an acidic phase in the second dimension. In two-dimensional chromatography it is important to evaporate the first phase thoroughly prior to proceeding to the second. This commonly is done by placing the plates into an empty TLC chamber, followed by placing a tube, through which dry nitrogen flows at a slow rate, into the chamber at one side and extending to the bottom. A slight opening is left at the top on the end of the chamber opposite where the tube is inserted for the nitrogen to pass out (expensive lids are available for this, but are not usually necessary). The plate is dried for approximately 30–60 min, after which the smell of the first solvent system should be dissipated. In cases where some fatty acid oxidation is not a problem, and where the laboratory humidity is 52% or less, it may be possible to dry the plates between dimensions in the air flow of a hood.

(c) *Subspecies separations.* Separation of phospholipid subspecies (different acyl compositions) has met with only limited success. The method in use has been argentation chromatography, and the best results have been obtained with diacylglycerol acetates after headgroup removal (Christie, 1982), with some improvements using phospholipids being recently described (Kennerly, 1986). On the other hand, Harwood (1976) separated an- and mono- through pentaenoic classes of PtdCho and PtdEtn, both with and without the headgroup. Despite this success, however, HPLC, as described below, may provide a better approach.

2. *High performance thin layer chromatography (HPTLC)*

A recent development in the field of thin layer chromatography is the use of small, higher resolution plates (10 cm × 10 cm as compared to 20 cm × 20 cm). Since these

TLC plates are half the height of standard plates the solvent front moves half the distance. In order to obtain adequate resolution they also have a support made from finer particles. Advantages of these chromatograms are high resolution in a short space, a faster development time, less space required for the development tanks, and less solvent and sample needed. The primary disadvantage is that the plates sometimes will not carry as much sample as needed for some assays, but this is not usually a problem. Methods with standard TLC do not necessarily transfer directly to HPTLC, and so adjustments may be needed (Touchstone et al., 1980).

Several solvent systems have been described for phospholipid separations by HPTLC using silica gel plates. Excellent separations have been obtained in one dimension. However, most of this work has been developmental, with some applications to animal systems, and the reported use of these plates for separations of plant lipid fractions has been minimal (Heape et al., 1985).

An example of one of the simplest of the systems is presented in Fig. 3.3 (Heape et al., 1985). These workers used Silica gel 60 F-254 10 cm × 10 cm plates, with a solvent system consisting of methyl acetate–n-propanol–chloroform–methanol–0.25% aqueous KCl (25:25:28:10:7; v/v). All the major phospholipids were separated, as well as DGDG and MGDG. Dugan (1985) also achieved separation of most phospholipid classes (cardiolipin was not reported) in one dimension using similar plates from another company. One difference was that these plates had a pre-absorbant strip at the bottom which allows for quicker spotting and concentration of the sample into a tight band as it enters the silica gel. This concentration feature can be of considerable benefit on these smaller plates. The solvent system in this latter case also gave separations which changed little with changes in relative humidity, an advantage to those of us who live in areas with widely fluctuating humidity and relatively poor building environmental control. The solvent system utilised was denatured ethanol–chloroform–ammonium hydroxide (50:6:6; v/v) and the plates were type HLF from Analtech (Newark, DE, USA). The system is sensitive to the amount of phospholipid applied, and should be kept to 1 µg or less of each (E. Dugan, pers. commun.).

For rapid, routine separations HPTLC has much to recommend it, and the techniques seem to be developing rapidly. It may be of great benefit in screening for lipid mutants.

3. Modification of silica layers

Some chromatographic applications require modification of the silica layer by addition of a cation or other component. If you prepare your own plates this may be accomplished by including it in the slurry. With commercially prepared plates the additional component may be sprayed onto the plates, or the plates lowered briefly into the appropriate solution.

4. Radioactivity detection

The end result of many experiments is radioactivity incorporated into one or more phospholipids. It is necessary to determine which fraction is radioactive, and usually the quantity of radioactivity incorporated.

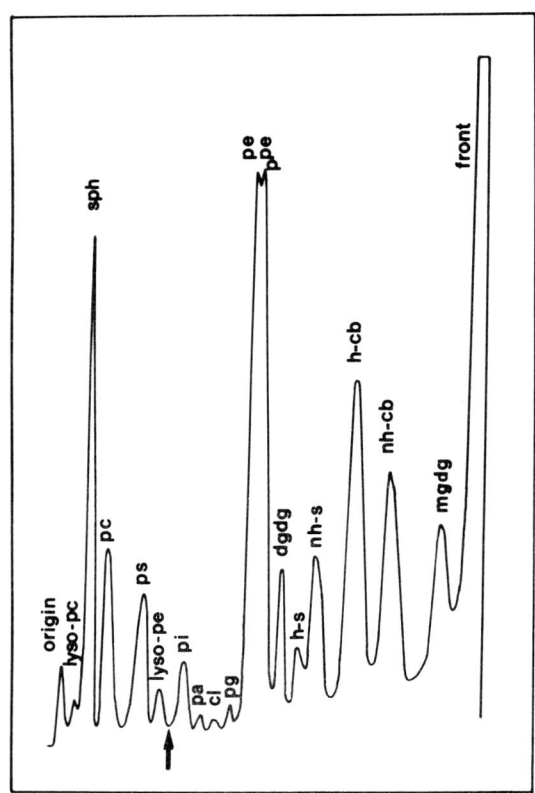

FIG. 3.3. High-performance thin layer chromatography of phospholipids according to Heape et al. (1985). Silica gel 60 F-254 HPTLC plates (10 cm × 10 cm) were developed with methyl acetate–n-propanol–chloroform–methanol–0.25% (w/v) aqueous KCl (25:25:28:10:7; v/v). The bands were visualised by copper acetate–phosphoric acid charring and scanned at 366 nm with a photodensitometer scanner. The sample was a mixture of a mouse sciatic nerve lipid extract plus lyso-PtdCho, lyso-PtdEtn, PtdOH, PtdGro, digalactosyldiacylglycerol (DGDG) and monogalactosyldiacylglycerol (MGDG). The peaks are: origin; lyso-PC, lyso-PtdCho; sph, sphingomyelin; pc, PtdCho + plasmalogen PtdCho; ps, PtdSer; lyso-pe, lyso-PtdEtn; pi, PtdIns; pa, PtdOH; cl, Bis-PtdGro; pg, PtdGro; pe, PtdEtn; ppe, plasmalogen PtdEtn; dgdg, DGDG; h-s, hydroxysulphatides; nh-s, non-hydroxysulphatides; h-cb, hydroxycerebrosides; nh-cb, nonhydroxycerebrosides; mgdg, MGDG; front, neutral lipids and solvent front. The arrow shows where soybean PtdIns would chromatograph in this system. The figure was modified by the original authors from Heape et al. (1985) and is used with permission.

(a) *Autoradiography.* Autoradiography has long been in use and the methods generally well established. Some advances in film speed and resolution have occurred and these are well advertised. Another useful technique of relatively recent development is the enhancement of the signal from the TLC plate by spraying a reagent onto it which will fluoresce in response to the radioactivity. Since the film is more responsive to this light than the radioactive particles, exposure times are greatly reduced. In practice, the plate is sprayed with the reagent, allowed to dry, and then X-ray film is laid onto the surface of the plate and held in place by a suitable holder. The holder may be kept at room temperature but more commonly is kept in a freezer (-20 or $-80°C$) to reduce background and sample degradation. Use of such enhancers of film exposure can reduce the exposure time from weeks to one or two days.

A different approach to viewing radioactivity on chromatograms is the 'spark imaging' system (Vose, 1980). The TLC plate is placed in an argon–methane mixture, which is ionised by the β-particles emitted from the sample. The ionisation is controlled by an electrode system mounted on a transparent plate. The flashes produced are recorded on polaroid film, and a mechanism is provided in these systems for aligning the TLC plate and film in order to mark the radiolabelled spot. Results can be obtained with as little as 20 min exposure.

(b) *Quantification.* Two methods are in general use for quantifying radioactivity on TLC plates. The first, and more common, is simply to scrape the spot(s) of interest into scintillation vials, add a scintillation cocktail, and measure the radioactivity in a scintillation counter. Special tools are sold for this, but scraping with a razor blade generally is adequate. The worker should be sure to wear plastic gloves and possibly a mask for protection from the dust. A revolution is occurring at this time in the scintillation cocktail field, as a result of increased awareness of hazards involved in the storage of nuclear waste. Often, the solution containing the radioactive compounds has been more hazardous than the radioactivity. Companies are responding to this, and now are producing ready-made, biodegradable cocktails to meet many needs. It is recommended that a suitable one be found and used in place of the older toluene- or hexane-based solutions. If it is necessary to use these latter solutions, care should be taken to dispose of them properly.

The second method for quantifying radioactivity on TLC plates is the use of radioisotope scanners. These instruments, manufactured by several companies, are more sensitive than earlier models and, with modern computer technology, are capable of providing high resolution two- or three-dimensional graphs of radioactivity on a TLC (or HPTLC) plate. These instruments are impressive; so also are the prices.

5. Spray reagents/quantification

A number of spray reagents have been developed for detection of phospholipids following thin layer chromatography. A great many of these have been compiled in an easily usable form by Vioque (1984). This compilation includes procedures for general lipid and phospholipid detection as well as more specific spray reagents for choline-, amino-, sugar-, etc. containing lipids.

A quick, sensitive way to detect lipid spots is to spray the plate with a 0.02% (w/v in methanol) solution of 8-anilino-4-naphthosulphonic acid. The lipid spots will fluoresce. A second rapid method relies on the fact that iodine reacts reversibly with double bonds to give a brown product. This can be taken advantage of for routine detection of lipid spots that contain sufficient quantities of unsaturated fatty acids. In practice, the plate may be 'sprayed' with iodine vapour produced by passing air over crystals in a Pasteur pipette (be careful! very low pressures and flow rates suffice; glass wool should be used to hold the crystals in place), by placing the plate in a chamber with an iodine-saturated atmosphere produced by vapours from iodine crystals in the bottom, or by a spray reagent (see Vioque, 1984, for a recipe). Finally, where large quantities of lipid are present, simply spraying the plate with water and observing the capacity for light reflection from the surface will allow for detection of lipid spots.

Quantification is most precisely done by scraping spots detected by a general spray reagent or iodine, extracting the lipids from the silica gel, and then subjecting the lipids to a colorimetric assay for the compound in question. Several methods for assaying the phosphate content of phospholipid fractions have been described, and some have been in long and routine use. Methods commonly used are based on the original methods published by Bartlett (1959) as described by Dittmer and Wells (1969). A more recent, very sensitive assay based on formation of a complex with a malachite green–ammonium molybdate reagent has been described and appears quite suitable (Duck-Chong, 1977). The method is highly sensitive, and so extreme care must be taken to assure that no phosphate contamination can arise from the glassware used. Residue from phosphate-containing detergents can be a particular problem.

One technique commonly used during method development utilises TLC or HPTLC plates sprayed with 10% (w/v) cupric sulphate in 8% (w/v) phosphoric acid followed by charring at 160°C (Touchstone et al., 1981). The concentration of these spots may be quantified by use of a scanning densitometer. This method is a promising one for rapid, on-plate determination of phospholipid concentrations.

6. Identification with hydrolysis products

Often it is necessary to proceed beyond simple chromatography comparisons for confirmation of identities of phospholipids. A procedure has been described for this involving alkaline hydrolyses of the acyl units, which may be separately analysed, followed by separation of the remaining water-soluble glycerophosphoryl moieties (Dawson, 1976).

B. High Performance Liquid Chromatography (HPLC)

Standard column chromatography has not proven capable of resolving all the phospholipid classes, and so its usefulness has been restricted to separations of the general lipid classes (see above). However, in recent years a new technique, which utilises finer particle sizes, with solvents being passed through under pressure, has been developed. This technique shows promise of being able not only to define the various classes of phospholipids adequately, but also of separating subspecies (different acyl compositions) of a particular class of phospholipid. This latter application has been used successfully in several cases with plant tissues (e.g. Rivnay, 1984; Demandre et al., 1985; Norman and St. John, 1986).

HPLC offers several advantages over TLC. It is safer, since all solvents are passed through a tightly sealed system and thereby vapours reduced, it is readily automated, quantitative results can be obtained without scraping, and sample exposure to oxidation in the air is greatly reduced.

HPLC in its simplest form can be described as a column of the appropriate properties, a pump capable of producing solvent pressures in the column at a maximum of about 250 to 300 bars, and a detector. The latter is not critical if a fraction collector is used and assays are performed manually. (In some cases fractions have been collected by hand!) At its most complex, the method can use multiple pumps to produce simple to complex solvent gradients, and multiple detectors.

Since the pressures routinely are moderately high, pumps are designed for low volume and high accuracy, and the system is equipped with very small internal diameter tubing.

It is imperative that great care be taken to assure that all solvents are degassed and filtered to remove dust, and that the sample also is free of particles. Many filtering systems for both sample and solvents are available, but in general filtration through a 0.2–0.45 µm filter (resistant to the solvents in use) under vacuum will meet these conditions. Filtration of the sample with similar filters under pressure with a syringe will work well. Make certain that all storage vessels are thoroughly cleaned and tightly sealed.

The major hindrance of rapid development of HPLC for use with phospholipids is the detectors. No simple, straightforward detector is capable of on-line detection of phospholipids at the sensitivity often required. Most researchers have relied on light absorbance at 200–210 nm or, more recently, a FID detector. These and other detectors are discussed below. Other drawbacks to HPLC are the initial cost, the cost of the relatively large quantities of solvent utilised per sample, and the extreme care which must be taken in order to assure a trouble-free run (see above). Problems such as these (discussed below) must be resolved before HPLC begins to supplant TLC as the method of choice for the routine separation of phospholipid classes. Despite these drawbacks, rapid strides are being made in HPLC technology and the researcher interested in work in the area of phospholipids should consider it as a potential tool.

1. Separations by headgroup

There are only two published reports of phospholipid mixture separations based on headgroup composition in plant tissues using HPLC only (Rivnay, 1984; Demandre *et al.*, 1985), but this technique has received considerably more attention with animal tissues. Unfortunately, almost every report has utilised different columns and different solvent mixtures, and none has reported on total separation of all headgroup classes (Christie, 1987; Sheely *et al.*, 1987; Shukla, 1988). This reflects the lack of a single system capable of clean separation of all phospholipid classes.

Rivnay (1984) separated phospholipids from soybean utilising HPLC, but found it necessary to apply TLC to some of the fractions in order to achieve adequate resolution. This approach may be necessary for some applications at this time.

The system used by Demandre *et al.* (1985) for separation of phospholipid classes was a 3.9 × 300 mm, 10 µm µPorasil (Waters) column, which was eluted with a linear gradient starting with 100% 2-propanol–hexane (4:3; v/v) to 100% 2-propanol–hexane–water (8:6:1.5; v/v) over a 20 min period, followed by a second 20 min period of the second solvent only (isocratic). The sample was dissolved in the first solvent (200–500 µg in 200 µl) for injection, and 20 µl fractions were applied to the column. The lipids were detected by UV absorbance at 205 nm (Fig. 3.4). Unfortunately these workers did not report on where phosphatidylinositol and phosphatidylserine appear in this profile and so the usefulness for a variety of studies is uncertain without additional experiments.

Please notice (Fig. 3.4) that a considerable decrease in base-line absorbance occurs during the course of development of the chromatogram. The high absorbance at the beginning is due to solvent absorbance of the UV light. This can be a major problem when only low lipid concentrations are possible or offset capabilities of the monitor are limited. This also, in many instances, requires constant monitoring of the equipment, thus reducing the value of the potential for automation.

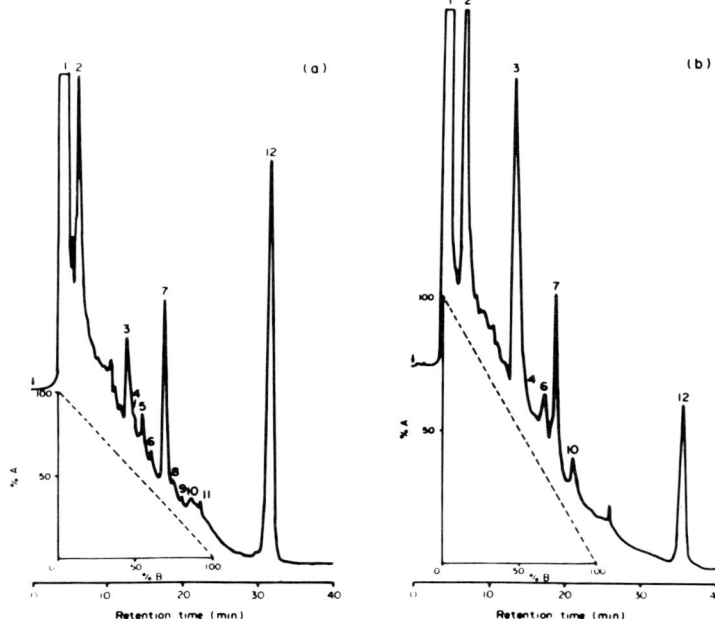

FIG. 3.4. High performance liquid chromatographic separation of major classes of plant phospholipids from potato tuber (a) and tobacco leaf (b) (Demandre et al., 1985). The column was 3.9 × 300 cm, 10 µm, µPorasil. The sample (20–50 µg) was applied in 200 µl 2-propanol–hexane (4:3; v/v) and eluted with a 20 min gradient ranging from 100% 2-propanol–hexane (4:3; v/v) to 100% 2-propanol–hexane–water (8:6:1.5; v/v), followed by 20 min isocratic elution with the latter solvent. Flow rates were 1 ml min^{-1}; detection was by absorbance at 205 nm. The lipids detected were: 1, neutral lipids; 2, MGDG; 3, DGDG; 4, sulpholipid; 5, PtdOH; 6, PtdEtn; 8, 9, 10, 11, unidentified; 12, PtdCho. See legends to Figs 3.1 and 3.3 for abbreviations. The figure is used with permission from Demandre et al. (1985).

2. Separation by acyl composition (subspecies)

High performance liquid chromatography shows considerable promise for resolving the many subspecies (acyl composition) of phospholipids extant in living tissue, and may well supplant argentation TLC. Some examples of the resolution obtainable are found in the plant research literature. Demandre et al. (1985) utilised HPLC-separated phosphatidylcholine, phosphatidylglycerol and phosphatidylethanolamine for analysis following separations on a µBondapak C_{18} column for reverse phase separations (Fig. 3.5). Several subspecies were obtained for each phospholipid, and detected by UV absorbance; quantification was by GLC of the methylated fatty acids after separation. Unfortunately, 18:2/18:2 molecular species did not separate from 16:0/18:3 species under the conditions used.

Norman and St. John (1986) have also separated molecular species of PC, but utilised the flame ionisation detector (FID) described below. This allowed them to make direct measurements straight from the column, thus eliminating the need to perform GLC separations for identification. Their column also was reverse-phase and did not separate the species mentioned above, nor 18:1/18:2 and 16:0/18:2.

A third approach to the detection of separated molecular species has been to replace the headgroup with a strong UV-absorbing group such as benzoate (Blank et al., 1984) or *p*-anisoyl (Kesselmeier and Heinz, 1985). This gives the distinct advantages of easier detection with the monitors most commonly available for HPLC systems. These results are capable of ready quantification. The added chromophores absorb in a region of the spectrum where the fatty acids would not (benzoate at 230 nm and *p*-anisoyl at 250 nm), and so the absorbance is constant for each molecule rather than dependent on the acyl composition. Blank *et al.* (1984) reported the separation of 29 molecular species of PE from beef brain. The major drawback of this approach to detecting subspecies would be when experiments are designed to determine the subspecies in which a newly added headgroup or a headgroup modification appears.

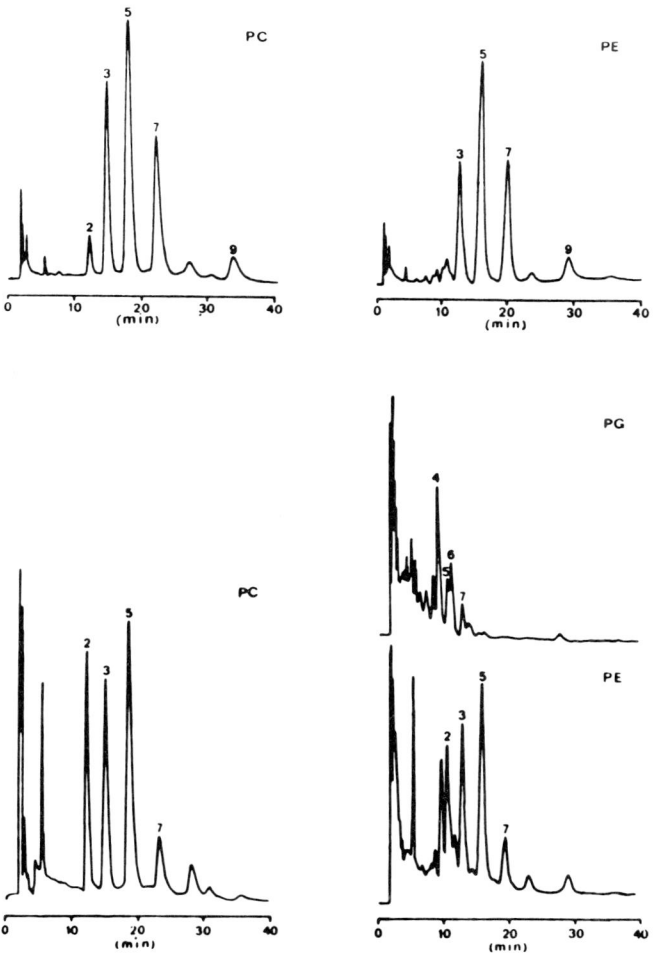

FIG. 3.5. High performance liquid chromatographic separations of plant phospholipid subspecies. Phospholipid species separated as described in Fig. 3.4 were further subfractionated on a 3.9 × 200 mm μBondapak C_{18} column in methanol–water–acetonitrile (90.5:7:2.5; v/v) at a flow rate of 1.5 ml min^{-1}. Detection was by absorbance at 205 nm. The peaks identified were: 2, C18:3/C18:3; 3, C18:2/C18:3; 4, C18:3/C16:1t; 5, C18:2/C18:2 and C16:0/C18:3; 6, C18:2/C16:1t; 7, C16:0/C18:2; 9, C18:0/C18:2. The figure is from Demandre *et al.* (1985) and is used with permission.

3. Detection and quantification

On-line detection and quantification of phospholipid fractions following HPLC remains a problem. The monitors most commonly utilised for plants have been UV-visible light absorbance and, more recently, flame ionisation detection (FID). These and other methods are discussed below.

(a) *Light absorbance monitors.* UV absorbance is the most commonly applied technique for detection of phospholipids eluting from a column. The wavelengths of value are from 195 to 210 nm, with 205 nm being most commonly utilised. The chosen wavelength normally is a compromise between elimination of solvent absorbance and maximising absorbance by the lipids. This lipid absorbance is dependent on the number of double bonds in the fatty acids of the phospholipids, with some small contribution from the ester linkage. Unfortunately, the absorbance is not very strong and is difficult to quantify due to varying numbers of double bonds in the phospholipids, making saturated phospholipids a particular problem. Nevertheless, despite these problems this method often is the only available detection procedure due to lesser sensitivity or greater expense of other methods. Replacement of headgroups with a chromophore has been discussed above. This approach provides a handy method for identifying molecular species within a particular phospholipid type.

(b) *Flame ionisation detector (FID).* Flame ionisation detection is the most common method of detection utilised in gas chromatography, but early efforts to adapt it to HPLC were unsuccessful and it has been slow to become commercially available. Recently, however, many early problems have been resolved and Tracor Instruments (Austin, TX) has marketed an instrument which is meeting with some success. The principle is straightforward. The sample is eluted onto a moving belt made of fibrous quartz, following which the solvent is evaporated by heating. The belt then moves through a hydrogen flame where the sample is ionised by combustion and the resulting current detected (Christie, 1987).

This instrument has met with success in plant research (Lynch *et al.*, 1983; Norman and St. John, 1986; Norman and Thompson, 1986) and may provide one solution to the detection problems, since it seems capable of detecting microgram quantities. Its use is not widespread, however, and the current high cost may keep it that way for some time. This problem might be alleviated somewhat if the instrument is truly successful in detecting a range of compounds and comes into greater demand. One additional problem is that solvent impurities may be detected, but this normally can be overcome by using high purity solvents.

(c) *Fluorescence.* Fluorescence often is useful for sensitive detection and quantification of compounds. Some limited use of fluorescing derivatives has been reported for phospholipid detection (Kaneko *et al.*, 1987; Postle, 1987), but the number of species is limited and further development will be required for widespread use.

(d) *Mass detection.* General detection of mass through changes in refractive index and light scatter have been utilised to detect lipids in the solvent stream, but in general these techniques are less than satisfactory due to the very low sensitivity.

Mass spectrometry also can be coupled to HPLC, and so identification of peaks may be achieved directly from the effluent. This technique utilises a moving belt similar to that with FID (see above) and the sample is ionised on the belt. This method has been used to separate phospholipid species and identify subspecies of phosphatidylcholine (Jungalwala et al., 1984).

(e) *Radioisotope detectors.* Flow-through radioisotope detectors recently have become available for HPLC uses from a variety of companies. Such detectors could be extremely useful for pulse-chase experiments. The best alternative to such detectors is to collect fractions and assay them in a scintillation counter.

4. Summary

The use of HPLC for purification of phospholipids and phospholipid molecular species remains in its infancy, but shows great promise. A major drawback has been the lack of suitable detectors. While the FID detector appears promising, the current high cost might be a problem for many laboratories. On the other hand, simply collecting fractions and running colorimetric assays might be suitable for many experiments (although obviously less automated and therefore prohibitive for mass screening).

There has been no report at the time of writing of a single column to separate all the phospholipid classes, yet the major classes can be separated well and a judicious choice of columns and solvent systems should allow separation of specific phospholipids of special interest. Some of the new developments in HPTLC appear to make this method a better choice for routine screening.

HPLC appears to be destined to becoming the method of choice for separation of phospholipid subclasses. The good separations being obtained offer real competition to the other major method in this field, argentation chromatography. Further development could allow superb separations of molecular species.

VI. SIDEDNESS OF PHOSPHOLIPIDS IN MEMBRANES

The distribution of phospholipids between the two layers (leaflets) of a bilayer membrane has been the subject of many investigations utilising microorganisms and animals. Unfortunately, such information is extremely limited for plant cells (Cheesbrough and Moore, 1980). Such data are useful for estimating the properties of membranes, the mechanisms of certain transport systems, etc. On the other hand, these estimates must be considered as first approximations since most techniques for estimating sidedness could affect flip-flop of phospholipids between the leaflets or the permeability of the membrane to the probe. Also, some of the phospholipids might be shielded from reacting by proteins with which they are associated (Cheesbrough and Moore, 1980). Thus, care must be taken to run proper controls and interpret the results cautiously. The primary methods utilised for sidedness studies are: (1) covalent bonding of probes; (2) enzymatic alterations of the phospholipids; and (3) specific transfer with phospholipid transfer (exchange) proteins.

A. Chemical Probes

Two effective chemical probes utilised for sidedness studies in plants are 2,4,6-trinitrobenzenesulphonic acid (TNBS) and 1,5-dinitro-3-fluorobenzene (DNFB) (Cheesbrough and Moore, 1980). Both probes react with free amino groups, and so would be restricted to estimations of phosphatidylethanolamine and phosphatidylserine. TNBS does not penetrate membranes, while DNFB penetrates readily (Marinetti and Love, 1976), thus providing a method whereby total reacting phospholipid may be compared with that reacting only with a probe on one side (TNBS). Treatment is in an iso-osmotic solution in order to avoid rupture of the organelle or cell being examined. The membranes are treated in 50 mM HEPES (with osmoticum) at pH 8.4 and 27°C with 2 mM dye for 20 min.

Detection of the amount of phospholipid accessible is estimated by determining how much phospholipid remains unreacted. This may be done following TLC and measuring either phosphate or radioactivity (from cells pre-labelled with radiolabelled precursors of the lipids) remaining in the spot as compared to untreated controls (see methods above; Cheesbrough and Moore, 1980).

B. Enzymatic Alterations

A widely used method, because of interest in phosphatidylcholine, the most abundant phospholipid, is treatment with phospholipase A_2. This enzyme normally exhibits a preference for PtdCho, but will remove the acyl unit from the 2-position of all phospholipids. The principle involves treating an intact cell or organelle with the enzyme, which does not pass through the membrane, and comparing the result with treatments of ruptured cells or organelles. An additional control may be necessary, since some of the phospholipids may be inaccessible to the probe (Cheesbrough and Moore, 1980). In this case, complete disruption of the membrane with detergent treatment will allow determination of whether this arises from such masking (perhaps by associations with proteins; Cheesbrough and Moore, 1980) or some problem with the assay.

Treatment with phospholipase A_2, usually from *Naja naja*, is done in an iso-osmotic solution containing 50 mM HEPES at pH 7.5 and 10 mM $CaCl_2$. Ten Iu (1 Iu = 1.0 µmol min^{-1}) of enzyme should suffice. The reaction is conducted at 0 and 37°C and the results compared. Unwanted rupture is more likely at the higher temperature. Following the reaction, the unreacted products are determined as described in Section V.

It is critical to determine the amount of time to treat the membranes in each situation. A time course of treatment might follow the theoretical profile diagrammed in Fig. 3.6, where the reaction reaches a plateau within a certain time, but shortly thereafter additional reactions occur. This usually is due to the limited stability of the membrane when a portion of the phospholipid is converted to the lysophospholipid, a very effective detergent.

C. Transfer Proteins

The preferred method for studies of this sort has not been utilised for plants. Advantage is taken of the existence of phospholipid transfer proteins (PLTP) which transfer phospholipids from one membrane to another across an aqueous medium. Such

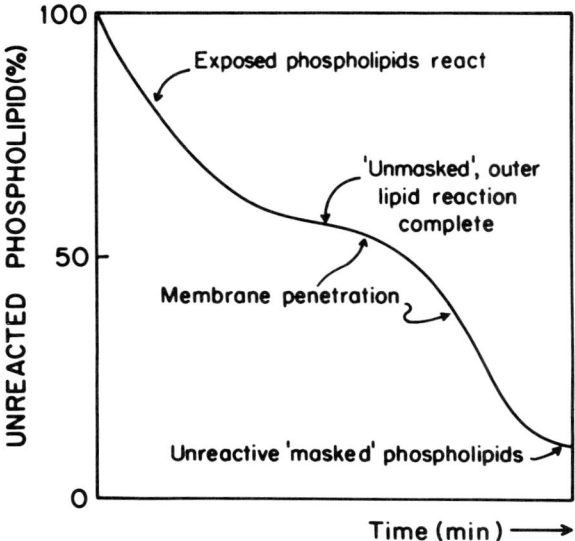

FIG. 3.6. Theoretical time-course of phospholipase A_2 treatment of intact cell or organelle membranes. The amount of a specific phospholipid declines upon treatment and conversion to the lyso-derivative until all the exposed phospholipids are acted on (plateau). As the membrane breaks down and the enzyme penetrates a second decrease is observed. If no plateau is obtained it is not possible to make conclusions concerning sidedness. A portion of the phospholipid may not react and may be protected by close association with proteins (Cheesbrough and Moore, 1980).

proteins have been highly purified and well defined from animal systems (Dawidowicz, 1987), but less so with plants (Kader et al., 1987). Advantage is taken of the general 'exchange' property of such enzymes under most conditions; that is, they will carry a phospholipid from one membrane to another, but after releasing it to the new membrane they will pick up a like phospholipid from this second membrane. The overall net result is a two-way flow of phospholipids. These transfer proteins may be either specific for a single phospholipid or generalist in nature (Bell and Coleman, 1980).

A typical experiment would be designed for a PLTP to transfer phospholipids between natural membranes and an artificial membrane containing the phospholipid(s) of choice. The molecular species composition of the artificial membrane should be as similar as possible to that which occurs naturally. Non-transferable lipids may be included in the artificial membrane in order to estimate the amount of mass transfer which has occurred through random collisions (Kader et al., 1987). The transfer is allowed to go to equilibrium, and the amount of phospholipid transferred from the natural membrane is determined (generally through the use of radiolabelled phospholipids of known specific activity in the artificial membrane). This is corrected for mass transfer (utilising dual labels, the second being in the 'non-transferable' lipid). This allows calculation of the accessible phospholipid in the exposed leaflet of the natural membrane.

The advantage of this method over the chemical or lipase probes is the minimal effect on phospholipid composition of the membrane in question. Both the chemical probe and phospholipase approaches lead to changes in the properties of the phospholipid in question, thereby leading to alterations in phospholipid flip-flop (movement between the two layers of the membrane) or localised membrane disruption, as well as the massive disruptions described above. The PLTP method does not have this effect since removed phospholipids are immediately replaced.

D. Summary

Most investigations to date have led to the conclusion that phospholipids of biological membranes are asymmetrically distributed. There is considerable concern, however, that the environment of the membranes can affect these distributions. Therefore simply isolating organelles or producing protoplasts might affect the phospholipid distributions. Treatment with the probes themselves might have some effect. Therefore, it must be re-emphasised that results of such studies remain a first approximation. On the other hand, they do give us useful information that is not obtainable by other means.

VII. PHOSPHOLIPID PRECURSORS

Precursors of phospholipids may be divided into lipid and non-lipid. The former generally are thought to be produced in the membrane and provide the lipophilic portion of the molecule, while the headgroup precursors are probably produced in the cytosol.

A. Lipid Precursors

The primary lipid precursor is phosphatidic acid, which in turn yields diacylglycerol and CDPdiacylglycerol. Diacylglycerol is a neutral lipid formed by the dephosphorylation of the phosphatidate and the isolation and characterisation of neutral lipids has been dealt with elsewhere (Chapter 2). CDPdiacylglycerol is formed by a reaction utilising CTP and phosphatidate and occurs at very low concentrations in the membrane. It may be purified and characterised by the chromatography methods described above.

B. Headgroup Precursors

The metabolic pathways resulting in headgroup synthesis and modification are several, and the precursors of the headgroups (choline, serine, ethanolamine, inositol and glycerol-P) are known to play key roles in pathways other than those directly related to phospholipid biosynthesis. Even the various headgroups themselves are often interrelated through pathways linking them through synthesis or utilisation. Thus the study of these precursors is complex and multiple pools may exist in the cell. However, questions centering on the rate and regulation of flow through the various pathways are intriguing, and methods have become available which allow for the beginnings of such studies, even though interpretations remain difficult.

1. Serine, ethanolamine, choline

The metabolism of these headgroup precursors with respect to phospholipid synthesis is complex and highly interrelated. Potential pathways are presented in various publications (Mudd and Datko, 1986b; Kinney and Moore, 1987). Unravelling these interactions requires separation of a large number of intermediates with similar properties. This has discouraged research in this area with both plant and animal systems for a number of years. The situation appears to be changing, however, with the recent publication of some interesting results (Mudd and Datko, 1986a, b; Kinney and Moore, 1987). Much remains to be done, however.

(a) *Purification.* The key to this work is a proper separation of all the intermediate involved. Separation of several of the key intermediates may be obtained by TLC (Kinney and Moore, 1987). For such measurements, an aqueous–methanol extract is obtained, evaporated to dryness, and redissolved in HEPES buffer (20 mM, pH 7). This solution is passed through an ion-exchange column packed with AG1-X8 (200–400 mesh) in the formate form and equilibrated with 20 mM glycylglycine buffer (pH 9). The column retains about 70% of the free bases and more than 95% of the phosphoryl and nucleotide bases when eluted with the pH 9 buffer (Kinney and Moore, 1987). Compounds bound to the column may be eluted with 0.1 M NH_4HCO_3. Both the HEPES and ammonium carbonate elutions are then separated on 500 µm silica gel G plates using methanol–0.6% (w/v) KCl–conc. NH_4OH (50:50:5; v/v) in the first dimension and, after drying the plate in air, 1-butanol–propionic acid–water (2:1:1:3; v/v) in the second.

Obtaining a full balance sheet, and thereby complete understanding of the pathways involved, is more involved. The best approach to date requires a combination of column, thin layer and paper chromatography, and electrophoresis, and has been described by Mudd and Datko (1986a,b). A flowchart of their central purification scheme, used initially for *Lemna* (Mudd and Datko, 1986a), is diagrammed in Fig. 3.7. The methods are not difficult, but the number of compounds to be dealt with for a detailed picture are many and some of the specific purifications require several hours. A complete characterization can require days of careful work.

The scheme involves an initial separation of precursors, products and intermediates into methanol–water or chloroform–methanol soluble, and insoluble fractions. The methanol–water fraction contains such key intermediates as S-adenosyl-L-methionine, choline, ethanolamine, L-serine, etc. The compounds in this fraction of further interest are subsequently separated by paper electrophoresis. The chloroform–methanol fraction contains the lipids, which may be fractionated into polar and non-polar lipids by methods mentioned above (Section IV), and these may be separated according to chromatography methods described here (Section V) and elsewhere in this volume (Chapters 2 and 4). The insoluble pellet is primarily protein and nucleic acid, which may be measured directly or subjected to hydrolysis and chromatography of the resultant monomeric units in order to determine the quantities of particular amino acids, etc. (Mudd and Datko, 1986a).

While most of the procedures are rather straightforward, care must be taken to avoid loss of certain intermediates. Of particular concern are two potential intermediates in the biosynthesis of choline, monomethylethanolamine and dimethylethanolamine. These two compounds are somewhat volatile in the unprotonated state (Mudd and Datko, 1986b), and so care must be taken to keep them protonated whenever possible, particularly avoiding paper chromatography involving development with basic solvents (Mudd and Datko, 1986b).

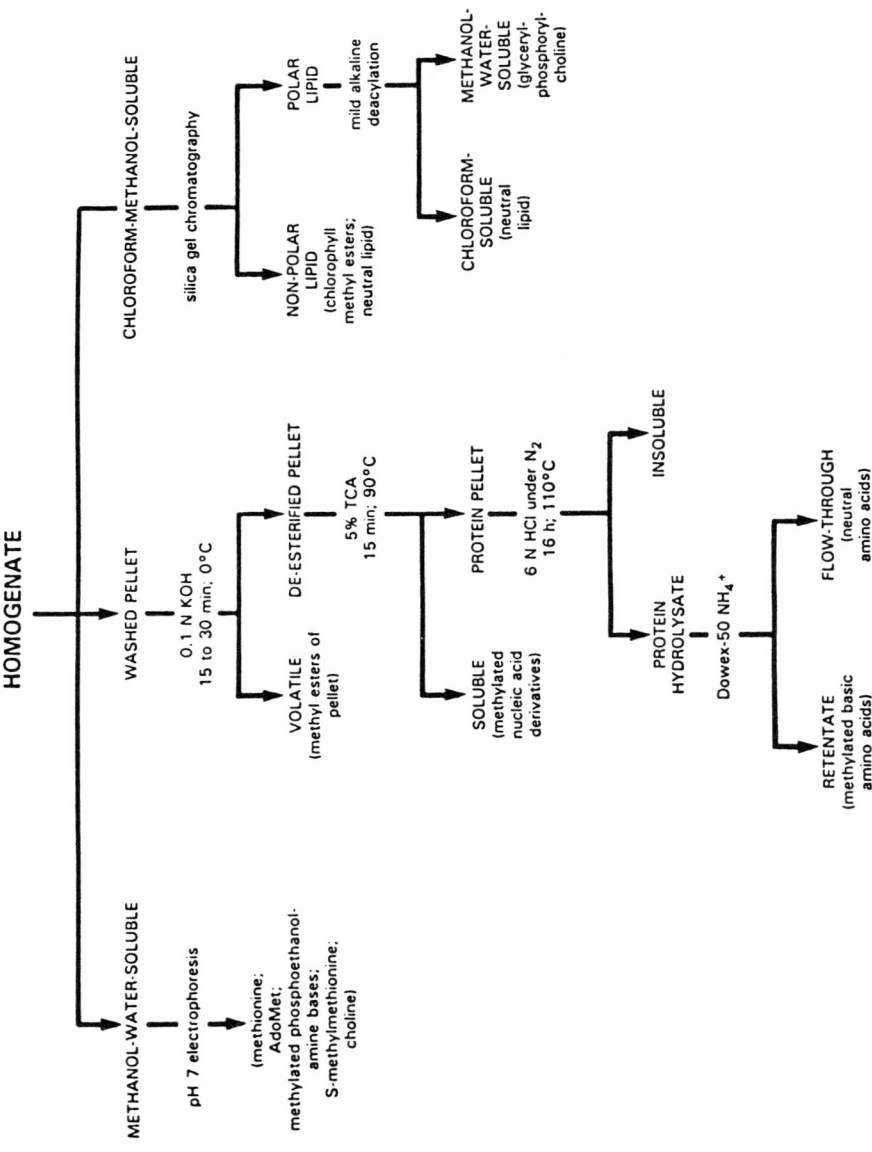

FIG. 3.7. Flow-chart of extraction and purification methods for products of serine, ethanolamine and choline incorporation. This sequence results in separation of these intermediates and their water-soluble derivatives potentially involved in PtdCho, PtdEtn and PtdSer synthesis, as well as other metabolic pathways. Details are presented in the text and by Mudd and Datko (1986a,b). The figure is from Mudd and Datko (1986a) and is used with permission.

Some effort has been made to utilise HPLC for similar investigations in animals (Mazzola and Kent, 1984), but detection is cumbersome and the technique has not been reported using plants.

(b) *Quantification.* Measurements of the amount of these intermediates in tissue is of importance in calculating possible separate pools and regulatory steps, yet most studies have dealt with radioactive pulse-chase techniques and reported only percentage of total radiolabel incorporated (Mudd and Datko, 1986a,b; Kinney and Moore, 1987). This is partly due to the difficulties in obtaining sufficient material for chemical assays, and the need to run several assays to cover all the intermediates. Methods are available for qualitative and quantitative analysis of most of the intermediates (Dittmer and Wells, 1969).

(c) *Summary.* Despite the difficulty of working with these intermediates, reliable techniques have now been developed and considerable progress made. Further developments in HPLC methodology should further inprove this separation.

VIII. OVERVIEW

Research on phospholipids has lagged behind many other areas, often due to the inadequacy or difficulty of available techniques. This is changing. Major improvements are occurring in the speed, resolution and ease of studying phospholipids, their precursors, and the pathways of their synthesis and degradation. Methods exist not only for the excellent separation of the major phospholipid species, but also for investigations of specific subspecies. With these techniques in hand, it may be anticipated that our understanding of how the synthesis and degradation of phospholipids occurs, and how these pathways are regulated, will improve dramatically over the next few years.

ACKNOWLEDGEMENT

Some of the research described was supported by grants from the National Science Foundation. Preparation of this manuscript was supported in part by grant DCB-8703739 for NSF.

REFERENCES

Bartlett, G. R. (1959). *J. Biol. Chem.* **234**, 466–471.
Bell, R. M. and Coleman, R. A. (1980). *Ann. Rev. Biochem.* **49**, 459–487.
Blank, M. L., Robinson, M., Fitzgerald V. and Snyder, F. (1984) *J. Chromatogr.* **298**, 473–482.
Bligh, E. G. and Dyer, W. J. (1959). *Can. J. Biochem. Physiol.* **27**, 911–917.
Cheesbrough, T. M. and Moore, T. S. (1980). *Plant Physiol.* **65**, 1076–1080.
Christie, W. W. (1982). "Lipid Analysis", 2nd edn. Pergamon Press, New York.
Christie, W. W. (1987). "HPLC and Lipids. A Practical Guide". Pergamon Press, New York.
Crosby, S. D. and Dale, G. L. (1985). *J. Chromatogr.* **323**, 462–464.
Dawidowicz, E. A. (1987). *Ann. Rev. Biochem.* **56**, 43–61.

Dawson, R. M. C. (1976). *In* "Lipid Chromatographic Analysis", 2nd edn. (G. V. Marinetti, ed.), Vol. 1, pp. 149–172. Marcel Dekker Inc., New York.
Demandre, C., Trémolières, A., Justin, A. M. and Mazliak, P. (1985). *Phytochemistry* **24**, 481–485.
Dittmer, J. C. and Wells, M. A. (1969). *Methods Enzymol.* **14**, 482–530.
Duck-Chong, C. G. (1977). *Lipids* **14**, 492–497.
Dugan, E. A. (1985). *LC Magazine* **3**, 126–128.
Folch, J., Lees, M. and Stanley, G. H. S. (1957). *J. Biol. Chem.* **226**, 497–509.
Garbus, J., DeLuca, H. F., Loomans, M. E. and Strong, F. M. (1963). *J. Biol. Chem.* **238**, 59–63.
Glaser, M., Ferguson, K. A. and Vagelos, P. R. (1974). *Proc. Natl. Acad. Sci. USA* **71**, 4072–4076.
Harwood, J. L. (1976). *Phytochemistry* **15**, 1459–1464.
Heape, A. M., Jugelin, H., Boiron, F. and Cassagne, C. (1985). *J. Chromatogr.* **322**, 391–395.
Horrocks, L. A. (1963). *J. Am. Oil Chem. Soc.* **40**, 235–236.
Hostetler, K. Y., Van Den Bosch, H. and Van Deenen, L. L. M. (1971). *Biochim. Biophys. Acta* **239**, 113–119.
Jungalwala, F. B., Evans, J. E. and McCluer, R. H. (1984). *J. Lipid Res.* **25**, 738–749.
Kader, J.-C., Vergnolle, C. and Mazliak, P. (1987). *Methods Enzymol.* **148**, 661–666.
Kaneko, T., Ohta, Y. and Machida, Y. (1987). *Agric. Biol. Chem.* **51**, 2023–2024.
Kennerly, D. A. (1986). *J. Chromatogr.* **363**, 462–467.
Kesselmeir, J. and Heinz, E. (1985). *Anal. Biochem.* **144**, 319–328.
Kinney, A. J. and Moore, T. S. (1987). *Plant Physiol.* **84**, 78–81.
Kolarovic, L. and Fournier, N. C. (1986). *Anal. Biochem.* **156**, 244–250.
Kolarovic, L. and Traitler, H. (1985). *J. High Resol. Chromatogr. Chromatogr. Commun.* **8**, 838–842.
Lynch, D. V., Gundersen, R. E. and Thompson, G. A. (1983). *Plant Physiol.* **72**, 903–905.
Marinetti, G. V. and Love, R. (1976). *Chem. Phys. Lipids* **16**, 239–254.
Mazzola, G. and Kent, C. (1984). *Anal. Biochem.* **141**, 137–142.
Mudd, S. H. and Datko, A. H. (1986a). *Plant Physiol.* **81**, 103–114.
Mudd, S. H. and Datko, A. H. (1986b). *Plant Physiol.* **82**, 126–135.
Nichols, B. W. (1964). *In* "New Biochemical Separations" (A. T. James and L. J. Morris, eds), pp. 321–337. Van Nostrand, New York.
Norman, H. A. and St. John, J. B. (1986). *J. Lipid Res.* **27**, 1104–1107.
Norman, H. A. and Thompson, G. A. (1986). *Plant Sci.* **42**, 83–88.
Postle, A. D. (1987). *J. Chromatogr.* **415**, 241–251.
Radwan, S. S. (1984). *In* "Handbook of Chromatography Lipids" (H. K. Mangold, ed.), Vol. I, pp. 481–508. CRC Press Inc., Boca Raton, FL.
Rivnay, B. (1984). *J. Chromatogr.* **294**, 303–315.
Rouser, G., Kritchevsky, G. and Yamamoto, A. (1976). *In* "Lipid Chromatographic Analysis", 2nd edn. (G. V. Marinetti, ed.), Vol. 3, pp. 713–776. Marcel Dekker, Inc., New York.
Sheeley, R. M., Hurst, W. J., Sheeley, D. M. and Martin, R. A. (1987). *J. Liquid Chromatogr.* **10**, 3173–3181.
Shukla, V. K. S. (1988). *Progr. Lipid Res.* **27**, 5–38.
Touchstone, J. C., Chen, J. C. and Beaver, K. M. (1980). *Lipids* **15**, 61–62.
Touchstone, J. C., Levin, S. S., Dobbins, M. F. and Carter, P. J. (1981). *J. High Resol. Chromatogr. Chromatogr. Commun.* **4**, 423–424.
Vioque, E. (1984). *In* "Handbook of Chromatography Lipids" (H. K. Mangold, ed.), Vol. II, pp. 309–329. CRC Press Inc., Boca Raton, FL.
Vose, P. B. (1980). "Introduction to Nuclear Techniques in Agronomy and Plant Biology". Pergamon Press, New York.

4 Glycolipid Analyses and Synthesis in Plastids

ROLAND DOUCE, JACQUES JOYARD,
MARYSE A. BLOCK and ALBERT-JEAN DORNE

Laboratoire de Physiologie Cellulaire Végétale, Département de Biologie Moléculaire et Structurale, Centre d'Etudes Nucléaires de Grenoble et Université Joseph Fourier, 85X, F-38041 Grenoble-cedex, France

I.	Introduction	71
II.	Isolation of intact plastids	72
	A. Isolation of intact chloroplasts	73
	B. Isolation of intact non-green plastids	75
	C. Thermolysin digestion of intact plastids	81
III.	Purification of plastid envelope membranes	82
	A. Purification procedures	82
IV.	Determination and characterisation of glycolipids	85
	A. Extraction of plant glycolipids	85
	B. Thin layer chromatography of glycolipids	86
	C. Column chromatography of glycolipids	88
	D. Analyses of fatty acids	90
V.	Biosynthesis of glycolipids	93
	A. Glycolipid synthesis by isolated intact plastids	93
	B. Glycolipid synthesis by isolated envelope membranes	94
	Acknowledgements	101
	References	101

I. INTRODUCTION

All plant tissues contain several polar lipids having a sugar moiety, i.e. glycolipids. The most widely distributed glycolipids among the plant kingdom are galactolipids and sulpholipids (Fig. 4.1). For instance, monogalactosyldiacylglycerol (MGDG) is probably the most abundant polar lipid on earth (Gounaris and Barber, 1983). Several reviews describe the expanding literature on galactolipid and sulpholipid structure, distribution and biosynthesis. The reader is referred to reviews recently published in the series *The Biochemistry of Plants* for a comprehensive survey of galactolipid and sulpholipid distribution and occurrence within plant tissues and cells, for a study of their biosynthesis and of the properties of the enzymes involved in their formation (Harwood, 1980; Douce and Joyard, 1980; Joyard and Douce, 1987; Mudd and Kleppinger-Sparace, 1987). Several other articles provide most interesting additional information on the structure, role and biosynthesis of plant glycolipids (Heinz, 1977; Roughan and Slack, 1982; Quinn and Williams, 1983; Frentzen, 1986; Murphy, 1986; Gounaris *et al.*, 1986; Heemskerk and Wintermans, 1987). All these studies demonstrate (a) that glycolipids are mostly concentrated within plastid structures, and (b) that plastids, which are the sites for fatty acid synthesis within plant cells, play a central role in glycolipid biosynthesis.

The purpose of this chapter is to provide methods for studies of the distribution and biosynthesis of plant glycolipids. Since plastids are of major importance for such studies, we will consider first methods for the purification of these organelles (chloroplasts and non-green plastids) and their fractionation to prepare envelope membranes. Then, we will describe the procedures we use for determination and characterisation of galactolipids and sulpholipids within plant membranes. Finally, we will describe methods for studying glycolipid biosynthesis by organelles, membrane fractions or purified enzymatic extracts. This chapter is largely based on our practical experience of higher plant organelles and glycolipids. The methods described here in detail are performed routinely in the authors' laboratory and include all the practical details necessary for their successful application.

II. ISOLATION OF INTACT PLASTIDS

Galactolipid biosynthesis is highly dependent upon the plant species analysed: for instance, Heinz and Roughan (1983) have clearly demonstrated that distinct patterns are obtained with spinach (a typical 16:3 plant*) or pea (a typical 18:3 plant) chloroplasts. The same is true for non-green plastids, as shown by Alban *et al.* (1988, 1989). Therefore, it is important to use 16:3 as well as 18:3 plant species for *in organello* studies of glycolipid biosynthesis.

The key to achieving optimal rates of glycerolipid biosynthesis by isolated organelles is the preparation of physiologically intact plastids. For instance, the highest rates of [1-^{14}C]acetate incorporation into chloroplast glycerolipids have always been obtained with chloroplasts having the highest rates for CO_2-dependent O_2 evolution. The development of procedures involving Percoll purification of plant cell organelles was indeed a breakthrough for the understanding of glycerolipid biosynthesis within chloroplasts. Percoll is a non-toxic silica sol gradient material [colloidal silica coated with poly(vinylpyrrolidone)] developed by Pharmacia. It has major advantages as a medium for density gradient, i.e. a low viscosity and a negligible osmotic potential, which have permitted the introduction of rapid purification procedures.

---

* Plants having only 18:3 fatty acid (linolenic acid) at both *sn*-1 and *sn*-2 positions of MGDG (see Fig. 4.1) are called 18:3 plants. A typical example is pea. Plants having both 18:3 and 16:3 (hexadecatrienoic acid) in MGDG are called 16:3 plants. In this case, whereas 18:3 can be at both *sn*-1 and *sn*-2 positions of the glycerol backbone of MGDG, 16:3 is restricted to the *sn*-2 position (see Fig. 4.1). A typical example is spinach. The situation is true for chloroplasts as well as for non-green plastids.

4. GLYCOLIPID ANALYSES AND SYNTHESIS IN PLASTIDS

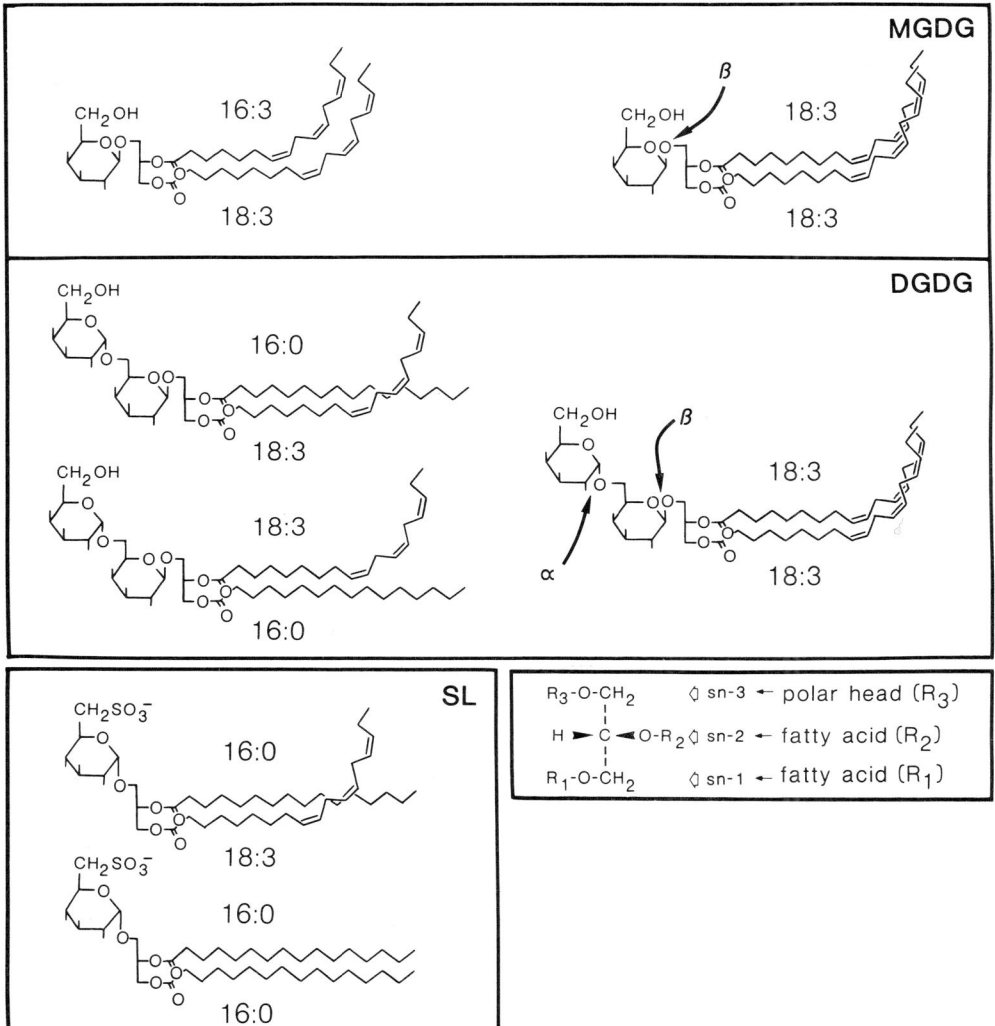

FIG. 4.1. Structure of plastid glycolipids. MGDG, monogalactosyldiacylglycerol; DGDG, digalactosyldiacylglycerol; SL, sulpholipid.

A. Isolation of Intact Chloroplasts

The method used for purification of intact chloroplasts is adapted from those described by Mourioux and Douce (1981) and Douce and Joyard (1982).

1. Purification method

(a) *Solutions*

Isolation medium: mannitol (or sorbitol), 330 mM; tetrasodium pyrophosphate, 50 mM, pH 7.8; $MgCl_2$, 1 mM; $MnCl_2$, 1 mM; EDTA, 2 mM; bovine serum albumin, 0.1% (w/v).

Washing medium: mannitol (or sorbitol), 330 mM; HEPES-KOH, 10 mM, pH 7.8; $MgCl_2$, 1 mM; $MnCl_2$, 1 mM; EDTA, 2 mM; bovine serum albumin, 0.1% (w/v).

Percoll solutions: Percoll 50% or 40% and 80% (v/v); mannitol (or sorbitol) 330 mM; HEPES-KOH, 10 mM, pH 7.8; $MgCl_2$, 1 mM; $MnCl_2$, 1 mM; EDTA, 2 mM; bovine serum albumin, 0.1% (w/v).

Note: For the preparation of large amounts of chloroplasts for further purification of envelope membranes, chloroplast stroma and thylakoids, the presence of $MgCl_2$, $MnCl_2$ and EDTA is not necessary. Furthermore, in this case, no bovine serum albumin is added to the washing medium and Percoll solutions.

(b) *Materials*

Nylon blutex (toile à bluter), 50 µm aperture (Tripette et Renaud, Sailly-Saillisel, Combles, France).

Muslin or cheesecloth or Miracloth, 50–60 cm large.

Blender (a 1-gallon Waring blender is most convenient for large quantities of leaves), the use of a Polytron (probe PT 35K) is recommended for pea leaves. Large leaves should be cut in small pieces prior to the grinding of the tissue.

Superspeed centrifuge, refrigerated (such as H-401 Kontron, RC5 Sorvall or J21 Beckman, for instance) with a whole set of fixed angle (Sorvall SS-34 for 50 ml tubes and Sorvall GS3 for 500 ml bottles) and swinging bucket rotors (Sorvall HB-4 for 50 ml tubes or HS-4 for 250 ml bottles). The use of a vertical rotor (Sorvall SV-288) is most convenient for rapid, efficient and elegant separation of low amounts of intact chloroplasts (Mourioux and Douce, 1981).

Note: For processing large amounts of leaves (0.5 to 3 kg) that are necessary for the preparation of envelope membranes, the use of large rotors (such as Sorvall rotors GS3 and HS-4) is strongly recommended.

(c) *Procedure*. The 6- to 8-week-old spinach leaves are harvested directly from the greenhouse about 2 h after the beginning of the light period. They are rinsed in deionised water, deribbed, placed in the grinding medium in the Waring blender (200 g leaves for 500 ml) and homogenised at low speed for 3 s. For pea leaves, the Polytron is set at 6 and homogenisation should be limited to a few s (less than 5). In both cases, longer homogenisation improves the yield of recovered chlorophyll but increases the proportion of broken chloroplasts (see Walker *et al.*, 1987). In addition, all operations are carried out in the cold (0–5°C) and are performed as quickly as possible. The homogenate is filtered rapidly through six layers of muslin and one layer of nylon blutex, and then centrifuged to purify intact chloroplasts, as described in Fig. 4.2(a). When large amounts of leaves are processed, the conditions for centrifugations are different due to the size of the rotors used (Fig. 4.2(b)).

The final pellets of intact chloroplasts are then carefully resuspended in a minimum volume of washing medium, with gentle agitation. From 200 g of spinach leaves, intact and purified chloroplasts corresponding to 3–5 mg chlorophyll (50–100 mg protein) are routinely prepared.

2. Chlorophyll determination

Mix 10 µl of chloroplast suspension in 10 ml acetone–water solution (80%, v/v). Spin the mixture for 10 min at $4500 \times g$ (in 30-ml Corex tubes, Sorvall SS-34 rotor with rubber adapters) in order to eliminate precipitated proteins. Absorbance of the supernatant is measured at 652 nm. The chlorophyll concentration of the original chloroplast suspension is estimated with the following formula: chlorophyll (mg ml^{-1}) = OD$_{652} \times 26$. More accurate measurements can be done using the formula described by Bruinsma (1961).

3. Chloroplast intactness

The procedures described above allow the preparation of highly intact chloroplasts, as judged by observation with phase contrast microscopy. As far as lipid synthesis by chloroplasts is concerned, it is recommended to assess the CO_2-dependent O_2 evolution using an oxygen electrode (Walker, 1980; Walker et al., 1987). The Hansatech oxygen electrode (Hansatech Ltd, Norfolk, UK) is extremely convenient for these studies. The experimental conditions for this assay are carefully described by Walker (1987).

B. Isolation of Intact Non-green Plastids

It is convenient to use cauliflower buds for the preparation of large amounts of intact and very pure non-green plastids. We also routinely use isolated sycamore cells for the preparation of amyloplasts. The methods used (conventional procedures for cauliflower, and via protoplast isolation for sycamore) can be generalised to other non-green tissues or isolated cells.

1. Purification of non-green plastids from cauliflower buds

The procedure used is adapted from that described by Journet and Douce (1985).

(a) *Solutions*

Isolation medium: mannitol, 300 mM; tetrasodium pyrophosphate, 20 mM, pH 7.6; EDTA, 1 mM; bovine serum albumin, 0.2% (w/v); cysteine, 4 mM (cysteine should be added just before the blending step).

Washing medium: sucrose, 300 mM; 3-(N-morpholino)propane sulphonic acid-KOH (MOPS-KOH), 10 mM, pH 7.2; EDTA, 1 mM; bovine serum albumin, 0.1% (w/v).

Percoll solution: Percoll, 35% (v/v); sucrose, 300 mM; MOPS-KOH, 10 mM, pH 7.2; EDTA, 1 mM; bovine serum albumin, 0.1% (w/v).

(b) *Materials.* See Section II.A.1(b) above.

(c) *Procedure.* All the experiment should be done at 0–4°C. The flower heads of cauliflower florets are cut, placed in a 1–gal Waring blender (1 kg of cut material for 2

Procedure A

Chop 200 g (150 mg chlorophyll) of deribbed spinach leaves.
Homogenise with 0.5 litre of isolation medium in a Waring blender at low speed for less than 5 s. Filter on muslin and nylon blutex. Centrifugations as follows:

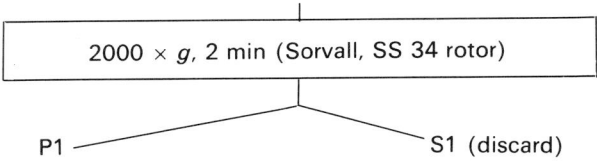

Resuspend in 8–15 ml washing medium, 2 ml
(4 mg chlorophyll) loaded on top of each
preformed Percoll gradient (4–8 tubes, 34 ml each);
prior to loading, linear gradients were preformed
by centrifugation for 100 min at $10\,000 \times g$
of a 50% Percoll solution (vertical rotor)

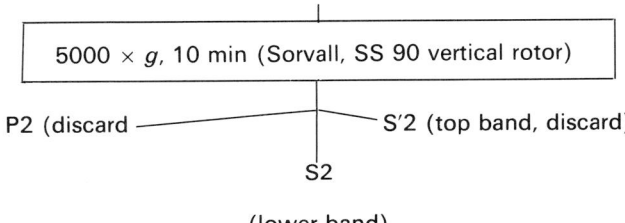

(lower band)

Recover with a pipette, dilute with 100 ml washing medium

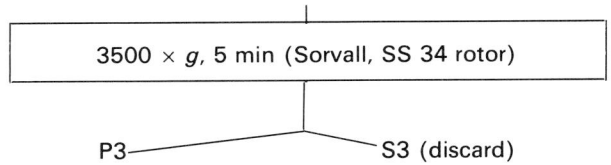

INTACT and PURIFIED CHLOROPLASTS

(a)

Procedure B

Chop 1 kg (750 mg chlorophyll) of deribbed spinach leaves.
Homogenise with 2 litres of isolation medium in a one gallon Waring blender at low speed for less than 5 s. Filter on muslin and nylon blutex. Centrifugations as follows:

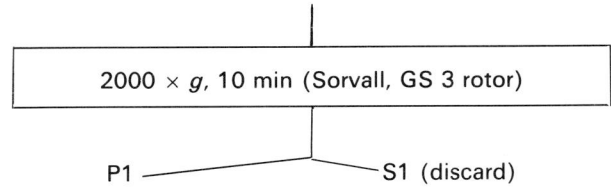

4. GLYCOLIPID ANALYSES AND SYNTHESIS IN PLASTIDS

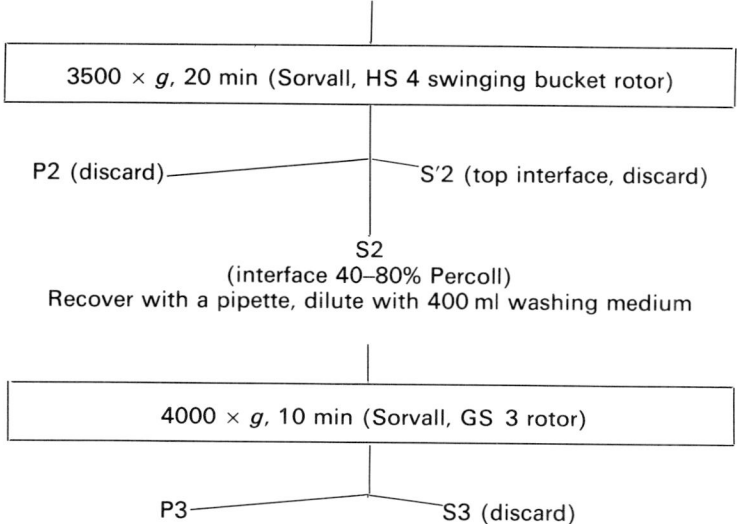

Resuspend in 60–80 ml washing medium, mix with a second batch, (from another kg of leaves), lay on top of Percoll gradients (4 tubes, 150 ml each) made of two layers (25 ml of 80% Percoll and 50 ml of 40% Percoll solutions)

3500 × g, 20 min (Sorvall, HS 4 swinging bucket rotor)

P2 (discard) — S'2 (top interface, discard)

S2
(interface 40–80% Percoll)
Recover with a pipette, dilute with 400 ml washing medium

4000 × g, 10 min (Sorvall, GS 3 rotor)

P3 — S3 (discard)

INTACT and PURIFIED CHLOROPLASTS

(wash this pellet again if envelope membranes are to be prepared)

(b)

FIG. 4.2. Protocol for isolation and purification of chloroplasts from spinach. Procedure A, adapted from Mourioux and Douce (1981), is suitable for processing about 200 g of spinach leaves, whereas Procedure B, adapted from Douce and Joyard (1982), is suitable for large amounts (1–2 kg) of spinach leaves.

litres of isolation medium) and ground at low speed for 3 s. Longer blending improves the yield of opened cells, but increases the proportion of damaged plastids. The homogenate is rapidly filtered through a nylon blutex. It is possible to grind again the material retained by the nylon blutex in 1 litre of fresh isolation medium. This procedure is a better way of improving both the yield and preventing breakage of plastids than a single long blending of the mixture. The suspension is then centrifuged to prepare first a crude pellet of plastids, and then to purify intact plastids by a series of differential and density gradient centrifugations, as described in Fig. 4.3. It is not possible to obtain intact plastids devoid of any contaminating extraplastidial structures by a single-step purification. The efficacy of the two-step method outlined in Fig. 4.3 was clearly demonstrated by Journet and Douce (1985). The final yield of intact plastids is, however, very low (0.5–1% of the total amounts of plastids present in the cauliflower tissue used). Starting from 1 kg of cut material, intact and purified plastids corresponding to approximately 10–15 mg protein are routinely prepared.

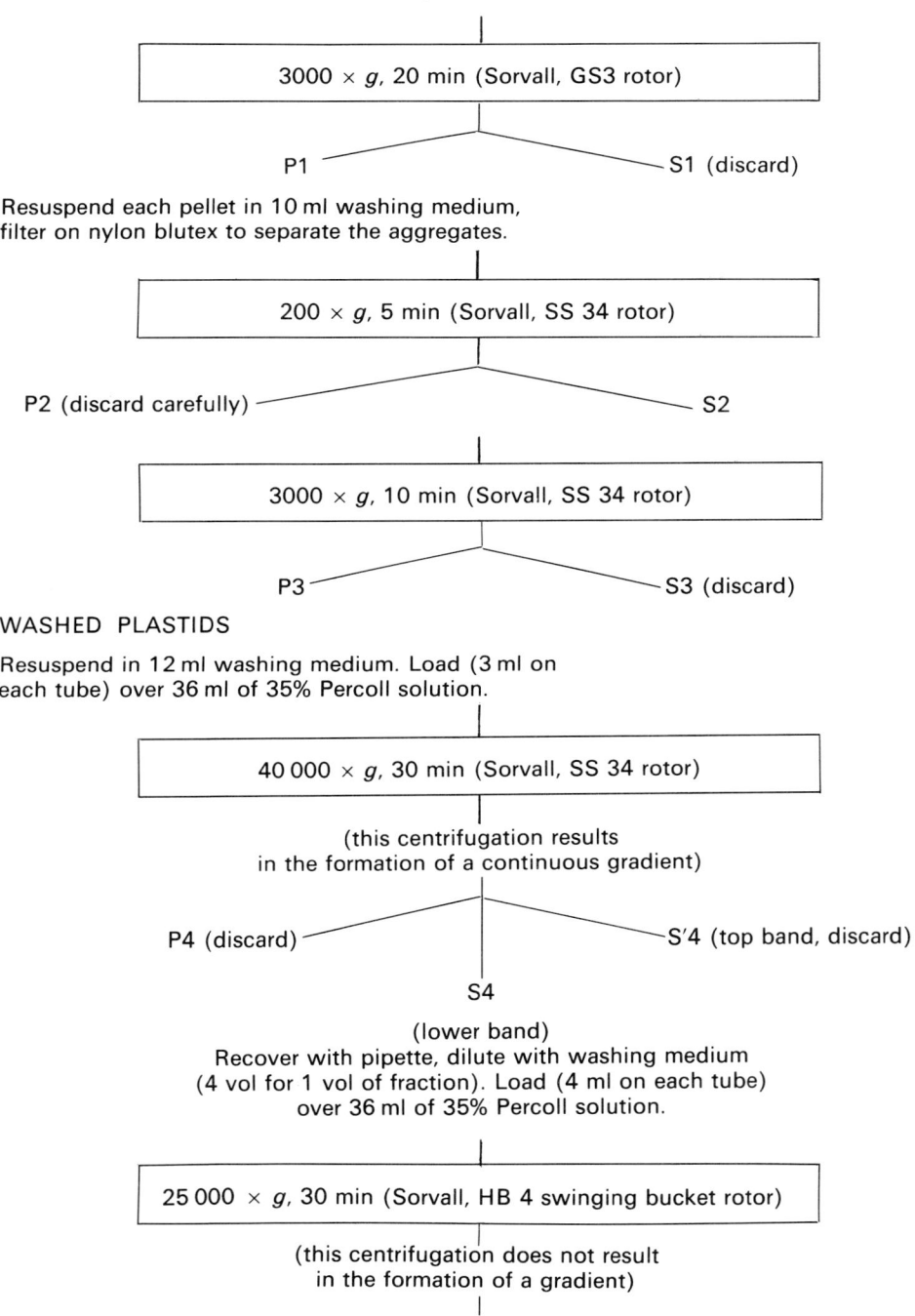

Recover with a pipette, dilute with washing medium
(10 vol for 1 vol of plastid suspension).

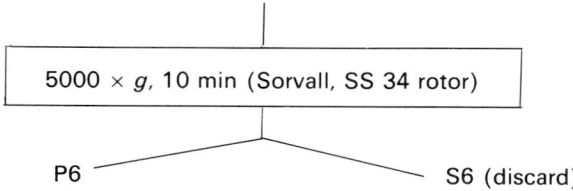

INTACT and PURIFIED NON-GREEN PLASTIDS

FIG. 4.3. Protocol for isolation of washed non-green plastids from cauliflower buds and their purification (adapted from Journet and Douce, 1985).

2. Purification of amyloplasts from sycamore cells

The procedure described is largely adapted from that described by Journet et al. (1986). A very important step in this preparation is the transfer of the sycamore cells to a sucrose-free medium about 15 h prior to the extraction. It is essential to decrease the size of the starch grains to get reasonable yields of intact plastids. A description of the events that occur within isolated cells after their transfer to a sucrose-free medium is given by Journet et al. (1986).

(a) *Solutions*

Culture medium: the complete composition was described by Bligny and Leguay (1987).

Cell digestion mixture: mannitol, 0.5 M; cellulase (Onozuka RS, Yakult Pharmaceutical Co., Nishinomiya, Japan), 1% (w/v); pectolyase Y-23 (Seishin Pharmaceutical Co., Nishinomiya, Japan), 0.1% (w/v); all these compounds are added to the cell culture medium, the final pH is adjusted to 5.7.

Protoplasts washing medium: mannitol, 0.5 M; in the cell culture medium.

Fractionation medium: mannitol, 0.5 M; MOPS-NaOH, 20 mM, pH 7.5; EDTA, 2 mM; bovine serum albumin, 0.1% (w/v); spermidine, 0.4 mM; β-mercaptoethanol, 7 mM; phenylmethylsulphonyl fluoride (PMSF), 1 mM; leupeptine, 10 μM; benzamidine-HCl, 1 mM; ε-aminocaproic acid, 5 mM; insoluble polyvinylpyrrolidone (K c.a. 25, Serva).

Suspension medium: mannitol, 0.5 M; MOPS-NaOH, 10 mM, pH 7.5; EDTA, 2 mM; PMSF, 1 mM; leupeptine, 10 μM; benzamidine-HCl, 1 mM; ε-aminocaproic acid, 5 mM.

Percoll solution: suspension medium plus Percoll, 50% (v/v).

(b) *Materials.* Shaker for digestion of cells at 25°C. Microscope. Other materials: see Section II.A.1(b) above.

(c) *Procedure.* Between 150 and 200 g of sycamore cells in exponential growth (3–5 days after the beginning of the culture) are used for protoplast preparation. The evening prior to the experiment, the cells are transferred to a sucrose-free culture medium. Then the cells are harvested and washed twice in culture medium containing 0.5 M mannitol. Sycamore cells are suspended in 150 ml of cell digestion mixture and incubated with constant shaking (20 cycles min^{-1}) for about 1 h at 25°C. Release of protoplasts from the cells is followed using a microsope. The suspension is filtered and protoplasts are recovered as described by Journet *et al.* (1986). The procedure for the gentle rupture of protoplasts, separation of the organelles and amyloplast purification is described in Fig. 4.4. Intact and purified amyloplasts thus obtained are recovered in a small volume of suspension medium. From 150–200 g of sycamore cells, the yield of purified intact amyloplasts is about 1 mg protein.

Sycamore protoplasts (from 150–200 g cells) are washed twice in fractionation medium. The suspension (100 ml) is loaded via the open end into the barrel of a 25 ml disposable syringe and protoplasts are forced out through a nylon blutex (20 µm). Repeat this step a second time. Control lysis of the protoplasts with a microscope. Centrifugations as follows:

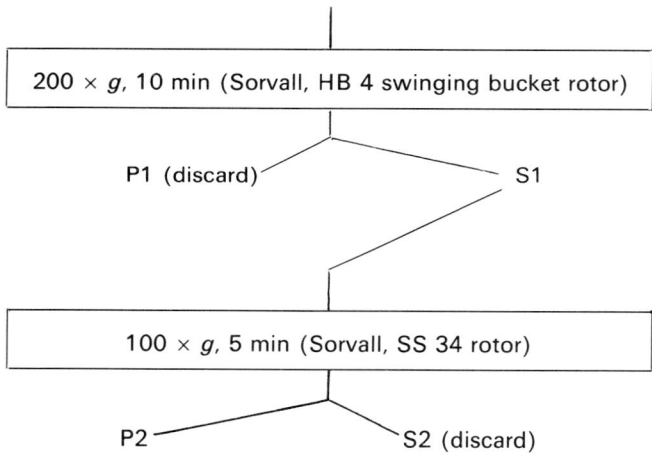

Resuspend carefully in fractionation medium,
filter on Miracloth to separate the aggregates.
Load (2 ml) on top of 50% Percoll solution (8 ml).

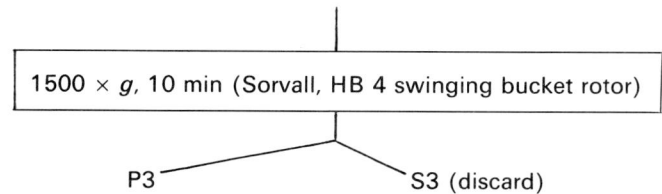

INTACT and PURIFIED AMYLOPLASTS

FIG. 4.4. Protocol for isolation and purification of amyloplasts from sycamore cells (adapted from Journet *et al.*, 1986).

3. Plastid intactness

The intactness of plastid preparations from cauliflower buds, as well as from sycamore cells, can be evaluated by measurement of phosphogluconate dehydrogenase (EC 1.1.1.44) latency in a quick and reliable spectrophotometric assay, as described by Journet and Douce (1985).

4. Plastid purity

The methods used to assess plastid purity are valid for chloroplasts as well as for non-green plastids. The reader is referred to the review by Quail (1979) for details on convenient marker enzymes. In addition, since we are concerned with lipids in this review, it is convenient to determine whether the preparation contains phosphatidylethanolamine (PE) and cardiolipin, which are absent from plastids and are two excellent markers for extraplastidial and inner mitochondrial membranes, respectively (Douce and Joyard, 1980).

C. Thermolysin Digestion of Intact Plastids

The presence of unnatural galactolipids (trigalactosyldiacylglycerol and tetragalactosyldiacylglycerol—Tri-GDG and Tetra-GDG) and of diacylglycerol in preparations of chloroplast envelope membranes is due to the functioning of a galactolipid:galactolipid galactosyltransferase (Van Besouw and Wintermans, 1978). This enzyme catalyses the intergalactolipid exchange of galactose and leads to the formation of diacylglycerol at the expense of MGDG during the course of envelope membrane preparation. When the *in vivo* polar lipid composition of envelope membranes is to be determined, or when envelope membranes with a low diacylglycerol content are required, it is important to inhibit fully the galactolipid:galactolipid galactosyltransferase. We have demonstrated that this enzyme is localised on the cytosolic side of the chloroplast (Dorne *et al.*, 1982) and of non-green plastids (Alban *et al.*, 1988), and therefore is accessible to proteolytic enzymes. Thermolysin is the most convenient proteolytic enzyme for such experiments (Joyard *et al.*, 1983, 1987a).

(a) *Solutions*
Incubation mixture: mannitol (or sorbitol, or sucrose), 330 mM; Tricine-NaOH, 10 mM, pH 7.8; $CaCl_2$, 1 mM; thermolysin (from *Bacillus thermoproteolyticus*, Calbiochem), 200 µg ml^{-1}.

Incubation solution for control: same as above plus EGTA, 10 mM.

Percoll solutions: Percoll, 35 or 40% (v/v); mannitol (or sorbitol, or sucrose), 330 mM; tricine-NaOH, 10 mM, pH 7.8; ε-aminocaproic acid, 5 mM; benzamidine-HCl, 1 mM; PMSF, 1 mM.

Washing solution: same as Percoll solution minus Percoll.

(b) *Materials.* Gyratory shaker (for instance New Brunswick, model G-2). Other materials: see Section II.A.1(b) above.

(c) *Procedure.* For reproducibility of data, it is essential to operate under standardised conditions. All the steps in this experiment are done at 4°C. Chloroplasts (1 mg chlorophyll ml^{-1}) or non-green plastids (10 mg protein ml^{-1}) are incubated for 1 h in the incubation mixture, under constant and gentle agitation. The digestion is terminated by addition of EGTA (10 mM). The suspension is then agitated for five additional minutes. Control experiments are carried out under the same conditions, except that at the same time, thermolysin, $CaCl_2$ and EGTA are added to the incubation mixture.

After incubation, it is important to remove both thermolysin and broken plastids from the mixture. This is achieved by centrifugation through a Percoll cushion. For spinach chloroplasts, centrifugation through a 40% Percoll layer for 5 min at $5000 \times g$ (HB-4 rotor) is used. For cauliflower bud plastids, centrifugation through a 35% Percoll layer for 30 min at $25\,000 \times g$ (HB-4 rotor) is necessary. The treated, repurified intact plastids are recovered as pellet in both conditions. They are resuspended in washing medium (10 vol for 1 vol of plastid suspension). Chloroplasts are centrifuged for 90 s at $3500 \times g$ (SS-34 rotor), whereas non-green plastids are centrifuged for 10 min at $3500 \times g$ (SS-34 rotor). An additional washing is necessary when envelope membranes have to be purified in order to remove all traces of Percoll.*

After thermolysin digestion, the integrity of the preparation can be monitored using the procedures described above (see Section II.A.3 for chloroplasts and Section II.B.3 for non-green plastids).

III. PURIFICATION OF PLASTID ENVELOPE MEMBRANES

Chloroplast envelope membranes are the major, if not the sole, site for galactolipid biosynthesis within the plant cell (Douce, 1974). In addition, envelope membranes from all non-green plastids analysed so far (from 16:3 as well as from 18:3 plants) contain the enzyme responsible for the formation of MGDG. The procedure described below is largely based on the method used to purify total envelope membranes from spinach chloroplasts originally described by Douce *et al.* (1973), extensively used in our laboratory (Douce and Joyard, 1982), and extended to non-green plastids (Alban *et al.*, 1988). Methods are available for the separation of outer and inner envelope membranes (Cline *et al.*, 1981; Block *et al.*, 1983), however, both methods yield low amounts of membranes and total envelope membranes are usually much more convenient for studies of galactolipid biosynthesis. Readers interested in such procedures are referred to methodological articles by Keegstra and Yousif (1986) and Joyard *et al.* (1987b).

A. Purification Procedures

(a) *Solutions*

Swelling medium: tricine-NaOH, 10 mM, pH 7.8; $MgCl_2$, 4 mM.

---

* It is necessary to remove Percoll from purified chloroplasts or non-green plastids when envelope membranes have to be prepared. Two successive washings (at least) are necessary. When only a partial washing is done, the Percoll remaining in the chloroplast pellet contaminates the different sucrose layers during high speed centrifugation, and this in turn prevents proper sedimentation of envelope vesicles (Douce and Joyard, 1982).

4. GLYCOLIPID ANALYSES AND SYNTHESIS IN PLASTIDS

Sucrose solutions for gradients: Sucrose, 0.6 and 0.93 M (for step gradient centrifugation) or 0.4 and 1.2 M (for continuous density gradients); tricine-NaOH, 10 mM, pH 7.8; $MgCl_2$, 4 mM.

Suspension medium: sucrose, 0.3 M; tricine-NaOH, 10 mM, pH 7.8.

(b) *Materials.* Ultracentrifuge (such as Beckman L65 or Sorvall OTD 2) with swinging-out bucket rotors (Beckman rotors SW-27, 6 × 38 ml; SW-40, 6 × 13 ml and SW-50, 6 × 5 ml, according to the amount of plastids used).

1. Procedure for the purification of envelope membranes from chloroplasts

The technique is based on gentle chloroplast swelling in the presence of Mg^{2+} and the use of discontinuous sucrose gradients (Douce *et al.*, 1973; Douce and Joyard, 1982). The buoyant density of the envelope membranes ($d = 1.12$) proved sufficiently different from that of the thylakoids ($d = 1.17$) to allow clean separation of the two membrane systems. A key point for the preparation of clean envelope membranes is the use of chloroplasts at least 90% intact. The presence of too many thylakoids in the chloroplast suspension susceptible to extensive disintegration during swelling will cause significant contamination of the envelope membranes with small pieces of thylakoids containing numerous plastoglobules.

Intact, purified chloroplasts (50–200 mg chlorophyll) prepared from Percoll gradients are suspended in 90 ml (final volume) of swelling medium. Under these conditions, water enters very quickly into the chloroplasts, the envelope membranes are unable to support the high pressure and burst in a few seconds with the liberation of stroma material. The suspension of broken chloroplasts is then layered on top of a discontinuous sucrose gradient (0.6 and 0.93 M) and centrifuged (Beckman SW-27 rotor) for 1 h at $72\,000 \times g$ (R_{max}), at 4°C. At the conclusion of this step, chloroplast components are clearly separated giving three distinguishable subfractions: a tightly-packed, dark-green pellet at the bottom of the tube (thylakoid subfraction), a yellow band at the interface of the two sucrose layers (envelope membrane subfraction) and a clear brown supernatant (soluble subfraction). All three subfractions are recovered:

(1) The soluble subfraction is used as a source of acyl-ACP(CoA):*sn*-glycerol 3-phosphate acyltransferase: this enzyme is essential for studies of glycerolipid biosynthesis with isolated envelope membranes (see Section V.B.1).
(2) The envelope membranes are recovered from the gradient and diluted with tricine-NaOH (10 mM, pH 7.8) to give a 0.2–0.3 M sucrose solution. The suspension is then centrifuged (Beckman SW-27 rotor) at $113\,000 \times g$ (R_{max}) for 40 min at 4°C. This fraction is used as a source of enzymes for the biosynthesis of phosphatidic acid, diacylglycerol and MGDG.
(3) The thylakoid fraction is contaminated by envelope membranes: we have found that about half of the envelope membranes are recovered together with the thylakoid pellet. Elimination of this contamination by envelope membranes (corresponding to less than 5% of the thylakoid proteins) is very difficult, but it is possible to lower its level by successive washings of the thylakoid pellet with suspension medium, followed by low speed centrifugation at $1500 \times g$ (Sorvall SS-34 rotor) for 5 min. Two to three successive washings reduce the contamination to less than 1%.

2. Procedure for the purification of envelope membranes from non-green plastids

The structure of non-green plastids from cauliflower buds is different from that of sycamore amyloplasts: cauliflower bud plastids have small starch grains and some internal membranes (Journet and Douce, 1985) whereas sycamore amyloplasts are mostly devoid of internal membranes but contain a larger starch grain inside. Therefore, the procedures for the preparation of their envelope membranes are slightly different. They are detailed by Alban et al. (1988).

(a) *Procedure for the purification of envelope membranes from cauliflower bud plastids.* Intact cauliflower plastids (0.5 ml, 50–80 mg protein) are frozen for 30 min at $-80°C$, then thawed to room temperature, treated with 3 ml of swelling medium and homogenised smoothly with a Potter–Elvehjem apparatus with a loose-fitting pestle. This ruptures the envelope, as judged by the lack of latency of gluconate 6-phosphate dehydrogenase (see Section II.B.3).

The broken plastid suspension (3.5 ml, 50–80 mg protein) thus obtained is layered on top of a linear sucrose gradient (total volume, 7.5 ml) made of 0.4–1.2 M sucrose, 10 mM tricine-NaOH (pH 7.8) and 4 mM $MgCl_2$. At the bottom of the gradient, a 2 ml sucrose (1.2 M) cushion, containing 10 mM tricine-NaOH (pH 7.8) and 4 mM $MgCl_2$, is layered prior to the sucrose gradient. Centrifugation (Beckman SW-40 rotor) for 15 h at $71\,000 \times g$ (R_{max}) results in the separation of three fractions: a supernatant, containing the soluble material, on top of the tube; a dense white starch pellet covered by a loose greenish pellet at the bottom of the tube; and, approximately in the middle of the tube, a large yellow band. The supernatant and the yellow band are removed successively from the top of the tube using a Pasteur pipette. In this case, the yellow fraction is diluted with swelling medium devoid of $MgCl_2$ (final volume, 13 ml) and centrifuged (Beckman SW-40 rotor) for 1 h at $218\,000 \times g$ (R_{max}). The pellets obtained are suspended in a minimal volume of suspension medium by using a Potter–Elvehjem homogeniser.

Using the procedure described above, the yield of purified envelope membranes is 1–2 mg protein (i.e. about 2–3% of the plastid protein), starting from 4–5 kg of cauliflower buds. Similar yields are obtained when using a discontinuous sucrose gradient (0.6 and 0.93 M, see Section III.A.1) to recover the envelope membrane fraction (at the 0.6 M/0.93 M interface).

(b) *Modification of the procedure for the preparation of envelope membranes from sycamore amyloplasts.* The differences from the method described above (see Section III.A.2(a)) are the following: (1) a simple osmotic shock is sufficient for the disruption of the envelope membranes, (2) the swelling medium as well as the different layers of the sucrose gradient are devoid of $MgCl_2$, (3) protease inhibitors are added in all media used and at all the steps of the purification.* With sycamore amyloplasts, a starch pellet (devoid of any membrane) is obtained at the bottom of the tube. From 150–200 g of isolated sycamore cells, the yield of purified envelope membranes is 0.2–0.3 mg, i.e. about 20–30% of the amyloplast proteins. The membrane to soluble protein ratio is much higher in amyloplasts than in cauliflower plastids, owing to the large starch grains.

---

* Addition of protease inhibitors is essential for the preparation of envelope membranes from sycamore cells. In absence of these inhibitors, the envelope polypeptide pattern is totally altered (Alban et al., 1988). This is true for almost any plastid prepared from protoplasts.

IV. DETERMINATION AND CHARACTERISATION OF GLYCOLIPIDS

Studies of plant cell glycolipids require their extraction from the membranes to be analysed, their separation [by thin layer chromatography (TLC), column chromatography, or high performance liquid chromatography (HPLC)], their characterisation and their quantitative determination.

A. Extraction of Plant Glycolipids

In this review, we wish to concentrate on lipid extraction from plant membranes or organelles. Although the Bligh and Dyer (1959) procedure is very efficient for most lipids, quantitative extraction of some glycerolipids, such as lysophosphatidic acid (Gardiner et al., 1984) and sulpholipid (Harwood, 1975), is not totally achieved. The procedure developed by Hajra (1974) is far more efficient for quantitative extraction of these glycerolipids; this method will be described too. The reader interested in lipid extraction from whole plant tissues or isolated cells is referred to the classical method of Folch et al. (1957), the book by Kates (1986) and to the interesting procedure of Browse et al. (1986). For plant tissues, enzymatic degradation of the lipids is highly possible and precautions to prevent such degradation should be taken (Douce, 1964).

(a) *Solutions*

Bligh and Dyer procedure: chloroform–methanol (1:2; v/v); chloroform; water.

Hajra procedure: chloroform–methanol (1:1; v/v); H_3PO_4–KCl solution, H_3PO_4, 0.2 M, KCl, 1 M.

(b) *Materials.* Low speed centrifuge (such as Kubota KN-70, with swinging bucket no. 53714).

1. Bligh and Dyer procedure

The membrane fraction to be extracted (200 μl) is mixed together with 750 μl chloroform–methanol (1:2; v/v) to form a single homogeneous phase. Then, 250 μl of chloroform and 250 μl of water are successively added to give a two-phase system containing the lipids in the lower (organic) phase. The two-phase system is quickly formed by centrifugation for 10 min at $400 \times g$ (Kubota KN-70), but the tubes can be left overnight at 4°C with the same result. The chloroform layer is recovered with a Pasteur pipette and evaporated to dryness under a stream of argon or nitrogen. The residue is dissolved in a given volume of chloroform for further use.

2. Hajra procedure

In order to form one liquid phase and simultaneously to extract the lipids, 10 ml of chloroform–methanol (1:1; v/v) are added to 1 ml of the membrane or organelle suspension with thorough mixing. 3.5 ml of the H_3PO_4–KCl solution are then added to form a two-phase system. Bubbling with nitrogen or argon before and after addition of the H_3PO_4–KCl solution prevents fatty acid oxidation. The two-phase system is quickly formed by centrifugation for 10 min at 400 × g (Kubota KN-70), but the tubes can be left overnight in a cold room (4°C) with the same result. The chloroform layer is recovered with a Pasteur pipette and evaporated to dryness under a stream of argon or nitrogen. The residue is finally dissolved in 1 ml of chloroform.

B. Thin Layer Chromatography of Glycolipids

1. Reagents

(a) *Solvent systems for TLC on silica gel 60 plates (Merck)*

System 1: chloroform–methanol–water (65:25:4; v/v).

System 2: chloroform–acetone–methanol–acetic acid–water (100:40:20:20:10; v/v).

Note: There are several other solvent systems that can be used for separation of glycolipids (see for instance Kates, 1986; Chapman and Barber, 1987; Kesselmeier and Heinz, 1987; Lem and Williams, 1987; Roughan, 1987). We found that for many studies on glycolipids (see Section V), solvent system 1 is simple and the most convenient. However, when optimal separation of plant glycerolipids is necessary, a two-dimensional procedure which combines both solvent systems 1 and 2, when carefully done, is far more reliable.

(b) *Solvent systems for TLC on cellulose pre-coated plates (Eastman Kodak)*

System 3: phenol–water (100:38; v/v).
System 4: methanol–formic acid–water (80:13:7; v/v).

(c) *Alkaline hydrolysis mixture:* KOH (0.2 N) in methanol; for neutralization, Amberlite IRC-50.

(d) *Reagents for characterisation of lipids:*

Anilinonaphthalene sulphonate (ANS), 0.2% (w/v) in methanol.

1-naphthol reagent: and 8 g 1-naphthol in 250 ml methanol, then 20 ml H_2O and finally 32 ml concentrated H_2SO_4.

2. Materials

Oven (for TLC plate activation).

Chromatographic chambers.

UV lamp (250 and 360 nm).

Nitrogen or argon.

3. Procedure

We successfully use silica gel 60 pre-coated TLC plates (Merck). In order to remove any traces of lipid, the plates (200 × 200 × 0.25 mm) can be washed by ascending chromatography overnight with solvent system 1. The solvent front, which contains the bulk of lipid contamination, is discarded. The washed plates are then activated in an oven at 110°C for at least 1 h.

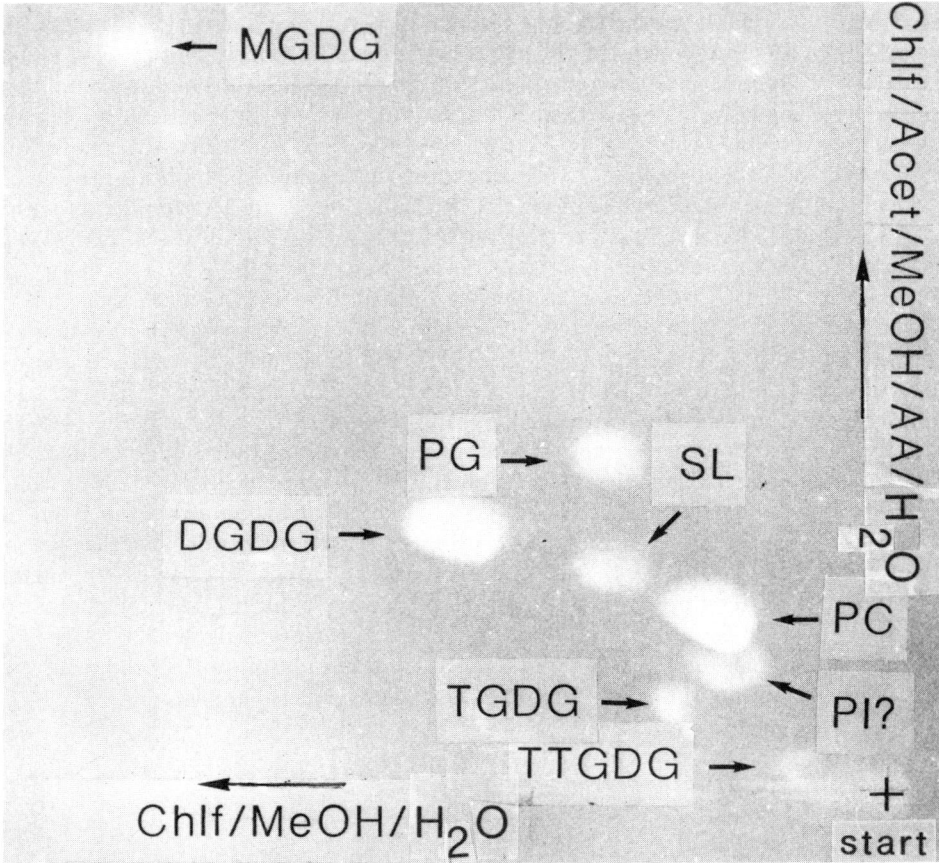

FIG. 4.5. Separation of envelope lipids by a two-dimensional TLC. Photography of the TLC plate under UV illumination, after spraying with ANS. The two solvent systems used are system 1 and 2, respectively, for the first and second dimension. The envelope membranes used for this separation were obtained from spinach chloroplasts obtained as described in Fig. 4.2, Procedure B. These chloroplasts were not treated with thermolysin, therefore the level of MGDG is low and the envelope membranes also contain unnatural galactolipids (Tri-GDG and Tetra-GDG). Abbreviations: Chlf, chloroform; MeOH, methanol; Acet, acetone; AA, acetic acid, TGDG, Tri-GDG; TTGDG, Tetra-GDG; PI, phosphatidylinositol; SL, sulpholipid; PG, phosphatidylglycerol; PC, phosphatidylcholine.

The lipid extract, dissolved in chloroform, is applied with a syringe or a Pasteur pipette to the lower right portion of the plate about 3 cm from each edge (200–600 μg of lipids are spotted). This amount allows the detection of lipids which make up even less than 1% of the lipid mixture. Solvent evaporation is hastened by a gentle stream of nitrogen. Immediately after spotting, plates are placed in rectangular chromatographic chambers lined on all sides with filter paper (Whatman no. 1) saturated with the first solvent system. About 150 ml of the solvent mixture is placed in the chamber. For the first direction, the solvent system is allowed to run to the front. The plate is removed and carefully dried at room temperature for at least 3 h under a slow stream of nitrogen, in a chromatographic chamber containing silica gel. This step is essential for a good further separation in order to eliminate water. The sample is then chromatographed in the second solvent system in the second dimension.

To bring out the lipid spots, the plates are sprayed with ANS and viewed under UV light at 360 nm. Figure 4.5 shows a photograph of a typical separation of envelope lipids as seen by UV illumination of a TLC plate sprayed with ANS. Glycolipids can also be identified by using a specific reagent such as 1-naphthol. Spray the reagent on the plate, place in an oven for 1 h at 110°C. Sterols and glycolipids will turn to either violet (sterols) or red/pink (glycolipids). Identification of the lipids can be done by co-chromatography with reference compounds (Roughan, 1987) or by comparison with a model separation. In this case, correlation of the R_f values on TLC plates with standards provide the basis for characterisation.

Chromatographic identification can be confirmed by mild alkaline deacylation, followed by thin layer chromatography on cellulose sheets for separation of hydrolysis products according to Benson and Maruo (1958). The samples to be analysed are first hydrolysed for 15 min in 1 ml KOH solution (0.2 N in methanol) at 37°C. The reaction is stopped and the solution neutralised by addition of beads of Amberlite IRC-50 (neutralisation is followed by using pH paper and the tip of a Pasteur pipette). The deacylated products are then chromatographed in a two-dimensional system on cellulose plates and developed first in solvent system 3. Phenol is then extracted from the cellulose sheet by washing with ether, prior to chromatography in the second dimension with solvent system 4. The polar head groups are identified by co-chromatography with standard compounds (glycerol, glycerol 3-phosphate, glycerol-galactose or glyceryl-quinovoside-6-sulphonate, for instance) obtained after alkaline hydrolysis of standard glycerolipids and glycolipids. Typical separations of polar headgroups of galactolipids, sulpholipid, phosphatidic acid and diacylglycerol are shown in Fig. 4.6.

C. Column Chromatography of Glycolipids

This procedure, which involves a low pressure liquid chromatography on silicic acid, is used when large amounts of glycolipids are to be prepared. This step is often used as a preliminary step prior to further separation of glycolipids by TLC or HPLC (Dorne *et al.*, 1987). Another procedure, described by Gounaris *et al.* (1983), and using acid-washed Florisil (magnesium silica) instead of silicic acid, can be used successfully.

(a) *Reagents*
Solvents: chloroform, acetone, methanol, and methanol–chloroform (100:10; v/v).

(b) *Materials.* Glass column (1 × 20 cm) filled with silicic acid (Biosil HA, <325 mesh, BioRad).

(c) *Procedure.* About 1–2 mg lipids (in a concentrated chloroformic solution) are layered onto a bed (100–200 mg) of silicic acid equilibrated with chloroform. Pigments and neutral lipids are first eluted by washing with 40 ml (about 5 column volumes) chloroform. A total polar lipid fraction can be eluted with 40 ml of chloroform–methanol (100:10; v/v). However, it is possible to extract first glycolipids with 40 ml acetone and then phospholipids with 40 ml methanol. The fraction containing glycolipids is then evaporated to dryness and the lipid residue redissolved in 1 ml chloroform for further analyses.

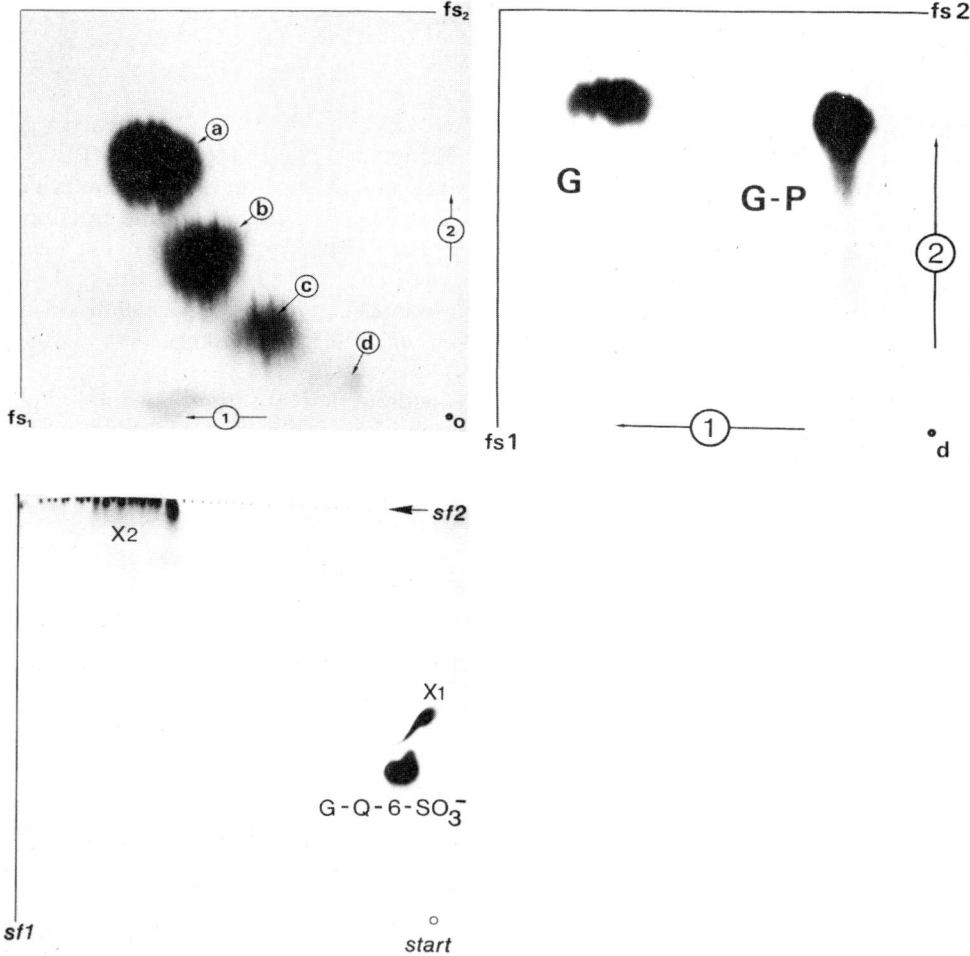

FIG. 4.6. Separation of polar headgroups after mild alkaline deacylation of plastid glycerolipids. Glycerolipids were hydrolysed according to Benson and Maruo (1958), and the polar headgroups separated by TLC. The solvent systems used are the same in the three examples, i.e. systems 3 and 4.
(A) Polar headgroups from galactolipids labelled with [^{14}C]galactose. (a), (b), (c) and (d) represent glyceryl-galactose with 1, 2, 3 and 4 galactoses deriving respectively from MGDG, DGDG, Tri-GDG and Tetra-GDG.
(B) Polar headgroups from glycerolipids labelled with sn-[^{14}C]glycerol-3-phosphate. Abbreviations: G, glycerol; G-P, sn-glycerol 3-phosphate. They were obtained after mild alkaline hydrolysis of diacylglycerol (for glycerol) and lysophosphatidic acid and phosphatidic acid (for sn-glycerol 3-phosphate). Reproduced with permission from Joyard and Douce (1977).
(C) [^{35}S]Sulphur-containing compounds obtained after incubation of isolated intact chloroplasts in the presence of labelled sulphate. Abbreviations: G-Q-6-SO$_3^-$, glyceryl-quinovosyl-6-sulphonate (which derives from sulpholipid), X$_1$ and X$_2$ were unknown compounds. Actually, compound X$_2$ does not derive from a sulpholipid but has been identified (Joyard et al., 1988) as the most stable form of elemental sulphur (S$_8$), which is hydrophobic. Reproduced with permission from Joyard et al. (1986).

D. Analyses of Fatty Acids

After preparation of individual lipids, the analyses of the molecular species, the determination of the positional distribution and the quantitative analyses of their fatty acids are important steps in glycolipid analyses.

1. Analyses of glycolipid molecular species

Molecular species of glycolipids can be separated rather easily by TLC using silica gel plates containing $AgNO_3$, as described by Roughan (1987). However, HPLC is probably the most convenient procedure to separate and analyse the different molecular species of glycolipids present in a given plant tissue or synthesised by plastid fractions.

Numerous procedures involving HPLC have been described in the last few years (Patton et al., 1982; Yamauchi et al., 1982; Lynch et al., 1983; Marion et al., 1984; Deleens et al., 1984; Demandre et al., 1985; Kesselmeier et al., 1985; Kesselmeier and Heinz, 1985, 1987; Kito et al., 1985; Smith et al., 1985; Heemskerk et al., 1986; Trémolières, 1986; Chapman and Barber, 1987).

It is possible to separate glycolipids by HPLC without derivatisation, but sulpholipid (which presents negatively charged residues) should be converted to the methyl ester derivatives with diazomethane for good resolution (Kesselmeier and Heinz, 1987). The detection (at 200 nm) is based upon the number of double bonds in the lipid species to be analysed. Therefore, detection of completely saturated species can only be achieved after the conversion of glycolipids into the corresponding *sn*-1,2-diacylglycerol by a chemical procedure (Heinze et al., 1984), followed by the formation of *p*-anisoyldiacylglycerols, containing a chromophoric acyl substituent at the *sn*-3 position of the glycerol (Kesselmeier and Heinz, 1987) which can be detected with a spectrophotometer at 250 nm after the separation by HPLC.

In our laboratory, we use routinely a Varian 5000 HPLC apparatus, equipped with a variable wavelength detector (UV 50) and a reverse-phase Spherisorb C6 column (5-μm particle size, 150 × 4.6 mm, Société Française Chromato Colonne, Gagny, France).

Finally, HPLC systems can be conveniently connected to a radioactivity detector, such as the Berthold HPLC radioactivity monitor (LB 506 D, equipped with GT 400 solid scintillation cell) or the Radiomatic Flow-One.

2. Analyses of fatty acid distribution

The method described below is largely based on the methods described by Fisher et al. (1973) and Tulloch et al. (1973). These analyses involve first a specific lipase from *Rhizopus arrhizus* treatment which removes fatty acids esterified to the *sn*-1 position of glycerolipids, as shown by Fisher et al. (1973). Then, free fatty acids and 2-acyl-*sn*-glycerolipid thus formed are separated by TLC and analysed by gas chromatography for identification. A similar procedure, using lipase from *Rhizopus* sp., is described by Roughan (1987).

(a) *Reagents*
Solvents: chloroform–methanol (2:1; v/v); methanol; diethylether; isopropanol.

Solvent systems for TLC: see solvent system 1 (Section IV.B.1(a)); chloroform–methanol–acetic acid–water (65:35:4:4; v/v).

Incubation mixture: Tris-HCl, 50 mM, pH 7.2; Triton X-100, 0.5% (v/v).

Lipase solution: lipase from *Rhizopus arrhizus* (Boehringer) is prepared by dissolving 30 µl lipase suspension in 1 ml Tris-HCl buffer (see above).

(b) *Materials*
Sonicator (such as Sonimass 250 T, Ultrasons, Annemasse, France).

Equipment for TLC: see Section IV.A(b).

Low speed centrifuge (Kubota KN-70).

(c) *Procedure.* Glycolipids (galactolipids, sulpholipid) to be analysed are scraped from TLC plates and transferred into test tubes. The gel is wetted with distilled water and the lipids extracted three times with chloroform–methanol (2:1; v/v) and once with pure methanol. Pooled extracts are taken to dryness under a stream of nitrogen. When radioactive glycolipids (which are generally in a low amount) are analysed it is recommended that unlabelled glycolipid (0.1–0.5 mg) is added as a carrier for optimal processing of the sample. The lipid samples are then dissolved in 1–2 ml diethylether and layered onto 0.7 ml incubation mixture. After removal of ether by a stream of argon and sonication (15 s, Sonimass 250 T, set at 60 W), a homogeneous dispersion is obtained. Next 0.3 ml of the *Rhizopus arrhizus* lipase solution is added and the incubation is conducted with shaking at room temperature for 15–30 min. The reaction is stopped by addition of 15 ml isopropanol. After solvent removal under a stream of argon, the residue is dissolved in 3 ml of chloroform–methanol (1:1; v/v) and, after vigorous shaking, centrifuged for 10 min at $400 \times g$ (Kubota KN-70). The supernatant contains hydrolysis products—*lyso*-compounds (2-acyl-*sn*-glycerolipid) and free fatty acids—which are then separated by TLC. The most convenient system for TLC separation of MGDG, *lyso*-MGDG and free fatty acids is described above (solvent system 1, see Section IV.B.1(a)). For sulpholipid derivatives, a mixture of chloroform–methanol–acetic acid–water (65:35:4:4; v/v) is recommended. *Lyso*-derivatives and unesterified fatty acids should be recovered in almost equal amounts. However, the method used is more accurate for the determination of fatty acids at the *sn*-2 position than for the *sn*-1 position (Fisher *et al.*, 1973; Tulloch *et al.*, 1973). This is due to the conversion of some 2-acyl-*sn*-glycerolipid into 1-acyl-*sn*-glycerolipid, which is almost immediately hydrolysed by the lipase. The free fatty acid fraction might therefore contain some fatty acids originating from the *sn*-2 instead of the *sn*-1 position.

The two lipid spots obtained from a given glycolipid (free fatty acids and *lyso*-derivative) are located (with ANS or after autoradiography) and scraped off the TLC plate. The fatty acids from each sample are transformed into fatty acid methyl esters and analysed by gas chromatography, as described below.

3. *Quantitative analyses of fatty acids*

The classical method used for quantitative determination of fatty acids involves the preparation of fatty acid methyl esters and their separation by gas chromatography, as described by Allen and Good (1971). HPLC can also be used for fatty acid analyses after derivatisation; the reader is referred to Kesselmeier and Heinz (1987) for a description of the procedure.

(a) *Reagents*

Transesterification mixture: methanol–sulphuric acid–benzene (100:5:5; v/v).

Solvents: hexane, methanol.

Internal standard: heptadecanoic acid (C17:0) or behenic acid (C22:0), 1 mg ml^{-1} in methanol.

(b) *Materials*
Oven set at 70°C.

Gas chromatograph equipped with a flame ionisation detector and an integrator. We routinely use a DELSI gas chromatograph 121 DFL containing a column packed with 10% diethylene glycol sulphonate on a Varaport 30 Chromosorb (or 10% Reoplex 400 on Chromosorb W) and a Delsi Integrator ENICA 21.

Gas proportional counter (such as Packard 894), for detection of radioactive fatty acids after gas chromatography.

(c) *Procedure.* Plastid suspensions containing free fatty acids and glycerolipids, or lipid areas scraped off TLC plates, are transferred to a 20-ml carefully washed screw-capped vial containing 4 ml of the transesterification mixture and an accurately measured quantity of the internal standard. For an optimal quantification of the fatty acids, the amount of internal standard to be added should be of the same order of magnitude as the linolenic acid content of the sample. The samples are bubbled with nitrogen (or argon), the cap of the vial slightly screwed, and the vials are placed for about 1 h in an oven set at 70°C. After 5 min, the cap is securely tightened. After the incubation, the vial is allowed to cool down and 5 ml of water are added. When cool, 2 ml of hexane are added to form two phases, the vial is vigorously shaken, and the upper phase containing the fatty acid methyl esters is recovered. The lower phase is extracted again with hexane and the organic phases are pooled together, and evaporated to dryness. Finally, a few drops of hexane are added to recover the methyl esters into a minimal volume for gas chromatography. Separations of fatty acid methyl esters are then carried out (at 175°C) using a gas chromatograph (with the equipment described above). Quantitative analyses are then made according to Allen and Good (1971) (see also Chapman and Barber, 1987).

The procedure described above can be combined with a radioactivity detector, which allows the simultaneous determination of radioactive fatty acids. However, a limitation of the method is the rather high amount of radioactive fatty acids necessary for accurate analyses.

4. Thin layer chromatography of fatty acid molecular species

It is often convenient to separate fatty acid methyl esters (prepared as described above) by TLC using silica gel containing silver nitrate. When used in combination with autoradiography, this procedure is more simple and more sensitive than radio gas chromatography, although (because it separates on the basis of numbers of double bonds) not all fatty acids can be separated.

(a) *Reagents*

Silica gel TLC plates sprayed with 5% $AgNO_3$ (in acetonitrile), and then activated at 110°C in an oven.

Benzene (beware of this very toxic solvent).

(b) *Materials.* Equipment for TLC and autoradiography (see Section IV.B.2).

(c) *Procedure.* Total fatty acid methyl esters are applied with a 50 μl Hamilton syringe to the lower position of the plate. The chromatogram is developed with benzene. This procedure allows separation of fatty acid methyl esters according to the number of double bonds: the mobility increases with the number of double bonds. It is then possible to locate the radioactive bands by autoradiography. The radioactive areas (saturated, monoenoic, dienoic and trienoic fatty acid methyl esters) are then scraped into scintillation vials for further counting.

When more detailed analyses of fatty acids are necessary, the use of paraffin-impregnated silica gel TLC plates could be convenient. The reader is referred to a description of this procedure by Roughan (1987).

V. BIOSYNTHESIS OF GLYCOLIPIDS

We will describe first glycolipid (galactolipid and sulpholipid) synthesis using intact plastids. Then, we will describe the procedures for studies of galactolipid synthesis using either isolated envelope membranes or an envelope fraction enriched in the enzyme responsible for MGDG synthesis.

A. Glycolipid Synthesis by Isolated Intact Plastids

The labelled substrates used to follow glycolipid (MGDG and sulpholipid) synthesis by isolated plastids (chloroplasts or non-green plastids) can be [^{14}C]acetate, sn-[^{14}C]glycerol 3-phosphate, UDP-[^{14}C]galactose (for MGDG) or [^{35}S]sulphate (for sulpholipid). In isolated plastids, glycolipid biosynthesis requires the coordinated functioning of: (a) the enzymatic complex (fatty acid synthetase) which catalyses the formation of acyl-ACP; (b) the enzymes of the Kornberg–Pricer pathway, responsible for the synthesis of lysophosphatidic acid, phosphatidic acid, diacylglycerol; (c) the enzymes responsible for the conversion of diacylglycerol into MGDG or sulpholipid. The procedures described below are adapted from those described by Roughan (1987), Kleppinger-Sparace *et al.* (1985) and Joyard *et al.* (1986).

1. Reagents

(a) *Incubation mixtures*

For MGDG synthesis: mannitol, 0.33 M; tricine-NaOH, 10 mM, pH 7.9; $NaHCO_3$, 10 mM; $MgCl_2$, 1 mM; $MnCl_2$, 1 mM; Na_3EDTA, 2 mM; Na_2HPO_4, 0.2 mM; Triton X-100, 0.13% (w/v); Na acetate, 0.2 mM; CoASH, 0.5 mM; ATP, 1 mM; sn-glycerol 3-phosphate, 1 mM; UDP-galactose (UDP-gal) 2 mM.

For sulpholipid synthesis: same as above, plus dithiothreitol (DTT), 0.5 mM; Na_2SO_4, 0.1 mM; minus UDP-gal.

Radioactive molecules: [^{14}C]acetate, 15 µCi µmol^{-1}; sn-[^{14}C]glycerol 3-phosphate, 5 µci µmol^{-1}; UDP-[^{14}C]gal, 5 µCi µmol^{-1}; [^{35}S]sulphate, 0.25 µCi nmol^{-1}.

(b) *Extraction mixture*: see Section IV.A.

2. Materials

Lamp: any system able to provide enough light to obtain a light intensity at the level of the reaction of at least 150 µmol m^{-2} s^{-1} is suitable. In addition, illuminated water baths provide a good illumination together with a controlled temperature and a thorough shaking of the mixture during the incubation.

Note: Illumination is not necessary for incubation with non-green plastids.

Equipment for TLC and autoradiography (see Section IV.B.2).

3. Procedure

Chloroplasts (corresponding to 120–180 µg chlorophyll) or non-green plastids (corresponding to 1.5–2 mg protein) are incubated with shaking at 25°C in 1 ml of incubation mixture. For chloroplasts, the incubation is done under illumination. Reactions (linear for at least 20 min) are stopped by the addition of the extraction mixture, as described above (see Section IV.A). The lipids are extracted and analysed by using the procedures described above. Two examples of such analyses are given below:

(1) Figure 4.7 presents a separation of chloroplast lipids by TLC after incubation in presence of [^{14}C]acetate to analyse sulpholipid synthesis; in addition, radioactive free fatty acids, lysophosphatidic acid, monoacylglycerol, phosphatidic acid and diacylglycerol are synthesised.
(2) Table 4.1 presents the fatty acid composition and positional distribution in the glycerolipids synthesised as shown in Fig. 4.7.

B. Glycolipid Synthesis by Isolated Envelope Membranes

In chloroplasts, as well as in non-green plastids, MGDG biosynthesis involves the functioning of the enzymes of the envelope Kornberg–Pricer pathway and of a MGDG synthase. It is convenient to follow MGDG formation in purified envelope membranes.

1. MGDG formation through the envelope Kornberg–Pricer pathway

The first enzyme of the Kornberg–Pricer pathway, the acyl-ACP: sn-glycerol 3-phosphate acyltransferase, is recovered in the plastid stroma after fractionation of the organelle (Joyard and Douce, 1977). Therefore, it is necessary to incubate the isolated envelope membranes in the presence of either the plastid stroma (to provide the enzyme) or lysophosphatidic acid (the substrate for the second acyltransferase). In addition, the true physiological acyl donors for the acyltransferases are acyl-ACPs (Frentzen et al.,

4. GLYCOLIPID ANALYSES AND SYNTHESIS IN PLASTIDS

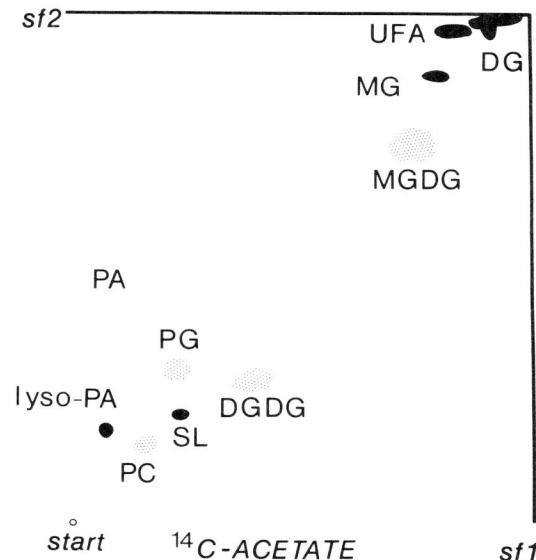

FIG. 4.7. Separation of chloroplast glycerolipids after incubation in the presence of [^{14}C]acetate and sulphate. Chloroplast lipids were extracted according to Hajra (1974), and separated by TLC using solvent systems 1 and 2. Radioactive lipids were revealed by autoradiography; dotted spots represent the position of non-radioactive plastid lipids. Reproduced with permission from Joyard et al. (1986). MG, monoacylglycerol; DG, diacylglycerol; PA, phosphatidic acid; PC, phosphatidylcholine; PG, phosphatidylglycerol; SL, sulpholipid; UFA, unesterified fatty acids.

TABLE 4.1. Fatty acid composition and positional distribution in glycerolipids synthesised by intact spinach chloroplasts incubated in the presence of sulphate and [^{14}C]acetate.

Lipid analysed	Radioactivity incorporated within the different fatty acids (%)											
	Total				sn-1 position				sn-2 position			
	<16:0	16:0	18:1	18:2	<16:0	16:0	18:1	18:2	<16:0	16:0	18:1	18:2
lyso-PA	0.5	4.6	94.9	—	0.5	4.6	94.9	—	—	—	—	—
PA	0.8	52.9	46.3	—	<0.5	12.0	87.9	—	1.5	93.0	5.5	—
DG	1.9	51.4	46.7	—	0.9	11.7	87.4	—	2.8	91.1	6.0	—
SQDG	—	57.4	41.6	1	—	21.3	76.7	2	<0.5	93.5	6.4	—

Experimental conditions for the incubation of chloroplasts are described in the text (Section V.A). After extraction of the lipids according to Hajra (1974), the chloroform–methanol soluble compounds are separated by two-dimensional TLC using solvent systems 1 and 2 (see Section IV.B). The radioactive spots are scraped for fatty acid analyses. Positional distribution was determined after lipase hydrolysis of the glycerolipids (see Section IV.D.2). Fatty acid methyl esters were prepared and then separated by gas chromatography associated with a gas proportional counter (see Section IV.D.3). Reproduced with permission from Joyard et al. (1986). SQDG, sulphoquinovosyldiacylglycerol or sulpholipid.

1983). However, it is quite difficult to synthesise large amounts of acyl-ACP for routine experiments. Therefore, it is convenient to use acyl-CoA instead of acyl-ACP, especially with envelope membranes which contain an acyl-CoA synthetase (Joyard and Douce, 1977). We usually restrict the use of acyl-ACP to experiments where analyses of the specificities and selectivities of acyltransferases have to be done. The procedures described below are adapted from those described by Joyard and Douce (1977), Douce and Joyard (1980), Frentzen et al. (1983) and Alban et al. (1988, 1989). A procedure to prepare and purify acyl-ACP is described by Rock and Garwin (1979).

(a) *Reagents*
Incubation mixtures:
  (1) Medium A for incubation with stroma: MOPS-NaOH, 10 mM, pH 7.9; $MgCl_2$, 1 mM; CoASH, 0.2 mM; ATP, 4 mM; sn-glycerol 3-phosphate, 1 mM; UDP-gal, 0.5 mM. The radioactive molecule is either sn-[^{14}C]glycerol 3-phosphate (5 µCi µmol^{-1}) or UDP-[^{14}C]-gal (5 µCi µmol^{-1}). In this experiment, it is convenient to use concentrated stroma (50 mg protein ml^{-1}), such as lyophylised chloroplast extract obtained from purified intact spinach chloroplasts (see Section III.A.1).
  (2) Medium B for incubation with lysophosphatidic acid: medium A minus sn-glycerol 3-phosphate; plus 1-oleoyl-sn-glycerol 3-phosphate (lysophosphatidic acid), 25 µM (synthesised as a radioactive compound from chloroplast stroma, using medium C, as described below).
  (3) Medium C for lysophosphatidic acid synthesis by chloroplast stroma: MOPS-NaOH, 10 mM, pH 7.1; sn-glycerol 3-phosphate, 1 mM; oleoyl-CoA, 0.1 mM, or oleoyl-ACP, 2 µM. The radioactive substrate can be either sn-[^{14}C]glycerol 3-phosphate (5 µCi µmol^{-1}), [^{14}C]oleoyl-CoA (5 µCi µmol^{-1}) or [^{14}C]oleoyl-ACP (50 µCi µmol^{-1}).
  (4) Medium D for incubation with acyl-ACP instead of acyl-CoA: medium A minus ATP and CoASH, plus palmitoyl-ACP, 1 µM; oleoyl-ACP, 1 µM; bovine serum albumin, 0.2 mg ml^{-1}. The radioactive substrate can be either sn-[^{14}C]glycerol 3-phosphate (5 µCi µmol^{-1}) or [^{3}H] or [^{14}C]acyl-ACP (50 µCi µmol^{-1}).

Miscellaneous: extraction mixture (see Section IV.A), solvent systems (see Section IV.B) for lipid extraction and chromatography.

(b) *Materials*
Equipment for TLC and autoradiography.

(c) *Preparation of lysophosphatidic acid.* About 10 mg chloroplast stroma are incubated for 1 h in 1.2 ml of medium C. The lipids are then extracted according to Hajra (1974), as described above (see Section IV.A). The organic phase containing the lipids is then chromatographed by TLC and recovered from the plates (see Section IV.B). The *lyso*-phosphatidic acid recovered is stored at −20°C under argon in chloroform.

(d) *MGDG synthesis.* Envelope membranes (corresponding to about 100 µg protein) and plastid stroma (corresponding to about 500 µg protein) are incubated at 20°C in 1.2 ml of medium A (or B or D). At given times, aliquots (200 µl) are taken and are extracted by the procedure of Hajra (1974), as described above (see Section IV.A). The organic phases are then recovered and are chromatographed as described above, using solvent system 1 (see Section IV.B). The lipids can be visualised after spraying with ANS and viewed under UV (see Section IV.B). Radioactive lipids can be revealed by autoradiography, the radioactive areas are scraped off and counted in scintillation vials. An example of such an experiment is presented in Fig. 4.8.

4. GLYCOLIPID ANALYSES AND SYNTHESIS IN PLASTIDS

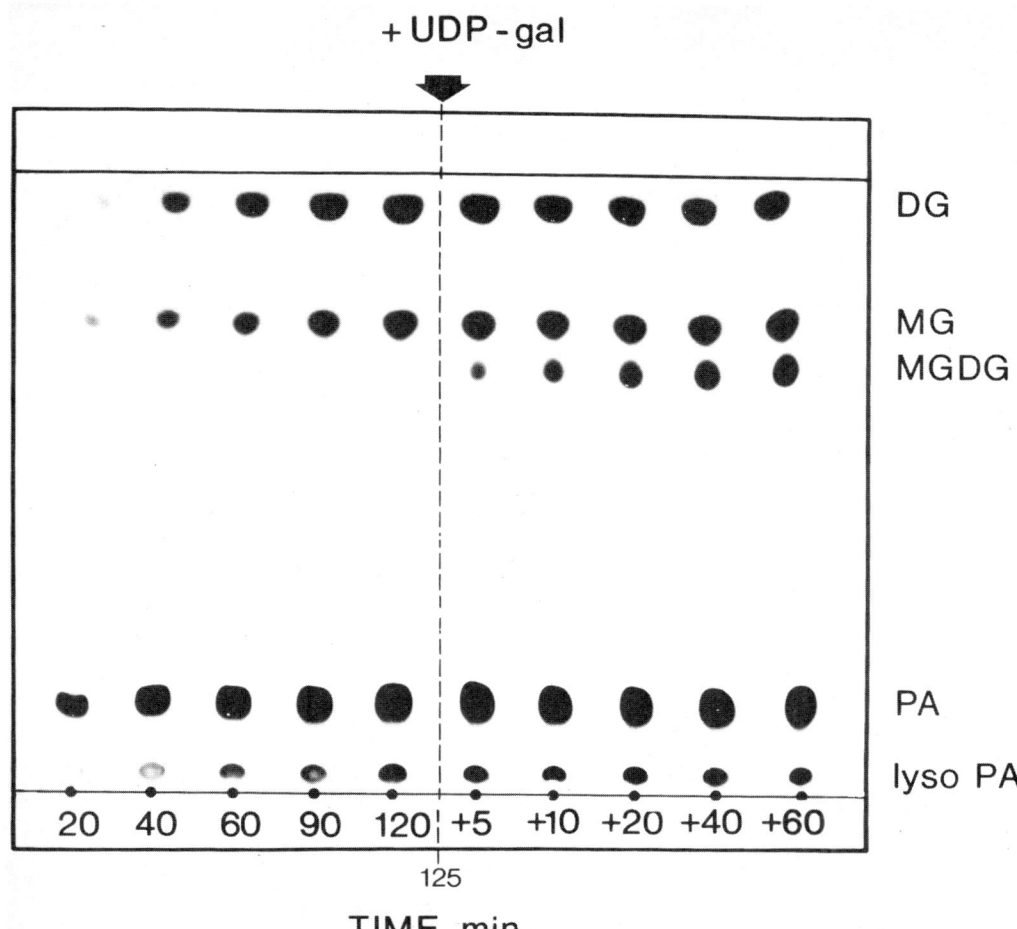

FIG. 4.8. Time course of glycerolipid synthesis by envelope membranes. Envelope membranes and stroma were incubated in presence of sn-[^{14}C]glycerol 3-phosphate. After 60 min incubation, UDP-gal is added for MGDG synthesis. This experiment demonstrates MGDG synthesis through the envelope Kornberg–Pricer pathway. Envelope lipids were separated by TLC using solvent system 1; radioactive lipids were revealed by autoradiography. Reproduced with permission from Joyard and Douce (1977).

2. Assay of MGDG synthase in isolated envelope membranes

All envelope membranes from chloroplasts or non-green plastids analysed so far contain the enzyme responsible for MGDG synthesis: the UDP-gal:1,2-diacylglycerol 3-β-D-galactosyltransferase (EC 2.4.1.46) or MGDG synthase. Interestingly, envelope membranes contain a galactolipid:galactolipid galactosyltransferase, which catalyses an intergalactolipid exchange of galactose and forms diacylglycerol during envelope preparation (Van Besouw and Wintermans, 1978; Heemskerk and Wintermans, 1987). This enzyme is responsible for the enrichment of envelope membranes in the substrate for MGDG synthase, i.e. diacylglycerol (Joyard and Douce, 1976). Therefore it is not necessary to add diacylglycerol to the incubation mixture. This procedure is adapted from the experiments described by Douce (1974), Joyard and Douce (1976) and Douce and Joyard (1980).

(a) *Reagents*

Incubation mixture: tricine-NaOH, 10 mM, pH 7.9; UDP-[^{14}C]gal, 0.2–0.6 mM (5 µCi µmol^{-1}). When formation of large amounts of digalactosyldiacylglycerol (DGDG), Tri-GDG or Tetra-GDG is needed, the addition of $MgCl_2$, 10 mM, and the use of a buffer at pH 7.2 is recommended.

Miscellaneous: extraction mixture (see Section IV.A) and solvent system 1 (see Section IV.B) are needed for lipid extraction and TLC chromatography.

(b) *Materials.* Equipment for TLC and autoradiography.

(c) *Procedure.* Envelope membranes (corresponding to about 50 µg protein) are incubated at 20°C in 1.2 ml of incubation mixture. At given times, aliquots (200 µl) are taken and extracted according to Bligh and Dyer (1959), as described above (see Section IV.A). The lipid phase is recovered and analysed by TLC, using solvent system 1, as described above (see Section IV.B). An example is presented in Fig. 4.9.

3. *Partial purification and assay of MGDG synthase after solubilization of envelope membranes*

The procedure used is based on that described by Covès et al. (1986, 1988). For these experiments, envelope membranes are stored concentrated (about 10 mg protein ml^{-1}) in a medium containing 50 mM MOPS-NaOH (pH 7.8) and 1 mM DTT, and in lipid nitrogen.

(a) *Reagents*

Detergent: 3-[(3-cholamidopropyl)dimethylammonio]-1-propane sulphonate (CHAPS) (Sigma).

Solubilisation and equilibration media: MOPS-NaOH, 50 mM; pH 7.8; DTT, 1 mM; CHAPS, 6 mM; KH_2PO_4, 50 mM.

Chromatography medium: same as above, except that KH_2PO_4 concentration is a linear gradient (50 to 162.5 mM) for the second part of the chromatography (see Fig. 4.10), and finally 275 mM, instead of 50 mM.

Washing medium: same as above, except that KH_2PO_4 concentration is 500 mM instead of 50 mM.

Incubation medium: MOPS-NaOH, 50 mM; pH 7.8; CHAPS, 6 mM; DTT, 1 mM; KH_2PO_4, 250 mM. To this medium are added: UDP-[^{14}C]gal, 600 µM, 45.8 MBq mmol^{-1}; and the following lipids: diacylglycerol and phosphatidylglycerol (PG).

Purified diacylglycerol (1,2-dioleoyl-*sn*-glycerol): it is prepared from a mixture of 1,2- and 1,3-diacylglycerol (Fluka) according to the procedure described by Covès et al. (1988).

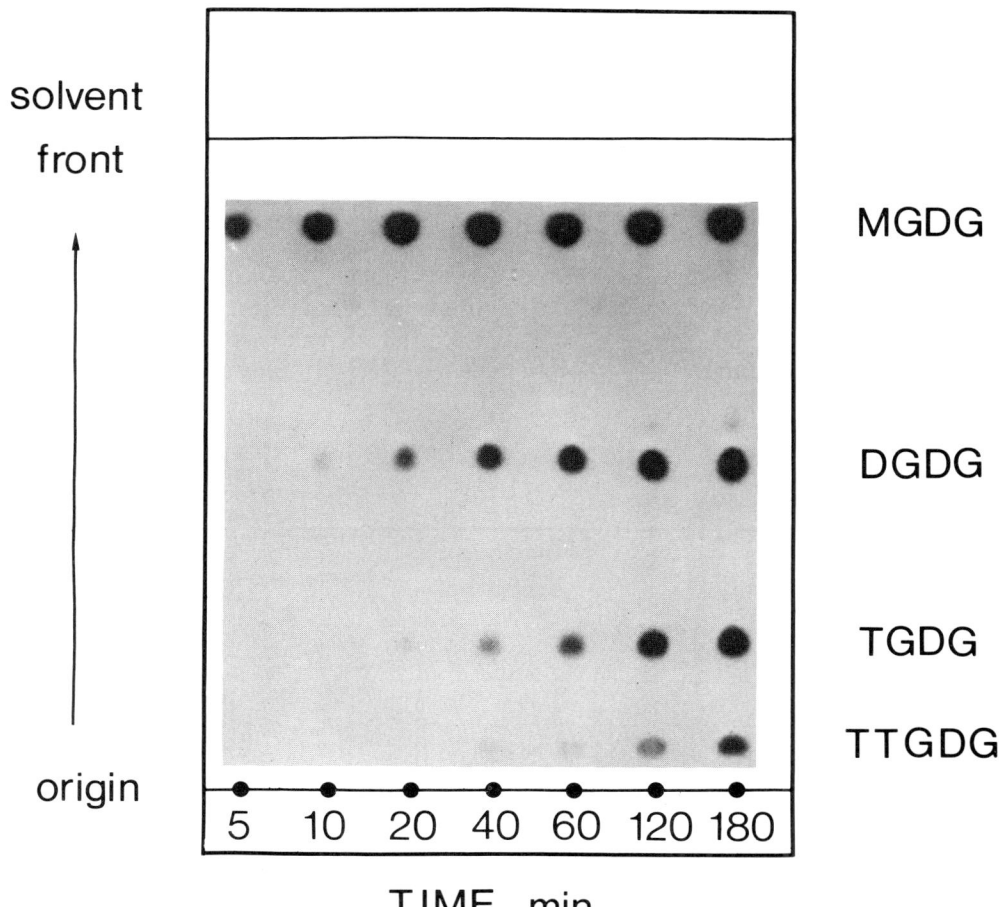

FIG. 4.9. Galactolipid synthesis by isolated envelope membranes incubated in the presence of UDP-[^{14}C]gal. Envelope lipids were separated by TLC using solvent system 1; radioactive lipids are revealed by autoradiography. It is possible to change the ratio between MGDG, DGDG, Tri-GDG and Tetra-GDG synthesis by isolated envelope membranes by modifying the cation concentration in the incubation mixture (by MgCl$_2$, for instance). At acidic pH, acyl-MGDG is also formed (Joyard and Douce, 1976). This compound migrates just above MGDG in solvent system 1.

PG: we use PG prepared from egg phosphatidylcholine (PC) by transphosphatidylation in the presence of free glycerol (Sigma).

Solvents: chloroform, methanol.

Extraction mixture: chloroform/methanol (1:2; v/v).

(b) *Materials*

Ultracentrifuge (Beckman L 65 or equivalent) with SW-50 rotor.

Chromatographic equipment including a 10 × 0.8 cm column containing hydroxyapatite. HA-Ultrogel (IBF, France) is recommended for good reproducibility.

(c) *Solubilisation and partial purification of MGDG synthase from chloroplast envelope membranes.* Chloroplast envelope membranes (about 10 mg protein) are incubated for 30 min at 0°C in 15 ml of solubilisation medium. The mixture is then centrifuged (Beckman, SW-50 rotor) for 15 min at 245 000 × g (R_{max}). The supernatant (0.5–0.6 mg protein ml^{-1}) containing all MGDG synthase is recovered. The solubilised envelope membranes are then loaded on top of the HA-Ultrogel column equilibrated in solubilisation medium prior to the experiment. The proteins are eluted, at a flow rate of 30 ml h^{-1}, by using an elution mixture containing first 50 mM KH$_2$PO$_4$ (peak 1), then a 50–162.5 mM KH$_2$PO$_4$ linear gradient (peak 2), and finally 275 mM KH$_2$PO$_4$, to elute peak 3 containing most of the MGDG synthase activity. Figure 4.10 presents the elution pattern obtained. The MGDG synthase-enriched peak is recovered and stored in liquid nitrogen for further use. The column is then washed with washing medium and equilibrated in solubilisation medium for further use.

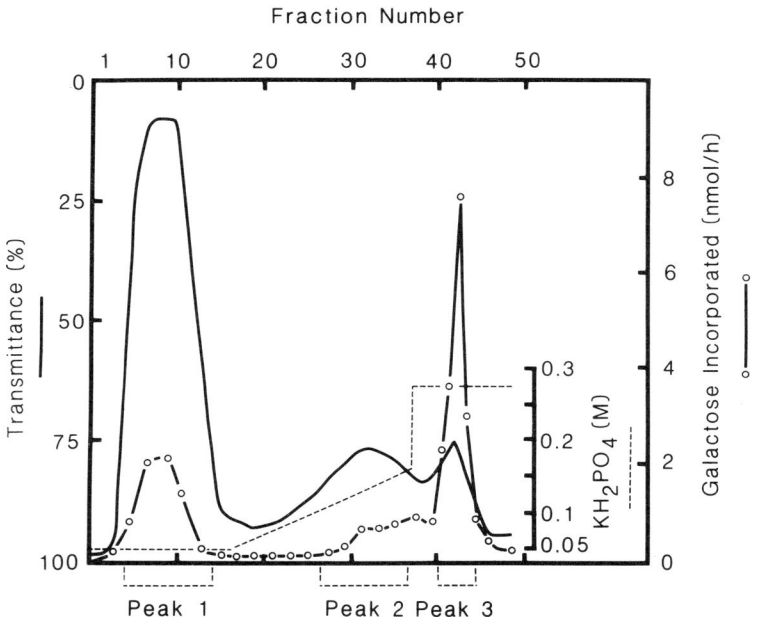

FIG 4.10. Fractionation of solubilised envelope by hydroxyapatite chromatography. Envelope membranes (10 mg protein) were solubilised in 15 ml (final volume) of solubilisation medium. After centrifugation (see text), the supernatant (15 ml, 8 mg protein) was loaded and peak 1 eluted with 15 ml solubilisation medium (30 ml h^{-1}). Peak 2 was eluted with a linear KH$_2$PO$_4$ gradient (36 ml of elution medium with 50 to 162.5 mM). Finally, 10 ml elution medium with 275 mM KH$_2$PO$_4$ was used to elute peak 3. The dotted line represents the variation of KH$_2$PO$_4$ concentration during the elution. MGDG synthase was assayed (galactose incorporated) in all fractions and the highest specific activity is recovered in peak 3. Reproduced with permission from Covès et al. (1986).

(d) *Assay of MGDG synthase after partial purification.* The MGDG synthase prepared as described above is highly delipidated (Covès et al., 1986, 1988). Addition of acidic lipids, and especially of PG, is essential to restore the activity (Covès et al., 1988). Diacylglycerol (100 µg) and PG (200 µg) dissolved in chloroform are introduced into glass tubes. After evaporation of chloroform under a stream of nitrogen, 150 µl of incubation mixture are added. The tubes are vigorously mixed to resuspend lipids in the incubation mixture (this mixture contains detergent). Then 50 µl (1.2–1.6 µg of protein) of the MGDG synthase-enriched fraction are added, and the tubes are vigorously mixed again. After 5 min at 20°C, the reaction is started by addition of UDP-gal. The reaction is stopped and the lipids extracted after 10 min incubation by adding 750 µl of extraction mixture, then 250 µl of chloroform and 250 µl water, according to Bligh and Dyer (1959). The organic phase is recovered, and the radioactivity in labelled MGDG determined by liquid scintillation counting. The reaction is linear for about 20 min under our experimental conditions. The activity is expressed as µmol of galactose incorporated per hour per mg protein.

ACKNOWLEDGEMENTS

We would like to thank Drs Claude Alban, Jacques Covès and Etienne-Pascal Journet for their collaboration in the development of some of the procedures described in this review.

REFERENCES

Alban, C., Joyard, J. and Douce, R. (1988). *Plant Physiol.* **88**, 709–717.
Alban, C., Joyard, J. and Douce, R. (1989). *Biochem. J.* **259**, 775–783.
Allen C. F. and Good, P. (1971). *Methods Enzymol.* **23**, 523–547.
Benson, A. A. and Maruo, B. (1958). *Biochim. Biophys. Acta* **41**, 328–333.
Bligh, E. G. and Dyer, W. J. (1959). *Can. J. Biochem. Physiol.* **37**, 911–917.
Bligny, R. and Leguay, J. J. (1987). *Methods Enzymol.* **148**, 3–16.
Block, M. A., Dorne, A. J., Joyard, J. and Douce, R. (1983). *J. Biol. Chem.* **258**, 13 273–13 280.
Browse, J. A., Warwick, N., Somerville, C. R. and Slack, C. R. (1986). *Biochem. J.* **235**, 25–31.
Bruinsma, J. (1961). *Biochim. Biophys. Acta* **52**, 576–578.
Chapman, D. J. and Barber, J. (1987). *Methods Enzymol.* **148**, 294–319.
Cline, K., Andrews, J., Mersey, B., Newcomb, E. H. and Keegstra, K. (1981). *Proc. Natl. Acad. Sci. USA* **78**, 3595–3599.
Covès, J., Block, M. A., Joyard, J. and Douce, R. (1986). *FEBS Lett.* **208**, 401–406.
Covès, J., Joyard, J. and Douce, R. (1988). *Proc. Natl. Acad. Sci. USA* **85**, 4966–4970.
Deleens, E., Schwebel-Dugue, N. and Trémolières, A. (1984). *FEBS Lett.* **178**, 55–58.
Demandre, C., Trémolières, A., Justin, A. M. and Mazliak, P. (1985). *Phytochemistry* **24**, 481–485.
Dorne, A. J., Block, M. A., Joyard, J. and Douce, R. (1982). *FEBS Lett.* **145**, 30–34.
Dorne, A. J., Joyard, J. and Douce, R. (1987). *Can. J. Bot.* **65**, 2368–2372.
Douce, R. (1964). *C.R. Hebd. Séances Acad. Sci. Paris* **259**, 3066–3069.
Douce, R. (1974). *Science* **183**, 852–853.
Douce, R. and Joyard, J. (1980). In "The Biochemistry of Plants. Lipids: Structure and Function" (P. K. Stumpf, ed.), Vol. 4, pp. 321–362. Academic Press, New York.
Douce, R. and Joyard, J. (1982). In "Methods in Chloroplast Molecular Biology" (M. Edelman, R. B. Hallick and N. H. Chua, eds), pp. 239–256. Elsevier Biomedical, Amsterdam.
Douce, R., Holtz, R. B. and Benson, A. A. (1973). *J. Biol. Chem.* **248**, 7215–7222.
Fisher, W., Heinz, E. and Zeus, M. (1973). *Hoppe-Seyler's Z. Physiol. Chem.* **354**, 1115–1123.
Folch, J., Lees, M. and Sloane-Stanley, G. H. (1957). *J. Biol. Chem.* **226**, 497–509.
Frentzen, M. (1986). *J. Plant Physiol.* **124**, 193–209.
Frentzen, M., Heinz, E., McKeon, T. and Stumpf, P. K. (1983). *Eur. J. Biochem.* **129**, 629–636.
Gardiner, S. E., Heinz, E. and Roughan, P. G. (1984). *Plant Physiol.* **74**, 890–896.

Gounaris, K. and Barber, J. (1983). *Trends Biochem. Sci.* **8**, 378–381.
Gounaris, K., Mannock, D. A., Sen, A., Brain, A. P. R., Williams, W. P. and Quinn, P. J. (1983). *Biochim. Biophys. Acta* **732**, 229–242.
Gounaris, K., Barber, J. and Harwood, J. L. (1986). *Biochem. J.* **237**, 313–326.
Hajra, A. K. (1974). *Lipids* **9**, 502–505.
Harwood, J. (1975). *Biochim. Biophys. Acta* **398**, 224–230.
Harwood, J. (1980). In "The Biochemistry of Plants. Lipids: Structure and Function" (P. K. Stumpf, ed.), Vol. 4, pp. 301–320. Academic Press, New York.
Heemskerk, J. W. M. and Wintermans, J. F. G. M. (1987). *Physiol. Plantarum* **70**, 558–568.
Heemskerk, J. W. M., Bögemann, G., Scheijen, M. A. M. and Wintermans, J. F. G. M. (1986). *Anal. Biochem.* **154**, 84–91.
Heinz, E. (1977). In "Lipids and Lipid Polymers in Higher Plants" (M. Tevini and H. K. Lichtenthaler, eds), pp. 102–120. Springer Verlag, Berlin.
Heinz, E. and Roughan, P. G. (1983). *Plant Physiol.* **72**, 273–279.
Heinze, F. J., Linscheid, M. and Heinz, E. (1984). *Anal. Biochem.* **139**, 126–133.
Journet, E. P. and Douce, R. (1985). *Plant Physiol.* **79**, 458–467.
Journet, E. P., Bligny, R. and Douce, R. (1986). *J. Biol. Chem.* **261**, 3193–3199.
Joyard, J. and Douce, R. (1976). *Biochim. Biophys. Acta* **424**, 126–131.
Joyard, J. and Douce, R. (1977). *Biochim. Biophys. Acta* **486**, 273–285.
Joyard, J. and Douce, R. (1980). *Methods Enzymol.* **69**, 290–301.
Joyard, J. and Douce, R. (1987). In "The Biochemistry of Plants. Lipids: Structure and Function" (P. K. Stumpf, ed.), Vol. 9, pp. 215–274. Academic Press, New York.
Joyard, J., Billecocq, A., Bartlett, S. G., Block, M. A., Chua, N. H. and Douce, R. (1983). *J. Biol. Chem.* **258**, 10 000–10 006.
Joyard, J., Blée, E. and Douce, R. (1986). *Biochim. Biophys. Acta* **879**, 78–87.
Joyard, J., Dorne, A. J. and Douce, R. (1987a). *Methods Enzymol.* **148**, 195–206.
Joyard, J., Block, M. A., Covès, J., Alban, C. and Douce, R. (1987b). *Methods Enzymol.* **148**, 206–218.
Joyard, J., Forest, E., Blée, E. and Douce, R. (1988). *Plant Physiol.* **88**, 961–964.
Kates, M. (1986). "Techniques of Lipidology. Isolation, Analysis and Identification of Lipids, 2nd edn. Elsevier, Amsterdam.
Keegstra, K. and Yousif, A. E. (1986). *Methods Enzymol.* **118**, 316–325.
Kesselmeier, J. and Heinz, E. (1985). *Anal. Biochem.* **144**, 319–328.
Kesselmeier, J. and Heinz, E. (1987). *Methods Enzymol.* **148**, 650–661.
Kesselmeier, J., Eichenberger, W. and Urban, B. (1985). *Plant Cell Physiol.* **26**, 463–471.
Kito, M., Takamura, H., Narita, H. and Urade, R. (1985). *J. Biochem.* **98**, 327–331.
Kleppinger-Sparace, K. F., Mudd, J. B. and Bishop, D. G. (1985). *Arch. Biochem. Biophys.* **240**, 859–865.
Lem, N. W. and Williams, J. P. (1987). *Methods Enzymol.* **148**, 346–350.
Lynch, D. V., Gundersen, R. E. and Thompson, Jr, G. A. (1983). *Plant Physiol.* **72**, 903–905.
Marion, D., Gandemer, G. and Douillard, D. (1984). In "Structure, Function and Metabolism of Plant Lipids" (P. A. Siegenthaler, and W. Eichenberger, eds), pp. 139–143. Elsevier, Amsterdam.
Mourioux, G. and Douce, R. (1981). *Plant Physiol.* **67**, 470–473.
Mudd, J. B. and Kleppinger-Sparace, K. F. (1987). In "The Biochemistry of Plants. Lipids: Structure and Function" (P. K. Stumpf, ed.), Vol. 9, pp. 275–289. Academic Press, New York.
Murphy, D. J. (1986). *Biochim. Biophys. Acta* **864**, 33–94.
Patton, G. M., Fasulo, J. M. and Robins, S. J. (1982). *J. Lipid. Res.* **23**, 190–196.
Quail, P. H. (1979). *Ann. Rev. Plant Physiol.* **30**, 425–487.
Quinn, P. J. and Williams, W. P. (1983). *Biochim. Biophys. Acta* **737**, 223–266.
Rock, C. O. and Garwin, J. L. (1979). *J. Biol. Chem.* **254**, 7123–7128.
Roughan, P. G. (1987). *Methods Enzymol.* **148**, 327–337.
Roughan, P. G. and Slack, C. R. (1982). *Ann. Rev. Plant Physiol.* **33**, 97–132.
Smith, L. A., Norman, H. A., Cho, S. H. and Thompson, Jr, G. A. (1985). *J. Chromatogr.* **346**, 291–300.
Trémolières, A. (1986). *Physiol. Vég.* **24**, 599–605.

Tulloch, A. P., Heinz, E. and Fisher, W. (1973). *Hoppe-Seyler's Z. Physiol. Chem.* **354**, 879–889.
Van Besouw, A. and Wintermans, J. F. G. M. (1978). *Biochim. Biophys. Acta* **529**, 44–53.
Walker D. A. (1980). *Methods Enzymol.* **69**, 94–104.
Walker, D. A. (1987). "The use of the oxygen electrode and fluorescence probe in simple measurements of photosynthesis." Research Institute for Photosynthesis, University of Sheffield (distributed by Packard Publishing Limited, 16, Lynch Down, Funtington, Chichester, West Sussex PO18 9LR, UK).
Walker, D. A., Cerovic, Z. G. and Robinson, S. P. (1987). *Methods Enzymol.* **148**, 145–157.
Yamauchi, R., Kojima, M., Isogai, M., Kato, K. and Ueno, Y. (1982). *Agric. Biol. Chem.* **46**, 2847–2849.

5 Waxes, Cutin and Suberin

T. J. WALTON

*Biochemistry Research Group, School of Biological Sciences,
University College, Singleton Park, Swansea, SA2 8PP, Wales, UK*

I.	Introduction	106
II.	The cuticular membrane	106
	A. Ultrastructure and histochemistry	106
	B. Isolation	107
	C. Physical properties	108
III.	Analysis of cuticular wax	109
	A. Extraction and isolation	109
	B. Chromatographic separation	110
	C. Identification of wax components	113
	D. Analysis of whole epicuticular wax	133
IV.	Analysis of cutin	135
	A. Isolation	135
	B. Depolymerisation	135
	C. Chromatography of monomer derivatives	136
	D. Identification of cutin monomers	138
	E. Molecular architecture of cutin	142
V.	Analysis of the suberin complex	145
	A. Isolation	145
	B. Ultrastructure	145
	C. Extraction and analysis of soluble lipid	146
	D. Depolymerisation and compositional analysis of the polymeric matrix	148
	E. Molecular architecture of the polymeric matrix	149
VI.	Concluding remarks	149
	Acknowledgements	152
	References	153

I. INTRODUCTION

The interface between plant tissue and the growth environment is provided by protective, extracellular envelopes in which lipids provide not only the waterproofing layer of soluble wax, but also provide the insoluble, polymeric structural matrices, cutin and suberin. Two types of structural organisation are found. On most aerial plant tissue, the continuous cuticular membrane, deposited exterior to the layer of epidermal cells, provides the protective barrier. The major structural component of the membrane is cutin, a biopolyester of hydroxyfatty acids, which is embedded in intracuticular wax and supports a waterproofing surface layer of epicuticular wax, typically composed of a complex mixture of highly hydrophobic aliphatic lipid classes. In tissues that undergo secondary growth, including woody stems, roots and underground storage organs, the external protective barrier is provided by the suberin complex, laid down between the plasmalemma and cell wall in the cork cells (phellem) of the periderm. The polymeric matrix, suberin, contains both aliphatic polyester domains similar to those of cutin and condensed aromatic domains resembling those of lignin, and is associated with a soluble wax fraction which confers resistance to water diffusion. In addition to those tissues which produce a surface periderm, suberisation also occurs at wound-healing sites in aerial plant tissue and provides the barrier layer at several diverse internal anatomical sites in root and aerial tissue.

The cuticular membrane and suberin complex and their associated waxes are important not only in restricting water vapour loss, controlling gas exchange and providing physical protection from pathogen and insect invasion, but also in limiting uptake of pesticides and reactive environmental pollutants. The organisation of the waxes and biopolymers of these surface barriers has been studied for over 150 years, but only since the mid-1960s has it been possible to establish details of ultrastructure and chemistry, and this knowledge has enabled systematic studies of the biosynthesis and physiology of these systems. In this chapter emphasis has been given to the analysis and chemistry of the cuticular wax fraction, which does not appear to have been extensively covered since the review of Tulloch (1976a), and some of the more recent advances in the understanding of ultrastructure and chemistry of the insoluble polymers arising from application of newer spectrometric techniques. Reviews of the biosynthesis of waxes and its genetic control (Kolattukudy, 1981, 1987; Bianchi, 1987; Wettstein-Knowles, 1987a,b), biosynthesis of the polymers cutin and suberin (Kolattukudy, 1978, 1984a; Kolattukudy and Espelie, 1985), the function of the barrier polymers (Schönherr, 1982) and the properties and molecular biology of fungal and bacterial cutinase systems (Kolattukudy, 1984b, 1987; Podila et al., 1988) have recently been published, whilst the earlier work in this field has been summarised by Kolattukudy (1970a,b) and Martin and Juniper (1970).

II. THE CUTICULAR MEMBRANE

A. Ultrastructure and Histochemistry

The fine structure of the cuticular membrane which covers the aerial parts of plants has been studied *in situ* and in isolated cuticles for over 150 years (for reviews see Martin and Juniper, 1970; Holloway, 1982a; Juniper and Jeffree, 1983). The depth of focus provided by scanning electron microscopy has been of particular value in studies of epicuticular wax morphology, and has allowed correlations between both chemical composition and environmental growth conditions and the ultrastructure of crystalline wax deposits in a range of species (Holloway *et al.*, 1977a,b; Hunt and Baker, 1979, 1982; Baker, 1974, 1982; Hemmers *et al.*, 1986; Sen, 1987), and observation of the erosion of waxes from leaf surfaces by environmental factors (Baker and Hunt, 1986; Crossley and Fowler, 1986; Guenthardt-Goerg and Keller, 1987; Grill *et al.*, 1987; Percy and Baker, 1987). In cuticles with few or no projecting wax structures, a platinum replica technique has been of value (Simpson and Wettstein-Knowles, 1980). Detailed ultrahistochemical investigations of all structural components of *Agave americana* leaf cuticular membrane both *in situ* and after isolation (Wattendorf and Holloway, 1980, 1982) have enabled earlier models of cuticular membrane structure to be refined and morphological terminology to be rationalised (Fig. 5.1), but since extremes of morphology and construction ranging from completely amorphous cuticular membranes to those in which all regions are reticulate or lamellate have been described (Holloway, 1982a), this schematic should not be regarded as a general model for cuticular membrane organization.

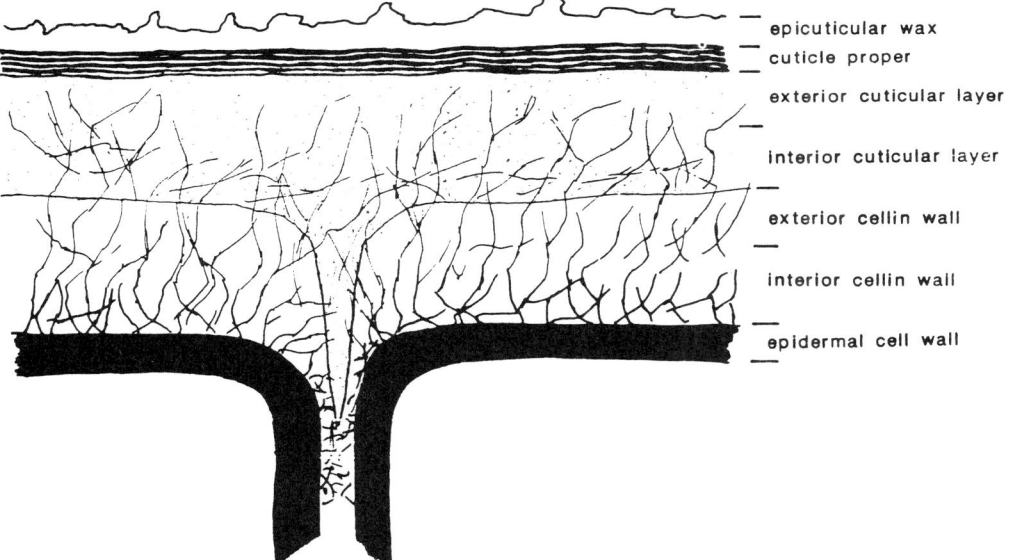

FIG. 5.1. Schematic of *Agave americana* cuticle. The cuticle proper, consisting of alternating lamellae of cutin and wax, together with the interior and exterior cuticle layers composed of carbohydrate fibres in an amorphous cutin matrix, comprise the cuticular membrane. It is 3–5 µm thick, and contains cutin and carbohydrate in a ratio of 2 : 1 (w/w). The fibrils in the cellin wall are pectocellulosic (from Wattendorf and Holloway, 1980, 1982).

B. Isolation

The cuticlar membrane may generally be detached from underlying tissue by treatment with ammonium oxalate–oxalic acid solution or by treatment with commercial pectinase (pectin glucosidase, poly-α-1,4-galacturonide: glycanhydrolase, EC 3.2.1.15), both procedures disrupting the lateral walls of the epidermal cells (Wattendorf and Holloway, 1982; Schönherr and Riederer, 1986). Temperatures above 44°C or high pH should be avoided in the isolation of cuticles for permeability studies (Schönherr et al., 1979; Haas and Schönherr, 1981; Holloway, 1982a), and for this application the milder enzymic procedure is generally preferred. Cuticles isolated by this method do, however, contain significant amounts of fatty acid and other lipophilic material derived from intracellular membranes and absorbed from the tissue slurry during enzymic digestion (Schönherr and Riederer, 1986). Whilst these extraneous lipids do not significantly affect the permeability properties of the cuticular membrane, they have almost certainly been erroneously included in earlier analyses of the soluble lipids extracted from enzymically isolated cuticles and identified as intracuticular wax, as discussed below.

The pectinocellulosic components may be selectively removed from isolated cuticular membranes by careful treatment of the preparation with cuprammonium ion (Matic, 1956) zinc chloride–HCl or sulphuric acid (Holloway and Baker, 1968). However, partial losses and chemical changes to cutin in the membrane may occur even with relatively short exposure to these reagents (Jones, 1978; Wattendorf and Holloway, 1982). Enzymic hydrolysis of the carbohydrate polymers with commercial cellulase (1,4-(1,3;1,4)-β-D-glucan-4-glucanohydrolase, EC 3.2.1.4) and pectinase preparations offers an effective, milder alternative to chemical digestion (Walton and Kolattukudy, 1972).

The procedures described above generally give satisfactory membrane preparations from fruit tissue or leaves with well developed cuticles such as those on xerophytes. Application of the procedures described above to leaves with thin cuticles yields a much less pure preparation, frequently contaminated with transport elements of the leaf. With such tissues careful mechanical stripping of the epidermal layer from the leaf is, when practicable, a valuable preliminary to chemical and enzymic treatment. In this context the *Pisum sativum* mutant *Argenteum*, in which the adaxial and abaxial epidermal layers are readily separated from leaf parenchyma (Hrazdina et al., 1982), appears to offer an attractive experimental system for ultrastructural, compositional and biosynthetic studies.

C. Physical Properties

1. Permeability

One of the main functions of the cuticle is to serve as a barrier to minimise water loss from plants to the atmosphere (Schönherr, 1982). The major role of the soluble cuticular lipid in determining water permeability has been demonstrated directly with cuticles isolated from *Clivia minata* leaves, when the water permeability coefficient was found to increase by two orders of magnitude following extraction of the cuticular wax (Schmidt et al., 1981). However, there is no straightforward relationship between the amount or composition of cuticular wax and water permeability of the cuticle (Haas and Schönherr, 1979; Riederer and Schneider, 1990), phase transition and ESR studies having indicated that permeability is much more dependent on the structural organisation of soluble lipids in the cuticular membrane (Schönherr et al., 1979). Exposure of a wide variety of cuticles to environmentally realistic ozone concentrations in combination with acid rain had no effect on their water and ion permeability (Garrec and Kerfown, 1989; Kerstiens and Lendzian, 1989). The structure and molecular organisation of the matrix polymer, cutin, appears to have only a small role in determining water permeability (Schönherr, 1976; Eckl and Gruler, 1980; Schmidt et al., 1981). Permeability of cuticles to lipophilic molecules is generally proportional to the oil–water partition coefficient of the permeating species (Price, 1982), whilst the permeation of individual lipophilic chemicals across the plant cuticle can be predicted from the octanol–water partition coefficient of the organic compound in conjunction with its molar volume (Kerler and Schönherr, 1988a,b).

Wettability of the leaf surface is governed by the physico-chemical characteristics of the epicuticular wax deposits (Holloway, 1970). Measurement of the contact angle of water droplets on leaf surfaces has shown that wettability is increased by erosion of the cuticular wax resulting from exposure to simulated acid rain of pH ≤ 4.6, leading to increased retention of rain droplets on the foliar surface and associated changes in the uptake and lateral movement of some inorganic ions (Percy and Baker, 1987).

2. Chemical reactivity

Reactive functional groups in the cuticles of some plant species can covalently bind chemical residues. In particular the covalent binding of compounds containing carboxylic acid functions by epoxide groups in cutin, which may lead to increased persistence of agrochemicals such as the phenoxyacetic acid herbicides in the environment, has been studied (Linskens et al., 1965; Riederer and Schönherr, 1986a). Methods have been developed for detection of epoxide groups in intact cuticular membranes by derivatisation as the iodohydrin (Holloway et al., 1981), and a quantitative method based on capillary gas–liquid chromatography (GLC) analysis of the chlorohydrins has been recently described (Riederer and Schönherr, 1988a).

III. ANALYSIS OF CUTICULAR WAX

A. Extraction and Isolation

1. Epicuticular wax

The amount of epicuticular wax produced is dependent on environmental growth conditions, increases in radiant energy flux or decreases in humidity and soil moisture content stimulating wax production (Baker, 1974; Baker and Procopiou, 1980; Hunt and Baker, 1982), whilst chemical composition is less influenced by environmental factors, but changes during tissue development (Tulloch, 1973; Baker et al., 1979; Baker and Hunt, 1981). Thus for reproducibility of wax yields and composition, the strict control of growth conditions and age of tissue harvested is necessary. Wax yields for leaves and fruits of many species have been summarised by Baker (1982).

The soluble lipids from the cuticular surfaces are generally isolated by brief (≤ 60 s) immersion (total surface wax) or washing (adaxial/abaxial surface wax) of fresh plant tissue with redistilled hexane or chloroform at room temperature (Salasoo, 1983; Holloway, 1984b). In waxes where there is a significant cyclic triterpenoid content a solvent of chloroform–diethylether (1 : 1; v/v) is preferrred (Baker et al., 1979). Selective removal of epicuticular wax has also been achieved mechanically, using films of cellulose acetate (Price, 1982) and collodion (Haas and Rentschler, 1984). Use of the latter film, which is insoluble in chloroform, allows the direct recovery and analysis of epicuticular wax without contamination, but neither technique is satisfactory for plants with hairy leaf surfaces.

Epicuticular wax can make a contribution to the aroma of plant tissue by trapping volatile compounds released from the plant. For analysis these may be recovered, together with the epicuticular wax, by immersion of plant tissue in diethylether, and the aroma components subsequently isolated by sublimation and trapping of the volatile fraction (Spence and Tucknott, 1983).

2. Intracuticular wax

In most studies intracuticular wax has been identified as the soluble lipid extracted from cuticular membranes prepared from plant tissue which had been briefly immersed in solvent to remove epicuticular wax (Baker and Procopiou, 1975; Hunt and Baker, 1980; Baker, 1982; Baker et al., 1982). This approach had been criticised, primarily because of the uncertainty of the specificity and efficiency of the initial solvent extraction step, and the collodion-film stripping method employed in a more recent comparative study of epi- and intracuticular wax (Haas and Rentschler, 1984). The demonstration that cuticles isolated with pectinase can contain significant amounts of fatty acids and other lipophilic material originating from intracellular membranes, further confirms that the composition and amount of intracuticular wax cannot be reliably determined from enzymatically isolated cuticular membranes (Schönherr and Riederer, 1986).

B. Chromatographic Separation

1. Column chromatography

As an initial step in purification, whole wax extracts are frequently fractionated by adsorption chromatography on columns of silicic acid or silica gel. Elution with stepwise gradients of chloroform or diethylether in hexane or by discontinuous elution with solvents of increasing polarity allows isolation of a relatively pure hydrocarbon fraction. Other fractions frequently contain more than one class of wax component (Table 5.1), and isolation of the individual classes generally requires further column chromatography or thin-layer chromatography. Alumina column chromatography has also been widely used where only the hydrocarbon fraction of cuticular wax is required (Gaskell et al., 1973; Tulloch, 1976a).

For preliminary fractionation of waxes containing large amounts of acidic wax components, DEAE Sephadex-A25 ion exchange column chromatography has been used as an alternative to adsorption chromatography (Franich et al., 1977; Table 5.1). More specialised column chromatography fractionation procedures have been described for the isolation of β-diketones via their copper complex (Horn et al., 1964; Netting and Wettstein-Knowles, 1976) and for the removal (Wettstein-Knowles, 1974; Wettstein-Knowles et al., 1984) or isolation (Avato et al., 1982) of fatty acids from whole wax extracts.

The low solubility of the surface lipids, together with the absence of a UV chromophore in saturated aliphatic components, has largely precluded the application of high

TABLE 5.1. Fractionation of total epicuticular wax by column chromatography.

Chromatography medium	Elution solvent	Components eluted	Wax source	Reference
Silicic acid (Biosil A)	Hexane	Hydrocarbons	Avena sativa leaf	Tulloch and Hoffman (1973a)
	CHCl$_3$:hexane (1:9; v/v)	Wax esters		
	CHCl$_3$:hexane (1:4; v/v)	β-Diketone + C22–C36 fatty acids		
	CHCl$_3$:hexane (1:3; v/v)	Alcohols		
	CHCl$_3$:hexane (1:1; v/v)	Alcohols, C14–C22 fatty acids		
	CHCl$_3$	Hydroxy-β-diketones		
Silica gel-G	Hexane	Hydrocarbons	Eragrostoid grasses	Tulloch and Hoffman (1974)
	Et$_2$O:hexane (1:49; v/v)	Alkyl esters, triterpenoid esters, aldehydes, diesters		Tulloch (1984)
	Et$_2$O:hexane (3:97; v/v)	Fatty acids + alcohols		
	Et$_2$O:hexane (2:23–1:4; v/v)	Alcohols, triterpenols, triacylglycerols		
Silica gel (Merck 60)	Pentane	Hydrocarbons	Euphorbia sp.	Hemmers and Gulz (1986)
	2-Chloropropane	Esters, aldehydes		
	Methanol	Alcohols, fatty acids		
DEAE-Sephadex 25	Et$_2$O:MeOH:H$_2$O (89:10:1; v/v/v)	Alkyl esters, secondary alcohols, alkane diols, estolides	Pinus radiata needle wax	Franich et al. (1977)
	CHCl$_3$:MeOH:H$_2$O:HCOOH (89:10:1:4; v/v/v/v)	Fatty acids, ω-hydroxyfatty acids, terpenoid acids		

performance liquid chromatography (HPLC) to cuticular wax analysis. An assessment of both adsorption and reverse-phase HPLC, underlining the limitations of HPLC in this particular application, is given by Zabkiewicz and Steele (1982).

2. Thin layer chromatography

Separation of most of the major classes of aliphatic wax components can be obtained by one-dimensional silica gel-G thin layer chromatography (TLC), and where small amounts of total wax are involved, as in biosynthetic experiments, preliminary fractionation by preparative TLC is frequently preferred to column chromatography. Benzene is often preferred as developing solvent for separation of components of low polarity such as hydrocarbons and wax esters, whilst chloroform-based solvent systems generally give better separation of the more polar wax classes. The relative mobility of most wax components is not affected by choice of solvent, but important exceptions to this are β-diketones, β-ketols and primary alcohols (Holloway and Brown, 1977; Tulloch, 1983). Separation of β-diketones from other aliphatic wax components by formation of the copper complex on a 'hybrid' TLC plate has been described (Mikkelsen and Wettstein-Knowles, 1978). Particular care is required with waxes containing triterpenoid wax components, as triterpenoid esters have very similar chromatographic behaviour to alkyl esters, whilst triterpenols generally overlap with the primary alcohol fraction on silica gel-G TLC (Tulloch, 1977; Hemmers et al., 1989), and resolution of aliphatic and terpenoid components requires reverse-phase TLC or GLC.

Each class of wax components on thin layer chromatograms can be visualised by sulphuric acid charring, whilst most triterpenoid-containing fractions can be selectively detected by their colour reaction following treatment with carbazole (Ghosh and Thakur, 1982). When recovery of the sample is required, the plate is sprayed with a fluorescent dye solution, such as 0.3% 2′7′-dichlorofluoroscein in ethanol and after air drying the chromatogram is inspected under UV light, when wax components appear as quenched areas on a fluorescent background. The R_f values of the principal classes of plant waxes in a range of chromatography systems are included in the comprehensive review of methodology for TLC analysis of surface lipids (Holloway, 1984b).

3. Gas–liquid chromatography

Gas–liquid chromatography (GLC) using columns packed with liquid phases stable at temperatures up to 400°C, in conjunction with temperature programming, allows routine separation and identification of individual components containing up to 70 carbon atoms, both in whole wax extracts and in fractions isolated by column and thin layer chromatography. Dexsil 300 has generally been found superior to OV- and SE-silicone phases, as the former carborane-siloxane phase has greater polarity and offers better resolution (Tulloch, 1976a). Free alcohol and carboxylic acid groups in the wax components must be derivatised prior to GLC analysis, generally by trimethylsilylation and methylation, respectively. Relative retention times for the principal constituents of plant wax on Dexsil 300 GLC have been presented (Holloway, 1984b).

Capillary GLC, with its inherent advantages of low sample requirement, high resolution and suitability for interfacing with mass spectrometry (MS), is now well established in studies of wax composition. Typically 15 m columns coated with OV-1 or

SE-52 have been used in the analysis of aliphatic constituents (Franich et al., 1985; Hemmers and Gülz, 1986; Scora et al., 1986; Audisio et al., 1987), whilst Tulloch (1983) employed a 30 m fused silica column chemically bonded with polymethyl (5% phenyl) siloxane. A 12 m quartz capillary coated with FFAP was used to separate saturated from unsaturated homologues (Hemmers and Gülz, 1986), whilst triterpenols have been characterised on 10 m OV-1 capillary columns (Hemmers et al., 1988).

C. Identification of Wax Components

1. Hydrocarbons

(a) *Aliphatic hydrocarbons.* The composition of the homologous series of odd carbon number alkanes (C17–C35, with C29 and/or C31 generally dominant, Table 2) is routinely established by packed column GLC on a non-polar stationary phase (Tulloch and Weenik, 1969; Tulloch, 1973). On such systems monoenoic hydrocarbons elute shortly before the alkane of the same chain length. Their provisional identification can be confirmed by GLC on a polar, polyester column, when the order of elution is the reverse of that on non-polar systems, and by GLC following catalytic hydrogenation of the hydrocarbon fraction, when the disappearance of putative alkenes is accompanied by a commensurate increase in the corresponding alkane component (Herbin and Robins, 1968a). Preliminary separation of the alkene fraction by preparative argentimetric TLC considerably facilitates subsequent GLC anaslysis (Nordby and Nagy, 1977). The geometry of the double bond in isolated alkenes can be determined from their infrared spectrum, whilst the position of the double bond is established from the composition of the fatty acid products yielded by permanganate-periodate oxidation (Tulloch and Hoffman, 1976).

In hydrocarbon fractions containing branched-chain alkanes, there is typically an homologous series of odd-carbon chain 2-methylalkanes (*iso*-C27–C33), isomeric with the *n*-alkane series, together with a series of even chain 3-methylalkanes (*anteiso*-C28–C34, Table 5.2); these three homologous series are well resolved by packed column GLC (Carruthers and Johnstone, 1959; Kaneda, 1967). However, isomeric 2- and 3-methylalkanes are not fully resolved by this procedure, but may be fully separated by capillary GLC (Wollrab et al., 1967; Streibl and Konečný, 1967). Prior to GLC analysis, the branched-chain components may be effectively concentrated by removal of *n*-alkanes from the hydrocarbon fraction, either by urea clathration (Coles, 1968; Baker and Holloway, 1975) or molecular seive chromatography (Kaneda, 1967). The characteristic fragmentation pattern in the electron impact mass spectrum (EIMS) of the individual alkanes from plant waxes allows each of the positional isomers to be distinguished, and provides direct confirmation of provisional identification based on chromatographic properties (Kolattukudy and Walton, 1972a).

(b) *Aromatic hydrocarbons.* Column chromatography on alumina resolved the hydrocarbons of banana (*Musa sapientum*) leaf wax into an *n*-alkane fraction and a second fraction whose constituents had chromatographic and spectrophotometric properties very similar to members of the phenanthrene, anthracene, naphthalene and pyrene families, and the identification of the wax components as a series of alkyl substituted polycyclic hydrocarbons was further supported by their EIMS fragmentation patterns (Nagy et al., 1965).

TABLE 5.2. Aliphatic components of epicuticular wax.

Wax fraction	Major or characteristic component	Chain length range	Good or only reported sources
Hydrocarbons			
n-Alkanes	$C_{31}H_{64}$, $C_{29}H_{60}$	C17–C37	Widespread
iso-Alkanes	$(CH_3)_2CH.(CH_2)_{27}$—CH_3	C29–C35	*Aeonium* sp. *Nicotiana tabaccum*
$anteiso$-Alkanes	CH_3CH_2—$CH(CH_3)$—$(CH_2)_{25}$—CH_3	C30–C34	*N. tabaccum*
Internally branched alkanes	9-, 11-, 13- and 15-methyl-triacontane	C29–C31	Walnut tree leaf
Dimethylalkanes	$(CH_3)_2$—CH—$(CH_2)_n$—CH—$(CH_3)_2$ $n = 8$–26	C14–C34	Montan wax, sugar cane, hops
Alk-1-enes	$CH_2\overset{c}{=}CH$—$(CH_2)_{25}$—CH_3	C26–C30	Sugar cane, walnut tree leaf
Alk-2-enes	$CH_3CH\overset{t}{=}CH(CH_2)_nCH_3$	C20–C33	Sugar cane
Alk-3-enes	CH_3—CH_2—$CH\overset{c}{=}CH$—$(CH_2)_{26}$—CH_3	C17–C33	Rose petal wax
	CH_3—CH_2—$CH\overset{t}{=}CH$—$(CH_2)_{24}$—CH_3	C19–C37	
Alk-5-enes	CH_3—$(CH_2)_3$—$CH\overset{c}{=}CH$—$(CH_2)_{22}$—CH_3	C27–C31	Rosaceae
Alk-7-enes	CH_3—$(CH_2)_5$—$CH\overset{c}{=}CH$—$(CH_2)_{20}$—CH_3	C17–C33	Rosaceae
Alk-9-enes	CH_3—$(CH_2)_7$—$CH\overset{c}{=}CH$—$(CH_2)_{20}CH_3$	C25–C33	Leaves and spikes of flowering *Agropyron intermedium*
Alk-10-enes	CH_3—$(CH_2)_8$—$CH=CH$—$(CH_2)_{21}$—CH_3	C31–C33	Sugar cane
Conjugated alkdienes	CH_3—$(CH_2)_n$—$CH=CH$—$CH=CH$—$(CH_2)_m$—CH_3 $n + m = 13$	C11–C33 (odd + even)	Rose flower wax
Aromatic hydrocarbons	Alkyl-phenanthrene/anthracene, naphthalenes, pyrenes		Banana leaf wax
Ketones			
Monoketones	$CH_3(CH_2)_8CO(CH_2)_{18}CH_3$	C24–C33	*Pyrus malus* *Brassica* sp.
α-Ketols	$CH_3(CH_2)_{12}.CHOH.CO.(CH_2)_{13}CH_3$	C29	*Brassica* sp.
β-Ketols	$CH_3(CH_2)_{11}.CHOH.CH_2.CO.(CH_2)_{13}CH_3$	C29	*Brassica* sp.
β-Diketones	$CH_3(CH)_{12}.CO.CH_2CO(CH_2)_{14}CH_3$	C27–C33	Barley spike, carnation leaf, *Brassica* sp., *Agropyron* sp.

Comments	References
Common, usually major. Odd chains predominate. Major components of tobacco and *Aeonium* species leaf wax; minor component of many waxes.	Tulloch (1976a) Eglinton et al. (1962, 1966), Kaneda (1961), Baker and Holloway (1975)
Minor component.	Stránský et al. (1970)
May be widespread minor alkane components.	Jarolinek et al. (1964), Wollrab et al. (1965a,b)
Odd and even chains often in equal abundance.	Sorm et al. (1964), Streibl and Stránský (1968), Stránský et al. (1970) Sorm et al. (1964)
Major olefin. of roses.	Wollrab et al. (1963), Wollrab (1967), Sorm et al. (1964)
	Tulloch and Hoffman (1976)
Very minor.	Sorm et al. (1964) Stoionova-Ivanova et al. (1971)
Equal in quality to *n*-alkanes.	Nagy et al. (1965)
Common, frequently but not invariably co-occur with alkane of same chain length. May also co-occur with secondary alcohol of same chain length and position of oxygen substituent.	Dodova-Anghelova and Ivanov (1973)
Rare minor component. 15-hydroxy-14-oxo-isomer also present.	Holloway and Brown (1977)
Rare minor component. 14-hydroxy-16-oxo-isomer also present.	Holloway and Brown (1977)
Frequent minor components of cereal waxes. C29–C33 generally dominant, positional isomers present.	Mikkelsen (1979) Tulloch (1983)

Continued

TABLE 5.2. continued

Wax fraction	Major or characteristic component	Chain length range	Good or only reported sources
Hydroxy-β-diketones	$CH_3(CH_2)_{12}CO.CH_2.CO$ $(CH_2)_8.CHOH.(CH_2)_5CH_3$	C31	*Agropyron smithii* leaves, *Poa ampla*
Oxo-β-diketones	$CH_3(CH_2)_{12}COCH_2.CO$ $(CH_2)_8CO(CH_2)_5CH_3$	C31	*Agropyron* sp. leaves
Oxo-β-ketol	$CH_3(CH_2)_5CO(CH_2)_8CO.$ $CH_2CHOH(CH_2)_{12}CH_3$	C31	*Agropyron elongatum*
Hydroxyoxo-β-diketones	$CH_3(CH_2)_2CHOH.(CH_2)_9$ $COCH_2CO(CH_2)_8$ $CHOH(CH_2)_5CH_3$	C31	*A. elongatum*
Dioxo-β-diketones	$CH_3(CH_2)_8CO(CH_2)_3COCH_2$ $CO(CH_2)_8CO(CH_2)_{14}CH_3$	C31	*A. intermedium*

Esters

Wax fraction	Major or characteristic component	Chain length range	Good or only reported sources
Long-chain alkan-1-ol acyl esters	$CH_3(CH_2)_{24}COOCH_2(CH_2)_{18}CH_3$	C32–C72	Widespread
Alkan-2-ol acyl esters	$CH_3-(CH_2)_{12}-CH-CH_3$ $\quad\quad\quad\quad\quad\quad\quad\;\; \mid$ $\quad\quad\quad\quad\quad OOC(CH_2)_{18}CH_3$	C31–C37	*Eucalyptus* sp. Barley (but not awn or leaf blade wax). Sorghum
7-Oxoalkan-2-ol acyl esters	$CH_3(CH_2)_{18}COOCH(CH_3)$ $(CH_2)_4CO(CH_2)_7CH_3$	C31–C37	Barley spike wax
Aromatic alcohol acyl esters	$C_7H_7-OOC(CH_2)_{26}CH_3$ $C_8H_9OOC(CH_2)_{24}CH_3$	C31–C37 (esters of benzylalcohol) C30–C38 (esters of 2-phenylethanol)	Jojoba leaf wax
Dioldiesters	$RCOO-CH_2-(CH_2)_{10}-$ CH_2OOCR'	C56–C60	Wheat and oat leaf wax
Polyesters (estolides)	$CH_3(CH_2)_{10}COOCH_2$ $(CH_2)_{12}CH_2OOC(CH_2)_{10}-$ $CH_2OOC(CH_2)_{10}CH_3$	M_r range 800–1500	*Pinus* sp. needle wax
Alkan-1-ol benzoyl esters	$C_7H_5-COOCH_2(CH_2)_{24}CH_3$	C29–C31	*Euphorbia dendriodes* *Citrus halimii*

Aliphatic alcohols

Primary alcohols

Wax fraction	Major or characteristic component	Chain length range	Good or only reported sources
n-Alkan-1-ols	$CH_3(CH_2)_nCH_2OH$ $n = 24, 26, 28$	C22–C34	Widespread
iso-Alkan-1-ols	$(CH_3)_2CH(CH_2)_{22}CH_2OH$	C24–C26	*Stellaria media*
anteiso-Alkan-1-ols	$CH_3CH_2(CH_3)CH(CH_2)_{21}CH_2OH$	C26 and C28	*Brassica* sp.

Comments	References
Generally substantial in *Agropyron* sp.	Tulloch (1976b)
Major component of *P. ampla* wax.	Tulloch (1978)
Rare and minor.	Tulloch (1976b, 1983)
Rare.	Tulloch (1983)
Rare, minor.	Tulloch (1983)
Very minor component.	Tulloch and Hoffman (1976)
Common major component. Common alcohols C26 and C28, common acids C18–C22. Unsaturated esters rare. C33 methyl esters with hexyl and octyl esters of C24–C36 acids in spruce and pine.	Tulloch (1984, 1987), Tulloch and Hoffman (1973a,b), Bianchi *et al.* (1989)
Rare.	Wettstein-Knowles and Netting (1976a), Lundqvist and Wettstein-Knowles (1983), Tulloch (1983), Wettstein-Knowles *et al.* (1984)
Very minor component.	Wettstein-Knowles and Madsen (1984)
Rare.	Gülz and Marner (1986)
Minor components in cereal wax. Major fatty acyl groups 22:1 (2*t*) and 24:1 (2*t*).	Tulloch, 1971, Tulloch and Hoffman (1973a)
Linear; ω-OH C12, C14, C16 fatty acid esterified to medium chain acid + diols. Cyclic estolides may also occur.	Herbin and Sharma (1969), Franich *et al.* (1978), Schulten *et al.* (1986)
Very minor components.	Gülz *et al.* (1987a) Gülz *et al.* (1987b)
Often several major components; in festucoid grasses C26 or C28 predominant, in panicoids, C32.	Tulloch (1976a, 1981)
Minor components of *Lolium perenne*, *Chenopodium album* and *Stelloria media* primary alkanols.	Allebone and Hamilton (1972)
May be minor components of many waxes.	Baker and Holloway (1975)

Continued

TABLE 5.2. continued

Wax fraction	Major or characteristic component	Chain length range	Good or only reported sources
Secondary alcohols	$CH_3(CH_2)_{13}CHOH(CH_2)_{13}CH_3$ $CH_3(CH_2)_{14}CHOH(CH_2)_{14}CH_3$	C29 and C31	Widespread, frequently same chain length distribution as alkane in same wax
Secondary diols	$CH_3(CH_2)_3CHOH-$ $(CH_2)_4CHOH(CH_2)_{18}$ CH_3	C29 and C31	*Pinus* sp.
Aliphatic aldehydes			
n-Alkanals	$CH_3(CH_2)_nCHO$ $n = 24, 26, 28$	C20–C34	Widespread
anteiso-Alkanals	$CH_3CH_2(CH_3)CH(CH_2)_{24}$ COOH	C27–C31	*Brassica oleraceae* gl_6 mutant
Fatty acids			
n-Alkanoic acids	$CH_3(CH_2)_nCOOH$ $n = 10–36$	C12–C38	Widespread minor component of many waxes
iso-Alkanoic acids	$(CH_3)_2CH(CH_2)_{20}COOH$	C24–C26	*Stellaria media*
anteiso-Alkanoic acid	$CH_3CH_2(CH_3)CH(CH_2)_{24}$ COOH	C25–C31	*Brassica oleraceae* gl_6 mutant
n-Alkenoic acids	$CH_3(CH_2)_nCH\overset{t}{=}CHCOOH$ $n = 18$ and 20	C22 and C24	*Triticum aestivum* var.

2. *Ketones*

(a) *Monoketones*. Components of the ketone fraction are readily identified by combined gas–liquid chromatography–mass spectrometry (GC-MS). Derivatisation is not necessary, as the EIMS of the ketone contains both a significant molecular ion to establish relative molecular mass, and intense ions formed by cleavage on either side of the carbonyl group (Fig. 5.2(I)), which together with a series of prominent, diagnostic rearrangement ions allow the position of the oxygen function to be unambiguously assigned (Dodanova-Anghelova and Ivanov, 1973).

(b) *Ketols*. Resolution of the α- and β-ketols which constitute this minor fraction of wax (Table 5.2) by preparative silica gel-G TLC in benzene yields an α-ketol fraction free from other wax components. However, the β-ketol fraction may be contaminated with primary alcohols and terpenols, and requires further preparative silica gel-G TLC purification using a chloroform–ethanol-based solvent (Holloway and Brown, 1977).

Comments	References
Common and major. Asymmetrical substitution patterns also found, e.g. C31-5-ol (*Rosa* petal wax), C31-9- and 10-ol (*Eragrostis trichoides*); presence of C29-10-ol correlated with occurrence of small crystalline wax tubes on plant surface.	Tulloch (1976a), Holloway *et al.* (1976), Mladenova *et al.* (1977), Baker and Hunt (1979), Tulloch (1984)
Range of C29 positional isomers present.	Franich *et al.* (1979) Tulloch (1987)
Common and minor, generally similar chain length to free primary alcohols in same wax. Octocosanal in polymeric form reported as the major component of sugar cane wax.	Tulloch (1976a)
	Netting *et al.* (1972)
Often a short chain homologous series (C12–C18) accompanied by a very-long-chain homologous series (C23–C36).	Tulloch (1976a) Baker (1982)
Minor components. Not present in normal *Brassica oleraceae*.	Allebone and Hamilton (1972) Netting *et al.* (1972)
Significant component of free acid fraction.	Tulloch and Hoffman (1973b)

Provisional identification of a ketol may be supported by increases in TLC R_f value following treatment with both 2,4-dinitrophenylhydrazine and with acetic anhydride (Holloway and Challen, 1966). Positional isomers of neither class of ketol are resolved by packed column GLC, and thus the identification and quantification of isomers was based on direct probe EIMS of isolated α- and β-ketol fractions, the diagnostic fragmentation pattern allowing provisional structural assignments for the isomers present in each class of ketol. These were substantiated and isomer content estimated by GC-MS of the trimethylsilylethers of each ketol fraction (Figs 5.2(II) and (III)) and of the sodium borohydride reduction products from each ketol fraction (Holloway and Brown, 1977).

(c)   β-*Diketones*. A useful indicator for the presence of β-diketones in wax fractions is their characteristic UV absorption at 273 nm ($E^{1\%}_{1cm}$ β-diketone = 260 (Tulloch, 1973)). Following separation from other wax components by copper complex formation discussed earlier, direct GC-MS analysis of the β-diketone fraction is possible. The

I CH₃−(CH₂)₇⫶CH₂⫶C⫶CH₂⫶(CH₂)₁₇−CH₃

with fragmentation labels: 155, 170 +H, 295, 310 +H, and C=O

II CH₃−(CH₂)₁₂−CH⫶C−(CH₂)₁₃−CH₃
         | ‖
        TMSO O

285 ← , → 225

III CH₃−(CH₂)₁₂⫶CH⫶CH₂⫶C⫶CH₂⫶(CH₂)₁₁−CH₃
              | ‖
             TMSO   O

→ 327 −90→ 237, 285, 211, 237←327 −90, 252←342 −90

IV CH₃−(CH₂)₁₂⫶CH₂⫶CH₂⫶C⫶CH₂⫶C⫶CH₂⫶CH₂⫶(CH₂)₁₀−CH₃
                      ‖   ‖
                      O   O

H→ , 253, 211, ←H
281, 268, 296, 309
192←250, 239, 281, 278→220

V CH₃−(CH₂)₁₂−CH₂⫶CH₂−C=CH−C⫶CH₂ CH₂−(CH₂)₁₀−CH₃
                    | ‖
                   TMSO   O

157, ←H, 325, 353

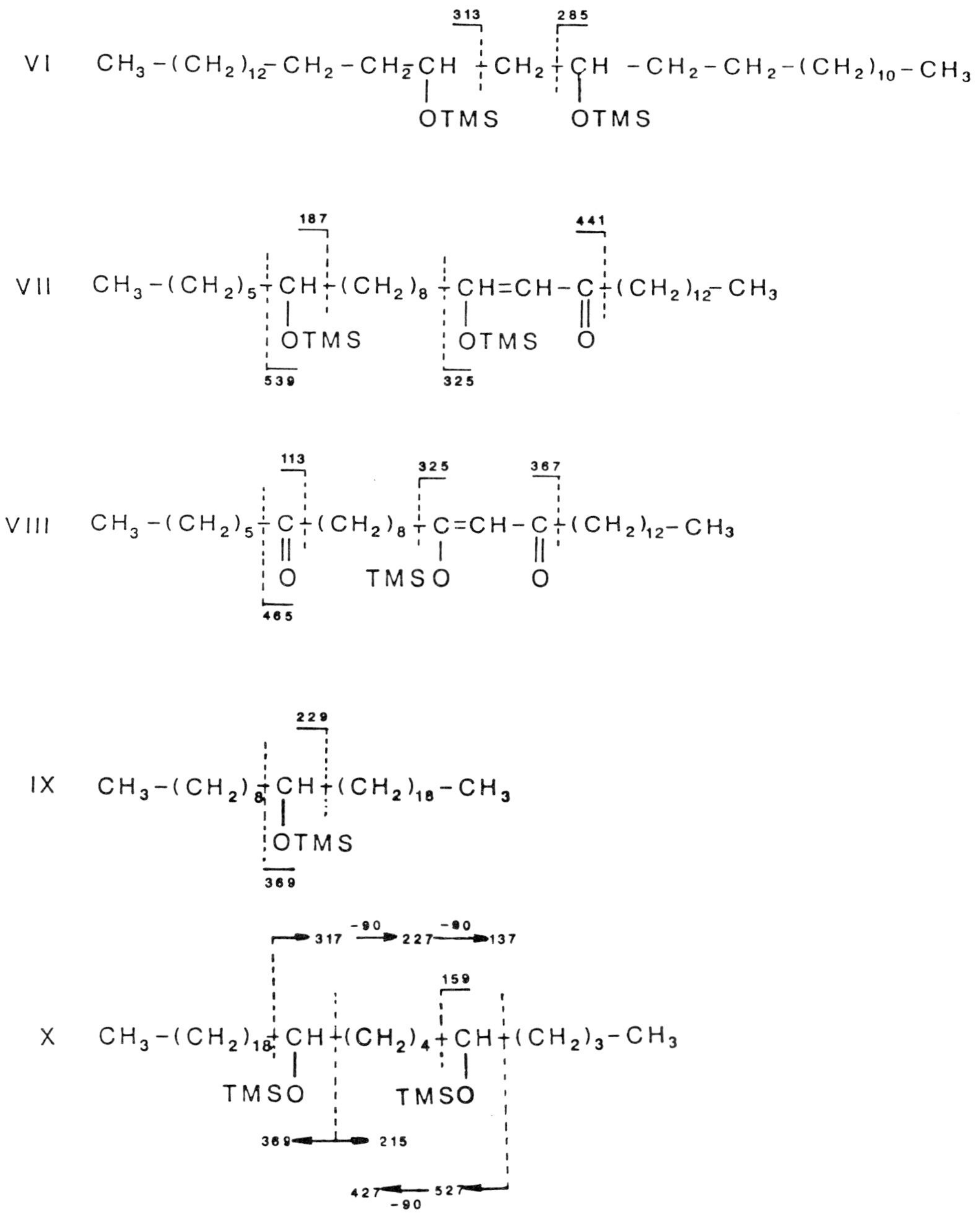

VI CH₃–(CH₂)₁₂–CH₂–CH₂–CH⫿CH₂⫿CH–CH₂–CH₂–(CH₂)₁₀–CH₃
                                    313   285
                           | |
                          OTMS   OTMS

VII CH₃–(CH₂)₅⫿CH⫿(CH₂)₈⫿CH=CH–C⫿(CH₂)₁₂–CH₃
                   187   441
              | | ||
             OTMS   OTMS   O
              539   325

VIII CH₃–(CH₂)₅⫿C⫿(CH₂)₈⫿C=CH–C⫿(CH₂)₁₂–CH₃
                  113   325   367
              || | ||
              O   TMSO   O
             465

IX CH₃–(CH₂)₈⫿CH⫿(CH₂)₁₈–CH₃
                229
              |
             OTMS
             369

       ┌─→ 317 —⁻⁹⁰→ 227 —⁻⁹⁰→ 137
                              159
X CH₃–(CH₂)₁₈⫿CH⫿(CH₂)₄⫿CH⫿(CH₂)₃–CH₃
              | |
             TMSO   TMSO
             369 ←── 215
                    427 ←── 527
                         ⁻⁹⁰

FIG. 5.2. Mass fragmentation of some wax ketones and secondary alcohols. I, nonacosan-10-one; II, 14-hydroxynonacosan-15-one TMS ether; III, 14-hydroxynonacosan-16-one TMS ether; IV, hentriacontane-14,16-dione; V, hentriacontane-14,16-dione TMS-enol ether; VI, hentriacontane-14,16-diol *bis*-TMS ether; VII, 25-hydroxyhentriacontane-14,16-dione TMS-enol ether; VIII, 25-oxohentriacontane-14,16-dione TMS-enol ether; IX, 10-hydroxynonacosanol TMS ether; X, nonacosane-5,10-diol *bis*-TMS ether. From: I, Dodenova-Anghelova and Ivanov (1973); II and III, Holloway and Brown (1977); IV, Tulloch (1976a); V, VI, VII, Tulloch and Hogge (1978); VIII, Mikkelsen (1979); IX, Holloway *et al.* (1976); X, Hunt and Baker (1979).

fragmentation pattern allows location of the β-diketone system (Jackson, 1971; Tulloch, 1976a), but the individual components are more readily characterised by GC-MS after conversion either to their TMS enol ethers (Tulloch and Hogge, 1978) or by analysis of the TMS ethers of the β-diols obtained by sodium borohydride reduction of the β-diketone fraction (Mikkelsen, 1979). Both TMS ether derivatives have major advantages in GC-MS studies, since, as exemplified in Figs 5.2(V) and 5.2(VI) for the case of hentriacontane-14,16-dione, their mass spectral fragmentation produces relatively few intense ions which are highly diagnostic of the pattern of substitution, whereas fragmentation of the β-diketone itself (Fig. 5.2(IV)) leads to a large number of ions which are of similar relative intensity, a disadvantage when mixtures of isomers are present. The position of substitution in the β-diketone may be confirmed by determination of the carbon chain lengths, either of the equimolar mixture of fatty acids and methyl ketones resulting from alkaline hydrolysis (Horn and Lamberton, 1962), or of the fatty acids yielded by iodoform oxidation (Netting and Wettstein-Knowles, 1976).

(d) *Hydroxy-β-diketones*. Isolation via the copper complex yields a fraction which can be characterised by direct mass spectrometry as containing hydroxy-β-diketones, although the fragmentation pattern does not allow the position of the hydroxyl group to be reliably assigned (Tulloch and Hoffman, 1973a). However, the structures and isomer compositions can in many cases be reliably assigned from the mass fragmentation pattern (Fig. 5.2(VII)) of the bis-TMS enol ether of the hydroxy-β-diketone fraction (Tulloch and Hogge, 1978). Alkaline hydrolysis of this class of β-diketone produces a mixture of methylketones, hydroxymethylketones, fatty acids and hydroxyfatty acids. The chain length and ^{13}C nuclear magnetic resonance (NMR) properties of the latter fraction allowed the position of the oxygen substituents in the wax component to be confirmed (Tulloch and Hoffman, 1973a). With two exceptions the hydroxy-β-diketones isolated to date are weakly dextrorotatory and have been assigned the *R* configuration (Tulloch and Hoffman, 1979; Tulloch, 1983).

(e) *Other substituted β-diketones*. These rare wax constituents isolated from *Graminae* spp. include oxo-β-diketones, dioxo-β-diketones, hydroxyoxo-β-diketones and oxo-β-ketols (Table 5.2). 25-Oxohentriacontane-14,16- and 10-oxohentriacontane-14,16-dione from *Agropyron intermedium* wax were originally characterised by proton magnetic resonance (PMR) and mass spectrometric analysis of the oxoalkanoic acid fractions obtained on alkaline hydrolysis. This allowed the position of the isolated carbonyl function to be located relative to the β-diketone system in the parent molecule, and the relative composition of isomers to be estimated (Tulloch and Hoffman, 1976). For routine identification and estimation of this class of β-diketones, including isomer composition, GC-MS analysis of the TMS enol ether (Fig. 5.2(VIII)) is the current method of choice (Tulloch and Hogge, 1978). 10,25-Dioxohentriacontane-14,16-dione was also identified as a very minor component of *A. intermedium* wax in the same study, and this was accompanied by larger amounts of hydroxyoxo-β-diketone. Characterisation of this as a mixture of 4-hydroxy-25-oxo-, 25-hydroxy-10-oxo- and 26-hydroxy-10-oxohentriacontane-14,16-dione in a relative proportion of c. 3 : 2 : 1 was based on GC-MS and ^{13}C-NMR analysis of the oxo- and hydroxyalkanoic acids yielded by alkaline hydrolysis of the hydroxyoxo-β-diketone fraction. More recently, Tulloch (1983) employed the iodoform reaction, previously used in characterisation of β-diketones

(Netting and Wettstein-Knowles, 1976), to degrade the hydroxyoxo-β-diketone fraction of *A. elongatum*, and by spectrometric analysis of the resulting hydroxy- and oxoalkanoic acids was able to confirm the structure as 4-hydroxy-25-oxohentriacontane-14,16-dione, which had been provisionally assigned from the mass spectral fragmentation of the TMS enol derivative. In the same wax sample, an unusual oxo-β-ketol was found as a minor component in the hydroxyoxo-β-diketone-containing fraction from column chromatography; GC-MS analysis of the TMS ether derivative of this minor component indicated its structure as 18-hydroxyhentriacontane-7,16-dione. This structure was largely confirmed by a comparative study of the ^{13}C-NMR properties of the naturally occurring oxo-β-ketol with those of model synthetic β-ketols and oxo-β-ketols.

3. Wax esters

(a) *Aliphatic monoesters.* The chain length range (C30–C56) of the homologous series of saturated aliphatic monoesters which commonly occurs in wax can be established directly by high temperature packed column or capillary GLC (Tulloch, 1973, 1976a; Audisio *et al.*, 1987). On columns of Dexsil-300 monoenoic esters have emergence temperatures slightly greater than the saturated ester of the same chain length, and their identity can be confirmed by catalytic hydrogenation and re-analysis by GLC. Presence of the much rarer homologous series (C31–C37) of odd carbon chain esters containing C13- and C15-alkan-2-ols can cause difficulties in GLC analysis, since the elution temperatures of these odd-chain esters are very similar to those of the even-chain ester containing one fewer carbon atom (Wettstein-Knowles and Netting, 1976a; Mikkelsen, 1984; Wettstein-Knowles, 1985). These ester classes can be readily distinguished by GC-MS, since the mass spectra of esters of primary alcohols obtained by electron impact ionisation mass spectrometry (EIMS) and chemical ionisation mass spectrometry (CIMS) contain a molecular ion, whilst those of secondary alcohols do not (Wettstein-Knowles and Netting, 1976b; Franich *et al.*, 1985). Combined GC-MS of the ester fraction not only differentiates between esters of alkan-1-ols and alkan-2-ols, but also allows ready identification and quantification of the alkyl and fatty acyl moieties which constitute each chain length in an homologous series of primary alcohol wax esters, since, as shown in Table 5.3, fragment ions arising from the fatty acid moiety are intense, whilst those arising from the fatty alcohol are relatively weak (Aasen *et al.*, 1971; Wettstein-Knowles and Netting, 1976b; Audisio *et al.*, 1987).

Tandem mass spectrometry (MS/MS) has also been applied in the compositional analysis of wax esters (Spencer and Plattner, 1984). Isobutane-CIMS of the crude wax ester generated a series of protonated molecular ions which established the chain length distribution in the alkyl ester fraction. Each of these ions was in turn selected using the first quadrupole mass filter and subjected to collision-induced dissociation with argon. For saturated esters the positively charged daughter ions obtained were found to be almost exclusively the protonated acid ion $[RCOOH_2]^+$, which thus directly defined the fatty acyl constituents in the ester and, since the relative molecular mass of the original ester was known, allowed the identity of the alkyl groups to be deduced. However, whilst interpretation was relatively straightforward for saturated esters, it was considerably more difficult for unsaturated esters. Not only were their fragmentation patterns more variable and complex, but in addition the presence of a single double bond doubles the number of structural isomers which may contribute to an observed protonated molecular ion. Prior to MS/MS analysis, therefore, the total ester fraction was saturated by tris(triphenylphosphine)chloro-rhodium (I) catalysed reduction with deuterium, resulting in the protonated molecular ions and their daughter protonated acid ions having m/z values 2 units greater per double bond, which allowed ready identification and quantification of isomeric unsaturated esters. When suitable MS facilities are not available, quantification and determination of the fatty acyl and alkyl moieties constituting wax ester fractions can also be readily achieved either by acid-catalysed alcoholysis (Tulloch and Weenik, 1969; Tulloch and Hoffman, 1973a) or, more rapidly, by transmethylation with tetramethylammonium hydroxide (Barta and Kömives, 1984) and, after derivatisation of the alcohol fraction obtained, GLC analysis of the products as discussed below.

TABLE 5.3. Diagnostic features in EI mass spectra of aliphatic wax esters.

Ion	Intensity	
	Alkan-1-ol esters	Alkan-2-ol esters
	^{a}R—C(=O)—O—R'^{b}	^{a}R—C(=O)—O—R'^{c}
$[RCO_2R]^+$	Present	Absent
$[RCO_2H_2]^+$	Base peak	High
$[RCO_2H]^+$	Intermediate	High/base
$[R'-1]^+$	Weak/intermediate	High
$[R'CO]^+$	Weak	High/base

^{a} R = $CH_3(CH_2)_{16-22}$—
^{b} R' = —$(CH_2)_{19-27}CH_3$
^{c} R' = —CH(CH$_3$)—$(CH_2)_{10-12}$CH$_3$

From Wettstein-Knowles and Netting (1976b).

The presence of trace amounts of the unusual wax esters containing the substituted secondary alcohol 7-oxopentadecan-2-ol in barley wax (Table 5.2) was initially indicated by their distinctive TLC, GLC and chemical properties; following borohydride reduction and methanolysis of the isolated ester fraction, the mass spectrum of the alcohol fraction was indistinguishable from that of synthetic 2,7-dihydroxypentadecane (Wettstein-Knowles and Madsen, 1984).

(b) *Higher aliphatic esters.* Aliphatic diesters (Table 5.2) composed of C8–C13-α,ω-alkanediols esterified by *trans*-2-docosenoic acid and *trans*-2-tetracosenoic acid, have been characterised as minor components of several waxes (Tulloch, 1971, 1973, 1983). Polyesters of greater chain length are insufficiently volatile for direct GLC analysis and therefore compositional studies of the linear and cyclic polyesters containing ω-hydroxyfatty acids, described as estolides, which characterise wax of *Pinus* and

Cupressaceae spp., relied upon analysis of their hydrolysis products (Rudloff, 1959; Herbin and Robins, 1968b; Herbin and Sharma, 1969). Subsequently, however, the structures of some of the linear estolides, which are substantial components of *Pinus radiata* secondary needle wax, were, after chromatographic fractionation, directly defined by EIMS, when at probe evaporation temperatures of 250–350°C satisfactory mass spectra for estolides up to M_r 1100 were obtained (Franich *et al.*, 1978). The mass spectrum of the estolide fraction of lowest M_r contained a base peak at m/z 183 and two sequences of peaks at m/z 876, 848, 820, 792, 764 and at m/z 902, 874, 846, 818, 790, which were identified as parent ions from two homologous series of estolides (Fig. 5.3(a)). Based on GLC compositional analysis of estolide hydrolysis products, it was possible to propose the two isomeric structures I and II (Fig. 5.3(b)) which would both give rise to the strongest ion in these series at m/z 792 ($[C_{50}H_{96}O_6]^+$). Both estolides terminate in esterified dodecanoic acid, but in estolide I, α,ω-C_{14}-diol is interesterified with C_{12}-ω-hydroxyfatty acid, whilst in structure II, α,ω-C_{12}-diol is interesterified with C_{14}-ω-hydroxyfatty acid. Studies on model esters indicated that fragmentation of the central ester linkage would give rise to diagnostic fragment ions which would allow the isomers to be distinguished and the linear sequence of the estolide to be established. In the mass spectrum (Fig. 5.3(a)), the base peak at m/z 183 and the strong ions at m/z 201 and m/z 592 are assigned to ions formed by cleavage of the terminal ester group, and thus confirm the presence of dodecanoic acid in this position but do not allow the isomers to be distinguished. However, the sequence of ions at m/z 381, 394, 399 and 439 can be assigned to cleavage of the ester linkage between α,ω-C_{14}-diol and C_{12}-ω-hydroxyfatty acid in estolide I (Fig. 5.3(b)), and the corresponding ions at m/z 409, 366, 427 and 411 to cleavage of a central ester linkage between C_{12}-α,ω-diol and C_{14}-ω-hydroxyfatty acid in estolide II (Fig. 5.3(b)), indicating the presence of both isobaric structures and establishing the linear sequence of the estolides.

Lack of volatility of high mass polyester fractions prevents application of this mass spectrometric sequencing procedure to estolides of $M_r > 1100$. However, carbon numbers of estolides up to M_r c. 2000 may be established by application of the newer ionisation techniques of field desorption (FD) and field ionisation (FI), which allow direct mass spectrometry of involatile fractions, and which are also of considerable value in establishing the total wax 'fingerprint' (see Section III.F) (Schulten *et al.*, 1987a,b). In the FD mass spectrum of the cuticular wax of *Picea abies* (Fig. 5.4), five homologous series of peaks were present in the mass range m/z 623 to m/z 1800, each of which (m/z 623–735; m/z 735–1017; m/z 991–1215; m/z 1245–1470; m/z 1501–1753) contained one major ion at m/z 679, m/z 933, m/z 1103, m/z 1358 and m/z 1613, respectively. This pattern was identified as sequences of protonated molecular ions derived from families of linear and cyclic estolides containing from two to seven ester units, and allowed for the first time relative molecular masses of estolides above M_r 1100 to be assigned. However, since the mass spectra obtained by FD or FI ionisation do not contain fragment ions derived from estolides, sequencing information is not provided by these techniques.

(c) *Aromatic esters*. Identification of acyl esters of aromatic alcohols in the leaf wax of Jojoba (*Simmondsia chinensis* (Link) Schneider) (Table 5.2) followed detection of two unusual ester fractions which co-chromatographed with aldehyde and ketone bands on silica gel-G chromatography. Each was resolved by capillary GLC into homologous

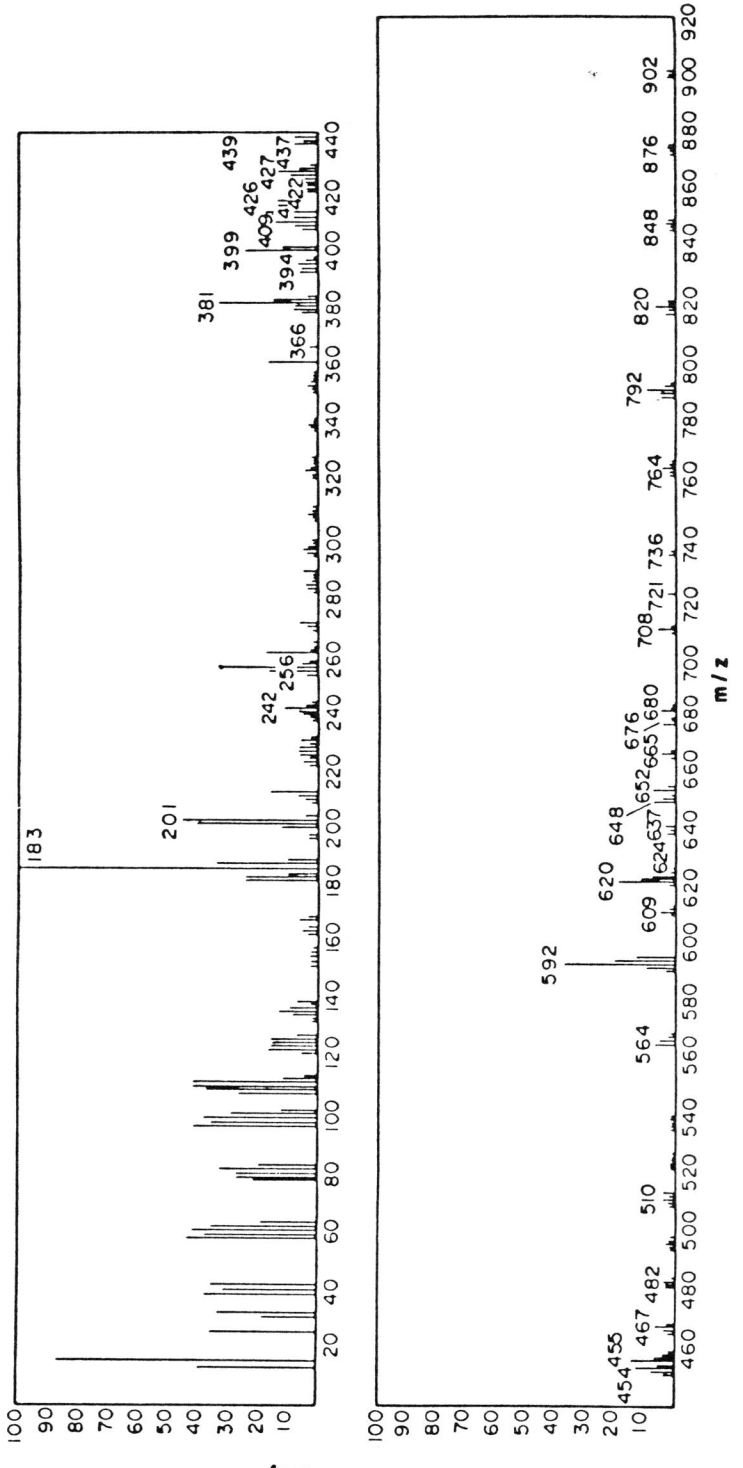

FIG. 5.3. (a) Electron impact mass spectrum of an estolide fraction from *Pinus radiata* needle wax.

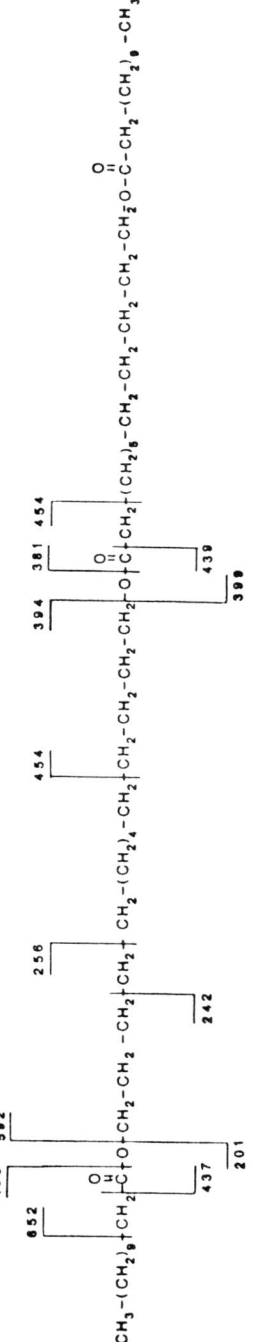

FIG. 5.3. (b) Structures and diagnostic mass fragmentation of isomeric linear estolides I and II. Reproduced with permission from Franich *et al.* (1978).

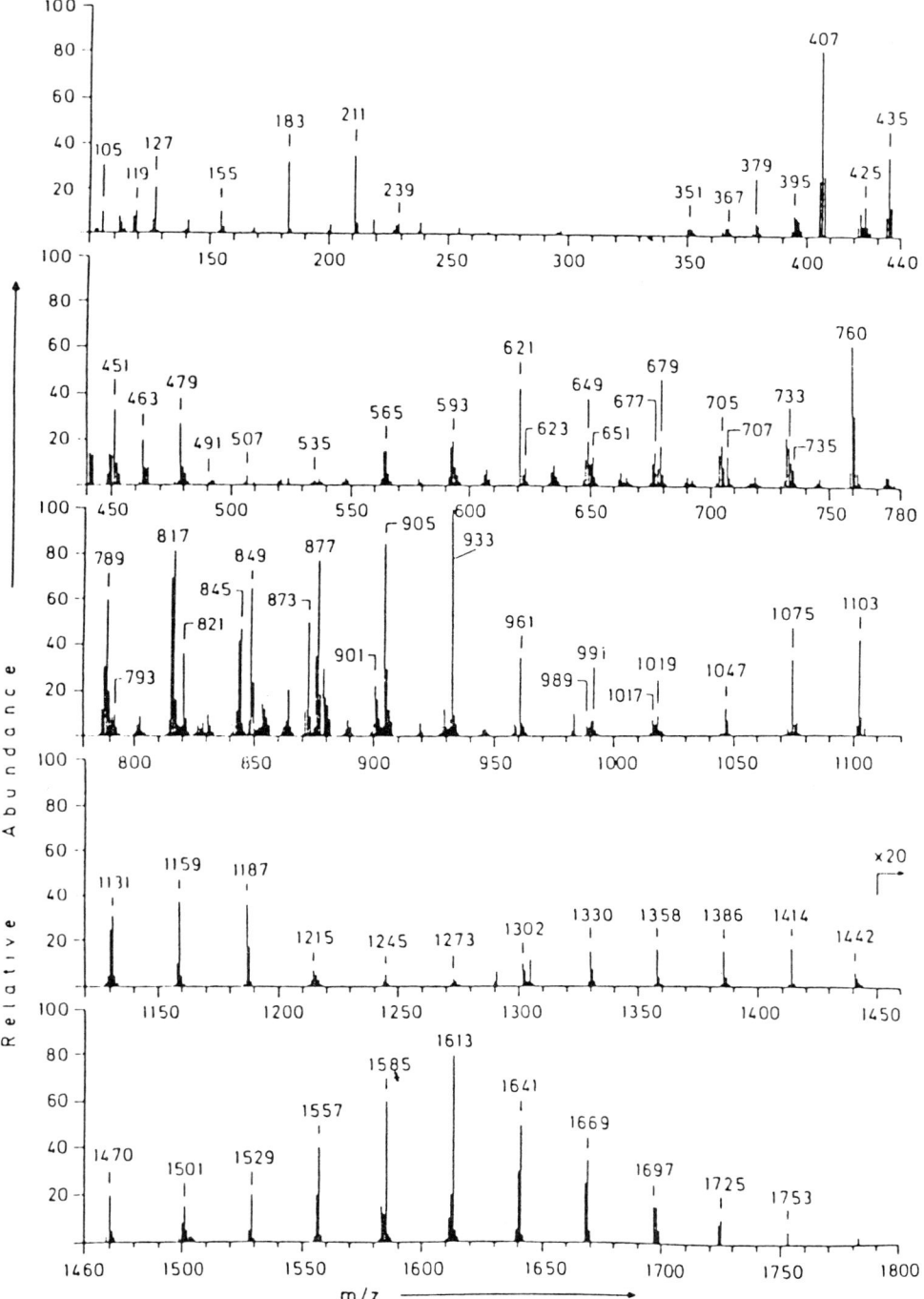

FIG. 5.4. Field desorption mass spectrum of whole epicuticular wax from *Picea abies* needles. Reproduced with permission from Schulten *et al.* (1986).

series of esters with carbon numbers between C30 and C38, and which yielded C22–C30 fatty acids on saponification. Identification of the alcohol moieties of the two series as benzyl alcohol and 2-phenylethyl alcohol was based on GC-MS analysis, when the Jojoba wax esters were found to have the same chromatographic properties and mass spectrum fragmentation patterns as synthetic C26- and C28-acyl esters of benzyl- and 2-phenylethyl alcohols (Gülz and Marner, 1986). The unusual class of esters detected by TLC as trace components of *Citrus halimii* and *Euphorbia dendroides* wax (Table 5.2), were characterised as acyl esters of benzoic acid by GC-MS (Gülz et al., 1987a,b).

4. Fatty alcohols

(a) *Primary alcohols.* For GLC analysis free fatty alcohols in wax and the alcohol moieties obtained by saponification or transesterification of wax ester fractions are converted to their acetate or fluoroacetate esters (Tulloch and Weenik, 1969); for preparation of the latter a very rapid derivatisation process involving 4-dimethylaminopyridine catalysed acylation has been described (Barta and Kömives, 1984). For analysis involving GC-MS of alcohols, the TMS ethers are the usual derivatives of choice, the fragmentation pattern readily allowing alkan-1-ols to be distinguished from alkan-2-ols. In the case of branched-chain primary alcohols, unequivocal structural assignments have been based on GC-MS analysis of the alkanes yielded by $LiAlH_4$ reduction of the mesylate derivatives (Baker and Holloway, 1975).

(b) *Secondary alcohols.* For structural characterisation of very-long-chain alcohols and diols with mid-chain hydroxyl substituents, mass fragmentation patterns (Figs 5.2(IX) and (X)) observed on GC-MS analysis of the trimethylsilyl (TMS) ether derivatives normally allows the facile and unequivocal location of in-chain substitutents (Holloway et al., 1976; Franich et al., 1979). Whilst wax esters do not normally contain secondary alcohols with mid-chain substituents, mixtures of alkan-2-ols and alkan-1-ols, which are difficult to separate by TLC or GLC as their acetate derivatives, are obtained from some wax esters. The components of such mixtures can be characterised either by GC-MS of the free alcohol (Vogt and Gülz, 1987) or by acid dichromate oxidation, yielding fatty acids and methyl ketones from primary alcohols and secondary alcohols, respectively, which may be readily separated by TLC (Lundqvist and Wettstein-Knowles, 1983; Mikkelsen, 1984).

5. Aldehydes

The composition of the fatty aldehyde fraction, which is generally of similar chain length distribution to the primary alcohols of the same wax, can be determined by direct GLC analysis (Radler and Horn, 1965). It is more often established by GLC and GC-MS analysis of the primary alcohols obtained by $NaBH_4$ reduction of the alkanal fraction, as discussed above (Kolattukudy, 1970b; Tulloch and Hoffman, 1977). Branched-chain fatty aldehydes were characterised as rare and minor components of *Brassica* by oxidation to the corresponding fatty acid and GLC analysis of their methyl ester derivatives (Netting et al., 1972).

6. Fatty acids

The unesterified branched-chain and *n*-fatty acids from wax are routinely determined, after preparation of their methyl esters, by GLC (Netting *et al.*, 1972; Tulloch, 1982b). A convenient alternative to preparation of methyl esters from free fatty acids via diazomethane of acid-catalysed esterification has recently been described. This alternative involves the direct acylation of the lithium aluminate product of fatty acids, aldehydes and esters from wax with trifluoroacetic anhydride, and GLC of the resulting trifluoroacetate esters (Barta and Kömives, 1984). When unsaturated fatty acids are also present, these may be fractionated by preliminary argentimetric-TLC, and identified by GLC on an appropriate polar liquid phase before and after catalytic hydrogenation (Avato *et al.*, 1982). The chain length distribution of the fatty acyl groups of wax esters is similarly determined by GLC of the ester fraction yielded by acid-catalysed methanolysis of the isolated wax component (Tulloch and Hoffman, 1973a). Characterisation of ω-hydroxyfatty acids found in hydrolysates of estolide fractions is discussed in Section IV.D below.

7. Terpenoids

(a) *Diterpenoids.* Two isomeric macrocyclic diterpenoid alcohols, α- and β-4,8,13-duvatriene-1,3-diol, have been identified as characteristic components of tobacco leaf wax, constituting the dominant wax fraction in immature leaves (Roberts and Rowland, 1962; Chang and Grunwald, 1976). Diterpenoid acids, of which abieta-8,11,13-trien-18-oic acid was the major component, were characterised by GC-MS analysis as substantial constituents of the acid fraction from *Pinus radiata* needle wax (Franich *et al.*, 1978).

(b) *Triterpenoids.* Pentacyclic triterpenoids are widely distributed components of plant epicuticular wax, and include triterpenoid acids, triterpenols, triterpenol esters and triterpenones (Fig. 5.5); phytosterols have only been identified in the epicuticular wax of a single species to date (Table 5.4). Analysis of the triterpenoid-containing fractions obtained by initial adsorption column chromatography of total wax extracts is often complicated by the presence of substantial amounts of aliphatic wax components of similar chromatographic properties. To facilitate the identification of individual components in the complex mixture of triterpenoids present in the epicuticular wax of *Euphorbia* spp., Gülz and his colleagues have recently developed systematic chromatographic procedures incorporating reverse-phase low pressure liquid chromatography on columns of silica gel-RP18 which allow the otherwise difficult separation of triterpenols from alkan-1-ols, as well as triterpenones, and the quantitative recovery of each fraction (Hemmers *et al.*, 1989). Reverse-phase TLC on the same chromatographic medium was a useful adjunct to the well established adsorption and argentimetric TLC procedures for qualitative analysis of cyclic triterpenoids (Gülz *et al.*, 1987a,b). Use of capillary columns with OV liquid phase allows both the direct GC-MS characterisation of triterpenols, triterpenones and sterols (Gülz *et al.*, 1987a,b; Hemmers *et al.*, 1989) as well as identification of triterpenols as their acetate ester and TMS ether derivatives (Tulloch, 1982a; Tulloch and Hoffman, 1982). The presence of the unusual 4,5-bis-methyl substitution pattern in ring A of filicanone (Fig. 5.5) has been confirmed by ^{13}C-NMR and PMR studies (Hemmers *et al.*, 1989).

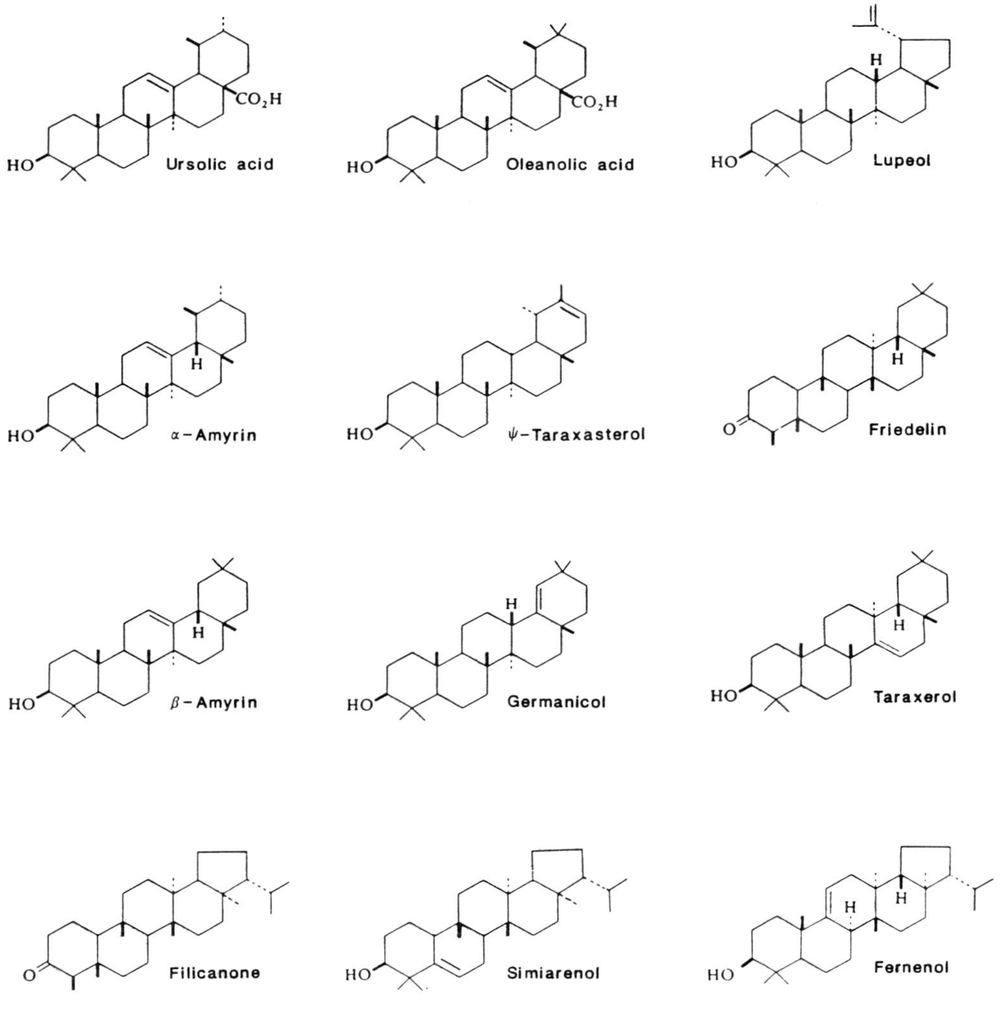

FIG. 5.5. Structures of representative epicuticular wax triterpenoids.

8. *Flavonoids*

Phenolics are widely distributed as minor components in the epicuticular leaf extracts of many species (Wollenweber and Dietz, 1981). Amongst the classes characterised are flavonoids, methylated flavonoids, flavonoid esters and flavonoid glycosides (Baker, 1982). In recent evaluations of the epicuticular flavonoid aglycones as chemotaxonomic markers for the genus *Cistus*, the identification of some 53 flavonoids by chromatographic and spectrometric procedures has been described (Vogt *et al.*, 1987a,b, 1988).

TABLE 5.4. Pentacyclic triterpenoids and sterols of epicuticular wax.

Compound	Wax source	Comments	References
Triterpenoid acids			
Ursolic acid	Apple	Major	Silva-Fernandes et al. (1964)
Oleanolic acid	Grape berry	Constitutes 70% of wax	Radler and Horn (1965)
11,12-Dehydroursolic acid lactone acetate	Euphorbia sp.	Minor	Horn and Lamberton (1964)
Triterpenols			
α-Amyrin	Euphorbia sp.	α- and/or β-amyrin frequently dominant component of triterpenoid fraction	Gülz et al. (1987a,b)
β-Amyrin			Hemmers et al. (1988, 1989)
Germanicol			Tulloch (1982a,b)
	Panicoid and eragrostoid grasses		
	Brassica napus		Holloway et al. (1977a)
Lupeol	Euphorbia sp.	Dominant triterpenol in E. characias	Hemmers et al. (1988)
α-Fernenol	Euphorbia sp.	Minor Simiarenol	Hemmers et al. (1988, 1989)
Simiarenol			
Taraxerol	Euphorbia lathyris	major, others present in traces	Hemmers et al. (1989)
Isomotiol			
ψ-Taraxasterol			
β-Glutinol	Panicum miliaceum	Traces	Tulloch (1982b)
Triterpenol acetates			
α-Amyrin acetate	Euphorbia sp.	β-Amyrin acetate dominant wax component of Tilia; minor class of Euphorbia sp. wax	Hemmers et al. (1988)
β-Amyrin acetate			Gülz et al. (1988b)
	Tilia sp.		
Lupeol acetate			
Triterpenol fatty acyl esters			
α-Amyrin esters	Euphorbia sp.	C16–C20 fatty acyl substituents. Minor terpenoid component in Euphorbia, major in grasses	Gülz et al. (1988a,b)
β-Amyrin esters	Tilia sp.		Tulloch (1982a,b)
Lupeol esters	Panicoid and eragrostoid grasses		
Ferneol esters			

TABLE 5.4. continued

Compound	Wax source	Comments	References
Germanicol esters			
Isoarborinol esters	Panicoid grasses		Tulloch (1982a,b)
Arborinol esters			
ψ-Taraxasterol- esters			
Triterpenones			
β-Amyrinone	Euphorbia sp.	Common in Euphorbia wax	Hemmers et al. (1988)
Lupenone			Gülz et al. (1988a)
Oleanone			
Taraxerone	Euphorbia lathyris	Taraxerone major, α-fernenone substantial, others in traces only	Hemmers et al. (1989)
Isomotione			
α-Fernenone			
β-Fernenone			
Simiarenone			
Filicanone			
Friedelanone	Citrus halimii	Substantial component of total wax	Gülz et al. (1987b)
Sterols			
Cholesterol			
Campesterol	Citrus halimii	Present in trace amounts	Gülz et al. (1987b)
Stigmasterol			
β-Sitosterol			

D. Analysis of Whole Epicuticular Wax

The methods described above allow the rigorous identification and quantification of each individual lipid component in fractions isolated by column and thin layer chromatography. Frequently, however, as in taxonomic surveys, assessment of environmental influences or screening of mutant varieties, there is a requirement for a straightforward procedure which involves minimal sample preparation and which yields the compositional fingerprint of whole wax samples. Until recently this has usually been achieved by temperature-programmed GLC analysis on Dexsil-300 liquid phase of whole wax extracts after acetylation. As shown in Fig. 5.6, this allows the hydrocarbon, free alcohol and wax ester components to be identified, and by use of appropriate internal standardisation procedures, quantified (Tulloch, 1973). This procedure can be extended to establish the chain length distribution of the acyl and alkyl groups in the total wax ester fraction by redetermination of the GLC pattern after acid methanolysis of whole wax, when increases in the fatty acid methyl ester and fatty alcohol acetate peaks allow the chain length distribution of the fatty acyl and fatty alkyl moieties of the original ester fraction to be deduced (Tulloch, 1976a).

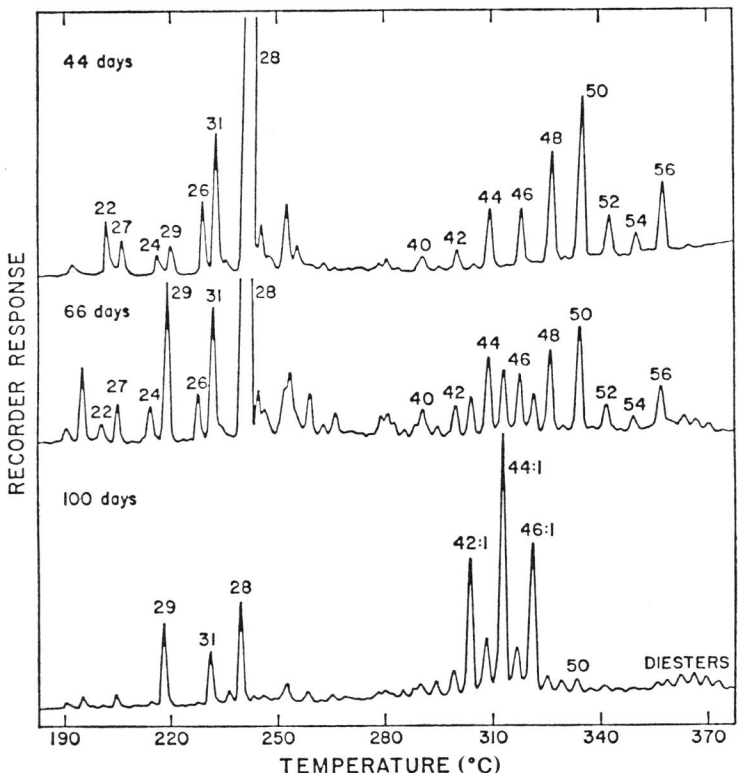

FIG. 5.6. Gas–liquid chromatographic analysis of leaf wax (after acetylation) from *Triticum aestivum* var. Selkirk collected at 44, 66 and 100 days after germination. Peaks with even numbers from 22 to 28 are free alcohol acetates, peaks with odd numbers from 27–31 are hydrocarbons, peaks with even numbers from 40 to 56 are long-chain esters and peaks with numbers 42:1, 44:1 and 46:1 are long-chain esters containing *trans* 2,3-unsaturation. Reproduced with permission from Tulloch, A. P. (1973).

The application of direct probe mass spectrometry using the 'softer' sample ionisation processes of FD and FI has recently enabled Schulten and his co-workers to develop a truly 'one-step' characterisation of whole wax extracts. These procedures have the further advantage of yielding information for very high mass ($M_r > 1100$) wax components, which are insufficiently volatile for direct GLC analysis. Because the mass spectra yielded by FD and FI ionisation of the major aliphatic components from epicuticular wax contain predominantly molecular ion species (Schulten *et al.*, 1987b), the mass spectra obtained from whole wax extracts contain relatively few ions, and these are highly diagnostic of the wax composition (Schulten *et al.*, 1986, 1987a). Thus in the FD mass spectrum of whole cuticular wax from *Picea abies* needles, a 'molecular ion fingerprint' up to m/z 1800 was obtained containing seven homologous series of ions (Fig. 5.4). The series between m/z 351 and m/z 491 was assigned to species arising by thermal elimination of water from protonated primary alcohols, of which the major components were octacosanol (represented by m/z 407) and triacontanol (represented by m/z 435). A second series of ions in the mass range m/z 536–929 was identified as containing molecular ions derived from an homologous series of alkyl esters containing up to 64 carbon atoms of which those of C42 and C56 chain length (m/z 621 and m/z

817, respectively) predominated. The five further homologous series of ions in the mass range m/z 623–1753 were assigned to homologous series of linear and cyclic estolides, as discussed in Section III.C.3 above. This technique has so far been used to investigate the effect of ageing and the impact of environmental factors on the composition of whole wax from needles of several coniferous species (Schulten et al., 1986, 1987a; Simmleit and Schulten, 1987), in which components containing 'in-chain' oxygen substituents are of relatively minor significance. Although more complex FD fragmentation patterns may be anticipated from angiosperm waxes containing relatively high proportions of secondary alcohols and ketones (Schulten et al., 1987b), it seems likely that the 'fingerprint' fragmentation patterns obtained in these cases will also be of value in comparative studies of whole wax composition.

IV. ANALYSIS OF CUTIN

A. Isolation

Sheets of cutin can be prepared by exhaustive extraction of isolated cuticular membrane preparations with chloroform and methanol (Walton and Kolattukudy, 1972; Holloway, 1984a), a recently described modification of this procedure allowing its application to very fragile cuticles (Riederer and Schönherr, 1988b). Prior to depolymerisation, the dried preparations are usually ground in a hammer mill to a fine powder. Frequently, however, when this approach is not practicable, as in many metabolic studies, the insoluble residue remaining after whole tissue has been physically disrupted and the epicuticular and intracellular lipid thoroughly extracted with appropriate solvent is used directly as a source of cutin (Kolattukudy and Walton, 1972b; Bowen and Walton, 1988). Several pre-treatments of cutin preparations prior to depolymerisation have been employed to facilitate identification of functional groups in the native polymer. Thus sodium borodeuteride and sodium borotritide reduction and semicarbazone and oxime derivatisation have been used in the characterisation of free carbonyl functionalities (Kolattukudy, 1972, 1974; Espelie et al., 1983), whilst expoxy acids in isolated cuticles can be converted quantitatively to the corresponding chlorohydrin with HCl–dioxane (Riederer and Schönherr, 1988a).

B. Depolymerisation

The ester bonds in cutin can be cleaved by alkaline hydrolysis to yield hydroxyfatty acids, by transesterification with methanol containing hydrochloric acid, sodium methoxide or boron trifluoride to yield hydroxyfatty acid methyl esters, or by exhaustive hydrogenolysis with $LiAlH_4$ to yield polyhydric fatty alcohols. The methodology and relative merits of each procedure have been critically reviewed (Holloway, 1984a). Methanolytic procedures have a number of advantages over hydrolytic methods, including direct production of monomer methyl esters and quantitative conversion of native expoxides to the corresponding methoxyhydrin, and where compositional analyses are to be based on GLC alone, over hydrogenolysis. When GC-MS facilities are available, either methanolysis or reductive cleavage with $LiAlD_4$ allows unambiguous identification and quantitative assessment of the original cutin components.

C. Chromatography of Monomer Derivatives

1. Thin layer chromatography

With the reduced sample requirement for structural characterisation by modern spectrometric methods, preparative TLC has now largely replaced column chromatography for the fractionation of cutin depolymerisation products. The products obtained by hydrolytic, methanolytic and hydrogenolytic depolymerisation of cutin can be largely resolved into classes by silica gel-G TLC, and comprehensive tables of R_f values for each class derivative, together with details of the TLC methodologies, have been presented (Holloway, 1984a). A single monomer class isolated or identified by this procedure may contain several members of an homologous series, saturated and unsaturated analogues or positional and stereoisomers. Separation of analogues of cutin monomers containing one or two double bonds from one another and from their saturated counterparts can be achieved by a single argentimetric TLC step (Walton and Kolattukudy, 1972), whilst diastereoisomeric *vic*-diols are readily separated by chromatography on layers of silica gel impregnated with boric acid (Holloway, 1972a).

2. Gas liquid chromatography

The analysis by GLC of either a monomer class isolated by thin layer chromatography or the total cutin depolymerisation product requires the prior derivatisation of any remaining free carboxyl or hydroxyl groups. Free acids from hydrolytic depolymerisation procedures are generally converted to their methyl esters by acid-catalysed esterification, since diazomethane methylation can lead to formation of methoxy methyl ester artifacts (Holloway and Deas, 1971a), although TMS esters have also been used (Holloway and Deas, 1971b). Acetylation, or trifluoroacetylation, of hydroxyl groups used in earlier compositional analysis (Baker and Holloway, 1970), has now largely been superseded by trimethylsilylation, due to the ease of derivatisation, the volatility of the resulting TMS ethers, and, in GC-MS analysis, the readily interpretable mass spectra data afforded by these derivatves. The TMS ethers of the hydroxyfatty acid methyl esters and of the polyhydric fatty alcohols released from cutin and suberin have been routinely analysed by packed-column GLC employing silicone-type liquid phases of differing polarity, and comprehensive tables of their GLC properties (Holloway, 1984a) and representative gas chromatograms (Holloway, 1982a) presented. Whilst good separation between major monomer homologues and classes is generally obtained, no single phase satisfactorily resolves all components. Positional isomers do not separate on packed-column GLC on any of these phases, so their identification and quantification requires GC-MS analysis.

Details of rather few analyses involving capillary GLC of cutin monomer derivatives have been published to date. Earlier reports indicated that a 20 m OV-101 column offered little overall improvement, co-elutions being comparable with packed SE-30 columns (Cardoso *et al.*, 1977; Holloway, 1984a). We have found that a 25 m capillary coated with OV-1 gave clear base-line separation of the TMS ether methyl ester derivatives of 7- and 8-hydroxy-hexadecanedioic acid and 9- and 10,16-dihydroxyhexadecanoic acid (Bowen and Walton, 1988), and this system also clearly resolved each of the unsaturated and saturated analogues derived from the major cutin acids on hydrogenolytic depolymerisation of apple fruit cutin (Fig. 5.7).

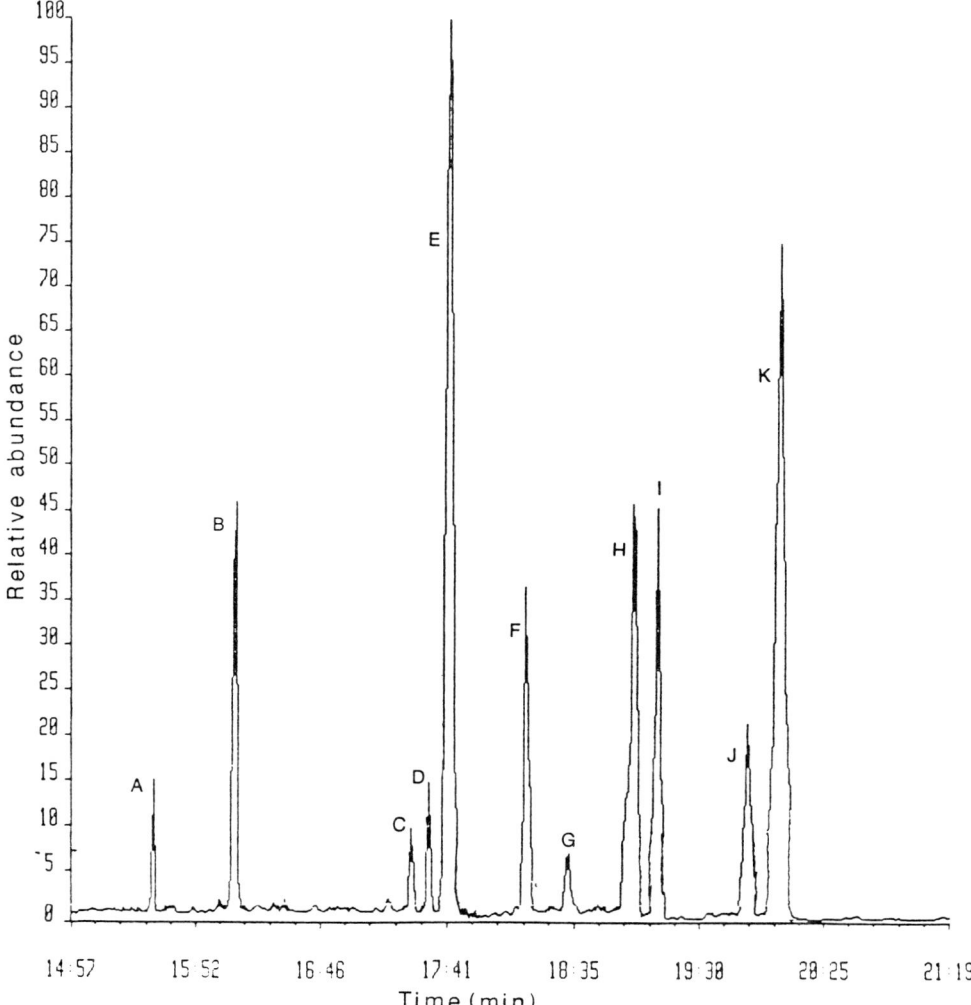

FIG. 5.7. Partial capillary gas chromatogram of the trimethylsilyl ether derivatives of the lithium aluminium hydride depolymerisation products from Golden Delicious apple fruit cutin. Identification of alcohols: A, nonadecanol; B, hexadecanediol; C, octadecadienediol; D, octadecenediol; E, hexadecanetriol; H, octadecenetriol; I, octadecanetriol; J, octadecenetetraol; K, octadecanetetraol. Column conditions: 25 m × 0.22 mm coated with OV-1 (0.1 μm), carrier (He) linear flow rate 60 cm s^{-1}, splitless injection at 100°C, programmed to 300°C at 10°C min^{-1}.

The relative composition and absolute amount of cutin has usually been based directly on the GLC data. However, a recent study has indicated that selective losses during the post-depolymerisation work-up procedure, together with differential sensitivity of the flame ionisation detection system, can account for underestimation of the

9,10,18-trihydroxyoctadecanoate content by 27% and of the 9,10-epoxy-18-hydroxyoctadecanoate fraction by 16%, and presents correction factors to allow quantitative gas chromatographic analysis of the major cutin acids (Riederer and Schönherr, 1986b).

D. Identification of Cutin Monomers

1. Combined GC-MS analysis

Since the initial application of GC-MS in the analysis of apple cutin (Eglinton and Hunnemann, 1968; Eglinton *et al.*, 1968), the technique has become established as the method of choice both for identification of monomers and quantification of the positional isomer composition. Interpretation of the EI mass spectra obtained from the TMS ether derivatives of the polyhydric fatty alcohols produced by hydrogenolytic and deuterolytic cleavage of cutin has been discussed previously (Walton and Kolattukudy, 1972; Kolattukudy and Walton, 1972a; Kolattukudy *et al.*, 1973; Kolattukudy, 1974). The characterisation of the methyl ester-, TMS ether-derivatives of cutin acids by mass spectrometry has been reviewed by Holloway (1982b).

Until very recently, all reported GC-MS analyses of cutin composition have employed EI ionisation. The first studies using chemical ionisation (CI) of cutin depolymerisation products, in conjunction with capillary GLC, have now been described, and suggest that the softer ionisation process can offer significant advantages in the identification of monomer derivatives. The CI mass spectra contain not only prominent α-cleavage ions which allow the location of mid-chain substituents and assessment of the relative proportion of positional isomers, but also contain intense ions which allow the facile assignment of relative molecular mass to the molecule, in contrast to EI mass spectra in which molecular and other ions of high mass that provide direct molecular weight information are frequently absent or of such low abundance that their identification can be difficult. Thus in the NH_3 CI mass spectra of the fatty acid methyl ester TMS ethers derived from the C16 cutin acids of *Pisum sativum* cutin, a protonated molecular ion ($[M + 1]^+$) and its ammonia adduct ($[M + 18]^+$) were prominent, and the ion formed by elimination of the in-chain trimethylsilanol group ($[M + 1 - 90]^+$) was the base peak in the spectra (Bowen and Walton, 1988), whilst, in the isobutane CI mass spectrum of the chlorohydrin methyl ester TMS ethers derived from 9,10-epoxy-18-hydroxyoctadecanoic acid from *Ficus elastica* cutin, the protonated molecular ion was the base peak (Riederer and Schönherr, 1988a). The spectra generated by both reagent gases also contained prominent fragment ions formed by cleavage of bonds immediately adjacent to the substituted mid-chain carbon atom, and whilst these α-cleavage ions were not of such high relative abundance as in the corresponding electron impact mass spectra, they readily identified the location of the mid-chain substituents in the positional isomers present, and since the pattern of α-cleavage ions in the CI mass spectrum was virtually identical to that in the corresponding EI mass spectrum, provided an equally reliable basis for determination of the relative proportions of positional isomers (Holloway and Deas, 1971b).

We have now extended the technique to TMS ether derivatives of the hydrogenolysis products from apple fruit cutin (T. J. Walton, unpubl. res.), and the NH_3 CI mass spectra of several of the characteristic components resolved by capillary GC (Fig. 5.7) are shown in Fig. 5.8. The provisional identification of hexadecanediol (Fig. 5.7, component B) was confirmed by the presence in the mass spectrum (Fig. 5.8(A)) of the intense ion at m/z 403, assigned to the protonated molecular ion $[M + H]^+$. This was the strongest peak in the spectrum above mass 200, and together with its ammonium adduct at m/z 420 $[M + NH_4]^+$, indicated an M_r of 402 as expected for the bis-TMS ether of hexadecanediol. Two minor components in the total ion chromatogram (Fig. 5.7, components C and D) were identified by their NH_3 CI mass spectra (Figs 5.8(B) and (C)) as an octadecadienediol and octadecenediol, respectively. In the case of component C in Fig. 5.7, the ion at m/z 427, assigned to the protonated molecular ion $[M + H]^+$, was the base peak in the spectrum and was accompanied by an intense ion at m/z 444 assigned to the ammonium adduct of the molecular ion $[M + NH_4]^+$ (Fig. 5.8(B)), indicating a relative molecular mass of 426, consistent with the proposed octadecadienediol structure. Similarly in the spectrum of component D in Fig. 5.7, the intense ions at m/z 429 ($[M + H]^+$) and m/z 446 ($[M + NH_4]^+$) (Fig. 5.8(C)) support the identification of octadecenediol. The ease with which molecular weights can be assigned to these cutin acids lacking mid-chain substituents indicates that CI mass spectrometry will be of particular value in analysis of the depolymerisation products from the aliphatic domains of suberin, which also generally lack in-chain functionalities. In the CI mass spectrum (Fig. 5.8(D)) of the peak provisionally identified as the hexadecanetriol fraction in the total ion chromatogram (Fig. 5.7(E)), the intense ions at m/z 491 and m/z 508 were assigned to the protonated molecular ion ($[M + H]^+$) and the ammonium adduct ion ($[M + NH_4]^+$), and together with the base peak at m/z 401, arising by elimination of trimethylsilanol from the protonated molecular ion ($[M + 1 - (CH_3)_3SiOH]^+$) indicated a relative molecular mass of 490, as expected for a hexadecanetriol. The pattern in the relative intensities of the α-cleavage ions at m/z 275 and 303 (hexadecane-1,7,16-triol) and m/z 289 and 317 (hexadecane-1,8,16-triol) was not significantly different from that in the corresponding EI mass spectrum (data not shown). The CI mass spectra of the octadecenetriol and octadecanetriol fractions, derived from 9,10-epoxy-18-hydroxy-C18 acids in cutin (components H and I, Fig. 5.7), showed similar patterns to those of the hexadecanetriol, in that the respective protonated molecular ion (m/z 517 and m/z 519) and ammonium adduct ions (m/z 534 and m/z 536) were intense (Figs 5.8(E) and (F)), and the ions formed by elimination of trimethylsilanol were the respective base peaks (m/z 427 and m/z 429). The pattern of α-cleavage ions (m/z 301, 303 and 317, octadecenetriol (Fig. 5.8(E); m/z 303 and 317, octadecanetriol, Fig. 5.8(F)) was very similar to that observed in the EI mass spectra of these derivatives (Walton and Kolattukudy, 1972), and allowed the location of the mid-chain substituent to be assigned, confirming the structures as octadecene-1,9,18-triol and octadecane-1,9,18-triol, respectively. In the CI mass spectra (Figs 5.8(G) and (H)) of the components provisionally identified as an octadecenetetraol and octadecanetetraol in the total ion chromatogram (Fig. 5.7, components J and K, respectively), the protonated molecular ion and accompanying ammonium adduct ion (m/z 605 and m/z 622 octadecenetetraol, m/z 607 and m/z 624 octadecanetetraol) were prominent and allowed ready assignment of relative molecular mass. In contrast to the C16 and C18 derivatives containing a single mid-chain TMS group, the ions formed by elimination of trimethylsilanol from the protonated molecular ion of the C18 tetraol TMS ethers, whilst prominent (m/z 515

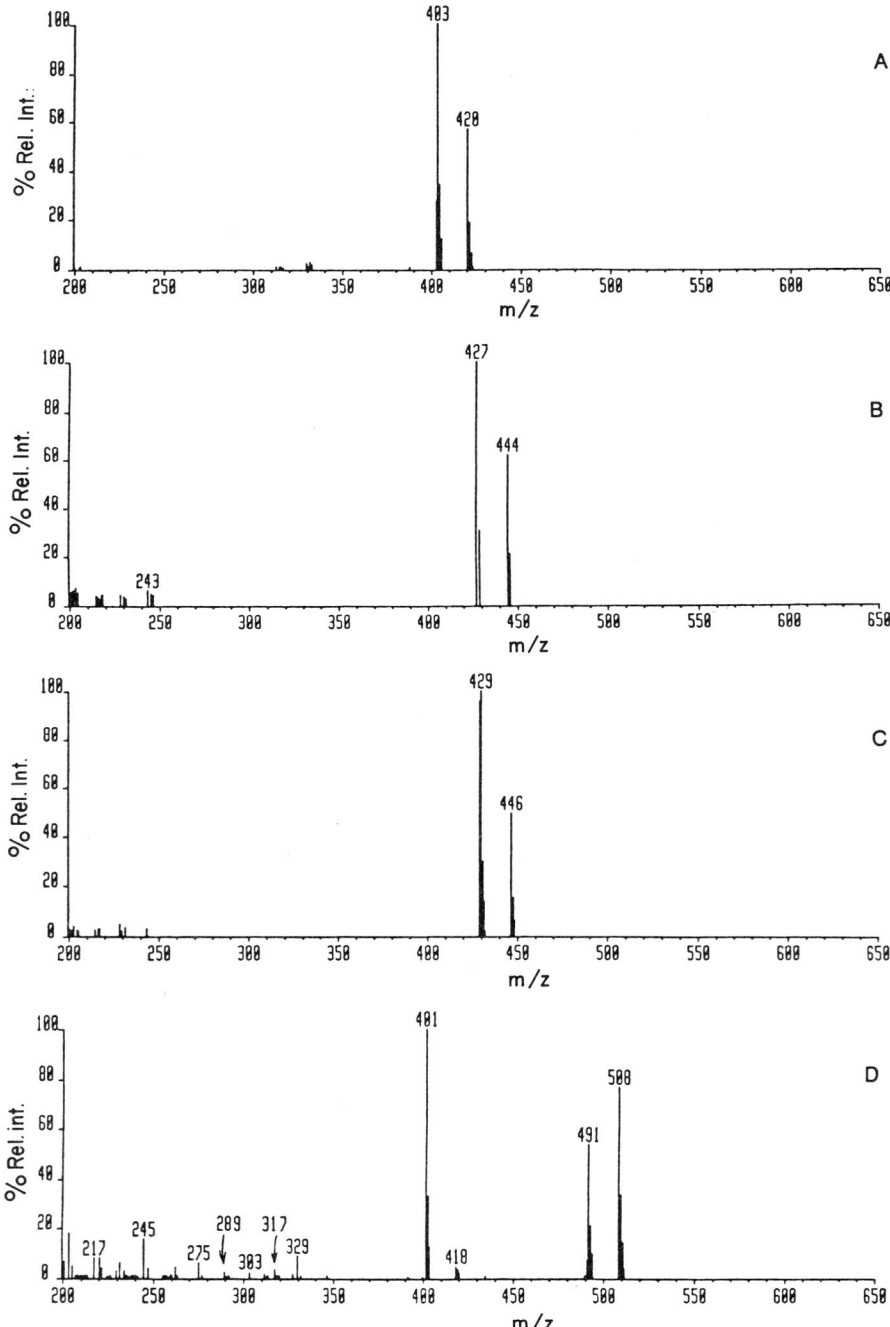

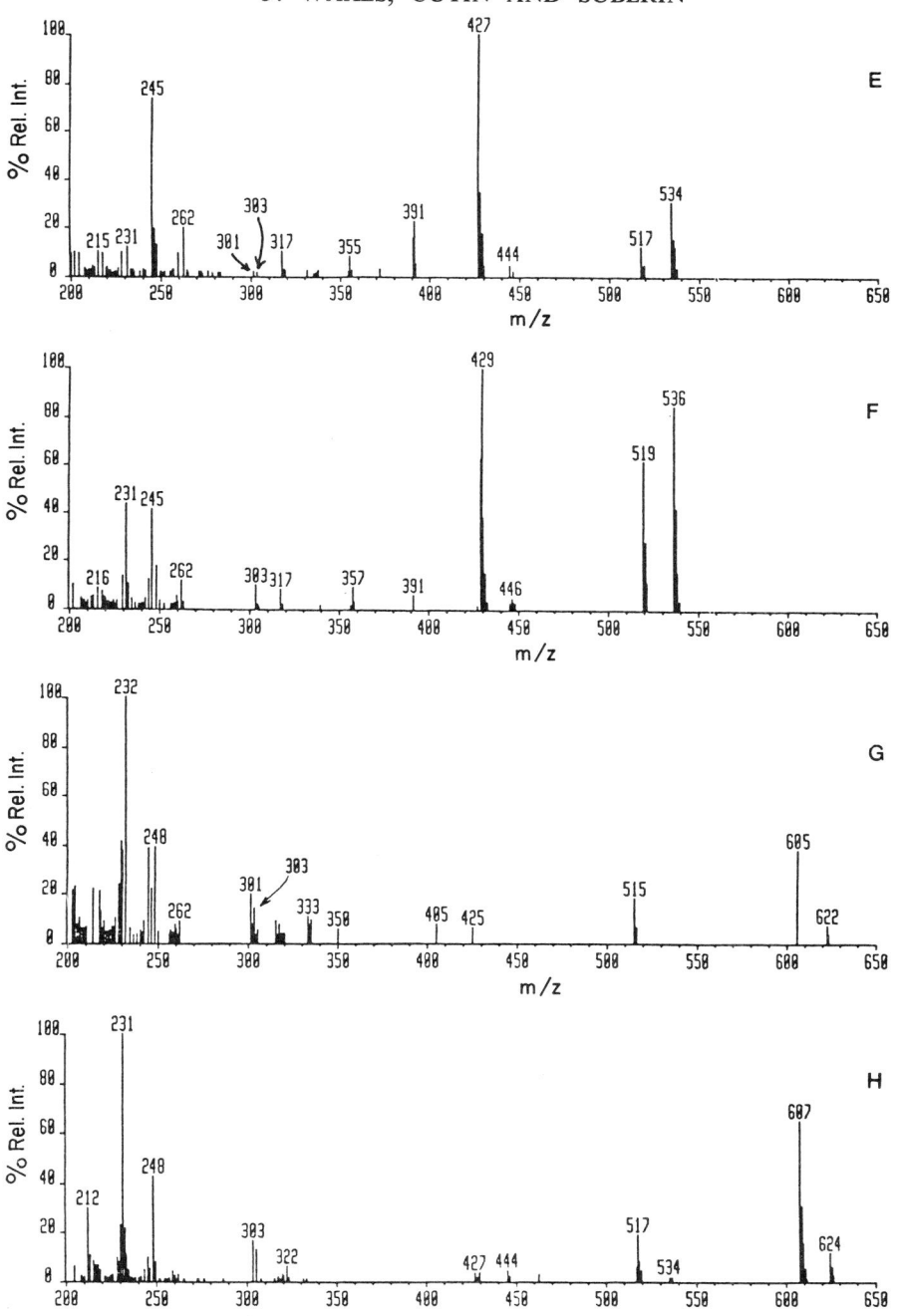

FIG. 5.8. NH$_3$ chemical ionisation mass spectra of the TMS ether derivatives of the lithium aluminium hydride depolymerisation products from Golden Delicious apple fruit cutin. Individual components were resolved by capillary GC (see Fig. 5.7) and mass spectra of the GC eluent determined by repetitive scanning of the 0–650 mass range (1.2 s cycle^{-1}). Identification of alcohols: A, hexadecanediol; B, octadecadienediol; C, octadecenediol; D, hexadecanetriol; E, octadecenetriol; F, octadecanetriol; G, octadecenetetraol; H, octadecanetetraol.

octadecenetetraol, m/z 517 octadecanetetraol) were of substantially lower relative intensity than the corresponding protonated molecular ion, and were of similar abundance to the major α-cleavage ions formed by cleavage of the central carbon–carbon bond between the *vic*-TMS groups (m/z 301 and 303 octadecenetetraol, m/z 303 octadecanetetraol). These fragmentation patterns allowed the identification of the C18 tetraols as octadecene-1,9,10,18-tetraol and octadecane-1,9,10,18-tetraol, derived from 9,10,18-trihydroxyoctadecenoic acid and 9,10,18-trihydroxyoctadecanoic acid, respectively, in cutin.

2. *Composition of cutin*

Compositional analyses employing GLC or GC-MS analysis have established that some 10 classes of substituted fatty acids occur in cutin. Despite their diverse pattern of substitution (Fig. 5.9), aliphatic monomers with carbon chains containing other than 16 or 18 carbon atoms are very rare, and thus the cutin acids can be regarded as belonging either to a C16 family or a C18 family. The depolymerisation products of many cutins are composed almost exclusively of derivatives of the C16 family of monomer acids, in which 10,16- and/or 9,16-dihydroxyhexadecanoic acids are the major components, accompanied in some species by smaller amounts of their positional isomers containing the mid-chain hydroxyl group at C-7 or C-8; in addition to this dihydroxyhexadecanoate fraction, much lower levels of monohydroxyhexadecane-1,16-dioic acid, 16-hydroxyhexadecanoic acid, hexadecane-1,16-dioic acid together with C14–C22 fatty acids are usually obtained. Relatively few cutins composed principally of C18 family monomers have been described, the majority of cutins containing significant amounts of both C16 and C18 family monomers, of which 9,10-epoxy-18-hydroxyoctadecanoic acid and 9,10,18-trihydroxyoctadecanoic acid are generally the most abundant, although substantial amounts of unsaturated C18 analogues are frequently encountered (Fig. 5.9). Fuller accounts of monomer structure, composition and distribution are found in the reviews of Kolattukudy and Espelie (1985) and Holloway (1982b, 1984a) and references therein.

E. Molecular Architecture of Cutin

1. *Polymer structure*

Understanding of the types of covalent linkages in cutin has been based largely on consideration of the chemical reactivity of the polymer. Depolymerisation with reagents that cleave ester linkages typically and consistently leaves at least 20–30% of the original cutin preparation as a solid residue. Further soluble products can be released by treatment of this insoluble material with sodium iodide and with hydriodic acid, suggesting, respectively, the presence of hydroperoxide-bound (10–20%) and ether-linked (1–2%) cutin monomers (Crisp, 1965; Croteau and Fagerson, 1972). However, even after exhaustive and repeated treatment of cutin with these reagents, a small refractory residue is generally obtained. A possible explanation for this observation is provided by the identification by pyrolysis GC-MS of a novel, highly aliphatic polymer in the cuticular membrane (Nip et al., 1986a,b; Tegelaar et al., 1989). This polymer was readily distinguished from polyester cutin by GC-MS analysis of the pyrolysis products, and was resistant to alkaline hydrolysis. It was present in *Agave americana* and *Clivia minata* leaf cuticle, which both contain substantial amounts of C18 family acids, and in *Beta vulgaris* leaf cutin preparations, which release no significant quantity of monomers on alkaline hydrolysis. It was not detected in *Lycopersicon esculentum* fruit and *Citrus limon* leaf, which possess polyester cutins in which C18 family acids are minor components. Ether linkages do not appear to be involved in cross-linking this polymer, and a polyethylene-type structure was suggested by ^{13}C-NMR studies.

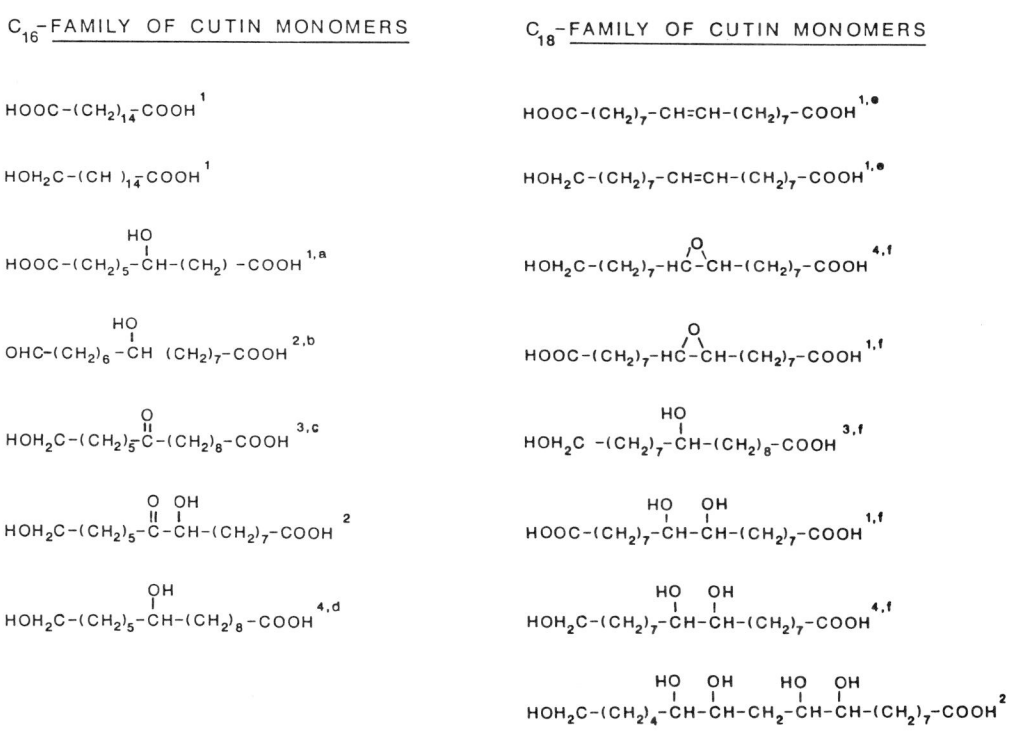

FIG. 5.9. Structures of cutin monomer acids. Notes: [1] common and minor; [2] rare; [3] uncommon; [4] common and abundant.
[a] 8-hydroxy isomer also occurs.
[b] 16-oxo-10-hydroxy isomer also occurs.
[c] 7-oxo-16-hydroxy, 8-oxo-16-hydroxy and 9-oxo-16-hydroxy isomers also occur.
[d] 7,16-dihydroxy, 8,16-dihydroxy and 9,16-dihydroxy isomer also occur.
[e] Δ9,12 dienoic and Δ9,12,15 trienoic analogues also occur.
[f] Δ12 monoenoic and Δ12,15 dienoic analogues also occur.
From Holloway and Deas (1971b), Kolattukudy (1974), Deas et al. (1974), Espelie et al. (1983), Holloway (1984a).

Knowledge of the intermolecular cross-linking between cutin monomers has been derived from analysis of the abundance of free primary and free secondary mid-chain alcohol groups, as well as unesterified carboxyl moieties, in several cutins composed primarily of the C16 family of monomer acids. Free hydroxyl groups have been 'labelled' by oxidation of cutin with CrO_3-pyridine complex prior to methanolytic depolymerisation (Deas and Holloway, 1977) and by mesylation followed by deuterolytic cleavage of the polymer (Kolattukudy, 1977). These approaches demonstrated that whilst only about half the mid-chain hydroxyl groups in the polymer are involved in cross-linking, most of the primary alcohol groups are involved in ester linkages, and it has been subsequently established that there are very few unesterified carboxyl groups (Kolattukudy, 1980). These studies, together with stoichiometry of the depolymerisation products, have enabled a structural model for cutins composed of C16 family monomers to be proposed (Kolattukudy, 1980). In this model, the minor phenolic acid component is shown esterified to the polymer matrix, although the evidence for this is not unequivocal, and the absolute configuration of the mid-chain hydroxyl group, established as L (Espelie and Kolattukudy, 1978), is not included in this proposal.

Physical studies of this biopolymer have until recently been restricted to X-ray diffraction, which suggested an amorphous structure and implied a single repeating unit in the polymer (Wilson and Sterling, 1976; Kolattukudy, 1980). Solid state ^{13}C-NMR studies of the polyester cutin from lime fruit have now provided the first direct spectroscopic data on the structure of the intact biopolymer, which, whilst generally consistent with the previously proposed model, suggested the presence of distinct polymer domains and indicated extensive cross-linking through secondary alcohol groups (Zlotnik-Mazori and Stark, 1988). However, since chemical analysis of lemon fruit cutin, which is of similar composition to that of lime (Espelie *et al.*, 1983), has indicated that less than 20% of secondary hydroxyl groups are esterified (Deas and Holloway, 1977), it appears essential to determine the esterified secondary alcohol content of the same cutin sample by both high resolution NMR analysis and by a chemical labelling method in order to establish the validity of the conclusions derived from the spectroscopic and chemical techniques. One approach to establish whether more than one type of polymeric repeat unit is present in cutin involves development of selective hydrolysis of the polymer with enzymes such as pancreatic lipase (Brown and Kolattukudy, 1978a,b) or fungal cutinase (Kolattukudy, 1984b), which preferentially hydrolyse primary alcohol esters, in combination with mass spectrometry of the high molecular weight, estolide-like oligomers released.

2. *Changes in cutin content and structure during ontogeny*

Generally cutin content of a cuticular membrane increases as the plant develops. In *Vicia faba* leaves, biosynthetic tracer experiments indicated that cutin monomer formation and incorporation of the monomers into the polymer essentially ceased in fully expanded tissue (Kolattukudy, 1970c; Kolattukudy and Walton, 1972b; Croteau and Kolattukudy, 1973), consistent with the view that cutin deposition is associated with active growth of tissue (Martin and Juniper, 1970). However, in *Clivia minata*, whose leaves grow at their base, cutin content increased after cell expansion had ceased (Riederer and Schönherr, 1988b).

During the normal ontogeny of many leaves and fruits, only minor changes in cutin composition have been observed (Deas and Holloway, 1977; Holloway, 1982b), and similarly no significant differences could be detected by either mass spectrometry or radiolabelling methods in the polymer formed during GA_3-induced extension of *Pisum sativum* stems (Bowen and Walton, 1988). On the other hand, embryonic apical buds of *Vicia faba* contain substantial amounts of 16-oxo-9- and 16-oxo-10-hydroxyhexadecanoic acids, monomers which cannot be detected in fully expanded tissue (Kolattukudy, 1974), and cutin of young rapidly expanding apple fruit contains significantly larger proportions of dienoic and trienoic C18 monomers than mature fruit (Kolattukudy *et al.*, 1973). Also in *Clivia minata*, there was a dramatic decrease in the number of monomer classes present with increasing age as the proportion of C18 monomers increased substantially, suggesting distinct phases of cutin synthesis, one associated with growing leaves and the other in leaves where expansion had ceased (Riederer and Schönherr, 1988b).

V. ANALYSIS OF THE SUBERIN COMPLEX

A. Isolation

Suberin of relatively high purity is available only from organs of those species which develop a thick superficial cork layer that can readily be stripped from underlying tissue (Holloway, 1984a). For isolation from other species, careful physical separation of the tissue of interest can be followed by treatment with ammonium oxalate/oxalic acid and by enzymic digestion to reduce carbohydrate polymer content, yielding a suberin-enriched preparation (Kolattukudy, 1978; Espelie *et al.*, 1979; Holloway, 1982c), but where physical separation of the suberised layer is impracticable, whole tissue has been frozen, lyophilised and used directly as a source for suberin (Pozuelo *et al.*, 1984; Sijmons *et al.*, 1985).

B. Ultrastructure

Suberised tissue from diverse anatomical sites, including tree barks, tuber periderm, root hypodermis and bundle-sheaths of grass, has a characteristic lamellar structure in electron micrographs (Kolattukudy, 1984a; Kolattukudy and Espelie, 1985). It is laid down between the plasmalemma and the cell wall, although the suberised Casparian band of root hypodermis is deposited at an early stage of ontogeny as an integral part of the radial and transverse cell walls (Clarkson and Robards, 1975). There is some controversy over the ultrastructure of the suberin complex and in the relationship of the chemical constituents to the laminated appearance typically seen in electron micrographs. Many workers believe that the dark, electron-opaque bands represent the polymeric components, whilst the electron-transparent regions represent the wax; specific inhibition of wax biosynthesis prevented development of the light bands in periderm of wound-healing potato tuber slices (Soliday *et al.*, 1979), although extraction of the wax does not change the appearance of the suberin complex.

C. Extraction and Analysis of Soluble Lipid

Thorough solvent extraction of the suberin preparation yields the soluble lipids, which can account for up to 40% of the sample dry weight, and the insoluble polymer matrix for subsequent analysis. Immersion of intact underground storage organs was used in the analysis of periderm suberin soluble lipids, but even relatively short exposures to solvent were found to extract substantial amounts of internal lipids, which must then be removed from the periderm wax chromatographically (Espelie et al., 1980b). To date the soluble lipids from tuber periderm, stem bark, seed tissue and root suberin have been characterised in relatively few species (Table 5.5). The major components have

TABLE 5.5. Soluble lipids isolated from suberin complexes.

Tissue	Component(s)	Comments	References
Underground storage organs			
Dioscorea deltiodea	C18–C25 alkane	C29, C30 and C31 chain lengths predominate	Hardman and Brain (1971)
Pastinaca sativa	C12–C34 fatty acids C12–C30 fatty alcohols	Major components, even chain lengths predominate	Espelie et al. (1980b)
Dauca carota			Espelie et al. (1980b)
Brassica napobrassica	C17–C35 alkanes	Even chain length alkanes substantial, several chain lengths predominate	Espelie et al. (1980b)
Beta vulgaris			Espelie et al. (1980b)
Ipomea batatas	Wax esters, ω-hydroxyfatty acids, dicarboxylic acids	Minor components	Espelie et al. (1980b)
Solanum tuberosum			Brieskorn and Binneman (1972), Soliday et al. (1979)
Barks			
Pinus contorta	C20–C22 fatty acids 13-epimanool	Major component Abundant	Rowe and Scroggins (1964)
Ailanthus altissima	C22 fatty acid	Major component	Chiarlo and Tacchino (1965)
Abies concolor	C22 fatty acid, C16–C22 hydroxyfatty acid	Hydroxy-C16 and C20 acids dominant. Wax saponified before analysis	Kurth (1967)
Tsuga mertensiana	C24 fatty acid	Major component in saponified wax isolates	Kurth (1967)
Aleurites montana	Friedelin, β-sitosterol, aleuritolic acid, betulinic acid	Aleuritolic acid present as acetate ester	Misra and Khastgir (1970)
Malus pumila	Friedelin C18–C28 fatty alcohols	Friedelin major component, widespread. Only traces of fatty alcohols	Sainsbury (1970), Chandler and Hooper (1979), Holloway (1982c)

TABLE 5.5. continued

Tissue	Component(s)	Comments	References
Pseudotuga menziessi	C16–C28 fatty acids C16–C24-dioic acids C16–C22-hydroxy-fatty acid	From analysis of saponified benzene-soluble wax. C22 hydroxyfatty acid dominant	Kurth (1967), Loveland and Laver (1972)
Saraca indica	C20–C35 alkanes C22–C30 alcohols C34–C60 wax esters 24-ethylcholest-5-en-3β-ol		Behari *et al.* (1977a,b)
Pinus sylvestris, Cerasus vulgaris, Platanus hydrida, Juglans regia, Betula pendula, Populus nigra, Aesculus hippocastanum, Syringa vulgaris	C15–C35 alkanes		Streibl *et al.* (1978)
Roots			
Phalaenopsis mariposa	C18–C33 alkane	C25–C27 predominant. May have contained alkane from internal tissue	Holman and Nichols (1972)
Malus pumila	Friedelin C18–C28 fatty alcohol	Friedelin major component, only traces of fatty alcohols	Holloway (1982c)
Seed tissue			
Gossypium sp. fibres	C18–C36 fatty alcohols C14–C36 fatty acids	C28 and C36 major fatty alcohols C16 major fatty acid	Ryser and Holloway (1985)
Citrus paradisi Chalazal region	*n*-alkanes fatty acids	C29 dominant	Espelie *et al.* (1980a)

generally been classes of aliphatic and terpenoid components typical of epicuticular waxes, and their analysis has involved similar chromatographic and spectrometric techniques to those described above. Important differences do occur between suberin and cuticular membrane soluble lipids, both in the relative proportion of classes present and in the chain length distribution of members of individual classes. In a study of the suberin-associated wax from the underground storage organs of seven species, fatty acids and fatty alcohols were major components, with smaller but substantial amounts of hydrocarbon, but only trace amounts of wax esters (Espelie *et al.*, 1980b). The fatty acids and fatty alcohols generally contained an even number of carbon atoms, and had chain length distributions similar to those covalently attached to the suberin polymer, but shorter than those in the leaf wax from the same species. The hydrocarbon fraction from these periderms contained a substantial proportion of even-chain length alkanes, and was further distinguished from the leaf alkanes by a wider range of chain lengths of which the average was significantly shorter than in the corresponding epicuticular wax. Since it is the soluble lipid waxes associated with suberin that confer the major barrier to water diffusion across the polymer (Soliday et al., 1979; Schönherr, 1982), more compositional data from a range of tissues and species would be of particular value in designing future studies of suberin complex physiology.

D. Depolymerisation and Compositional Analysis of the Polymer Matrix

1. Polyester domain

The aliphatic portion of the suberin polymer, which typically constitutes 5–30% of the insoluble matrix, can be depolymerised with the same ester-targeted reagents used in the cleavage of cutin discussed above. The monomer derivatives obtained from suberin from a diverse range of tissues (Table 5.6) have been identified by similar chromatographic and GC-MS analysis to that described earlier for cutin analysis, and are generally characterised by a high proportion of ω-hydroxyfatty acids and dicarboxylic acids of chain length range C16–C24, in which monoenoic C18 components are frequently dominant, together with smaller amounts of very long chain (C20–C30) *n*-fatty acids and *n*-fatty alcohols (Table 5.6). Although substantial levels of monomers containing mid-chain substituents are present in suberin from several tree bark corks (Holloway, 1972a, 1982c), such components are only minor constituents of suberin from other sources (Table 5.6), whereas they are dominant components of most cutin structures. Thus the characteristic GLC profile of the depolymerisation products normally allows a suberin-type polyester to be readily distinguished from a cutin-type polyester (Kolattukudy, 1978; Espelie et al., 1979; Ryser and Holloway, 1985), whilst the amount of C18 diol present in the hydrogenolysis products provides a useful measure of the extent of suberisation (Kolattukudy and Dean, 1974). As discussed earlier, GC-CIMS will be of particular value in future analyses of the aliphatic suberin monomers.

2. Aromatic domain

The aromatic matrix of suberin, which is resistant to methanolytic and hydrogenolytic depolymerisation, is partially cleaved by nitrobenzene oxidation to yield a mixture of phenolic aldehydes. These can be fractionated by preparative TLC and, after acetylation, analysed by GLC and GC-MS (Cottle and Kolattukudy, 1982; Espelie et al., 1982). Under these depolymerisation conditions the release of *p*-hydroxybenzaldehyde and vanillin appears to be characteristic of the aromatic domain of suberin (Table 5.6), whereas syringaldehyde is obtained as a major component from nitrobenzene oxidation of lignin; this pattern was also observed following CuO oxidation and thioglycolic acid derivatisation of suberin formed in potato wound periderm (Hammerschmidt, 1985).

The extent of root suberisation, as indicated by quantification of the depolymerisation products from both aromatic and aliphatic domains, can be significantly affected by deficiency of specific metal ions during growth, magnesium depletion causing increased root suberisation in *Zea mays* (Pozuelo *et al.*, 1984) and iron deficiency effecting reduced suberisation in *Phaseolus vulgaris* roots (Sijmons *et al.*, 1985). The yields of aromatic aldehydes released from a range of suberised tissues by nitrobenzene oxidation has been presented (Kolattukudy and Espelie, 1985). In addition to the condensed phenolics of the aromatic matrix, periderm suberin of tubers from several species has been found to contain covalently bound ferulic acid as a minor component, probably esterified to a hydroxyl group of the polymer (Riley and Kolattukudy, 1975).

E. Molecular Architecture of the Polymeric Matrix

Details of the structure within the aromatic domains of the matrix remain largely unknown, but the composition of the nitrobenzene oxidation products, together with the histochemical properties of suberin (Brisson *et al.*, 1976), indicate cross-linking in the condensed aromatic domains via ether and carbon–carbon linkages in a polymer structure similar to, but distinct from, that of lignin. By analogy with lignified tissue, it has been postulated that the aromatic domain may be attached to the cell wall, and this model also proposes that the aromatic domain is linked via ester bonds to the aliphatic domain (Kolattukudy, 1980). Direct experimental evidence of the structure of the suberin complex and its relationship to the cell wall polymers consistent with this earlier model has recently been obtained from ^{13}C-NMR studies of suberised cell wall tissue from wound-healing potato tubers (Garbow *et al.*, 1989; Stark *et al.*, 1989). The spectroscopic data of the matrix indicated that aromatic domains of suberin are juxtaposed to, rather than embedded in, the cell wall polymers and that the bulk methylenes of aliphatic domains of suberin, which had more motional freedom than those in cutin, were outnumbered two to one by aromatic carbons. The changes in the NMR spectra following BF_3:MeOH depolymerisation of the aliphatic polyester domains of the suberin complex indicated selective removal of the relatively mobile methylene carbons, although no substantial change in the aliphatic to aromatic carbon ratio was apparent in the residual polymer. This may imply the presence of a substantial amount of non-ester linked aliphatics in this system, but may also indicate that refractory, highly aliphatic polymers reported to be present in the isolated periderm layer from beech (Nip *et al.*, 1986b) may be of more widespread occurrence in suberins.

VI. CONCLUDING REMARKS

The application of recently developed spectrometric techniques to plant waxes, cutin and suberin has initiated a new phase in extending our fundamental knowledge of their composition, and particularly their molecular architecture and ultrastructure. The development of analytical techniques which allow complete analysis of wax or polymer composition on small sections of a single leaf has already made possible detailed studies of cuticular membrane development during ontogeny, and should allow wider experimental evaluation of the effects of specific modifications of composition and structure on the water and solute transport properties of the cuticular membrane and suberin complex. The capacity to fingerprint directly large numbers of wax samples has many applications, including assessment of environmental pollution on wax composition,

TABLE 5.6. Components of the polymeric suberin matrix.

Source of suberin	Characteristic aliphatic monomers	Major chain length[a]
Bark corks		
Quercus suber	C16–C26 ω-hydroxyfatty acids 9,10-dihydroxyoctadecane-1,18-dioic acid	C18:1
Betula pendula	C16–C26 ω-hydroxyfatty acids 9,10-dihydroxyoctadecane-1,18-dioic acid	C18:1
Ribes sp.	C16–C24 ω-hydroxyfatty acids C16–C24-dioic acids	C16:0 and C18:1 C16:0 and C18:1
Malus pumila	C16–C28 fatty acids and alcohols C16–C26 ω-hydroxyfatty acids 9,10-, 18-trihydroxyoctadecanoic acid 9,10-epoxy-18-hydroxyoctadecanoic acid	C22–C24 C18:1 and C22
Underground storage organ periderm		
Solanum tuberosum		
Tuber periderm	C16 and C18:1 dioic acids	C18:1
Tuber wound periderm	C16–C28 ω-hydroxyfatty acids	C18:1
Daucus carota	C16–C34 fatty acids	C18:1, C20–C24
Pastinaca sativa	C16–C25 ω-hydroxyfatty acid	C16 and C18:1
Brassica napobrassica Brassica rapa Beta vulgaris Ipomomea batatas	C15–C24 dioic acid	C16, C18 and C18:1
Root suberins		
Vicia faba	C16 and C18:1 diol[c]	C16
Malus pumila	C16–C28 fatty acid and alcohols C16–C26 ω-hydroxyfatty acids 9,10,18-trihydroxyoctadecanoic acid 9,10-epoxy-18-hydroxyocta-decanoic acid	C22 and C24 C18:1 and C20
Zea mays	C15–C26 fatty alcohols	C17 (hypodermis)
	C16–C26 ω-hydroxyfatty acids	C16 (endodermis) C24 (hypodermis)
	C16–C26 dioic acids	C18:1 (both sources)
Phaseolus vulgaris	C18:1–C26 fatty acids and alcohols C16 and C18:1 ω-hydroxyfatty acids C16 and C18:1 dioic acids	

5. WAXES, CUTIN AND SUBERIN

Ratio in nitrobenzene oxidation products of p-hydroxybenzaldehyde: vanillin : syringaldehyde	References
2.6 : 1 : 1	Holloway (1972a) Kolattukudy and Espelie (1985)
—	Holloway (1972a)
—	Holloway (1972b)
—	Holloway (1982c)
8.9 : 33.3 : 1 1.3 : 1 : — — : 1 : —[b] — : 1 : —[b] 1 : 1.7 : 1	Kolattukudy and Agrawal (1974) Kolattukudy and Dean (1974) Brieskorn and Binneman (1975) Cottle and Kolattukudy (1982) Kolattukudy and Espelie (1985) Kolattukudy et al. (1975) Kolattukudy and Espelie (1985)
— —	Espelie et al. (1979) Holloway (1982c)
19.1 : 16.4 : 1	Pozuelo et al. (1984)
4.6 : 1.1 : 1	Sijmons et al. (1985)

TABLE 5.6. continued

Source of suberin	Characteristic aliphatic monomers	Major chain length[a]
Seed tissue suberin		
Gossypium sp.	C16–C24 ω-hydroxyfatty acids	C22
Cotton fibres and seed coat epidermis	C16–C24 dioic acids	C22
Citrus paradisi	C16–C24 ω-hydroxyfatty acids	C18:1
Seed chalazal region	C16–C24 dioic acids	C16 and C22
Vegetative and fruit tissue		
Crassula argenta	C16–C24 fatty alcohols[c]	
Leaf wound-polymer	C16–C22 α,ω-diols	C18:1
Phaseolus vulgaris	C16–C22 fatty alcohols	C20
Bean pod wound-polymer	C16 and C18:1 α,ω-diols	
Sorghum bicolor	C16–C26 fatty acids	C18:1
Shoot endodermis	C16–C24 ω-hydroxyfatty acids	C16
	C16–C24 dioic acid	C18:1
Agave americana	C16–C24 fatty acids	
Crystal idioblasts	C16–C24 ω-hydroxyfatty acids	C18:1
	C16–C22 dioic acid	C18:1
Zea mays	C17–C30 fatty alcohols	C17 and C19 alkan-2-ols
Leaf bundle sheaths	C16–C24 ω-hydroxyfatty acids	
	Dihydroxyhexadecanoic acid	9,16-diOH
Lycopersicon esculentum	C18–C22 fatty alcohols	C20 and C22
Fruit wound-polymer	C16 and C18:1 ω-hydroxyfatty acids	C18:1
	C16 and C18:1 dicarboxylic acids	C18:1
Fruit protoplasts	C12–C24 fatty acids	C18 and C20
	C12–C22 dioic acids	C18 unsat.

[a] *n*-Saturated unless otherwise indicated: fatty alcohols are alkan-1-ols.
[b] Only vanillin released.
[c] LiAlH$_4$ depolymerisation product, dicarboxylic acids and ω-hydroxyfatty acids not differentiated.
[d] Cotton seed coat suberin.

characterisation of wax mutant varieties and in chemotaxonomic surveys of wax. Thus the chemistry, biochemistry and physiology of the barrier polymers and associated waxes of higher plants is entering an exciting new era.

ACKNOWLEDGEMENTS

I thank Drs G. Bianchi and P.-G. Gülz, Professor Dr H.-R. Schulten, and Dr R. Stark, for making available preprints of unpublished material.

Ratio in nitrobenzene oxidation products of p-hydroxybenzaldehyde: vanillin : syringaldehyde	References
0.4 : 6.0 : 1.0 [d]	Yatsu et al. (1983) Ryser et al. (1983) Ryser and Holloway (1985) Kolattukudy and Espelie (1985)
0.6 : 8.7 : 1.0	Espelie et al. (1980a) Kolattukudy and Espelie (1985)
—	Dean and Kolattukudy (1976)
—	Dean and Kolattukudy (1976)
—	Espelie and Kolattukudy (1979)
0.2 : 2.2 : 1.0	Espelie et al. (1982) Kolattukudy and Espelie (1985)
—	Espelie and Kolattukudy (1979)
—	Dean and Kolattukudy (1976)
20.8 : 1.8 : 1	Rao et al. (1984, 1985a,b)

REFERENCES

Aasen, A. J., Hoffstetler, H. H., Iyengar, B. T. R. and Holman, R. T. (1971). *Lipids* **6**, 502–507.
Allebone, J. F. and Hamilton, R. J. (1972). *J. Sci. Food Agric.* **23**, 777–786.
Audisio, G., Rossini, A., Bianchi, G. and Avato, P. (1987). *J. High Res. Chromatogr. Chromatogr. Commun.* **10**, 594–597.
Avato, P., Mikkelsen, J. D. and Wettstein-Knowles, P. von (1982). *Carlsberg Res. Commun.* **47**, 377–390.
Baker, E. A. (1974). *New Phytologist* **73**, 955–966.
Baker, E. A. (1982). *In* "The Plant Cuticle" (D. F. Cutler, K. L. Alvin and C. E. Price, eds), pp. 139–165. Academic Press, London.

Baker, E. A. and Holloway, P. J. (1970). *Phytochemistry* **9**, 1557–1562.
Baker, E. A. and Holloway, P. J. (1975). *Phytochemistry* **14**, 2463–2467.
Baker, E. A. and Hunt, G. M. (1979). *Phytochemistry* **18**, 1059–1060.
Baker, E. A. and Hunt, G. M. (1981). *New Phytologist* **88**, 731–747.
Baker, E. A. and Hunt, G. M. (1986). *New Phytologist* **102**, 161–173.
Baker, E. A. and Procopiou, J. (1975). *J. Sci. Food Agric.* **26**, 1347–1352.
Baker, E. A. and Procopiou, J. (1980). *J. Hort. Sci.* **55**, 85–87.
Baker, E. A., Bukovac, M. J. and Flore, J. A. (1979). *Phytochemistry* **18**, 781–784.
Baker, E. A., Bukovac, M. J. and Hunt, G. M. (1982). *In* "The Plant Cuticle" (D. F. Cutler, K. L. Alvin and C. E. Price, eds), pp. 33–44. Academic Press, London.
Barta, I. Cs. and Kömives, T. (1984). *J. Chromatogr.* **287**, 438–441.
Behari, M. Andhiwal, C. K. and Ballantine, J. A. (1977a). *Ind. J. Chem. Sect. B* **15**, 765–766.
Behari, M., Andhiwal, C. K. and Streibl, M. (1977b). *Collect. Czech. Chem. Commun.* **42**, 1385–1388.
Bianchi, G. (1987). *Gaz. Chim. Ital.* **117**, 707–716.
Bianchi, G., Avato. P., Scarpa, O., Murelli, C., Audisio, G. and Rossini, A. (1989). *Phytochemistry* **28**, 165–171.
Bowen, D. J. and Walton, T. J. (1988). *Plant Sci.* **55**, 115–127.
Brieskorn, C. H. and Binneman, P. H. (1972). *Tetrahedron Lett.* 1127–1130.
Brieskorn, C. H. and Binneman, P. H. (1975). *Phytochemistry* **14**, 1363–1367.
Brisson, J. D., Robbs, J. and Peterson, R. L. (1976). *Microsc. Soc. Can.* **3**, 174–175.
Brown, A. J. and Kolattukudy, P. E. (1978a). *Arch. Biochem. Biophys.* **190**, 17–26.
Brown, A. J. and Kolattukudy, P. E. (1978b). *J. Agric. Food Chem.* **26**, 1263–1266.
Cardoso, J. N., Eglinton, G. and Holloway, P. J. (1977). *Adv. Org. Geochem.* **7**, 273–287.
Carruthers, W. and Johnstone, R. A. W. (1959). *Nature* **184**, 1131–1132.
Chandler, R. F. and Hooper, S. N. (1979). *Phytochemistry* **18**, 711–724.
Chang, S. Y. and Grunwald, C. (1976). *Phytochemistry* **15**, 961–963.
Chiarlo, B. and Tacchino, E. (1965). *Riv. Ital. Sostanze. Grasse* **42**, 122–124.
Clarkson, D. T. and Robards, A. W. (1975). *In* "The Development and Function of Roots" (J. C. Torrey and D. T. Clarkson, eds), pp. 415–436. Academic Press, London.
Coles, L. (1968). *J. Chromatogr.* **32**, 657–661.
Cottle, W. and Kolattukudy, P. E. (1982). *Plant Physiology* **69**, 393–399.
Crisp, C. E. (1965). Ph.D. Thesis, University of California, Davis, CA.
Crossley, A. and Fowler, D. (1986). *New Phytologist* **103**, 207–218.
Croteau, R. and Fagerson, I. S. (1972). *Phytochemistry* **11**, 353–363.
Croteau, R. and Kolattukudy, P. E. (1973). *Biochem. Biophys. Res. Commun.* **52**, 863–869.
Dean, B. B. and Kolattukudy, P. E. (1976). *Plant Physiol.* **58**, 411–416.
Deas, A. H. B. and Holloway, P. J. (1977). *In* "Lipids and Lipid Polymers in Higher Plants" (M. Tevini and H. K. Lichtenthaler, eds), pp. 293–299. Springer-Verlag, Berlin.
Deas, A. H. B., Baker, E. A. and Holloway, P. J. (1974). *Phytochemistry* **13**, 1901–1905.
Dodova-Anghelova, M. S. and Ivanov, C. P. (1973). *Phytochemistry* **12**, 2239–2242.
Eckl, K. and Gruler, H. (1980). *Planta* **150**, 102–113.
Eglinton, G. and Hunneman, D. H. (1968). *Phytochemistry* **7**, 313–322.
Eglinton, G., Gonzalez, A.G., Hamilton, R. J. and Raphael, R. A. (1962). *Phytochemistry* **1**, 89–102.
Eglinton, G., Hamilton, R. J., Kelly, W. B. and Reed, R. J. (1966). *Phytochemistry* **5**, 1349–1352.
Eglinton, G., Hunneman, D. H. and McCormick, A. (1968). *Org. Mass Spectrom.* **1**, 593–611.
Espelie, K. E. and Kolattukudy, P. E. (1978). *Lipids* **13**, 832–833.
Espelie, K. E. and Kolattukudy, P. E. (1979). *Plant Sci. Lett.* **15**, 225–230.
Espelie, K. E., Dean, B. B. and Kolattukudy, P. E. (1979). *Plant Physiol.* **64**, 1089–1093.
Espelie, K. E., Davies, R. W. and Kolattukudy, P. E. (1980a). *Planta* **149**, 498–511.
Espelie, K. E., Sadek, N. Z. and Kolattukudy, P. E. (1980b). *Planta* **148**, 468–476.
Espelie, K. E., Wattendorf, J. and Kolattukudy, P. E. (1982). *Planta* **155**, 166–175.
Espelie, K. E., Köller, W. and Kolattukudy, P. E. (1983). *Chem. Phys. Lipids* **32**, 13–26.
Franich, R. A., Wells, L. G. and Barnett, J. R. (1977). *Ann. Bot. (Lond.)* **41**, 621–626.
Franich, R. A., Wells, L. G. and Holland, P. T. (1978). *Phytochemistry* **17**, 1617–1623.
Franich, R. A., Gowar, A. P. and Volkman, J. K. (1979). *Phytochemistry* **18**, 1563–1564.

Franich, R. A., Goodin, S. J. and Volkman, K. (1985). *Phytochemistry* **24**, 2949–2952.
Garbow, J. R., Ferrantello, L. M. and Stark, R. E. (1989). *Plant Physiol.* **90**, 783–787.
Garrec, J. P. and Kerfown, C. (1989). *Environ. Exp. Bot.* **29**, 215–218.
Gaskell, S. J., Eglinton, G. and Brunn, T. (1973). *Phytochemistry* **12**, 1174–1176.
Ghosh, P. and Thakur, S. (1982). *J. Chromatogr.* **240**, 515–517.
Grill, D., Pfeifhofer, H., Hartwig, H. and Gottfried, W. H. (1987). *Eur. J. Pathol.* **17**, 246–255.
Guenthardt-Goerg, M. S. and Keller, T. (1987). *Trees (Berlin)* **1**, 145–150.
Gülz, P.-G. and Marner, F.-J. (1986). *Z. Naturforsch.* **41c**, 673–676.
Gülz, P.-G., Hemmers, H., Bodden, J. and Marner, F.-J. (1987a). *Z. Naturforsch.* **42c**, 191–196.
Gülz, P.-G., Scora, R. W., Müller, E. and Marner, F.-J. (1987b). *J. Agric. Food Chem.* **35**, 716–720.
Gülz, P.-G., Bodden, J., Müller, E. and Marner, F.-J. (1988a). *Z. Naturforsch.* **43c**, 19–23.
Gülz, P.-G., Müller, E. and Moog, B. (1988b). *Z. Naturforsch.* **43c**, 173–176.
Hammerschmidt, R. (1985). *Potato Res.* **28**, 123–127.
Haas, K. and Rentschler, I. (1984). *Plant Sci. Lett.* **36**, 143–147.
Haas, K. and Schönherr, J. (1979). *Planta* **146**, 399–403.
Haas, K. and Schönherr, J. (1981). *Biochem. Physiol. Pflanzen.* **176**, 101–104.
Hardman, R. and Brain, K. R. (1971). *Phytochemistry* **10**, 1115–1119.
Hemmers, H. and Gülz, P.-G. (1986). *Phytochemistry* **25**, 2103–2107.
Hemmers, H., Gülz, P.-G. and Hangst, K. (1986). *Z. Naturforsch.* **41c**, 521–525.
Hemmers, H., Gülz, P.-G. and Marner, F.-J. (1988). *Z. Naturforsch.* **43c**, 799–805.
Hemmers, H., Gülz, P.-G., Marner, F.-J. and Wray, U. (1989). *Z. Naturforsch.* **44c**, 193–201.
Herbin, G. A. and Robins, P. A. (1968a). *Phytochemistry* **7**, 239–255.
Herbin, G. A. and Robins, P. A. (1968b). *Phytochemistry* **7**, 1325–1337.
Herbin, G. A. and Sharma. K. (1969). *Phytochemistry* **8**, 151–160.
Holloway, P. J. (1970). *Pesticide Sci.* **1**, 156–163.
Holloway, P. J. (1972a). *Chem. Phys. Lipids* **9**, 158–170.
Holloway, P. J. (1972b). *Chem. Phys. Lipids* **9**, 171–179.
Holloway, P. J. (1982a). *In* "The Plant Cuticle" (D. F. Cutler, K. L. Alvin and C. E. Price, eds), pp. 1–32. Academic Press, London.
Holloway, P. J. (1982b). *In* "The Plant Cuticle" (D. F. Cutler, K. L. Alvin and C. E. Price, eds), pp. 45–85. Academic Press, London.
Holloway, P. J. (1982c). *Phytochemistry* **21**, 2517–2522.
Holloway, P. J. (1984a). *In* "CRC Handbook of Chromatography, Lipids" (H. K. Mangold. G. Zweig and J. Sherma, eds), Vol. 1, pp. 321–345. CRC Press, Boca Raton, FL.
Holloway, P. J. (1984b). *In* "CRC Handbook of Chromatography, Lipids" (H. K. Mangold, G. Zweig and J. Sherma, eds), Vol. 1, pp. 347–380. CRC Press, Boca Raton, FL.
Holloway, P. J. and Baker, E. A. (1968). *Plant Physiol.* **43**, 1878–1879.
Holloway, P. J. and Brown, G. A. (1977). *Chem. Phys. Lipids* **19**, 1–13.
Holloway, P. J. and Challen, S. B. (1966). *J. Chromatogr.* **25**, 336–346.
Holloway, P. J. and Deas, A. H. B. (1971a). *Chem. Ind.*, 1140.
Holloway, P. J. and Deas, A. H. B. (1971b). *Phytochemistry* **10**, 2781–2787.
Holloway, P. J., Jeffree, C. E. and Baker, E. A. (1976). *Phytochemistry* **15**, 1768–1770.
Holloway, P. J., Brown, G. A., Baker, E. A. and Macey, M. J. K. (1977a). *Chem. Phys. Lipids* **19**, 114–127.
Holloway, P. J., Hunt, G. M., Baker, E. A. and Macey, M. J. K. (1977b). *Chem. Phys. Lipids* **20**, 141–155.
Holloway, P. J., Brown, G. A. and Wattendorf, J. (1981). *J. Exp. Bot.* **32**. 1051–1066.
Holman, R. T. and Nichols, P. C. (1972). *Phytochemistry* **11**, 333–337.
Horn, D. H. S. and Lamberton, J. A. (1962). *Chem. Ind.*, 2036–2037.
Horn, D. H. S. and Lamberton, J. A. (1964). *Aust. J. Chem.* **17**, 477–480.
Horn, D. H. S., Kranz, Z. H. and Lamberton, J. A. (1964). *Aust. J. Chem.* **17**, 464–476.
Hrazdina, G., Marx, G. A. and Hoch, H. C. (1982). *Plant Physiol.* **70**, 745–748.
Hunt, G. M. and Baker, E. A. (1979). *Chem. Phys. Lipids* **23**, 213–221.
Hunt, G. M. and Baker, E. A. (1980). *Phytochemistry* **19**, 1415–1419.
Hunt, G. M. and Baker, E. A. (1982). *In* "The Plant Cuticle" (D. F. Cutler, K. L. Alvin and C. E. Price, eds), pp. 279–292. Academic Press, London.

Jackson, L. L. (1971). *Phytochemistry* **10**, 487–490.
Jarolinek, P., Woolrab, U., Streibl, M. and Sorm, F. (1964). *Chem. Ind.* 237–238.
Jones, J. H. (1978). *Plant Physiol.* **62**, 831–832.
Juniper, B. E. and Jeffree, C. E. (1983). "Plant Surfaces". Edward Arnold, London.
Kaneda, T. (1967). *Biochemistry* **6**, 2023–2032.
Kerler, F. and Schönherr, J. (1988a). *Arch. Environ. Contam. Toxicol.* **17**, 1–6.
Kerler, F. and Schönherr, J. (1988b). *Arch. Environ. Contam. Toxicol.* **17**, 7–12.
Kerstiens, G. and Lendzian, K. J. (1989). *New Phytologist* **112**, 21–27.
Kolattukudy, P. E. (1970a). *Ann. Rev. Plant Physiol.* **21**, 163–192.
Kolattukudy, P. E. (1970b). *Lipids* **5**, 398–402.
Kolattukudy, P. E. (1970c). *Plant Physiol.* **46**, 759–760.
Kolattukudy, P. E. (1972). *Biochem. Biophys. Res. Commun.* **49**, 1040–1046.
Kolattukudy, P. E. (1974). *Biochemistry* **13**, 1354–1363.
Kolattukudy, P. E. (1977). *Recent Adv. Phytochem.* **11**, 185–246.
Kolattukudy, P. E. (1978). In "Biochemistry of Wounded Plant Tissue" (G. Kahl, ed.), pp. 43–84. Walter de Gruyter & Co., Berlin.
Kolattukudy, P. E. (1980). *Science* **208**, 990–1000.
Kolattukudy, P. E. (1981). *Ann. Rev. Plant Physiol.* **32**, 539–567.
Kolattukudy, P. E. (1984a). *Can. J. Bot.* **62**, 2918–2933.
Kolattukudy, P. E. (1984b). In "Lipases" (B. Borgstrom and H. Brockman, eds), pp. 471–504. Elsevier/North Holland Biomedical. Amsterdam.
Kolattukudy, P. E. (1987). In "The Biochemistry of Plants", Vol. 9—Lipids (P. K. Stumpf, ed.), pp. 291–314. Academic Press, New York.
Kolattukudy, P. E. and Agrawal, V. P. (1974). *Lipids* **9**, 682–691.
Kolattukudy, P. E. and Dean, B. B. (1974). *Plant Physiol.* **54**, 116–121.
Kolattukudy, P. E. and Espelie, K. E. (1985). In "Biosynthesis and Biodegradation of Wood Components" (T. Higuchi, ed.), pp. 161–207. Academic Press, New York.
Kolattukudy, P. E. and Walton, T. J. (1972a). *Progr. Chem. Fats Other Lipids* **13**, 119–175.
Kolattukudy, P. E. and Walton, T. J. (1972b). *Biochemistry* **11**, 1897–1907.
Kolattukudy, P. E., Walton, T. J. and Kushwaha, R. P. S. (1973). *Biochemistry* **12**, 4488–4498.
Kolattukudy, P. E., Kronman, K. and Poulouse, A. J. (1975). *Plant Physiol.* **55**, 567–573.
Kurth, E. F. (1967). *Tappi* **50**, 253–258.
Linskens, H. F., Heinen, W. and Stoffers, A. L. (1965). *Residue Rev.* **8**, 137–178.
Loveland, P. M. and Laver, M. R. (1972). *Phytochemistry* **11**, 430–432.
Lundqvist, U. and Wettstein-Knowles, P. von (1983). *Carlsberg Res. Commun.* **48**, 321–344.
Martin, E. A. and Juniper, B. E. (1970). "The Cuticles of Plants". Edward Arnold, London.
Matic, M. (1956). *Biochem. J.* **63**, 168–178.
Mikkelsen, J. D. (1979). *Carlsberg Res. Commun.* **44**, 133–147.
Mikkelsen, J. D. (1984). *Carlsberg Res. Commun.* **49**, 391–416.
Mikkelsen, J. D. and Wettstein-Knowles, P. von (1978). *Arch. Biochem. Biophys.* **188**, 172–181.
Misra, D. R. and Khastgir, H. N. (1970). *Tetrahedron* **26**, 3017–3021.
Mladenova, K., Stoianova-Ivanova, B. and Camaggi, C. M. (1977). *Phytochemistry* **16**, 269–272.
Nagy, B., Modzeleski, V. and Murphy, M. T. J. (1965). *Phytochemistry* **4**, 945–950.
Netting, A. G. and Wettstein-Knowles, P. von (1976). *Arch. Biochem. Biophys.* **174**, 613–621.
Netting, A. G., Macey, M. J. K. and Barker, H. N. (1972). *Phytochemistry* **11**, 579–585.
Nip, M., Tegelaar, E. W., Brinkhuis, H., De Leeuw, J. W., Schenck, P. A. and Holloway, P. J. (1986a). *Org. Geochem.* **10**, 769–778.
Nip, M., Tegelaar, E. W., De Leeuw, J. W., Schenck, P. A. and Holloway, P. J. (1986b). *Naturwissenschaften* **73**, 579–585.
Nordby, H. E. and Nagy, S. (1977). *Phytochemistry* **16**, 1393–1397.
Percy, K. E. and Baker, E. A. (1987). *New Phytologist* **107**, 577–589.
Podila, G. K., Dickman, R. B. and Kolattukudy, P. E. (1988). *Science* **242**, 922–925.
Pozuelo, J. M., Espelie, K. W. and Kolattukudy, P. E. (1984). *Plant Physiol.* **74**, 256–260.
Price, C. E. (1982). In "The Plant Cuticle" (D. F. Cutler, K. L. Alvin and C. E. Price, eds), pp. 237–252. Academic Press, London.
Radler, F. and Horn, D. H. S. (1965). *Aust. J. Chem.* **18**, 1059–1069.
Rao, G. S. R., Willison, J. H. M. and Ratnayake, W. M. N. (1984). *Plant Physiol.* **75**, 716–719.

Rao, G. S. R., Willison, J. H. M. and Ratnayake, W. M. N. (1985a). *Can. J. Bot.* **63**, 2177–2180.
Rao, G. S. R., Willison, J. H. M., Ratnayake, W. M. N. and Ackman, R. G. (1985b). *Phytochemistry* **24**, 2127–2128.
Riederer, M. and Schneider, G. (1990). *Planta* **180**, 154–165.
Riederer, M. and Schönherr, J. (1986a). *Arch. Environ. Contam. Toxicol.* **15**, 97–105.
Riederer, M. and Schönherr, J. (1986b). *J. Chromatogr.* **360**, 151–161.
Riederer, M. and Schönherr, J. (1988a). *Arch. Environ. Contam. Toxicol.* **17**, 21–25.
Riederer, M. and Schönherr, J. (1988b). *Planta* **174**, 127–138.
Riley, R. G. and Kolattukudy, P. E. (1975). *Plant Physiol.* **56**, 650–654.
Roberts, D. L. and Rowland, R. L. (1962). *J. Org. Chem.* **27**, 3989–3994.
Rowe, J. W. and Scroggins, J. H. (1964). *J. Org. Chem.* **29**, 1554–1562.
Rudloff, E. von (1959). *Can. J. Chem.* **37**, 1038–1042.
Ryser, U. and Holloway, P. J. (1985). *Planta* **163**, 151–163.
Ryser, U., Meier, H. and Holloway, P. J. (1983). *Protoplasma* **117**, 196–205.
Sainsbury, M. (1970). *Phytochemistry* **9**, 2209–2215.
Salasoo, I. (1983). *Biochem. Syst. Ecol.* **11**, 17–20.
Schmidt, H. W., Merida, T. and Schönherr, J. (1981). *Z. Pflanzenphysiol.* **105**, 41–51.
Schönherr, J. (1976). *Ecol. Studies* **19**, 148–159.
Schönherr, J. (1982). *In* "Encyclopedia of Plant Physiology" (O. L. Lange, P. S. Noble, C. B. Osmond and H. Ziegler, eds), Vol. 12B, pp. 153–179. Springer-Verlag, Berlin.
Schönherr, J. and Riederer, M. (1986). *Plant Cell Environ.* **9**, 459–466.
Schönherr, J., Eckl, K. and Gruler, H. (1979). *Planta* **147**, 21–26.
Schulten, H.-R., Simmleit, N. and Rump, H. H. (1986). *Chem. Phys. Lipids* **41**, 209–224.
Schulten, H.-R., Murray, K. E. and Simmleit, N. (1987a). *Z. Naturforsch.* **42c**, 178–190.
Schulten, H.-R., Simmleit, N. and Murray, K. E. (1987b). *Fresenius Z. Anal. Chem.* **327**, 235–238.
Scora, R. W., Mueller, E. and Gülz, P.-G. (1986). *J. Agric. Food Chem.* **34**, 1024–1026.
Sen, A. (1987). *Z. Naturforsch.* **42c**, 1153–1158.
Sijmons, P. C., Kolattukudy, P. E. and Bienfait, H. F. (1985). *Plant Physiol.* **78**, 115–120.
Silva-Fernandes, A. M., Baker, E. A. and Martin, J. T. (1964). *Ann. Appl. Biol.* **53**, 43–58.
Simmleit, N. and Schulten, H. R. (1987). *In* "Acid Rain: Scientific and Technical Advances" (R. Perry, R. M. Harrison, J. N. B. Bell and J. N. Lester, eds), pp. 546–553. Selper Ltd., London.
Simpson, D. and Wettstein-Knowles, P. von (1980). *Carlsberg Res. Commun.* **45**, 465–481.
Soliday, C. L., Kolattukudy, P. E. and Davis, R. W. (1979). *Planta* **146**, 607–614.
Sorm, F., Wollrab, U., Jarolinek, P. and Streibl, M. (1964). *Chem. Ind.*, 1833–1834.
Spence, R.-M. M. and Tucknott, O. G. (1983). *Phytochemistry* **22**, 2521–2523.
Spencer, G. F. and Plattner, R. D. (1984). *J. Am. Oil Chem. Soc.* **61**, 90–94.
Stark, R. E., Zlotnik-Mazori, T., Ferrantello, L. M. and Garbow, J. R. (1989). *Am. Chem. Soc. Symp. Ser.* **399**, 214–229.
Stoianova-Ivanova, B., Mladenova, K. and Malova, I. (1971). *Phytochemistry* **10**, 2525–2528.
Stránský, K., Streibl, M. and Kubelka, V. (1970). *Collect. Czech. Chem. Commun.* **35**, 882–891.
Streibl, M. and Konečný, K. (1967). *Chem. Ind.*, 546.
Streibl, M. and Stránský, K. (1968). *Fette, Seiffen, Anstrichmittel* **70**, 343–348.
Streibl, M., Stránský, K. and Herout, V. (1978). *Collect. Czech. Chem. Commun.* **43**, 320–326.
Tegelaar, E. W., De Leeuw, J. W., Largeau, C., Dereune, S., Schulten, H. R., Mueller, R., Boon, J. J., Nip, M. and Sprenkels, J. C. M. (1989). *J. Anal. Appl. Pyrolysis* **15**, 29–54.
Tulloch, A. P. (1971). *Lipids* **6**, 641–644.
Tulloch, A. P. (1972). *J. Am. Oil Chem. Soc.* **49**, 609–610.
Tulloch, A. P. (1973). *Phytochemistry* **12**, 2225–2232.
Tulloch, A. P. (1976a). *In* "Chemistry and Biochemistry of Natural Waxes" (P. E. Kolattukudy, ed.), pp. 235–287. Elsevier, Amsterdam.
Tulloch, A. P. (1976b). *Phytochemistry* **15**, 1153–1156.
Tulloch, A. P. (1977). *Lipids* **12**, 233–234.
Tulloch, A. P. (1978). *Phytochemistry* **17**, 1613–1615.
Tulloch, A. P. (1981). *Can. J. Bot.* **59**, 1213–1221.
Tulloch, A. P. (1982a). *Phytochemistry* **21**, 661–664.
Tulloch, A. P. (1982b). *Phytochemistry* **21**, 2251–2255.

Tulloch, A. P. (1983). *Phytochemistry* **22**, 1605–1613.
Tulloch, A. P. (1984). *Phytochemistry* **23**, 1619–1623.
Tulloch, A. P. (1987). *Phytochemistry* **26**, 1041–1043.
Tulloch, A. P. and Hoffman, L. L. (1973a). *Lipids* **8**, 617–622.
Tulloch, A. P. and Hoffman, L. L. (1973b). *Phytochemistry* **12**, 2217–2223.
Tulloch, A. P. and Hoffman, L. L. (1974). *Phytochemistry* **13**, 2535–2540.
Tulloch, A. P. and Hoffman, L. L. (1976). *Phytochemistry* **15**, 1145–1151.
Tulloch, A. P. and Hoffman, L. L. (1977). *Can. J. Bot.* **55**, 853–857.
Tulloch, A. P. and Hoffman, L. L. (1979). *Phytochemistry* **18**, 267–271.
Tulloch, A. P. and Hoffman, L. L. (1982). *Phytochemistry* **21**, 1639–1642.
Tulloch, A. P. and Hogge, L. R. (1978). *J. Chromatogr.* **157**, 291–296.
Tulloch, A. P. and Weenik, R. O. (1969). *Can. J. Chem.* **47**, 3119–3126.
Vogt, T. and Gülz, P.-G. (1987). *Z. Naturforsch.* **42c**, 157–158.
Vogt, T., Proksch, P. and Gülz, P.-G. (1987a). *J. Plant Physiol.* **131**, 25–36.
Vogt, T., Proksch, P., Gülz, P.-G. and Wollenweber, E. (1987b). *Phytochemistry* **26**, 1027–1030.
Vogt, T., Gülz, P.-G. and Wray, V. (1988). *Phytochemistry* **27**, 3712–3713.
Walton, T. J. and Kolattukudy, P. E. (1972). *Biochemistry* **11**, 1885–1897.
Wattendorf, J. and Holloway, P. J. (1980). *Ann. Bot.* **46**, 13–28.
Wattendorf, J. and Holloway, P. J. (1982). *Ann. Bot.* **49**, 769–804.
Wettstein-Knowles, P. von (1974). *FEBS Lett.* **42**, 187–191.
Wettstein-Knowles, P. von (1985). *Carlsberg Res. Commun.* **50**, 239–262.
Wettstein-Knowles, P. von (1987a). *In* "The Metabolism, Structure and Function of Plant Lipids" (P. K. Stumpf, J. B. Mudd and W. D. Nes, eds), pp. 489–498. Plenum Publ. Corp., New York.
Wettstein-Knowles, P. von (1987b). *In* "Plant Molecular Biology" (D. von Wettstein and N.-H. Chua, eds), pp. 305–314. Plenum Publ. Corp., New York.
Wettstein-Knowles, P. von and Madsen, J. O. (1984). *Carlsberg Res. Commun.* **49**, 57–67.
Wettstein-Knowles, P. von and Netting, A. G. (1976a). *Carlsberg Res. Commun.* **41**, 225–235.
Wettstein-Knowles, P. von and Netting, A. G. (1976b). *Lipids* **11**, 478–484.
Wettstein-Knowles, P. von, Mikkelsen, J. D. and Madsen, J. O. (1984). *Carlsberg Res. Commun.* **49**, 611–618.
Wilson, L. A. and Sterling, C. (1976). *Z. Pflanzenphysiol.* **77**, 359–371.
Wollenweber, E. and Dietz, V. H. (1981). *Phytochemistry* **20**, 869–932.
Wollrab, V. (1967). *Collect. Czech. Chem. Commun.* **32**, 1304–1308.
Wollrab, V., Streibl, M. and Sorm, F. (1963). *Collect. Czech. Chem. Commun.* **28**, 1904–1913.
Wollrab, V., Streibl, M. and Sorm, F. (1965a). *Collect. Czech. Chem. Commun.* **30**, 1654–1659.
Wollrab, V., Streibl, M. and Sorm, F. (1965b). *Collect. Czech. Chem. Commun.* **30**, 1670–1675.
Wollrab, V., Streibl, M. and Sorm, F. (1967). *Chem. Ind.*, 1872–1873.
Yatsu, L. Y., Espelie, K. E. and Kolattukudy, P. E. (1983). *Plant Physiol.* **73**, 521–524.
Zabkiewicz, J. A. and Steele, K. D. (1982). *Chromatographia* **16**, 92–97.
Zlotnik-Mazori, T. and Stark, R. E. (1988). *Macromolecules* **21**, 2412–2417.

6 Polyacetylenes and Related Compounds: Analytical and Chemical Methods

JØRGEN LAM[1] and LENE HANSEN[2]

[1]University of Aarhus, 8000 Aarhus C, Denmark
[2]BST-Center, 7000 Fredericia, Denmark

I.	Introduction	160
II.	Experimental	160
	A. Collection of plant material	160
	B. Preparation of plant material for extraction	161
	C. Column separation	161
	D. Initial observations by ultraviolet spectroscopy and TLC	162
	E. Methods for isolation of single compounds	163
	F. Detection by GLC and HPLC	166
	G. Characterisation of isolated and purified compounds	167
	H. Chemistry	175
III.	Differences and similarities within groups and tribes of the family Asteraceae and other families	179
	A. Asteraceae	179
	B. Polyacetylenes in Umbelliferae and Araliaceae	180
	C. Polyacetylenes in Campanulaceae	180
	D. Acetylenes in fungi	180
IV.	Biosyntheses of acetylenes and related compounds	181
V.	Biological activities	181
	Acknowledgements	181
	References	182
	Recommended additional reading: chemical, biogenetical and biological aspects	183

I. INTRODUCTION

The so-called polyacetylenes are a large group of compounds existing in plants, fungi and insects, which normally contain conjugated systems of double and triple bonds. It should be noted that the term polyacetylenes is generally used to include natural compounds, which may possess only one or two triple bonds in their chemical structure, or some compounds related to this group but with chemically distinct structures, for example α-terthienyl (**1**). Bithienylbutynene (**2**) is another example of a compound containing only one triple bond and thus it is not a true 'poly'-acetylene.

However, both of these compounds are thought to be biogenetically related to the polyacetylenes and most likely produced along with the polyacetylenic compounds or from polyacetylenes as described by Bohlmann *et al.* (1973), Bu'Lock (1964) and Bohlmann and Zdero (1985).

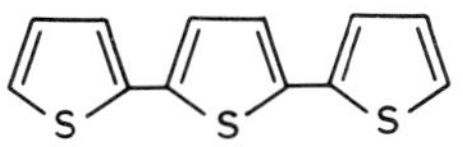

FIG. 6.1. (**1**) α-Terthienyl, (α-T); (**2**) bithienylbutynene (BBT).

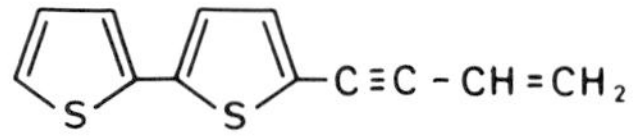

About 1000 compounds belonging to the group of acetylenes and related compounds are known today. Many of these compounds have been isolated from various tribes of Asteraceae (Compositae). The tribes Anthemideae, Astereae, Heliantheae, Inuleae, Helenieae and Cynareae are the most acetylene-rich and the diversity of compounds in these tribes is very great. The compounds range in polarity from almost non-polar hydrocarbons to polar diol acetylenes. In Campanulaceae, Umbelliferae and Araliaceae alcohols and/or ketones and aldehydes are the most frequently occurring. Fungi contain alcohols, esters, lactones and even several free acids; thus generally the most polar acetylenes found in nature are present in fungi. Of course, the different structures and the relative instability of a large number of these compounds makes it impossible to give a complete description for the handling of these compounds, but some general outlines may be given.

II. EXPERIMENTAL

A. Collection of Plant Material

Plant material may be collected from nature or obtained from a botanical garden. Fresh material is preferred, as some of the polyacetylenes are so unstable that drying may lead to a reduction of recovery. If dried material is used, a small fresh sample should be obtained to compare the proportions found in fresh and dry material. The material should be correctly identified and a voucher specimen should be deposited at a botanical herbarium every time a new plant is to be investigated. In cases where the plant materials are obtained from a botanical garden, it is still very important to check the naming of the plant and the authority.

B. Preparation of Plant Material for Extraction

As root material very often differs from that of the aerial parts of the plant in its content of acetylenes and other compounds, the usual procedure is to extract the root material separately from the rest of the plant. For similar reasons it is preferable to separate the flowers or the flower heads from the leaves and stems.

1. Root material

Root material is usually crushed in a suitable crushing machine. In order to keep the temperature low during the crushing procedure, dry ice may be added to prevent decomposition of the polyacetylenes which may occur at more elevated temperatures. Carbon dioxide is thought to prevent unnecessary oxidation at the same time.

The root material will often contain variable amounts of water, which may decrease the efficiency of extraction with an organic solvent. Therefore dry sodium sulphate in suitable amounts may be added to the crushed material before extraction is performed with diethylether. The material is covered completely with solvent and is shaken or stirred frequently during the extraction at room temperature. The flasks are stoppered and covered with tin foil to reduce decomposition by ultraviolet irradiation. Repeated extraction is carried out at room temperature until the last extraction no longer produces significant amounts of acetylenes in the solution, determined by means of ultraviolet spectroscopy. The extraction procedure is usually finished after three or four extractions.

2. Aerial parts

For extraction of the aerial parts of the plant a non-polar solvent is usually chosen for the preliminary extractions in order to keep the extraction of chlorophylls at a minimum. The chlorophylls may be troublesome to remove from the extracts, and hence pentane, hexane, heptane or light petroleum ether (b.p. 30–60°C), in which the chlorophylls are only slightly soluble, are convenient solvents for leaves and stems and also for flower heads. A secondary extraction may be performed with more polar solvents and thus the chlorophylls will be a part of the extract. If the plant material contains very polar substances like acetylenic alcohols with one or more alcohol groups this procedure may be required to obtain a quantitative extraction of the plant material.

C. Column Separation

The combined extracts either from roots, leaves and stems or from flower heads are separately reduced to small volumes and from each combined extract an aliquot is evaporated to dryness in order to obtain information about what size of column should be chosen for further separation.

Whereas a time-consuming procedure was previously applied (3–5 days of chromatography on the column) a more rapid chromatographic method described by Still and co-workers (1978) is now used in our laboratory. The silica gel used is 40–63 µm (400–230 mesh) silica gel 60 (E. Merck No. 9385). The method allows separation of grams of extract within an hour or two. Moreover, the method has the advantage over the previously applied method in that a small pressure of nitrogen is used to protect against oxidation of the plant material on the column. In addition the shorter time on the column means that hydrolysis may also be reduced, a factor which is not easy to control. The solvents used for development on the column are distilled light petroleum ether (b.p. <50°C) at the beginning with an increasing amount of diethylether up to 100% of the mixture. Finally, methanol is often used to elute the most polar compounds from the column.

The method has been slightly modified, as we usually apply a mixture of diethylether and light petroleum ether. This is used especially when extracts are not fully soluble in light petroleum ether alone. In order to avoid development of bubbles in the column (due to the presence of diethylether) a water-cooled mantle is used around the column.

By following the described procedure a series of fractions will be obtained. As very unstable compounds may be present in the solutions, the flasks should be stoppered immediately and protected against ultraviolet light by use of tin foil around each flask, which again should be stored in a refrigerator.

Caution: To avoid the danger of explosion, a spark-proof refrigerator is required for storing diethylether fractions.

D. Initial Observations by Ultraviolet Spectroscopy and TLC

In order to obtain information about the purity and the polarity of each single fraction obtained by column chromatography, an ultraviolet spectrum is recorded for each fraction and a semiquantitative determination is carried out in cases where this is possible. Thin layer chromatography (TLC) is performed for all the fractions on silica gel plates (Kieselgel 60 G, Art.nr. 7731, Merck). Usually 20 µl of each fraction are applied to thin layer plates, so that 20 µl of the first 11 fractions are placed on plate number one and the plate (20 × 20 cm) is placed in a chamber saturated with light petroleum ether vapour and developed with pure light petroleum. The second plate is first spotted with 20 µl of fraction 11 which is also applied to the first plate. On the second plate 20 µl of fractions 12–21 are applied subsequently. The second plate may now be developed with 10% diethylether in light petroleum ether. By using increasing amounts of diethylether for development of the successive plates, it is possible to get information about the increasing polarity of the compounds, which combined with the information about the ultraviolet spectra may form the basis for the further analytical treatment of the fractions obtained by column chromatography.

Several possibilities for visualising the spots on the plates are available, but usually a neutral solution of potassium permanganate (0.32%) is sprayed onto the plates. This oxidises most organic compounds, leaving yellow spots. For highly unsaturated compounds the oxidation appears to be very fast, whereas saturated compounds are oxidised very slowly, an observation which may give preliminary information. Another method of visualisation that may be used for the pure acetylenic compounds is application of Vanillin or *p*-dimethylaminobenzaldehyde in organic solvents. The method is described by Picman *et al.* (1980) for a series of polyacetylenic compounds. It

6. POLYACETYLENES AND RELATED COMPOUNDS

yields a number of different colour reactions, which may be informative in the characterisation of unsaturated compounds.

In a few cases compounds may crystallise from a certain fraction and an almost pure compound may be obtained. Such a compound may then be subjected to various spectroscopical methods directly (UV, IR, NMR and MS). Unfortunately, pure compounds are not obtained readily and therefore repeated chromatography is the most usual procedure.

E. Methods for Isolation of Single Compounds

1. Repeated column or preparative thin layer chromatography using silica gel

From the ultraviolet spectroscopy and the thin layer chromatography carried out to get preliminary information about the individual column fractions, a decision may be made that further chromatography is to be carried out, either by repeated column chromatography or by use of preparative thin layer chromatography. If, for example, the preliminary thin layer chromatogram reveals that a certain fraction or a series of fractions were overloaded with respect to one or two or even more compounds, it may be worthwhile to repeat the separation on a column (see Section II.C) or to perform a separation by preparative thin layer chromatography.

2. Caffeine-impregnated plates

The use of 5–10% of caffeine added to silica gel may yield a good separation of two compounds of comparable structure which differ slightly in the number of π-electrons present. The electronic system of caffeine has the effect of interacting with π-electrons in unsaturated compounds. Examples of otherwise poorly separable compounds where separations have been achieved by the addition of 10% of caffeine to silica gel on plates or columns are given in Fig. 6.2. Here (3) is the most strongly bound compound and thus may be separated from (4). Compounds (3) and (4) occur frequently in a mixture in the plant group of Coreopsedinae of the tribe Heliantheae, e.g. in *Bidens* and *Dahlia* species.

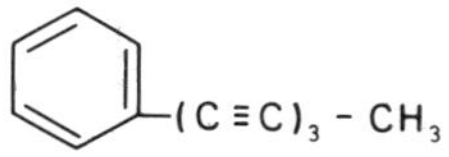

$$\text{Ph}-(C\equiv C)_3-CH_3$$

(3)

$$\text{Ph}-(C\equiv C)_2-CH=CH-CH_3$$

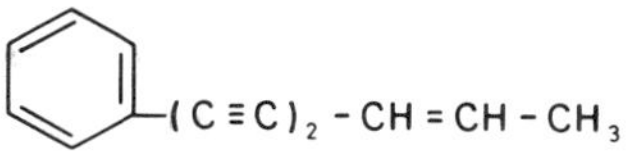

(4)

FIG. 6.2. (3) 1-Phenylhepta-1,3,5-triyne, PHT; (4) 1-phenylhepta-1,3-diyn-5-ene.

The thiophene compounds previously mentioned (1 and 2) may also be more readily separated on caffeine–silica plates than on silica without caffeine. *Echinops* and *Tagetes* are two species where both (1) and (2) usually occur together.

Two types of acetylenes from *Chrysanthemum leucanthemum* as described by Wrang and Lam (1975) are shown in Fig. 6.3. They are not easily separated on preparative silica plates. By using silica gel enriched with 5% of caffeine it is possible to separate the two types of acetylenes, whereas (**5a**) and (**5b**) or (**6a**) and (**6b**), respectively, could not be separated by means of caffeine as there is no difference in the number of π-electrons in their structures.

$$CH_3-(C\equiv C)_3-CH_2-CH\overset{Z}{=}CH-(CH_2)_n-CH=CH_2$$

(**5a**) $n = 4$ (**5b**) $n = 5$

$$CH_3-(C\equiv C)_3-\overset{E}{CH=CH}-\overset{E}{CH=CH}-(CH_2)_m-CH=CH_2$$

(**6a**) $m = 3$ (**6b**) $m = 4$

FIG. 6.3. Two types of acetylenes from *Chrysanthemum leucanthemum*.

The use of caffeine may be advantageous for separating compounds which differ slightly in the number of π-electrons, but its use is restricted to apolar or less polar compounds, as solvents used for development of thin layer plates or columns with added caffeine cannot be very polar. Caffeine is dissolved by diethylether and more polar solvents, and would be moved ahead of the compounds to be separated by these organic solvents so that the separating effect would be negligible. As many of the polyacetylenic compounds in nature are hydrocarbons and very often have slight differences in their π-electron systems, the method may often be used effectively. Even with slightly more polar compounds such as esters, ketones, aldehydes and thiophene derivatives, the method may be used with good results. As a general rule up to 10% of diethylether in light petroleum ether is a solvent mixture which does not dissolve the caffeine too much from a TLC plate. The separated compounds should preferably be eluted from preparative TLC plates impregnated with caffeine with light petroleum ether, as a greater amount of diethylether will, of course, dissolve caffeine together with the compound looked for. However, caffeine is soluble in water, which is usually not the case for the acetylene or the related compound. A partitioning between water and an organic solvent may be used for separation of caffeine from the compound to be analysed in case of contamination.

Tetramethyluric acid, which is less soluble in diethylether and other polar solvents than caffeine is, may be used as a complexing agent. However, the interaction seems to be too strong, so that during chromatography the material to be separated is not so easily eluted off the column or the plate.

Silver nitrate has been proposed for separation of unsaturated compounds containing different numbers of π-electrons for some purposes as described by Stahl (1967), but the method has not yet attracted attention for polyacetylene separations, although it may prove useful.

For preparative chromatography an application of a fraction or a solution may be performed by placing a great number of spots on a TLC plate, or a continuous band from a pipette may be drawn out slowly. After development in a chamber saturated with the solvent also used for the development on the plate, the plate is partially covered, so that only 10% of the developed material at either edge of the plate is kept free for spraying with potassium permanganate. The position of the unsprayed bands may, therefore, be visualised (Fig. 6.4). The bands are scraped off into filters and the compounds are eluted with a convenient solvent. The fractions obtained are monitored by ultraviolet spectroscopy, and further procedures may be carried out according to the results obtained.

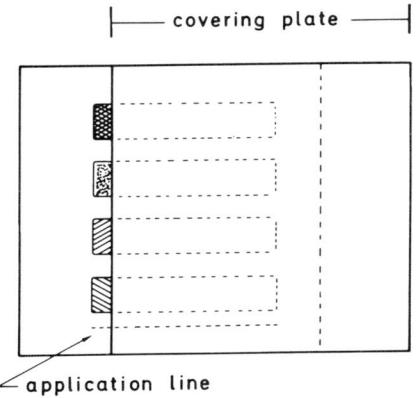

FIG. 6.4. Preparative TLC. About 10% of the plate is visualised.

The possibility of using alumina for separation has been considered; indeed it has been tried in the past. However, the high rate of decomposition has led to the abandonment of use of alumina. The greater degradation of polyacetylenes by use of alumina than by silica was shown by Lam (1973).

The procedure for making preparative thin layer plates has been described by Stahl (1967). The procedure for making silica gel plates containing 10% of caffeine is as follows: 2.5 g of caffeine are dissolved in water by heating until the caffeine has been totally dissolved. The solution is cooled to a temperature of 40–60°C and 25 g of silica gel (Kiesel gel 60G Merck Art. 7731) are added to the solution of caffeine and a slurry at about 35°C is distributed on glass plates (20 × 20 cm). This amount is enough for five analytical plates of about 0.3 mm. For making preparative plates about three times this quantity is required, depending on the thickness of the layer required. The plates are air-dried for no more than 5 min at room temperature and then oven-dried at 120°C for 30 min.

3. Centrifugal preparative chromatography

An alternative separation method is based on centrifugation (chromatotron). This method has been used for various purposes, for instance as described by Zeng *et al.* (1983). In this case the separation was performed on silica GF starting with a chloroform–methanol mixture (100:5) and after a gradient elution ending with an elution using a mixture (100:20). We have modified the disc coverage to a caffeine-containing silica (10% of caffeine). The preparative layer is 1 mm thick and the preparative plate is produced from a slurry placed on the disc under a rotation of $\frac{1}{2}$–1 revolution per second. The preparation is similar to that described in Section II.E.2.

A solution of the mixture to be separated is placed on the central part of the disc under rotation (700 r.p.m.). Elution under the influence of both centrifugal force and interaction between electrons in the caffeine and compounds in the mixture yields a separation of the compounds. A mixture of 11 mg of 1-phenylhepta-1,3-diyne-5-ene (**3**) and 6 mg of 1-phenylhepta-1,3,5-triyne (**4**) is eluted with light petroleum containing 0.1% of methanol at a flow of about 3–5 ml min^{-1}. The separation is completed in a very short time and we usually collect 5 ml fractions. Four fractions contained 9.6 mg of pure (**3**), one fraction contained 1.36 mg of a mixture of (**3**) and (**4**), and the five succeeding fractions contained 5.6 mg of (**4**). The unseparated part accounted for only about 8% of the total mixture applied. Smaller fractions than 5 ml could perhaps yield even better results; however, the result was quite acceptable.

Two other compounds, α-terthienyl (**1**) and bithienylbutynene (**2**), which are separated well on caffeine-impregnated silica plates, could also be separated well by using centrifugal preparative chromatography on caffeine-impregnated silica discs.

The advantages of using this method are: (1) rapid separation; and (2) a restricted use of solvents for this kind of chromatography.

F. Detection by GLC and HPLC

Minute amounts of polyacetylenes are exuded from roots into the surrounding soil. It has been possible to detect these various acetylenes by use of gas–liquid chromatography (GLC) after development of standard gas chromatograms with pure compounds from stock solutions. Gas–liquid chromatography has also been used to advantage in cases where *cis-* and *trans-* forms of the same molecular structure turn out to be difficult to separate. However, some decomposition may occur at the high temperatures often required to allow a compound to pass through the heated column of a gas chromatograph.

A more gentle method for separation of acetylenic structures is high performance liquid chromatography (HPLC). This method has often been used for separation of acetylenes and thiophenes, as for instance by Porter *et al.* (1979), Mansfield *et al.* (1980) and Sütfeld (1987).

In our laboratories a separation has been carried out of a mixture consisting of α-terthienyl (**1**), bithienylbutynene (BBT) (**2**), a dimer of BBT (iso-cardopatine), and the following compound

$$CH_3-(C{\equiv}C)_2-\underset{S}{\langle\!\!\!\langle\,\,\rangle\!\!\!\rangle}-C{\equiv}C-CH(Cl)-CH_2OAc$$

By using diode array detection the ultraviolet spectra are drawn after separation on reversed phase column material. By using this method it was revealed that the chlorine compound consists of two components very similar in polarity and in ultraviolet spectrum, which may indicate the presence of previously unknown isomers. The method has the advantage of being gentle, as heating is not required, and also benefits from its use of reversed phase chromatography as well as straight phase chromatography. Both methods of chromatography may be applied to a mixture of a plant extract with acceptable results.

In some cases pre-treatment of an extract with a short silica column or a Sep-Pak cartridge may retain the most polar material while allowing a more easily separable mixture through. This generally leads to better separations than are possible with crude extracts.

G. Characterisation of Isolated and Purified Compounds

A combination of spectroscopic methods can elucidate the structure of acetylenic and related compounds. It may be easy to elucidate the structure of a compound so well that full characterisation is possible for many structures of this group of organic substances, especially where no asymmetric carbon atom is present.

1. Absorption spectroscopy

The acetylenes occurring in nature are very often highly conjugated substances with one or more triple bond(s) in conjugation with double bonds. Such compounds usually exhibit ultraviolet spectra with characteristic patterns. They absorb in the region between 200 and 410 nm. The most highly oxidised carbon compound known from nature is the trideca-3,5,7,9,11-pentayn-1-ene, CH_3—$(C{\equiv}C)_5$—$CH{=}CH_2$. It possesses, among other absorption bands, its most red-shifted band at 410 nm. This compound very frequently occurs in plants of the family Asteraceae (Compositae).

In nature there are some other compounds with fewer triple bonds in conjugation and often with one or more triple bond(s) replaced by double bonds. This kind of conjugation gives a series of acetylenic compounds with characteristic UV patterns for the different compounds, as the number of electrons in the conjugated system play a major role in determining the spectrum. Some frequently occurring compounds and their ultraviolet spectra are given in Fig. 6.5.

Related to these compounds are alcohols and esters or both, co-occurring with the hydrocarbons in a plant. As an oxidation of a methyl group to the corresponding alcohol will not change the chromophore, the ultraviolet spectra of alcohols or esters will remain the same as for the hydrocarbons.

A great number of other acetylenes may show characteristic UV patterns, whereas triple and double bonds in conjugation with unsaturated ring structures such as thiophenes and furans or enol ethers may yield less characteristic ultraviolet spectra, as shown in Fig. 6.6.

An unsaturated alcohol with the OH group in the allyl position may be oxidised to an aldehyde by the use of MnO_2. This oxidation may be performed in an organic solvent, e.g. diethylether. It usually takes about 10–15 min to oxidise a few milligrams of an alcohol to the corresponding aldehyde under stirring. The resulting carbonyl group is a part of the prolonged chromophore, which gives rise to a shift of the ultraviolet spectrum towards longer wavelengths. Thus oxygen or sulphur atoms may play a decisive role in the appearance of an absorption spectrum.

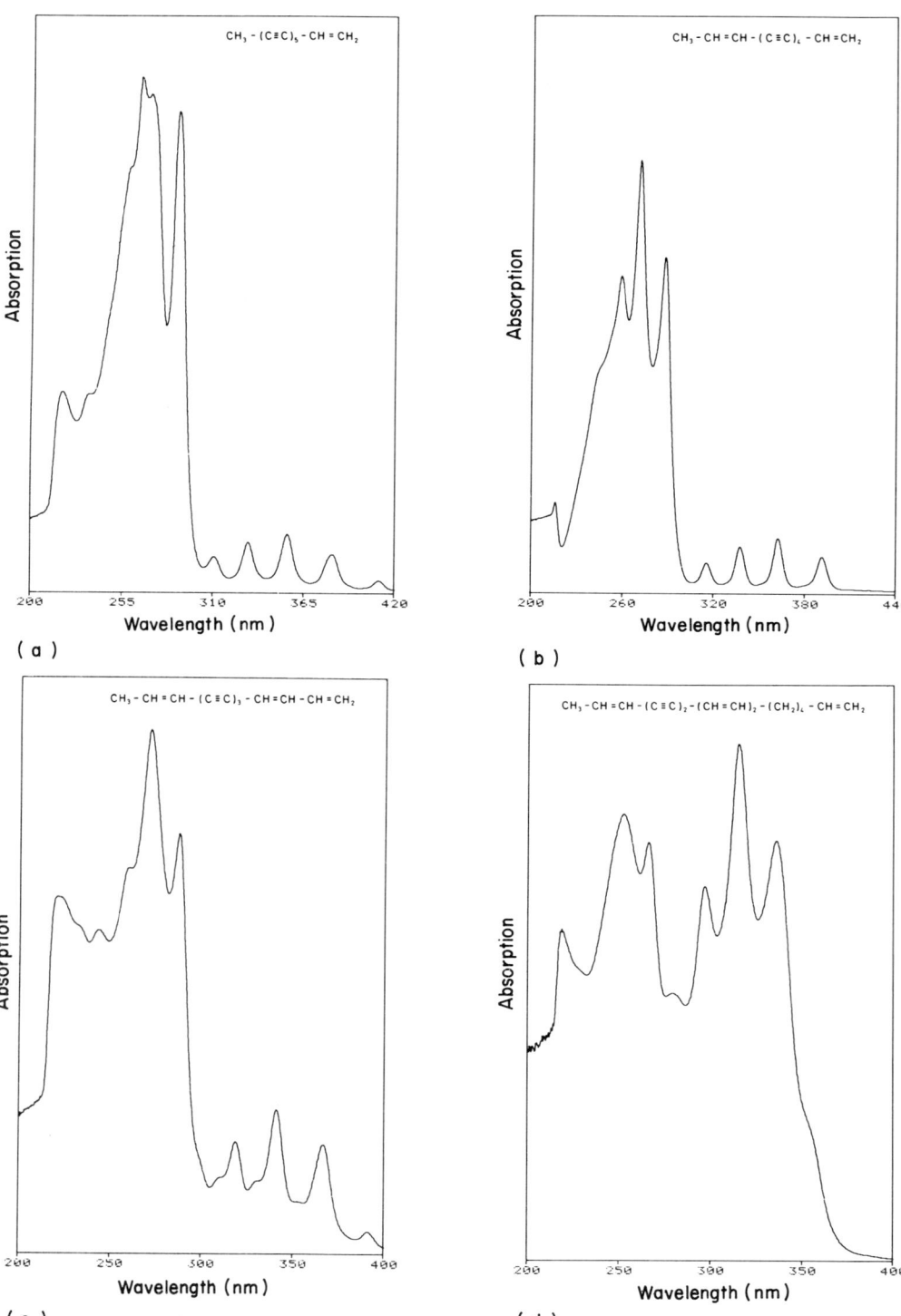

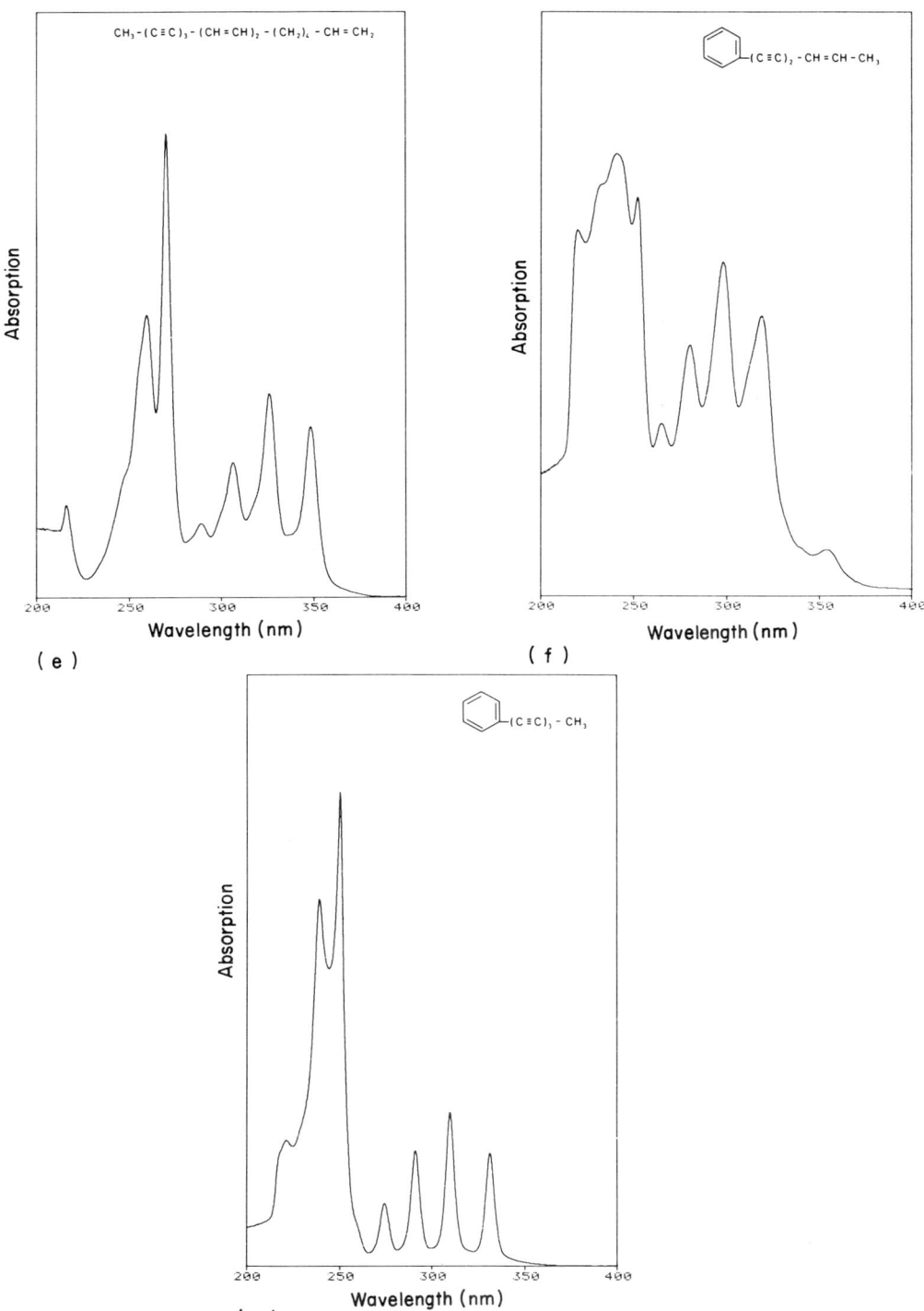

FIG. 6.5. Ultraviolet spectra of some frequently occurring compounds. Characteristic patterns can be seen.

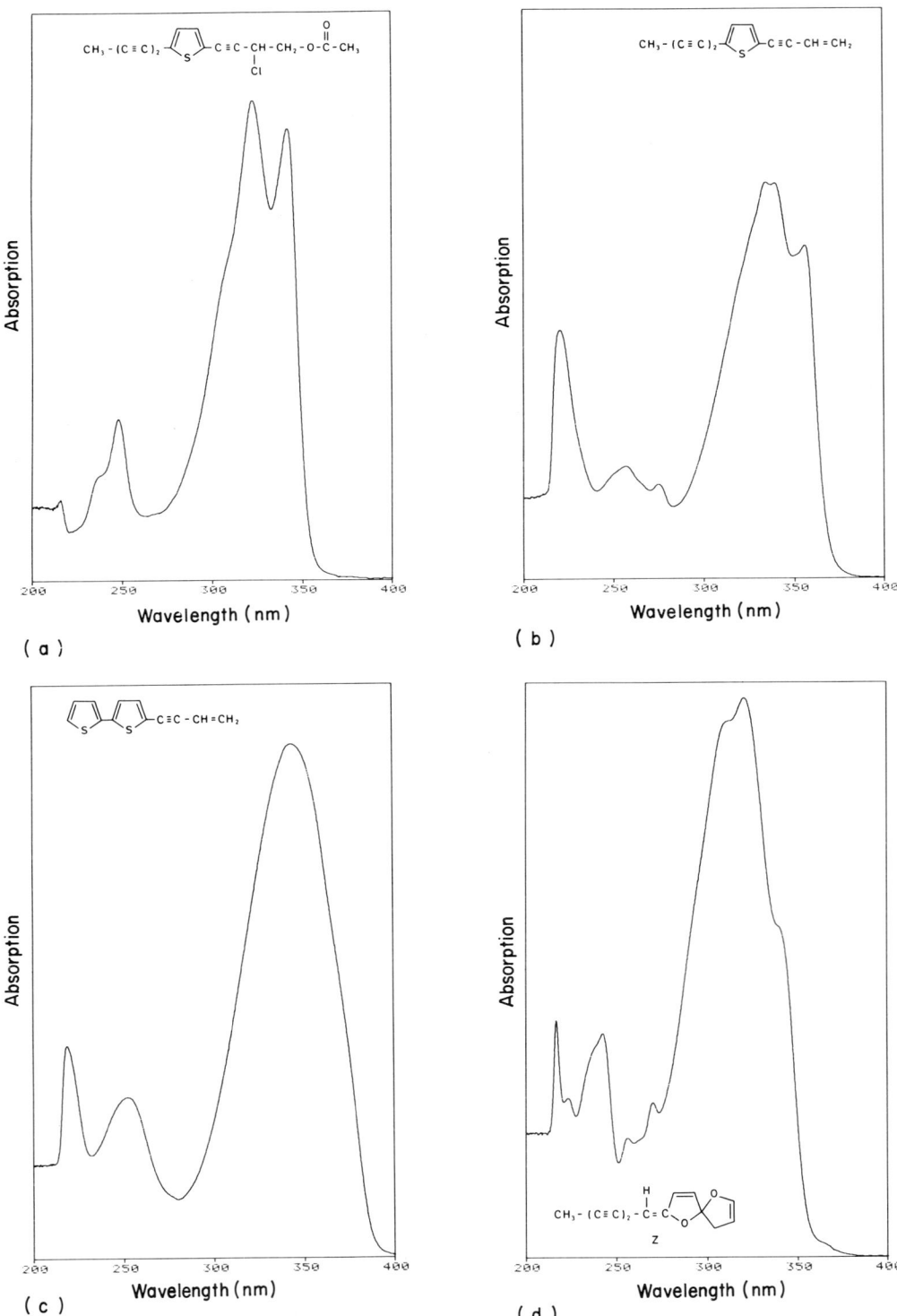

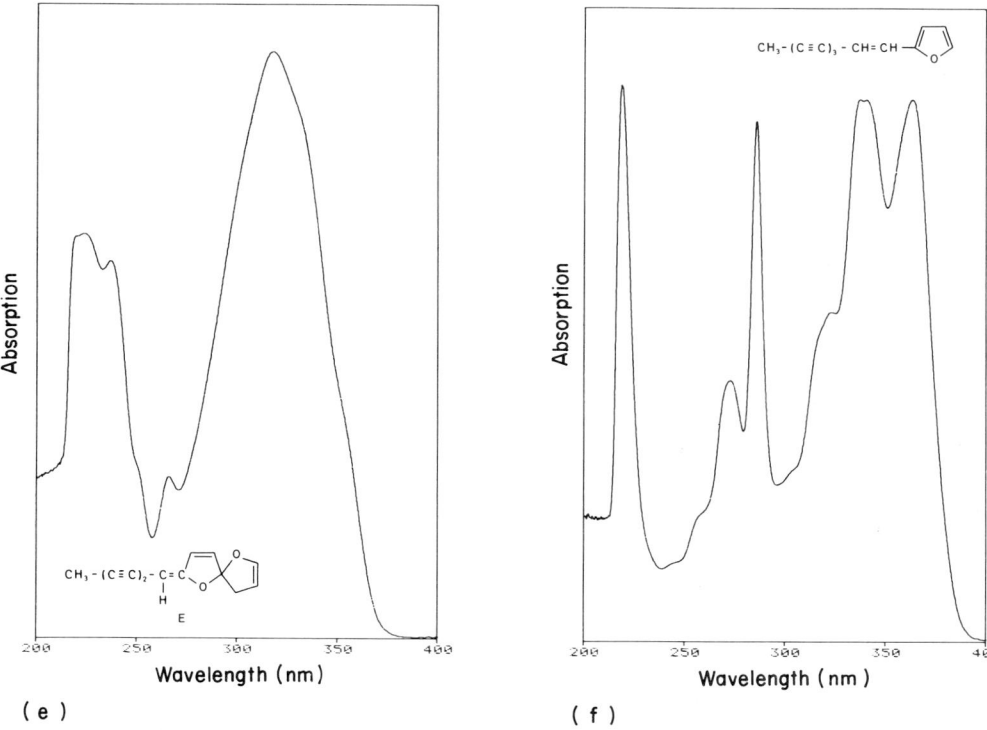

FIG. 6.6. Ultraviolet spectra of compounds with unsaturated ring structures show less characteristic patterns.

An example of a polyacetylenic alcohol oxidised to a conjugated aldehyde is given in Fig. 6.7. The spectrum may be recorded in diethylether. As soon as the solution of the aldehyde is evaporated and the aldehyde again dissolved in acidic methanol, the formation of an acetal may occur. The ultraviolet spectrum will be similar to that of the alcohol before oxidation. The chromophore will now be shortened as the aldehyde oxygen atom takes part in the acetal formation.

It is thus possible to obtain information about the chromophore by application of ultraviolet spectroscopy to the compound in solution, but in order to obtain information about a saturated part of the molecule or about the functional groups (e.g. alcohol or ester groups), infrared spectroscopy is required to provide additional information.

2. Infrared spectroscopy

Infrared spectroscopy can give very valuable information about a structure containing triple bonds. A triple bond may be end positioned when a sharp signal for the H—C≡ stretching at 3300 cm^{-1} will be present. For an internal triple bond there will usually be a band at about 2200 cm^{-1}. Under very symmetric conditions the band of the triple bond may not be observed, but in some cases more than one signal will be seen in the region between 2170 and 2250 cm^{-1}. A very strong signal may occur if the triple bond is conjugated with a carbonyl group. This may be a very informative observation. Very often the triple bonds are conjugated with one or two *trans*-double bonds, thus giving rise to bands in the region between 935 and 965 cm^{-1}, whereas a *cis*-double bond can be observed at about 700 cm^{-1}. Furthermore, a vinyl group, which is often a part of an acetylenic compound, may be observed at 915 and 985 cm^{-1}. These features are the most general for polyacetylenes.

Some naturally occurring acetylenes are liquid substances at room temperature and the infrared spectrum may be recorded from a film of the compound placed between sodium chloride plates. Solid compounds may be recorded either in potassium bromide tablets or in solution.

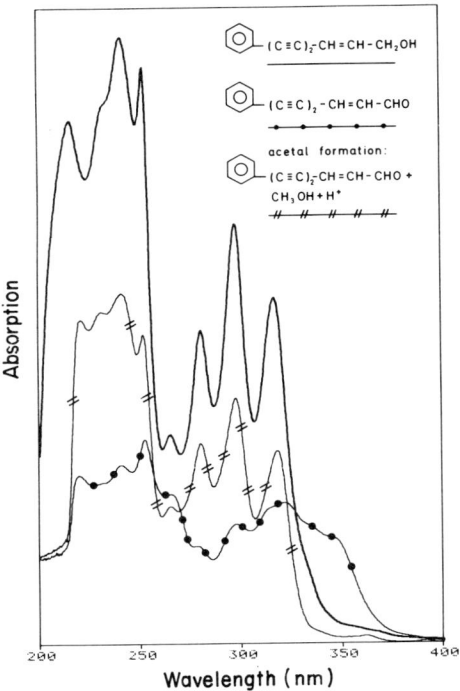

FIG. 6.7. Oxidation of a polyacetylenic alcohol followed by UV spectroscopy. The sequence is allylic alcohol → aldehyde → acetal.

3. *Nuclear magnetic resonance spectroscopy*

(a) *^{1}H-NMR spectroscopy*. Proton magnetic resonance spectroscopy (^{1}H-NMR spectroscopy) has long been of enormous value in the characterisation of acetylenes and related compounds. The CH$_3$—C≡C— and the CH$_3$—CH=CH—C≡C— end structures are quite characteristic, the former showing a singlet at about 2 δ and the latter double doublet at about 1.85 δ. The vinyl end of many acetylenic compounds may be detected by its characteristic pattern. *Cis*- or *trans*-double bonds (Z- or E-) may be distinguished in some uncomplicated cases, as for instance for the ^{1}H NMR spectra of *cis*- or *trans*-dehydromatricariaester, which both occur in *Chrysanthemum leucanthemum*. The ^{1}H-NMR spectra of these compounds are shown in Fig. 6.8.

6. POLYACETYLENES AND RELATED COMPOUNDS

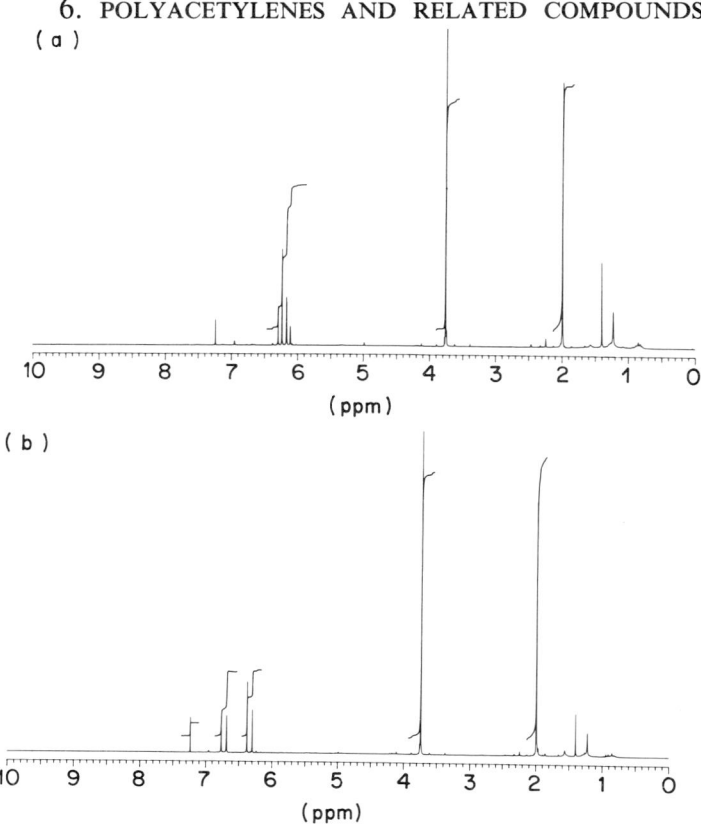

FIG. 6.8. ¹H-NMR spectra of (a) Z-dehydromatricariaester and (b) E-dehydromatricariaester.

The ¹H-NMR spectra of 1-phenylhepta-1,3,5-triyne (PHT) (**3**), and 1-phenylhepta-3,5-diyn-1-ene (**4**), are presented in Figs 6.9(a), (b), and (c), and are different in two respects. In Fig. 6.9(a) the methyl group neighbouring a triple bond yields a singlet at about 2 δ, while in (b) the methyl group connected to a double bond with two hydrogen atoms involved results in a double doublet at about 1.82 δ. In addition the structure (b) shows a pattern from the double bond at 5.6 and 6.35 δ, indicating a *trans*-double bond. Signals in the region 0.85–1.55 δ and in the region between 3 and 4 δ are due to solvent impurities. The signal at 7.24 δ comes from the solvent $CDCl_3$. The integrals at about 1.82 δ and 2 δ may give the proportional value of the relative amounts of the two compounds (a) and (b) in a mixture, in this case approximately 1:6.

Much information may be obtained about the groups neighbouring the double bonds or in between double bonds from their shifts and their resonance patterns obtained by ¹H-NMR spectroscopy. A comprehensive description of ¹H-NMR spectra of several naturally occurring thiophene compounds has been given by Bohlmann and Zdero (1985).

(b) *¹³C-NMR spectroscopy.* For the most complicated acetylenes carbon-13 nuclear magnetic resonance (¹³C-NMR) spectra may reveal useful information, as for instance the shifts of carbonyl groups in aldehydes, ketones and esters will give information about the presence of such groups in a molecule. ¹³C-NMR spectra of several thiophenes have been described by Bohlmann and Zdero (1985).

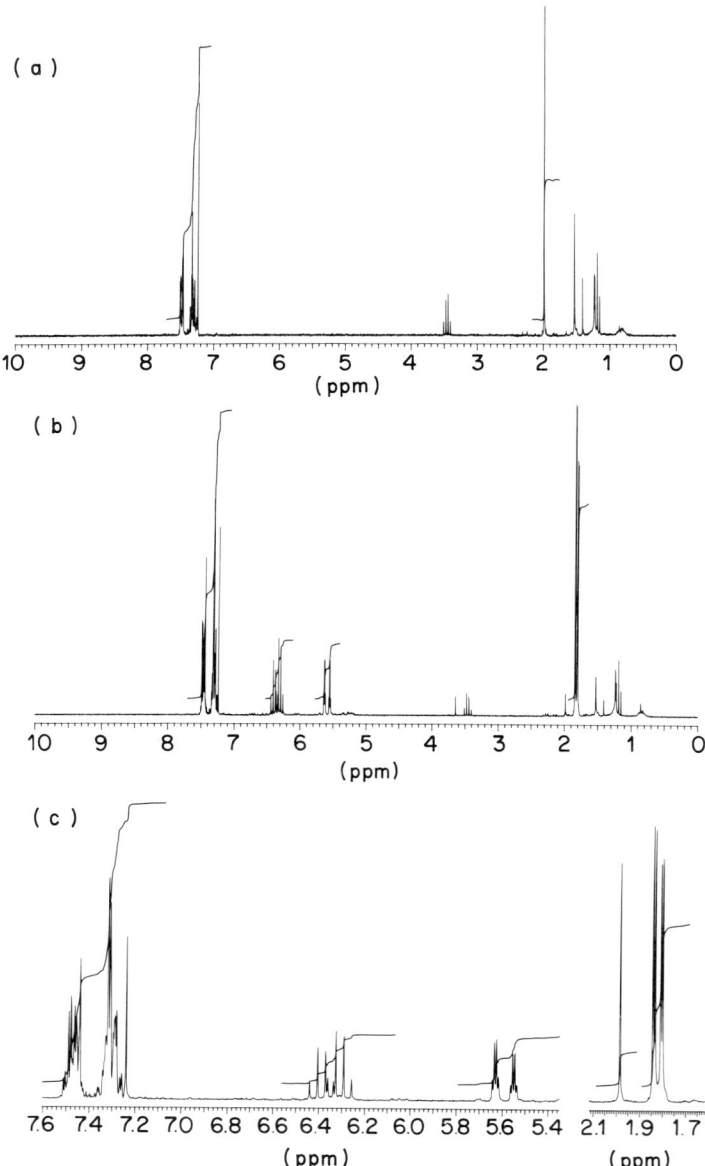

FIG. 6.9. ^{1}H-NMR spectra of (a) 1-phenylhepta-1,3,5-triyne, (**3**), (b) 1-phenylhepta-1,3-diyne-5-ene, (**4**) and (c) 1-phenylhepta-1,3-diyne-5-ene (6 parts) and 1-phenylhepta-1,3,5-triyne (1 part).

4. Mass spectrometry

The mass spectra of polyacetylenes containing several triple bonds and in many cases conjugated with double bonds reveal a fragmentation pattern similar to that of benztropylium cation and indenyl cation with fragments at 141 amu and 115 amu, as described by Aplin and Safe (1967) and later by Allen and Thomas (1971). Allen and Thomas have shown by exact mass determination that the observed masses 141.0713

6. POLYACETYLENES AND RELATED COMPOUNDS

and 115.0555 agree well with calculated masses for the proposed benztropylium cation (141.0704) and the indenyl cation (115.0548). This is a general feature for the highly conjugated acetylenes, which may be of great value for the indication of the presence of polyacetylenes in a mixture when determined by combined gas–liquid chromatography with mass spectrometry (GC-MS).

5. *Optical rotation*

Where an asymmetric carbon is present rotational measurements may be of great value.

H. Chemistry

A great diversity of functional groups and a few heteroatoms may be associated with acetylenes or compounds biosynthetically formed from acetylenes. Thus alcohol, ester, aldehyde, ketone, acid and lactone groups are among the most common; besides these groups epoxides, spiroenolethers, furans and thiophenes containing heteroatoms in ring structures frequently occur. Phenyl rings are also present in several naturally occurring acetylenes.

The conjugated systems of the polyacetylenes may be the basis for the study of many chemical aspects, some of which will be mentioned as examples below.

1. *Ultraviolet irradiation*

Cardopatine and isocardopatine are dimer products of BBT [5-(3-buten-l-ynyl)-2,2'-bithienyl, or bithienylbutynene] isolated from *Cardopatium corymbosum* by Selva *et al.* (1978). This plant belongs to the tribe Cynareae of the family Asteraceae. From another species of the same tribe, *Echinops bannaticus* Rochel, the isocardopatine was isolated in our laboratory (unpubl. data). The compound was characterised by UV, ^{1}H-NMR, ^{13}C-NMR and MS.

After irradiation of bithienylbutynene with ultraviolet light at 350 nm three dimer products could be separated on a TLC plate. The compounds were characterised by UV, NMR and MS data. One of the compounds had the same R_f value as that of isocardopatine isolated from the *E. bannaticus*.

Selva *et al.* (1978) point to the possibility of photochemical and/or enzyme-catalysed formation of bithienylbutinene in the roots of *Cardopatium corymbosum*. The enzyme catalysed formation is perhaps the most likely possibility as the dimer has been isolated repeatedly from the roots of *E. bannaticus* and only in trace amounts from some other *Echinops* species, whereas there has been no sign of the presence of any dimer product in *E. sphaerocephalus* or *E. persicum*, although these species also contain bithienylbutynene. As the treatment of the root material from the different plants has been the same in all cases, there is no reason to believe that a photochemical process is significant, although irradiation of bithienylbutynene may lead to a mixture of dimer products in very minute amounts. The dimer, which we have isolated, was present in the root material, but could not be detected in the aerial part of *E. bannaticus*.

Another photochemical process, which may be more extensive than presently recognised, is a Z- to E-form conversion. The compound (the Z-form, Fig. 6.10) has been transformed experimentally by irradiation with ultraviolet light at 350 nm (unpubl.

data) into the E-form (Fig. 6.11). These compounds (Z- and E-matricarialactones) are both present in the roots of *Solidago virgaurea* (Astereae). The plants usually grow in shade (woods), but they may also be found in open areas. Although the Z-form is the more abundant in plants from both habitats, the plants collected on open areas contained more of the E-form than plants collected in woods. This finding came from a seasonal study of *Solidago virgaurea* described by Lam (1971).

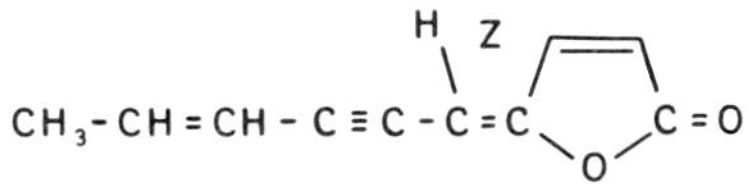

$$CH_3-CH=CH-C\equiv C-C=C\overset{H}{\underset{}{\diagup}}\overset{Z}{\diagdown}\cdots C=O$$

FIG. 6.10. Z-matricarialactone.

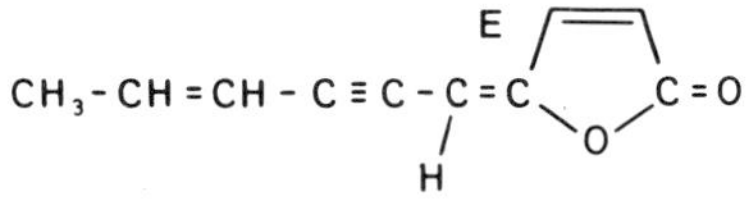

$$CH_3-CH=CH-C\equiv C-C=C\overset{E}{\underset{H}{\diagup}}\cdots C=O$$

FIG. 6.11. E-matricarialactone.

In *Chrysanthemum leucanthemum* and in *Santolina chamaecyparissus*, both plants of the tribe Anthemideae, the Z- and E-spiroethanol ethers (Fig. 6.12) are present in the proportions: $Z:E = 1:9$ in the roots of *C. leucanthemum* and $Z:E = 3:1$ in the aerial parts of *S. chamaecyparissus*. A solution of the E-form could be changed to the Z-form by irradiation at 350 nm. Irradiation of the Z-form did not lead to the E-form. The presence of increased amounts of the Z-form in the aerial parts of plants agree well with the change of the E-form to the Z-form under the influence of ultraviolet light.

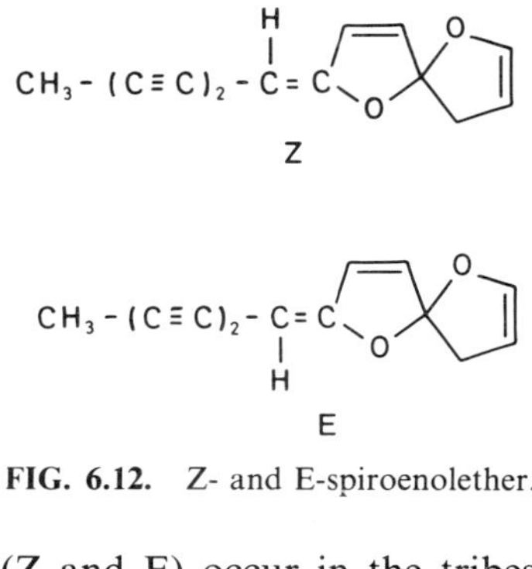

$$CH_3-(C\equiv C)_2-\overset{H}{\underset{Z}{C}}=C\cdots$$

$$CH_3-(C\equiv C)_2-\overset{}{\underset{H}{C}}=C\cdots\overset{E}{}$$

FIG. 6.12. Z- and E-spiroenolether.

Dehydromatricariaesters (Z and E) occur in the tribes Astereae and Anthemideae (Fig. 6.13). Both compounds occur in the roots of *Artemisia vulgaris* in the proportion $Z:E = 6:1$ and in the flower heads of *C. leucanthemun* in the proportion $Z:E = 1:\infty$, which means that in most experiments almost complete conversion to the E-form could be detected.

Irradiation of E-dehydromatricariaester and Z-dehydromatricariaester at 350 nm leads to the formation of an equilibrium state in both cases. The rearrangement may be followed by ultraviolet spectroscopy and by separation. Quantitative determination is then performed of the separated compounds after the reaction. After a certain period of irradiation in a solution of cyclohexane the maxima of the ultraviolet spectra for both

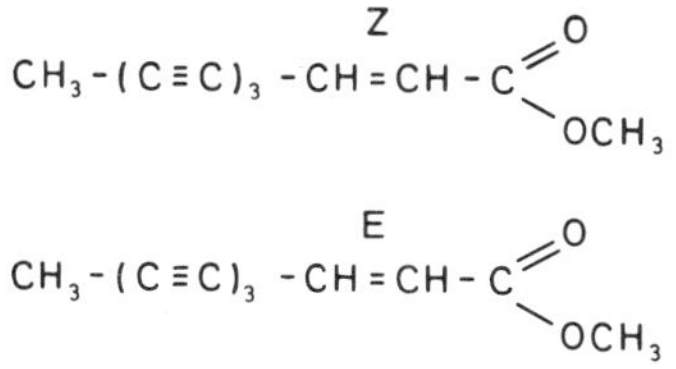

6. POLYACETYLENES AND RELATED COMPOUNDS

$$CH_3-(C\equiv C)_3-CH\overset{Z}{=}CH-C\overset{O}{\underset{OCH_3}{\diagdown}}$$

$$CH_3-(C\equiv C)_3-CH\overset{E}{=}CH-C\overset{O}{\underset{OCH_3}{\diagdown}}$$

FIG. 6.13. Z- and E-dehydromatricariaester.

E- and Z-forms in the equilibrium state occur at the same wavelength, 345.7 nm. The ultraviolet spectra for the pure Z-form and E-form exhibit maxima at 348.2 and 345.2 nm, respectively. After about 1 h of irradiation the solutions of each compound reach the equilibrium state. It is thought that a decomposition occurs at the same time because a loss of material can be observed. By use of TLC the two forms are well separated.

We have not been able to detect dehydromatricariaesters from the roots of *C. leucanthemum*, although they may be present in minor amounts. The presence of E-dehydromatricariaester as the dominant form in the aerial parts of *C. leucanthemum* and the Z-form as the prevailing form in the root material of *A. vulgaris* may be explained partly by a conversion of the Z-form to the corresponding E-form under the influence of ultraviolet light.

The experimental reactions carried out with BBT, the Z-matricarialactone, the Z- and E-spiroenolethers, and the Z- and E-dehydromatricariaesters clearly show that special precautions should be taken in order to avoid an inappropriate influence of ultraviolet light on the compounds in question. Presumably many other highly conjugated compounds should be protected against the influence of ultraviolet irradiation in order to avoid photodegradation.

(a) *Formation of thiophenes from polyacetylenes.* A very comprehensive description of naturally occurring thiophenes, including spectroscopic data of acetylenic thiophenes, is given by Bohlmann and Zdero (1985). Feeding experiments with labelled precursors reveal that thiophenes are formed from acetylenes in nature. The metabolic pathway is still unclear. Some early experiments with addition of H_2S to polyacetylenes were carried out by Schulte *et al.* (1962, 1965). Thiophenes corresponding to the naturally occurring ones could be produced synthetically.

(b) *Diels–Alder reaction.* Maleic acid anhydride can be used for determination of the position of double bonds in a conjugated system as described by Bohlmann *et al.* (1965) and by Kaufmann and Lam (1965) for compounds isolated from *Dahlia merckii* and *D. scapigera*, respectively. The chromophore in conjugated compounds is reduced, resulting in a change in the ultraviolet spectrum.

(c) *Epoxides.* Epoxides often occur in such large amounts that it is not reasonable to believe they are artifacts. The presence of pontica epoxide

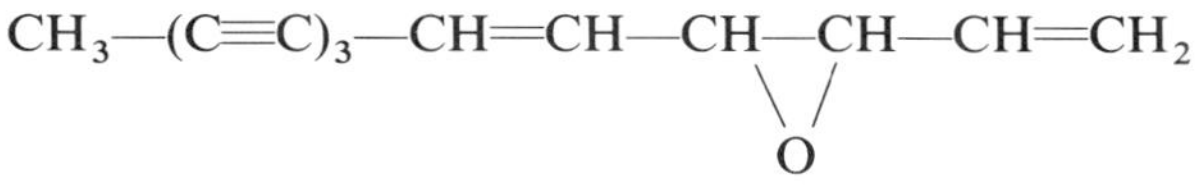

$$CH_3-(C\equiv C)_3-CH=CH-\underset{\underset{O}{\diagdown\diagup}}{CH-CH}-CH=CH_2$$

in the amount of 100–400 mg kg^{-1} in the root material of *Artemisia pontica* indicates that this compound is present in the plant with an as yet unknown function. The compound may be analysed by spectrophotometry. In addition, cleavage of the epoxy group to a diol, as well as a sodium periodate reaction, can lead to a further degradation resulting in known aldehydes, as described by Bohlmann *et al.* (1960).

(d) *Alcohols*. Acetylenes with one or two alcohol groups are frequently encountered in nature. The corresponding acetates or diacetates often co-occur with the alcohols in a given plant. In the laboratory the alcohols may be transformed into the esters by mild treatment with acetic anhydride or acetyl chloride. In addition, the esters can be transformed into the corresponding alcohols by mild hydrolysis with an alcoholic potassium hydroxide solution at room temperature.

3. Photoprotection

Many of the naturally occurring acetylenes and related compounds possess the property of being photosensitisers with different biological activities resulting. This light-activated toxicity is thought to act as an efficient defence against plant pests. At the same time some of the polyacetylenes are decomposed. Some of the most highly conjugated compounds, for instance the trideca-1-en-3,5,7,9,11-pentayne hydrocarbon, $CH_3-(C\equiv C)_5-CH=CH_2$, the trideca-1,11-dien-3,5,7,9-tetrayne hydrocarbon, $CH_3-CH=CH-(C\equiv C)_4-CH=CH_2$ and other compounds are very unstable in the presence of ultraviolet light. How can they 'survive' temporarily in the plant? In plants a great number of compounds may interact with electrons in, for instance, polyacetylenes, if the compounds taking part in the interaction contain a conjugated electron system.

Such a system could be involved in the interaction with the highly conjugated electron system of an acetylenic compound with the result that the acetylene will be more or less protected against the influence of ultraviolet light. An investigation which confirms this has been carried out with caffeine and different naturally occurring compounds, all with a stabilising effect, that gives the acetylenes a somewhat longer life. The observation that polyacetylenes were better protected on a silica plate containing caffeine compared to a plate covered with silica without caffeine was described by Lam (1973). This result gave us the idea of investigating the relative protection exerted by other compounds to polyacetylenes and related thiophene compounds. For this purpose some purines other than caffeine, some lignans and flavones were chosen. They all had some effect, described in a paper given at the NOARC Conference held in Aarhus, Denmark, July 1987 (see Lam *et al.*, 1988).

The protective and loose binding effects of caffeine in a mixture with silica gel are thus of great importance for the separation and characterisation procedures applied to unstable polyacetylenes. The lignans have a limited effect, whereas the lignan-related compound dillapiol and the flavones provide a certain degree of protection to acetylenic and related compounds.

III. DIFFERENCES AND SIMILARITIES WITHIN GROUPS AND TRIBES OF THE FAMILY ASTERACEAE AND OTHER FAMILIES

A. Asteraceae

Acetylenic and related compounds seem to be confined to certain groups and tribes, but overlaps are also seen.

The tribe Cynareae has several genera containing unsaturated compounds with an aldehyde group isolated from the unsaturated part of the molecule. The compounds may be polyenes or polyacetylenes in the unsaturated part. For instance $CH_3-CH_2-(CH=CH)_4-(CH_2)_n-CHO$ (where n is a number from 4 to 7) occurs in *Centaurea scabiosa* L., in *C. montana* L. and in other genera, especially in the leaves and stems of these plants.

Another compound named Aplotaxen, which occurs frequently in the roots of several genera within the tribe Cynareae, has the following structure:

$$CH_3-CH_2-[(CH=CH)-CH_2]_3-(CH_2)_4-CH=CH_2$$

These compounds are not acetylenic, but due to their metabolic relationship they may be considered as being related to the naturally occurring acetylenes with which they co-occur in the plants of Cynareae. The structure of Aplotaxen is closely related to that of linoleic acid. Within Cynareae and Inuleae chlorine-containing compounds occur more frequently than in other tribes of Asteraceae. The chain length is usually between 13 and 17 carbon atoms for the compounds present in Cynareae.

In the tribe Astereae C10 acetylenes and methylesters of acids or lactones corresponding to the acids frequently occur (e.g. matricaria esters, dehydromatricaria esters, dihydromatricaria esters and the corresponding lactones). There is some overlap between the tribes Astereae and Anthemideae in the presence of compounds with C10 acetylenes. Thus *Solidago altissima* (Astereae) and other genera of the same tribe contains Z-dehydromatricaria ester, which is also found in *Artemisia vulgaris*. Z- and/or E-dehydromatricaria ester are present in *Chrysanthemum leucanthemum* and other plants of the same tribe (Anthemideae). The tribe is also well-known for its spiroenol ethers with 13 carbon atoms in the basical structure.

α-Terthienyl and bithienylynene (BBT) and related derivatives occur in *Tagetes* species (Helenieae) and in *Echinops* species (Cynareae), showing an overlap between these tribes.

Although about 1000 different acetylenic compounds are known to occur in nature, certain patterns are characteristic of individual species and it is very important to recognise them when new plants are considered for investigation. Of course, the knowledge of what is to be expected in a plant may be of importance in choosing the isolation and characterisation methods to be applied, but it should also be emphasised that pre-conceived ideas should be avoided.

As an example, we mention here the differences we found between plants of a single genus when we investigated four subspecies of *Dahlia coccinea* Cav. There is a general pattern of compounds present in these subspecies, but one of the *Dahlias* contained relatively large amounts of 1-phenylhepta-1,3-diyne-5-ene and derivatives thereof, both in the root material and in the aerial parts. Only one of the other subspecies contained trace amounts of this hydrocarbon (Lam et al., 1968). A horticultural species also contained the 1-phenylhepta-1,3-diyn-5-ene. Two of the four subspecies showed trace amounts of the hydrocarbon 1-phenylhepta-1,3,5-triyne (PHT), whereas the horticultural species had a high content of 63 mg of PHT per 1000 g of fresh root material. This compound is well known from other plants within Coreopsedinae, to which group the genera *Dahlia, Cosmos, Coreopsis* and *Bidens* of the Heliantheae belong. In the aerial parts of *Bidens alba* var. *radiata* 0.05–0.1% of PHT is present calculated on basis of fresh material.

Acetylenic alkamides seem to be restricted to certain genera within Anthemideae and Heliantheae; both are tribes belonging to Asteraceae, whereas pure olefinic alkamides have been isolated from Piperaceae, Aristolochiaceae and Rutaceae as well. The last three families are not, however, considered to contain acetylenic compounds.

From *Achillea* species (Anthemideae) a great number of both acetylenic and pure olefinic alkamides have been reported. Other genera within Anthemideae and Heliantheae contain both acetylenic and olefinic amides, whereas only an olefinic amide was found in the third tribe (Senecioneae).

About 130 alkamides have been characterised up to the present. Examples of amine groups taking part in the amide formation are given in the review article by Greger (1984). One of the most frequently encountered amines is isobutylamine, but ring structures are also met, for instance piperidines, pyrrolideines, piperideines and phenylethylamines.

B. Polyacetylenes in Umbelliferae and Araliaceae

Both families are rich in alcohols and ketones such as falcarinol, falcarinone and related compounds with two or more alcohol groups or with two ketone groups constitute the most frequently encountered acetylenes in the two families. A recent review article about polyacetylenes in Araliaceae has been published by Hansen and Boll (1986a).

C. Polyacetylenes in Campanulaceae

As described by Bentley et al. (1969) and Lam amd Kaufman (1969), the plant family Campanulaceae has several members of the genus *Campanula* which contain a number of C14 polyacetylenes with a tetrahydropyranylether part in the structure, but C14 straight-chain compounds are also present in several cases. Most commonly represented are the ene-diyne and ene-diyn-ene chromophores, but other acetylenic patterns may also occur as described by Bohlmann et al. (1973). With the ene-diyne and the ene-diyn-ene chromophores represented, the Campanulaceae bear some relationship to Compositae plants, especially when compared to plants of the tribe Astereae, except that the chain length is most often reduced to C10 in this tribe.

D. Acetylenes in Fungi

Bu'Lock (1964) describes a series of polyacetylenes that are essentially different from those usually isolated from Asteraceae, Araliaceae, Umbelliferae and other highly developed plant families. However, dehydromatricaria acid, present as the methyl ester in Asteraceae, is also present in fungi. Matricarianol and dehydromatricarianol are related compounds present in fungi as well. A series of allenic-diacetylenes with an end-positioned acetylenic group was described by Bew et al. (1966a,b). Similar compounds were described earlier by Jones et al. (1960) from other members of Basidiomycetes. Expoxides and amides are present in Asteraceae and in Basidiomycetes. End-positioned diacetylenes are common in fungi as mentioned above. Another type of end-positioned diacetylene occurs among acetylenic amides found to be present in genera of Anthemideae and Heliantheae (Asteraceae).

IV. BIOSYNTHESES OF ACETYLENES AND RELATED COMPOUNDS

A great number of biosynthetic experiments have been carried out in order to throw light upon the mechanisms of synthesis in nature. The greater part of the acetylenes and related compounds are produced via an acetyl-CoA/malonyl-CoA based biosynthesis route with several side reactions and chain shortenings occurring. It is commonly accepted that oleic acid and the related crepenynic acid both play an important role in the further route of desaturation and formation of polyacetylenic and related compounds such as monothiophenic, dithiophenic, as well as trithiophenic compounds. Also, the polyacetylenic and polyenic alkamides are supposed to be derived from unsaturated fatty acid structures. Many other unsaturated compounds related to polyacetylenes could just as well be included. It is therefore not feasible to reference all the proposed schemes published in literature, and only a few of these as described by Anchel (1967), Bu'Lock and Smith (1967), Bohlmann et al. (1973), Jones and Thaller (1974), Wrang and Lam (1975), Sørensen (1977), Greger (1984), and Hansen and Boll (1986a) will be referred to.

V. BIOLOGICAL ACTIVITIES

Although a description of biological activities is beyond the scope of the present chapter dealing with analytical and chemical methods, it should be briefly mentioned that many of the naturally occurring acetylenes and related compounds often possess more than one biological property. Some of the compounds are photoactive, a property that might lead to generation of toxic singlet oxygen; in other cases a dark effect may be exerted. Other compounds may be growth regulators, retarding germination of seeds or the growth of roots, as described by Numata et al. (1974).

Known activities exerted by acetylenes and related compounds are of fungicidal, antibiotic, antiviral, nematicidal, insecticidal, or algicidal character. Some compounds may provoke allergy, as for example described by Hansen and Boll (1986b). A synergetic effect may be brought about by interaction between acetylenes or related compounds and other naturally occurring systems like lignans. This is an expanding area of research which requires further study.

ACKNOWLEDGEMENTS

The critical reading by Professor J. T. Arnason, Ottawa, is greatly appreciated by the authors.

REFERENCES

Allen, E. H. and Thomas, C. A. (1971). *Phytochemistry* **10**, 1579–1582.
Anchel, M. (1967). *In* "Antibiotics", Vol. II, Biosynthesis (D. Gottlieb and P. D. Shaw, eds), pp. 189–215. Springer-Verlag, Berlin.
Aplin, R. T. and Safe, S. (1967). *Chem. Commun.*, 140–142.
Bentley, R. K., Jenkins, J. K., Jones, E. R. H. and Thaller, V. (1969). *J. Chem. Soc. (C)*, 830–832.
Bew, R. E., Chapman, J. R., Jones, E. R. H., Lowe, B. E. and Lowe, G. (1966a). *J. Chem. Soc. (C)*, 129–135.
Bew, R. E., Cambie, R. C., Jones, E. R. H. and Lowe, G. (1966b). *J. Chem. Soc. (C)*, 135–138.
Bohlmann, F. and Kleine, K.-M. (1965) *Chem. Ber.* **98**, 872–875.
Bohlmann, F. and Zdero, C. (1985). *In* "Heterocyclic Compounds. Thiophene and Its Derivatives" (S. Gronowitz, ed.), Vol. 44, pp. 261–323. John Wiley and Sons, Chichester.
Bohlmann, F., Arndt, C. and Bornowski, H. (1960) *Chem. Ber.* **93**, 1937–1944.
Bohlmann, F., Burkhardt, T. and Zdero, C. (1973). *In* "Naturally Occurring Acetylenes", pp. 70–73. Academic Press, London and New York.
Bu'Lock, J. D. (1964). "Progress in Organic Chemistry. Polyacetylenes and Related Compounds in Nature", pp. 86–134. Butterworth, London.
Bu'Lock, J. D. and Smith, G. N. (1967). *J. Chem. Soc. (C)*, 332–336.
Greger, H. (1984). *Planta Med.* **50**, 366–375.
Hansen, L. and Boll, P. M. (1986a). *Phytochemistry* **25**, 285–293.
Hansen, L. and Boll, P. M. (1986b). *Phytochemistry* **25**, 529–530.
Jones, E. R. H. and Thaller, V. (1974). *Anales Quimica* **70**, 1009–1014.
Jones, E. R. H., Leeming, P. R. and Remers, W. A. (1960). *J. Chem. Soc.*, 2257–2263.
Kaufmann, F. and Lam, J. (1965). *Acta Chem. Scand.* **19**, 1267–1268.
Lam, J. (1971). *Phytochemistry* **10**, 647–653.
Lam, J. (1973). *Planta Med.* **24**, 107–111.
Lam, J. and Kaufmann, F. (1969). *Chem. and Ind.*, 1430.
Lam, J., Kaufmann, F. and Bendixen, O. (1968). *Phytochemistry* **7**, 269–275.
Lam, J., Breteler, H., Arnason, J. T. and Hansen, L. (eds) (1988). Proceedings of the 1st NOARC Conference, Aarhus, Denmark, July 1987.
Mansfield, J. W., Porter, A. E. A. and Smallman, R. V. (1980). *Phytochemistry* **19**, 1057–1061.
Numata, M., Kobayashi, A. and Ohga, N. (1974). *In* "Studies in Urban Ecosystems" (M. Numata, ed.), pp. 22–25.
Picman, A. K., Ranieri, R. L., Towers, G. H. N. and Lam, J. (1980). *J. Chromatogr.* **189**, 187–198.
Porter, A. E. A., Smallman, R. V. and Mansfield, J. W. (1979). *J. Chromatogr.* **172**, 498–504.
Schulte, K. E., Reisch, J. and Hörner, L. (1962). *Chem. Ber.* **95**, 1943–1954.
Schulte, K. E., Rücker, G. and Meinders, W. (1965). *Tetrahedron Lett.* **11**, 659–661.
Selva, A., Arnone, A., Mondelli, R., Sprio, V., Ceraulo, L., Petruso, S., Plescia, S. and Lamartina, L. (1978). *Phytochemistry* **17**, 2097–2100.
Stahl, E. (1967). "Dünnschichtchromatographie". Springer-Verlag, Berlin, Heidelberg and New York.
Still, W. C., Kahn, M. and Mitra, A. (1978). *J. Org. Chem.* **43**, 2923–2925.
Sütfeld, R. (1987). *In* "Modern Methods of Plant Analysis", Vol. 5, pp. 104–113. Springer-Verlag, Berlin, Heidelberg, New York, London, Paris and Tokyo.
Sørensen, N. A. (1977). *In* "The Biology and Chemistry of the Compositae" (V. H. Heywood, J. B. Harborne and B. L. Turner, eds), Vol. I, pp. 385–409. Academic Press, London, New York and San Francisco.
Wrang, P. A. and Lam, J. (1975). *Phytochemistry* **14**, 1027–1035.
Zeng, M. Y., Li, L.-N., Chen, S.-F., Li, G.-Y., Liang, X.-T., Chen, M. and Clardy, J. (1983). *Tetrahedron* **39**, 2941–2946.

RECOMMENDED ADDITIONAL READING:
Chemical, Biogenetical and Biological Aspects

Bohlmann, F. (1965). *Fortsch. Chem. Forsch.* **6**, 65–100.
Bohlmann, F. (1967). *Fortsch. Chem. Organisch. Naturstoffe* **25**, 1–62.
Bohlmann, F. and Mannhardt, H. J. (1957). *Fortsch. Chem. Organisch. Naturstoffe* **16**, 1–70.
Bohlmann, F., Bornowski, H. and Arndt, C. (1962). *Fortsch. Chem. Forsch.* **4**, 138–272.
DiCosmo, F. and Towers, G. H. N. (1984). *In* "Phytochemical Adaptions to Stress" (B. N. Timmerman, C. Steelink and F. A. Loewus, eds), Vol. 18 of Recent Advances in Phytochemistry, pp. 97–175. Plenum, New York.
Engvild, K. C. (1986). *Phytochemistry*, **25**, 781–791.
Heitz, J. R. and Downum, K. R. (eds) (1987). "Light-Activated Pesticides", ACS Symposium Series, No. 339. American Chemical Society, Washington, DC.

7 Phytochrome and Other Photoreceptors

LEE H. PRATT[1], HORST SENGER[2]
and PAUL GALLAND[2]

[1] *Botany Department, University of Georgia, Athens, GA 30602, USA*
[2] *Fachbereich Biologie-Botanik, Philipps-Universität, 3550 Marburg, FRG*

I.	Introduction	186
II.	Radiometric units and optical equipment	187
	A. Radiometric units	187
	B. Light sources	189
	C. Light detectors	190
	D. Filters	190
III.	Action spectroscopy	191
IV.	Phytochrome: isolation and purification	193
	A. General considerations	194
	B. Purification from etiolated tissue	195
	C. Purification from green tissue	197
V.	Phytochrome: molecular characterisation	199
	A. SDS PAGE	199
	B. Spectral characterisation	201
	C. Quantitation	203
VI.	Phytochrome: phototransformations	208
	A. Quantum yields and photoequilibrium values for Pfr	208
	B. Molar extinction spectra for Pr and Pfr	212
	C. $[Pfr]_x$ as a function of wavelength ($[Pfr]_x^\lambda$)	212
VII.	Phytochrome: molecular biology	213
	A. Phytochrome control of transcription of its own genes	214
	B. Epitope mapping with fusion proteins	215
VIII.	UV/blue-light photoreceptors	219
	A. Problems of UV/blue-light photoreceptor identification	219

	B. Possible methods for photoreceptor identification	220
	C. Photoactivation of enzymes: nitrate reductase	223
IX.	DNA-photolyases	223
	A. Action spectra for *in vitro* photoreactivation	224
	B. Assays for DNA-photolyases	225
	C. *In vivo* transformation assay	226
	References	227

I. INTRODUCTION

Light is the obligatory prerequisite for all plant life and thus life on Earth. Its function as a photosynthetic energy source is almost self-evident. Less appreciated, though no less important, is its role in the regulation of numerous biological processes including the regulation of photosynthesis itself (Fig. 7.1). These light-regulated processes include directional movement (phototaxis, phototropism, photonasty), photomorphogenesis and photorepair of UV-damaged DNA (photoreactivation). The term photomorphogenesis, which pertains *in sensu strictu* to morphogenetic changes, is often applied in a wider sense that also includes changes in specific metabolites, enzymes or pigments.

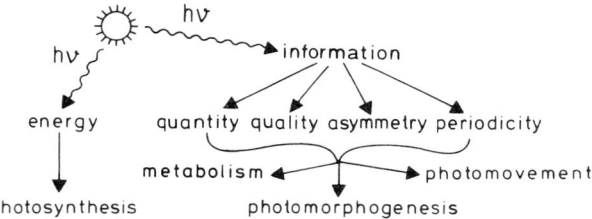

FIG. 7.1. The role of light in plant life.

Light is absorbed by special photoreceptor(s), which consist of a protein moiety and chromophore(s) with conjugated π-systems. Photon absorption and the concomitant transition from the S_0 ground state to one of the higher excited singlet states constitutes the first step in a chain of events that includes signal transformation, possible amplification and signal transduction to an effector site. In some cases this signal transduction is of short range, for example, when the receptor molecule is a photoactivatable enzyme. Examples are the DNA-photolyases (DNA repair enzymes) of some organisms. In most other cases signal transduction is more complicated, for example, when gene expression is involved.

The range of light fluence rate (casually, but incorrectly called 'intensity') in which plants can operate is enormous. UV/blue-light receptors and also phytochrome receptors can operate often in a range that spans more than seven orders of magnitude. The sporangiophore of the lower fungus *Phycomyces*, for example, is sensitive between 10^{-9} and $10 \, \text{W m}^{-2}$ of blue light, which is similar to the range of vertebrate vision. Inside this range plants can recognise fluence rate ('intensity') and, provided that multiple or photochromic receptors are present, they can also discriminate between different wavelengths, a property that is analogous to colour vision in vertebrates. The information passed on from the site of perception to the effector site can vary in quantity, quality and periodicity. The signal distribution in the plant can be symmetric or can have a specific asymmetry, causing in this way directed responses of localised morphogenetic events. For none of the known photoreceptors do we know the actual signal transduction pathway, i.e. the reaction steps that occur between photoreceptor excitation and the final response. This field is thus wide open for research and represents one of the most attractive challenges for investigators in photophysiology.

Currently we distinguish between two major groups of photoreceptors. The phytochromes, which absorb most efficiently in the red region of the visible spectrum, have been extensively investigated during the past 40 years and are, as a result, well described with respect to their physiology, molecular structure and their molecular biology (general reviews in Kendrick and Kronenberg, 1986). The so-called UV/blue-light photoreceptors are a heterogeneous group of pigments absorbing in the near-UV and blue region (general reviews: Senger, 1980, 1984, 1987; Galland and Senger, 1988a). With the exception of the DNA-photolyases, these photoreceptors have not yet been isolated and chemically identified. There is, however, a general consensus that flavins and in some cases also rhodopsins are involved in blue-light reception. The participation of pterins in the reception of near-UV light is presently under investigation (Galland and Senger, 1988b). The greatest difficulty that one has to overcome before a UV/blue-light photoreceptor can be isolated is the present lack of a photoreceptor assay. As long as such a photoreceptor assay is unavailable these receptors must still be studied and characterised by their physiological effects. Our knowledge about their function and mode of action still depends, therefore, on action spectroscopy and other physiological methodology.

As a result of the widely differing advances that have been made in the fields of phytochrome and UV/blue-light photoreceptors, this chapter is necessarily very heterogeneous with respect to available biochemical methods. Most of this chapter reflects the rapid advances that the biochemistry and molecular biology of phytochrome has experienced during the past 15 years. We have nevertheless included the description of non-biochemical methods such as action spectroscopy and general photophysical methods that are also of relevance for *in vitro* experiments.

II. RADIOMETRIC UNITS AND OPTICAL EQUIPMENT

Photoreceptor analysis necessitates in one way or another the availability of monochromatic or polychromatic light. To familiarise biochemists with the technical aspects of photobiology, we give here some basic information about commonly used light sources, light calibration and radiometric units. Recommended references providing more detailed technical information include Björn (1971), Jagger (1967, 1977), Montagnoli (1980), Hartmann (1982) and Holmes (1984a,b). The technical literature from manufacturers of lamps, light detectors and optical equipment is easily accessible and provides most of the theoretical and technical information that a newcomer requires.

A. Radiometric Units

For most photobiological or photochemical experiments the variable of interest is the amount of light falling on a unit surface area per unit time. This is the energy fluence rate (for short: fluence rate), which is given in units of $W\,m^{-2}$ ($=10^7\,erg\,m^{-2}\,s^{-1}$). If the photon content of the light is emphasised the energy fluence rate can be converted to the photon fluence rate, which is given in units of $mol_{(photons)}\,m^{-2}\,s^{-1}$. The term 'intensity' which is often casually used in place of fluence rate should be avoided, because it represents according to strict definition the energy output of the light source rather than the energy at the surface of light application. If light pulses rather than continuous irradiation are applied one uses the units of energy fluence ($J\,m^{-2} = W\,m^{-2}\,s$) or photon fluence [$mol_{(photons)}\,m^{-2} = (mol_{(photons)}\,m^{-2}\,s^{-1})\,s$].

TABLE 7.1. Conversion factors for commonly occurring units.

To convert from	to	Multiply by
ångström	nanometre	1.000×10^{-1}
calorie (thermochemical)	joule	4.184
calorie $cm^{-2}\,s^{-1}$	watt m^{-2}	4.184×10^4
calorie $cm^{-2}\,s^{-1}$	mol $m^{-2}\,s^{-1}$	$3.499 \times 10^{-4}\,\lambda^a$
calorie $cm^{-2}\,s^{-1}$	quanta $m^{-2}\,s^{-1}$	$2.107 \times 10^{20}\,\lambda$
calorie $cm^{-2}\,min^{-1}$	watt m^{-2}	6.973×10^2
calorie min^{-1}	watt	6.973×10^{-2}
calorie s^{-1}	watt	4.148
electron volt	joule	1.602×10^{-19}
erg	joule	1.000×10^{-7}
erg	quanta	$5.036 \times 10^8\,\lambda$
erg $cm^{-2}\,s^{-1}$	mol $m^{-2}\,s^{-1}$	$8.362 \times 10^{-12}\,\lambda$
erg $cm^{-2}\,s^{-1}$	quanta $m^{-2}\,s^{-1}$	$5.036 \times 10^{12}\,\lambda$
erg $cm^{-2}\,s^{-1}$	watt m^{-2}	1.000×10^{-3}
erg s^{-1}	calorie min^{-1}	1.434×10^{-6}
erg s^{-1}	watt	1.000×10^{-7}
joule	mol	$8.362 \times 10^{-9}\,\lambda$
langley	calorie cm^{-2}	1.000
micrometer (micron)	nanometre	1.000×10^3
mol $m^{-2}\,s^{-1}$	quanta $m^{-2}\,s^{-1}$	6.022×10^{23}
mol $m^{-2}\,s^{-1}$	watt m^{-2}	$1.196 \times 10^8\,\lambda^{-1}$
quanta $m^{-2}\,s^{-1}$	mol $m^{-2}\,s^{-1}$	1.661×10^{-24}
quanta $m^{-2}\,s^{-1}$	watt $m^{-2}\,s^{-1}$	$1.987 \times 10^{-16}\,\lambda^{-1}$
watt cm^{-2}	watt m^{-2}	1.000×10^4
watt m^{-2}	mol $m^{-2}\,s^{-1}$	$8.362 \times 10^{-9}\,\lambda$
watt m^{-2}	quanta $m^{-2}\,s^{-1}$	$5.036 \times 10^{15}\,\lambda$

Reprinted with permission from Holmes (1984b).
[a] λ equals wavelength in nm.

Table 7.1 lists the most common units found in the literature and conversions among units. The conversion from energy content (in $J\,m^{-2}$ or $W\,m^{-2}$) to photon content (quanta m^{-2}, $mol\,m^{-2}$, or quanta $m^{-2}\,s^{-1}$, $mol\,m^{-2}\,s^{-1}$) of a light source can be made by Planck's equation: $E = h\nu = hc/\lambda$, where h is Planck's constant ($6.626 \times 10^{-34}\,J\,s^{-1}$), c is the speed of light ($2.998 \times 10^8\,m\,s^{-1}$), and λ is the wavelength (nm). For 1 mole of photons, we obtain: $E = N_A hc/\lambda$, where E is the energy per mol of photons ($J\,mol^{-1}$) of wavelength λ (nm) and N_A is Avogadro's number ($6.022 \times 10^{23}\,mol^{-1}$).

B. Light Sources

1. Incandescent lamps

Incandescent lamps (filament lamps) can be used in the visible (above 400 nm) and infrared range of the electromagnetic spectrum. Besides ordinary light bulbs from household lamps, incandescent tungsten lamps or tungsten–halogen lamps are recommended. The latter have a greater power output, a greater lifetime and greater stability than regular tungsten lamps. Regular slide projectors are relatively inexpensive light sources that provide collimated light of rather high fluence rate. Because the emission spectrum of an incandescent lamp depends on the filament temperature there is a rapid drop of energy output below 600 nm. Below 400 nm these lamps provide very little power. If, however, the organism is very light sensitive, this small power output in the near-UV can be sufficient to achieve response saturation. Before more expensive near-UV or UV lamps are purchased one should, therefore, take into account the sensitivity of the organism. In most biological experiments the excess of infrared irradiation is undesirable and appropriate infrared filters should be used in combination with incandescent light. A 10–15% solution of copper sulphate serves perfectly this purpose. If only small areas need to be irradiated one or two heat-absorbing filters, for example type KG-1 (Schott Optical Glass), should be placed into the light path. If the power output of the lamps is controlled by a dimmer one should be aware that the emission spectrum shifts towards the infrared region as lamp output is reduced. This does not necessarily create problems as long as interference filters, which provide monochromatic light, are used.

2. Fluorescent lamps

If rather large areas need to be irradiated, for example for the maintenance and growth of biological material, fluorescent lamps are the most economical and practical choice. There are a large variety of fluorescent-light tubes that can provide light of different spectral compositions (cool-white, daylight, blue, blacklight, blacklight-blue, green, gold, red and far-red). Fluorescent lamps are often in demand as laboratory safelights: gold fluorescent lamps to avoid photoreactivation (photorepair below 500 nm) of UV-irradiated organisms, green fluorescent lamps to avoid photodegradation of chlorophylls or phototransformation of phytochrome, and red fluorescent lamps for work with blue-light sensitive organisms. One caveat: all of these 'safety' lights contain some spectral contamination at wavelengths below or above their emission maxima, which may still be perceived by very sensitive organisms. In such cases one should adjust the emission of these lamps by the addition of green, red or other appropriate acrylic plastic filters that remove the residual contaminating wavelengths. By combining white fluorescent lamps with far-red fluorescent lamps one can effectively control the phytochrome status of plants, i.e. the ratio of far-red-absorbing phytochrome (Pfr) to total phytochrome. For work on UV damage of DNA, germicidal lamps (30-WT8 from General Electric) are very convenient. Because more than 85% of their power output occurs at 254 nm (mercury line), they can be considered for all practical purposes as monochromatic light sources that do not require further very costly UV-interference filters.

3. Arc lamps

High-pressure xenon and high-pressure mercury arc lamps are recommended for experiments that require very high fluence rates. These lamps are expensive as they require special housings and power supplies. They can be provided with modular sets of light guides (fibre optics), filter holders, focusing lenses and automated shutters, which makes these lamps suitable for high-fluence-rate irradiations of small areas (spectrophotometry, fluorescence studies, small biological objects). High-pressure xenon arc lamps are suited for work in the visible region. They have a rather flat power output between 400 and 800 nm and little emission below 300 nm. Mercury arc lamps are useful for work below 400 nm, where much of the power output occurs in sharp, discrete emission lines. Above 380 nm these lamps have only four major mercury lines so that they are only of limited use between 400 and 600 nm.

C. Light Detectors

A silicon photodiode attached to an electrometer can measure fluence rates over a wide range of wavelengths. Such photodiodes are optimised for different spectral regions. Examples are the PIN-10DP/SB (250–1100 nm; optimum 300–500 nm) or the UDT-UV100 (200–1100 nm; optimum 200–400 nm) (both from United Detector Technology, Hawthorne, CA; 1 cm^2 active area). Photodiodes can be purchased either calibrated or uncalibrated as a function of wavelength. It is desirable to calibrate photodiodes once or twice a year. Because fluence rates above $10\,\mathrm{W\,m^{-2}}$ may render the photodiodes temporarily light insensitive, they should never be exposed to very bright light.

Thermopiles are less light sensitive than photodiodes and they, therefore, cannot always be used for the measurement of dimly illuminated areas. Because the sensitivity of thermopiles is wavelength independent, all other measuring devices can be calibrated with them. The thermopile itself can be calibrated with a standard lamp that is provided by the US National Bureau of Standards or other internationally accepted Bureaux.

D. Filters

1. Neutral density filters

Neutral density filters are an easy way to lower the fluence rate of the actinic light. They can be provided with different optical densities ranging from 0.3 (or even less) to 5. Many of these filters are spectrally rather flat. They can be purchased either uncalibrated or calibrated as a function of wavelength. Even if calibrated filters are bought, it is recommended to check the transmission of these filters in different spectral regions, especially in the near-UV and the infrared regions. Neutral density filters can be made from different absorptive materials such as gelatin (Wratten filters from Eastman Kodak, Rochester, NY), gray acrylic plastic (Röhm and Haas Plexiglas), or glass (Schott type NG) and glass with reflective coating. When a stack of reflective filters is used, the total transmission is higher than the product of the individual filters. Equations to calculate the transmission of a stack of reflective filters have been published (Poff et al., 1984). One needs in any case to be cautious when using combinations of these filters and should if possible measure the transmission of the combined stack. A stack of white paper sheets is often sufficient to attenuate the light if neutral density filters are unavailable. Wire mesh can often be used for light attenuation. The advantage of wire mesh is that it is spectrally flat. When several layers of wire mesh are used one has to be careful about local periodicity of fluence rates (Moiré pattern).

2. Interference filters, bandpass filters, cutoff filters

For the measurement of action spectra either a monochromator or narrow-band interference filters (for example: Balzers; Corion Corporation; Schott Optical Glass) are required to obtain monochromatic light. Interference filters have two unequal sides: one appears coloured and the other has a reflective coating (mirror side). The mirror side of the filters should face towards the light source. Unnecessary heating should be avoided and it is recommended to place a heat-absorbing filter (Schott type KG) between the light source and the interference filter. Often they need to be used in combination with cutoff filters because of spectral impurities from other wavelength regions. This is especially a problem with UV and near-UV interference filters that still transmit some blue light. It is therefore advisable to use them in combination with an ultraviolet-transmitting black glass, which blocks visible light (Schott type UG). The use of chemical filters is often the most practical and economical way of excluding unwanted spectral contamination. A 10 cm layer of water removes practically all radiation above 1200 nm. The half-maximum transmission of an aqueous solution of copper sulphate occurs between 330 and 550 nm, and that of nickelous–cobaltous sulphate occurs between 235 and 330 nm (Jagger, 1977). Glass bandpass filters are often a reasonable alternative to the more costly interference filters if the requirement for spectral purity is not very high. They can be provided for UV, near-UV and visible light (manufacturers: Corning; Kodak; Schott). These filters have typically half-maximal bandwidths of 50–100 nm. Glass cutoff filters are also available for the UV, near-UV and visible region. To eliminate unwanted UV radiation from the actinic light one can use window glass (cutoff near 330 nm), polystyrene film (cutoff near 280 nm), Pyrex glass #774 (cutoff near 300 nm), or Mylar polyester film (cutoff near 310 nm) (Jagger, 1977).

III. ACTION SPECTROSCOPY

Action spectroscopy is a method used to determine *in vivo* (or *in vitro*) the absorption spectrum of the photoreceptor (receptor pigment) that mediates a biological response. The technique consists basically of studying the dependence of the response on wavelength. When the photoreceptor for a particular response is not yet isolated and characterised *in vitro*, an action spectrum gives the first essential information, i.e. the shape of its absorption spectrum. The action spectrum thus becomes the prerequisite for the future isolation of the photoreceptor.

The rate constant, k (s^{-1}), with which a photoreceptor goes into an excited state after photon absorption is given by:

$$k = N_\lambda \varepsilon_\lambda \Phi_\lambda \qquad (7.1)$$

where N_λ is the photon fluence rate (mole$_{(photons)}$ m^{-2} s^{-1}), ε_λ the molar extinction coefficient (m^2 mole$_{(receptor)}^{-1}$) and Φ_λ the quantum effectiveness (dimensionless), the probability that the receptor goes into the excited state after photon absorption. The left and the right side of Eq. (7.1) have the dimension s^{-1}. In action spectroscopy the quantum effectiveness (also called quantum efficiency or quantum yield) is defined as the probability that an absorbed photon causes a primary photochemical reaction, i.e. the first reaction in a (perhaps rather long) chain of biochemical reactions that results in a measurable biological response. In order to perform action spectroscopy the nature of this primary reaction does not need to be known. The magnitude of the quantum effectiveness for a photoresponse is the product of the quantum effectiveness for phototransformation as defined in Eq. (7.1) and of the probability with which the primary photochemical reaction can compete with the other two processes of de-excitation, i.e. fluorescence (or phosphorescence if the primary reaction occurs from the triplet state) and radiationless de-excitation.

Action spectroscopy is based on the fundamental assumption that the system must produce identical responses to different wavelengths when the rates of the primary reactions are identical. The rate of the primary reaction must be proportional to the rate constant of the phototransformation in Eq. (7.1). Therefore, if two responses to wavelengths λ_1 and λ_2 are identical, the rate constants of phototransformation must also be identical: $k_{\lambda 1} = k_{\lambda 2}$ or $N_{\lambda 1}\varepsilon_{\lambda 1}\Phi_{\lambda 1} = N_{\lambda 2}\varepsilon_{\lambda 2}\Phi_{\lambda 2}$ or $N_{\lambda 1}\sigma_{\lambda 1} = N_{\lambda 1}\sigma_{\lambda 2}$, where σ_λ is the cross-section of phototransformation. This leads to the relation:

$$\frac{\sigma_{\lambda 1}}{\sigma_{\lambda 2}} = \frac{(1/N_{\lambda 1})}{(1/N_{\lambda 2})} \tag{7.2}$$

This basic equation of action spectroscopy states that the cross-sections of phototransformation are proportional to the reciprocals of the photon fluence rates.

In practice, an action spectrum is measured with the following two-step procedure. First, one measures the response as a function of N_λ, the photon fluence rate (mol m^{-2} s^{-1}). When light pulses rather than continuous irradiation are used to elicit the response, N_λ should be replaced by $F_\lambda = N_\lambda \Delta t$, the photon fluence (mol m^{-2}) where Δt is the duration of the light pulse with the photon fluence rate N_λ. The equations that are used to construct an action spectrum apply equally well to F_λ and N_λ. For any given wavelength one measures these photon fluence rate–response curves. A criterion response (for instance a 50% response level) is chosen and the photon fluence rate $N_{\lambda i}$ at λ_i that is required to elicit the response is determined. A plot of $1/N_\lambda$ against wavelength is the action spectrum. Alternatively, one may normalise all values of $1/N_{\lambda i}$ to the value of $1/N_{\lambda \text{ref}}$ for a reference wavelength, usually the wavelength where the photon requirement is the lowest. Because $1/N_\lambda$ is proportional to $\sigma_\lambda = \varepsilon_\lambda \Phi_\lambda$ the action spectrum represents the relative spectrum of the cross-section of the photoreceptor. To obtain the absolute spectrum of the cross-section, one would have to know the quantum effectiveness for the primary reaction. In practice it is useful to plot photon fluence–response curves on a logarithmic abscissa. Besides the obvious advantage of enabling one to plot a wide range of photon fluence rates, this method is particularly useful, because for any two wavelengths that bring about the same response the curves must be shifted parallel to each other by the interval $N_{\lambda 1}/N_{\lambda 2}$, as $\ln N_{\lambda 1} - \ln N_{\lambda 2} = \ln(N_{\lambda 1}/N_{\lambda 2})$. Finding parallel photon fluence–response curves thus provides a test for the underlying assumptions of classical action spectroscopy. If the photon fluence rate–response curves (or photon fluence–response curves) are not parallel, one deals most likely with a more complex photoreceptor system. This is the case for responses controlled by phytochrome or other photochromes (receptors that exist in two photo-interconvertible states with different, but overlapping absorption spectra). The procedures outlined above do not hold in such cases, because two responses rather than only one primary response (the phototransformations in the forward and backward directions) are involved. Classical action spectroscopy cannot, therefore, be applied for the prediction of the absorption spectra of photochromes. This does not mean that action spectroscopy is useless. As a matter of fact, in the case of phytochrome the photoreversibility was an excellent tool for its identification and a prerequisite for its isolation. The interpretation of action spectra involving complex photoreceptors or photochromes requires specific mathematical models. This is a domain of analytical action spectroscopy (Hartmann and Cohnen-Unser, 1972; Hartmann, 1982).

IV. PHYTOCHROME: ISOLATION AND PURIFICATION

Phytochrome was first isolated by Butler *et al.* (1959), who not only identified it unequivocally as a chromoprotein, but also provided a rapid spectrophotometric assay suitable for following its purification. Present methods of phytochrome isolation and purification derive from these initial efforts of Butler *et al.*, and from subsequent improvements in methodology introduced by Siegelman and Firer (1964), Mumford and Jenner (1966), Correll *et al.* (1968), Rice *et al.* (1973), and Pratt (1973). A review of this early work is available elsewhere (Pratt, 1980).

Phytochrome was first purified as either a 60 kDa monomer (Mumford and Jenner, 1966), or as an apparent tetramer or hexamer of 42 kDa monomers (Correll *et al.*, 1968). Gardner *et al.* (1971) eventually documented that proteolytic degradation products were purified in both cases. Nevertheless, until the demonstration by Vierstra and Quail (1982) and Kerscher and Nowitzki (1982) that the monomer of full-size oat phytochrome was 124 kDa, it continued to be purified as a slightly degraded, *c.* 120 kDa monomer (e.g. Pratt, 1973; Rice *et al.*, 1973). It is now evident that phytochrome is unusually susceptible to the endoproteolytic activity that is ubiquitous in crude plant extracts. Phytochrome subsequently has been shown to exist in solution as a dimer (Hunt and Pratt, 1980; Jones and Quail, 1986), while its undegraded monomer size has been documented to range from 118 to 127 kDa (Vierstra *et al.*, 1984; Cordonnier *et al.*, 1986a,b).

It has also become evident in recent years that phytochrome from a single organism can be heterogeneous. For example, the bulk of the phytochrome from etiolated and light-grown oats, referred to here as etiolated-oat and green-oat phytochrome, respectively, differ from one another immunochemically and spectrophotometrically (Tokuhisa *et al.*, 1985; Shimazaki and Pratt, 1985; Pratt and Cordonnier, 1987), as well as biochemically (Cordonnier *et al.*, 1986b).

It is also evident that etiolated- and green-plant phytochromes may themselves be heterogeneous. Hershey *et al.* (1985) documented the existence of at least four etiolated-oat phytochrome genes that are expressed at the mRNA level and that differ by about 2% in sequence, while Grimm *et al.* (1987) have in part confirmed this heterogeneity at the protein level. Recent evidence that there are single genes for etiolated-pea (Sato,

1988) and etiolated-rice (Kay et al., 1989a) phytochrome, however, indicate that the situation in oats cannot be generalised. Nevertheless, recent biochemical and immunochemical observations indicate that green-oat phytochrome exhibits greater heterogeneity than is observed for etiolated-oat phytochrome (Cordonnier and Pratt, 1989; Wang et al., 1989). Consistent with these observations is the report that there are two genes in *Arabidopsis* whose characteristics are consistent with their being green-*Arabidopsis* phytochrome genes (Sharrock and Quail, 1989). These two genes exhibit relatively low homology to each other, consistent with the marked differences between the two types of green-oat phytochrome that have been found. While it is exciting to speculate about the potential implications of the heterogeneity seen for phytochrome, its significance, if any, remains to be elucidated.

A. General Considerations

1. Controlled illumination

Because phytochrome is a photochemically active chromoprotein, and because its two forms exhibit differential reactivity *in vitro*, it is best to work under illumination that is not absorbed significantly by either form of this pigment. Suitable illumination is provided by green fluorescent bulbs (e.g. Sylvania F40G) wrapped with plastic filters (e.g. one layer each of Roscolene No. 874 [medium green] and No. 877 [medium blue-green], Rosco Laboratories, Inc., Port Chester, NY 10573) (see Pratt, 1984, for spectrum).

2. Starting tissue

Phytochrome is considerably more abundant in fully dark-grown, as opposed to light-grown, tissue. Consequently, the overwhelming majority of work with phytochrome *in vitro* has been done with etiolated-plant phytochrome. When beginning with etiolated tissue, however, even infrequent exposure of etiolated seedlings to other than green light dramatically depletes its phytochrome content. It is therefore important to grow etiolated tissue in absolute darkness.

3. Phytochrome assay

The unique photoreversibility of phytochrome, which was first utilised as an *in vitro* assay for this chromoprotein by Butler et al. (1959), is best measured with a dual-wavelength spectrophotometer. Because this assay (a) is highly specific for phytochrome, (b) is sufficiently sensitive, and (c) can be performed reasonably quickly, it is ideal for following phytochrome during the initial stages of its purification. For the purposes of this chapter, one unit (U, *c.* 1 mg) of phytochrome is the quantity that in 1 ml and for a 1 cm light path yields by this assay a $\Delta\Delta A$ of 1 when assayed in the red and far-red at wavelengths corresponding to the absorbance maxima for red-absorbing phytochrome (Pr) and Pfr, respectively (see Pratt, 1983, for detailed discussion). If a dual-wavelength spectrophotometer is unavailable, a conventional split-beam, single-wavelength spectrophotometer can be adapted to this application (Jung and Song, 1979). Of course, once phytochrome is well separated from other pigments and has been sufficiently concentrated, it can be monitored directly at either c. 666 nm (for Pr) or c. 730 nm (for Pfr).

4. General precautions

As already noted, phytochrome is especially susceptible to endoproteolytic attack. All work should therefore be done as rapidly as possible and at the lowest practical temperature. Whenever possible, appropriate endoprotease inhibitors should be utilised. Phenylmethylsulphonylfluoride (PMSF) generally works well when beginning with etiolated tissue (e.g. Vierstra and Quail, 1982), but is not always sufficient (Cordonnier and Pratt, 1982). In particular, it is useless when beginning with green oat leaves (Cordonnier *et al.*, 1986b), in which case iodoacetamide or leupeptin may be adequate replacements (Tokuhisa and Quail, 1989).

Phytochrome also contains reactive sulphydryls that require protection by a relatively strong reductant (Smith and Cyr, 1988). While β-mercaptoethanol is most commonly used for this purpose, it has been reported to stimulate endoprotease activity in crude extracts, in which case addition of sodium bisulphite has been recommended (Litts *et al.*, 1983; Vierstra and Quail, 1983a).

B. Purification from Etiolated Tissue

No attempt will be made here to choose among the variety of protocols available for purification of full-sized phytochrome to homogeneity (Litts *et al.*, 1983; Vierstra and Quail, 1983a; Datta and Roux, 1985; Grimm and Rüdiger, 1986; Chai *et al.*, 1987; Smith and Cyr, 1988). Rather, a generally applicable, rapid, and simple procedure for partial purification of phytochrome will be outlined. This abbreviated procedure, which is adapted from Vierstra and Quail (1983a), constitutes the initial steps in several of the protocols that have been used for its purification to homogeneity (Vierstra and Quail, 1983a; Datta and Roux, 1985; Grimm and Rüdiger, 1986; Chai *et al.*, 1987).

1. Standard protocol for Avena

With a 4-1 Waring blender, extract 1 kg of tissue into 750 ml of 0.1 M Tris-Cl, 0.14 M $(NH_4)_2SO_4$, 10 mM EDTA, 50% (v/v) ethylene glycol (pH 8.3 at 4°C), to which 1.56 g of powdered sodium bisulphite and 15 ml of 0.2 M PMSF in dimethylsulphoxide (stored at $-20°C$) are freshly added. The extraction buffer is brought to $-20°C$ prior to use. It does not freeze because of the ethylene glycol. If the tissue to be extracted is freshly harvested, it should be chilled to 4°C, pre-irradiated with red light (R), and added to the blender. If it is frozen, the initial homogenate should be irradiated with R. This irradiation converts all phytochrome to Pfr, which minimises proteolytic degradation near the *N*-terminus (Vierstra and Quail, 1983a). A suitable R source is the output of unfiltered Sylvania Standard (but not Wide Spectrum) Gro-lux fluorescent lamps. The crude homogenate may be filtered through three layers of cheesecloth. Alternatively, it may be clarified more quickly and easily by centrifugation through a Braun Model MP50 juice extractor, in which a piece of Miracloth is used as the filter.

Add 10 ml of 10% (v/v) poly(ethylenimine) per litre of crude homogenate, stir for 15 min, centrifuge for 15 min at $25\,000 \times g$, and discard the pellet. Add 10 ml l^{-1} of the

0.2 M PMSF stock solution to the supernatant. To prepare the 10% poly(ethylenimine), mix 100 ml of pure reagent with 700 ml H_2O, cool to 4°C, adjust pH to 7.8 with 3 N HCl, allow to stand overnight at 4°C, readjust pH to 7.8 with 3 N HCl, and bring to 1 l.

Per litre of clarified extract, add *rapidly* 250 g of powdered $(NH_4)_2SO_4$ and stir until dissolved (c. 10 min). Keep pH near 7.8 with 2.5 M Tris. Centrifuge for 15 min at $30\,000 \times g$ in a 6×250-ml rotor, discard the supernatant, dissolve the pellet in 170 ml of 50 mM Tris-Cl, 5 mM EDTA, 14 mM β-mercaptoethanol, 2 mM PMSF (freshly added from the 0.2 M stock), 25% (v/v) ethylene glycol (pH 7.8 at 4°C), and clarify by centrifugation. This volume of buffer provides an appropriate concentration of residual $(NH_4)_2SO_4$ for application of the sample to the hydroxyapatite column.

Add the clarified sample to a 50-ml bed volume (2.5×10 cm) hydroxyapatite column. While self-made hydroxyapatite is satisfactory (Pratt, 1984), commercially obtained hydroxyapatite has also been used successfully. Prior to sample application, the column is equilibrated with 50 mM Tris-Cl, 5 mM EDTA, 70 mM $(NH_4)_2SO_4$, 14 mM β-mercaptoethanol, 25% ethylene glycol (pH 7.8 at 4°C), until the eluant is stable at pH 7.8. It is operated throughout at a flow rate of $8\,\text{ml min}^{-1}$. Add the phytochrome-containing sample, wash with two column volumes of the preceding buffer minus the $(NH_4)_2SO_4$ and ethylene glycol, then with five column volumes of 5 mM K-phosphate, 5 mM EDTA, 14 mM β-mercaptoethanol (pH 7.8). Elute with a linear gradient made with 150 ml each of 5 and 200 mM K-phosphate, each containing 5 mM EDTA and 14 mM β-mercaptoethanol at pH 7.8. Monitor the eluant at 280 and 730 nm. Phytochrome purity may be estimated from the ratio, A_{730}/A_{280}.

Pool phytochrome-containing fractions of the desired purity, add $(NH_4)_2SO_4$ from a saturated solution to a final concentration of 30%, stir for 5 min, centrifuge for 10 min at $50\,000 \times g$, and discard the supernatant. Dissolve the pellet in 0.1 M Na-phosphate, 1 mM EDTA (pH 7.8) to give a phytochrome concentration near 0.5–$1\,\text{mg ml}^{-1}$, clarify by centrifugation for 10 min at $50\,000 \times g$, irradiate with far-red light (FR), and store at or below -70°C. Phytochrome is unstable *in vitro* at -20°C. Data from a satisfactory purification by this protocol are summarised in Table 7.2.

TABLE 7.2. Partial purification of phytochrome from 1 kg of 4-day-old etiolated *Avena* shoots grown at 25°C.

	Volume (ml)	Phytochrome		Yield (%)
		(U ml^{-1})	(U)	
Clarified homogenate	1540	0.0176	27.1	100
Poly(ethylenimine) supernatant	1180	0.0163	19.2	71
First $(NH_4)_2SO_4$ fractionation	174	0.084	14.6	54
Hydroxyapatite pool	72	0.096	6.9	25
Second $(NH_4)_2SO_4$ fractionation	3.3	0.71	2.3[a]	8.5

[a] Much of the phytochrome remained in the pellet after clarification. This sample had an A_{667}/A_{280} ratio as Pr of 0.16 ($\approx 15\%$ pure). Greater recovery after this second $(NH_4)_2SO_4$ fractionation is normally obtained if the pellet is dissolved in a larger volume of buffer, in which case the purity may also be significantly reduced.

2. Comments

With minor modifications, the preceding protocol can be used with etiolated tissue from a wide variety of sources. For example, if the plant tissue has a high lipid content, as in the case of soybean shoots, then lyophilisation followed by solvent extraction of the lipid is important in order to obtain optically clear extracts. Moreover, the appropriate $(NH_4)_2SO_4$ concentration needs to be determined empirically for each starting tissue, while the use of poly(ethylenimine) may in some cases lead to unacceptably low recovery of phytochrome at this initial step (Shimazaki *et al.*, 1981).

As already indicated, several strategies exist for further purification of phytochrome. Vierstra and Quail (1983a) complete the purification of full-size etiolated-oat phytochrome by subsequent chromatography with Affi-Gel Blue and Bio-Gel A-1.5, while Datta and Roux (1985) utilise hydroxyapatite a second time, followed by Fractogel. In contrast, Grimm and Rüdiger (1986) and Chai *et al.* (1987) rely largely upon back-extraction of the $(NH_4)_2SO_4$ pellet that is obtained following hydroxyapatite chromatography.

C. Purification from Green Tissue

The following protocol yields approximately 0.5 U of phytochrome from 4 kg of 9- to 11-day-old oat leaves harvested from greenhouse-grown plants. Based upon the assumptions that green-oat phytochrome represents about 1 part in 50 000 of the soluble protein from green oat leaves, and that it has the same extinction coefficients as etiolated-oat phytochrome, then this protocol yields a purification of approximately 250-fold. Since green-oat phytochrome is even more labile to endoproteases than is etiolated-oat phytochrome (Cordonnier *et al.*, 1986b; Tokuhisa and Quail, 1989), it is essential to maintain phytochrome at the lowest possible temperature, and to complete the procedure as rapidly as possible.

1. Standard protocol for Avena

Cut green oat leaves, either frozen or freshly harvested and chilled to 4°C, into 3–4-cm long pieces with a sharp razor blade. Homogenise in a 4-litre Waring blender with 375 ml of 0.1 M Tris-Cl, 0.14 M $(NH_4)_2SO_4$, 10 mM EDTA, 50% (v/v) ethylene glycol (pH 8.3 at 4°C), to which 6 ml of 2.5 M Tris, 37.5 ml of 0.2 M iodoacetamide, and 0.87 g of sodium bisulphite are freshly added. The additional Tris base prevents the pH from dropping too low in the crude homogenate. Grind for as brief a period as possible, but until the tissue is thoroughly extracted. Filter the homogenate as for etiolated-oat phytochrome. Spray with antifoam (e.g. Dow Corning #316 Silicone Release Spray) to eliminate foam. Add 10 ml of 10% (v/v) poly(ethylenimine) per litre of crude homogenate, stir for 1 min, centrifuge for 15 min at 25 000 × g and discard the chlorophyll-containing pellet. Per ml of supernatant, add 0.61 ml of 95% saturated $(NH_4)_2SO_4$ neutralised to pH 7.8 with Tris base and containing 10 mM iodoacetamide, stir for 1 min, centrifuge for 15 min at 25 000 × g and discard the supernatant. Dissolve the pellet in 30 ml of HA buffer [50 mM Tris-Cl, 5 mM EDTA, 10 mM iodoacetamide, 25% (v/v) ethylene glycol (pH 7.8 at 4°C)], clarify by centrifugation at 40 000 × g for 15 min, freeze the supernatant rapidly with liquid N_2, and store at or below −70°C. This initial sample preparation is repeated eight times prior to hydroxyapatite chromatography. With practice, it is possible to complete each such preparation within 100–110 min.

Prepare a hydroxyapatite column of 200 ml bed volume in a column with inside diameter of c. 5 cm. Equilibrate with approximately 3 litres of HA buffer, to which $(NH_4)_2SO_4$ has been added to 70 mM, until the pH of the eluate is stable at 7.8. Operate throughout at a flow rate of 8 ml min^{-1}. Thaw eight aliquots prepared by poly(ethylenimine) and $(NH_4)_2SO_4$ fractionation as above. Clarifiy by centrifugation for 10 min at 40 000 × g, dilute each aliquot to 170 ml with HA buffer, and apply to the column. To minimise proteolysis, it is desirable to thaw the aliquots a few at a time, so that none of the sample waits too long before being applied to the column. Wash the column with 400 ml of 50 mM Tris-Cl, 5 mM EDTA, 10 mM iodoacetamide (pH 7.8 at 4°C), followed by 400 ml of 5 mM K-phosphate, 5 mM EDTA, 10 mM iodoacetamide (pH 7.8). Elute with a linear gradient prepared with 600 ml each of 5 mM and 200 mM K-phosphate, each 5 mM in EDTA, 10 mM in iodoacetamide, and at pH 7.8. Monitor phytochrome spectrophotometrically and protein by Bradford assay (Bradford, 1976).

Pool the purest fractions and the two 'side' fractions separately (pools I and II, respectively). Precipitate each by addition of 95% saturation $(NH_4)_2SO_4$, 10 mM iodoacetamide (pH to 7.8 with tris base) to a final $(NH_4)_2SO_4$ concentration of 30%, centrifuge for 15 min at 25 000 × g, and discard the supernatant. Dissolve each pellet in sufficient 0.1 M Na-phosphate, 1 mM EDTA (pH 7.8) to make a phytochrome concentration of c. 100 or 60 mU ml^{-1} for pools I and II, respectively, clarify by centrifugation for 10 min at 40 000 × g, freeze with liquid nitrogen, and store at or below −70°C. Typical results obtained with this procedure are summarised in Table 7.3.

TABLE 7.3. Partial purification of phytochrome from 4 kg of 11-day-old, greenhouse-grown, green *Avena* leaves. Entries are averages of five sequential purifications.

	Volume (ml)	Phytochrome (mU ml^{-1})	(mU)	Protein (mg ml^{-1})[a]	(mg)	Purity[b]	Yield (%)
First $(NH_4)_2SO_4$ cut							
Before dilution[c]	220	5.75	1265	5.40	1190	1.07	100
After dilution	1360	0.70	952	0.92	1251	0.76	75
Hydroxyapatite							
Pool I	89	4.53	403	1.45	129	3.1	32
Pool II	66	2.97	196	1.42	94	2.1	15
Second $(NH_4)_2SO_4$ cut							
Pool I	3.8	107	407	18.3	70	5.8	32
Pool II	3.1	56	174	19.7	61	2.8	14

[a] Protein content was estimated by Bradford assay (Bradford, 1976), with reference to bovine serum albumin.
[b] Phytochrome purity is expressed as mU ml^{-1} phytochrome divided by mg ml^{-1} protein.
[c] Sum of 8 preparations after resuspension and clarification, but before freezing.

2. Comments

Significant purification of phytochrome from fully light-grown, green plant tissues has seldom been reported. In addition to the partial purification of green-plant phytochrome from oats (Tokuhisa *et al.*, 1985; Shimazaki and Pratt, 1985; Cordonnier *et al.*, 1986b), its partial purification has been reported from pea (Shimazaki *et al.*, 1981) and its complete purification from the alga, *Mesotaenium* (Kidd and Lagarias, 1990). As in the case of etiolated-oat phytochrome, it is likely that minor modifications of the above protocol will be required for its successful extrapolation to other plant tissues. Recent observations indicate that a four-fold increase in the amount of poly(ethylenimine) added to the filtered crude homogenate has no effect on recovery, but yields a two-fold increase in purity.

V. PHYTOCHROME: MOLECULAR CHARACTERISATION

Because phytochrome is a chromoprotein, it is possible to utilise for its characterisation any methods that are applicable to characterisation of proteins in general. The emphasis here will be on only those methods that are of special interest with respect to phytochrome, either because they are uniquely applicable to phytochrome, or because they deal with a standard characterisation that is essential to document the status of a preparation.

A. SDS PAGE

1. Extent of degradation

Given the sensitivity of phytochrome to endoproteolytic attack, it is essential that each preparation be characterised with respect to monomer size and homogeneity. While inferences concerning monomer size can be made based upon spectrophotometric measurements (see below), the most direct, general method for size characterisation is SDS PAGE (sodium dodecyl sulphate polyacrylamide gel electrophoresis), followed by immunoblotting if the sample is not pure. Because these methods are not specific to phytochrome, and because they are widely available, only those points that are important relative to phytochrome are highlighted here.

The relatively large size of the phytochrome monomer requires the use of a porous gel. Good results have been obtained with 5–10% (e.g. Cordonnier *et al.*, 1985) and with 7.5–15% (Fig. 7.2) linear-gradient, as well as with 6% (Vierstra and Quail, 1983a) and 7% (Litts *et al.*, 1983) polyacrylamide gels. Otherwise, with one exception, routine conditions of electrophoresis (Laemmli, 1970), electroblotting (Towbin *et al.*, 1979), and immunostaining (e.g. Pratt *et al.*, 1986) are satisfactory. The one exception is the use of a modified sample buffer for the extraction of fresh or lyophilised tissue, in which case the sample buffer is 125 mM Tris-Cl, 10% (v/v) β-mercaptoethanol, 4% (w/v) SDS, 20% (v/v) glycerol, and 0.002% (w/v) bromophenol blue (pH 6.8 at 25°C) (Vierstra *et al.*, 1984). Ratios of 2 ml per g fresh weight tissue and 1 ml per 45 mg lyophilised powder work effectively.

Because the mass difference between full-size and slightly degraded phytochrome may be no more than 2 to 4 kDa out of *c.* 120 kDa, it is essential that an absolute size reference be included in an immunoblot. When possible, an SDS sample buffer extract from lyophilised tissue, of the same species as that used for the phytochrome preparation being characterised, is ideal (e.g. Pratt *et al.*, 1986). To minimise, if not eliminate, degradation during sample preparation, freeze freshly harvested tissue rapidly by immersion in liquid N_2, grind to a powder in a mortar under liquid N_2, and

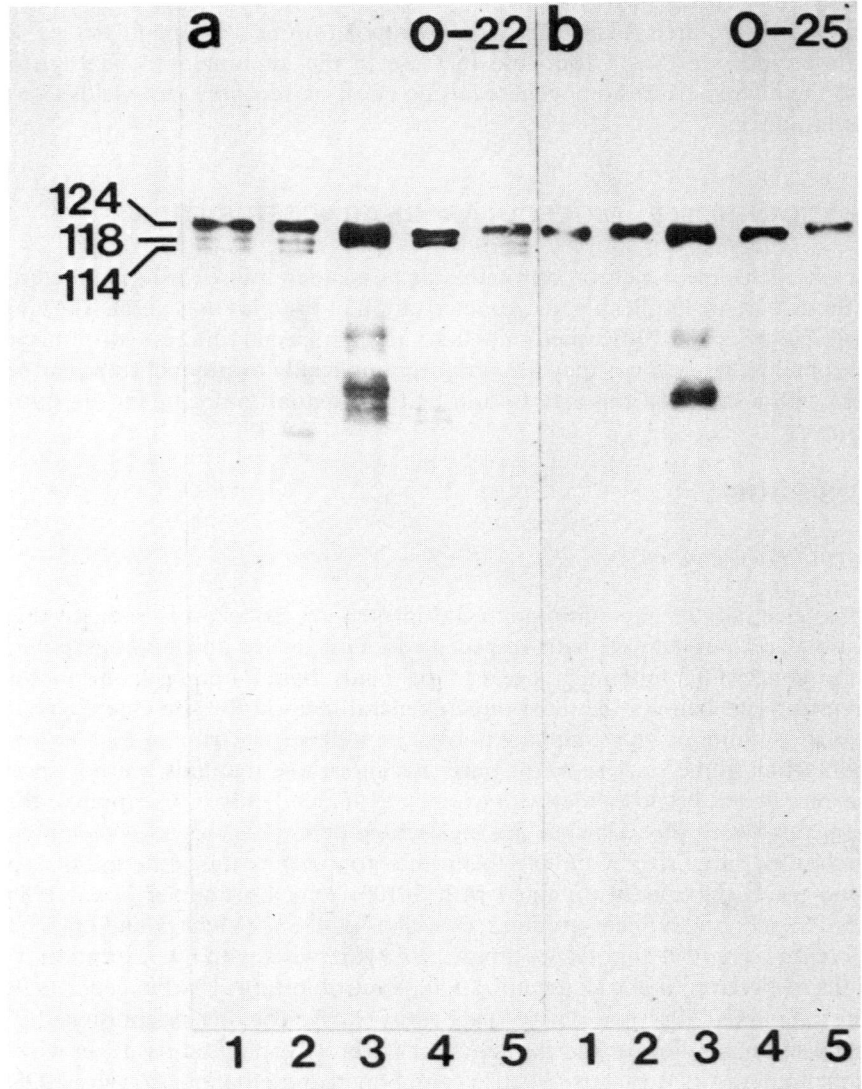

FIG. 7.2. Immunoblot analysis of phytochrome documenting the outcome of proteolytic cleavage at its N-terminus. Samples were prepared for electrophoresis, separated by SDS PAGE in a 7.5–15% linear-gradient gel, electroblotted, and immunostained as described in detail elsewhere (Pratt et al., 1986). Lanes 1 and 5, 1 μl of SDS sample buffer extract from lyophilised, 4-day-old etiolated oat shoots; lanes 2, 25 ng of etiolated-oat phytochrome purified as described in this chapter; lanes 3, 50 ng of etiolated-oat phytochrome purified as described in this chapter, but allowed to degrade by incubation at room temperature for 6 h at 25°C in a clarified homogenate of 4-day-old etiolated oat shoots; lanes 4, 50 ng of immunopurified, etiolated-oat phytochrome (Hunt and Pratt, 1979). Blots were immunostained with 1 μg ml^{-1} of MAb Oat-22 (a) or 1 μg ml^{-1} MAb Oat-25 (b). Estimated sizes of phytochrome peptides are indicated in kDa.

transfer to a lyophilisation apparatus without permitting the sample to thaw at any time. Extract the lyophilised powder by addition of SDS sample buffer (see above) that is pre-heated to 100°C, and incubate for 5 min. An undegraded phytochrome preparation should exhibit an immunostaining profile identical to that provided by the reference sample (Fig. 7.2, compare lanes 1 and 2). Degradation is immediately obvious, as in the case of a crude extract permitted to incubate at room temperature in the absence of protease inhibitor (Fig. 7.2, lane 3), or in the case of an immunopurified phytochrome sample (Fig. 7.2, lane 4) prepared with less care taken to prevent proteolysis (Hunt and Pratt, 1979b).

If suitable monoclonal antibodies (MAbs) are available, they can assist in determining the extent of degradation at the *N*-terminus of phytochrome. While MAb Oat-22, which detects an epitope 37 ± 12 kDa from the *N*-terminus (Pratt *et al.*, 1988), immunostains all high molecular weight phytochrome peptides (Fig. 7.2, panel a), MAb Oat-25, which detects an epitope at *c.* 6 kDa from the *N*-terminus (Cordonnier *et al.*, 1985; Pratt *et al.*, 1988), immunostains only those peptides that still possess this *N*-terminal region (Fig. 7.2, panel b).

2. Visualisation with Zn^{2+}

Because visible extinction of the phytochrome chromophore is lost upon denaturation of the protein moiety, the position of phytochrome in a denaturing gel cannot be determined by absorbance of its chromophore. The chromophore becomes, however, intensely fluorescent when complexed with Zn^{2+}. Berkelman and Lagarias (1986) have taken advantage of this property to develop a sensitive method for the rapid and specific visualisation of phytochrome following SDS PAGE.

To utilise the method, SDS PAGE is performed in the usual manner (Laemmli, 1970), with the following exceptions. Prepare buffers for the gel and for the electrophoresis chambers with 1 mM zinc acetate. Following electrophoresis, visualise fluorescence with a *c.* 300-nm UV transilluminator. Photograph the gel through a yellow filter (e.g. Kodak Wratten Filter #8). With Type 57 Polaroid Land Film, an exposure of *c.* 0.5 s at an aperture of f 4.5 can be satisfactory. Exposure to the UV illuminator should be minimized since the Zn^{2+}-induced fluorescence bleaches relatively quickly. The gel may then be stained for total protein as desired. As little as 100 ng of phytochrome may be detected in this way.

It is also possible to detect Zn^{2+}-induced fluorescence of phytochrome following its electrotransfer to a medium such as nitrocellulose (Parks *et al.*, 1987). Rinse the blot in water and incubate for 1 h in 1.3 M zinc acetate, 20% ethanol. Photograph as described above.

B. Spectral Characterisation

Absorption of light by the phytochrome chromophore is strongly dependent upon its interaction with the protein moiety. In addition to the obvious differences between spectra for Pr and Pfr, spectra of green- and etiolated-plant phytochromes can differ. Moreover, the absorbance spectrum of phytochrome is a function of the extent to which it has been degraded, and of the extent to which it is denatured. Thus, since one can deduce much about the properties of phytochrome in a given sample from its spectral characteristics, a careful spectral analysis is important as part of a routine characterisation of any preparation.

1. Etiolated-oat versus green-oat phytochrome

Absorbance spectra of etiolated- and green-oat phytochrome purified as described here are indicative of relatively undegraded, undenatured phytochrome samples (Fig. 7.3). Note that for the etiolated-oat phytochrome preparation, it is pure enough that essentially all extinction beyond 500 nm can be ascribed to phytochrome (Fig. 7.3(a)). The same cannot be said, however, for the green-oat phytochrome preparation (Fig. 7.3(a)). Both the absorbance maxima and absorbance difference maxima and minima differ between etiolated- and green-oat phytochrome (Fig. 7.3(b)).

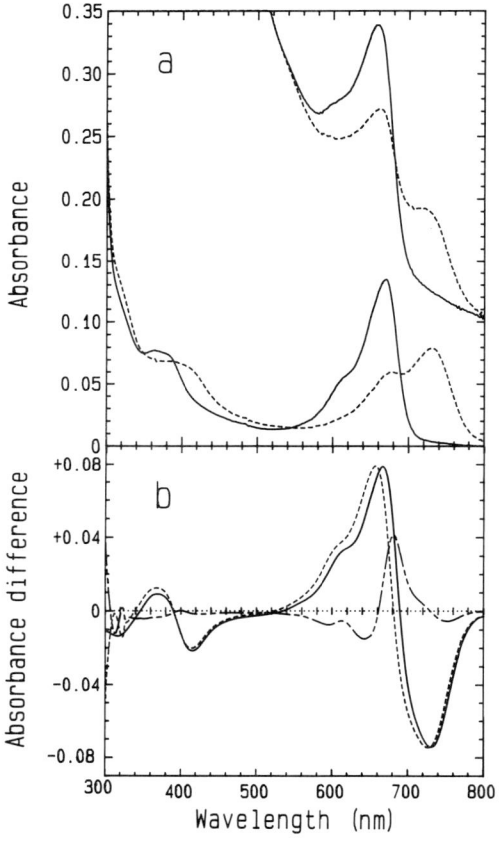

FIG. 7.3. Absorbance spectra of etiolated- and green-oat phytochrome purified as described in this chapter. (a) Absorbance spectra of etiolated-oat phytochrome (0.16 U phytochrome per mg protein; lower spectra) and green-oat phytochrome (0.0075 U phytochrome per mg protein; upper spectra) after saturating irradiation with far-red (———) or red (- - -) actinic light. The green-oat phytochrome spectra are offset by 0.1 A for clarity. (b) Pr – Pfr difference spectra for etiolated-oat (———) and green-oat (- - -) phytochrome from (a). The etiolated-oat difference spectrum minus the green-oat difference spectrum is also given (— - - —). For etiolated-oat phytochrome, absorbance maxima are at 669 nm and 730 nm, while absorbance difference peaks are at 370 nm, 415 nm, 667 nm and 731 nm and isosbestic points are at 391 nm and 690 nm. For green-oat phytochrome, absorbance maxima (corrected for the base-line ramp) are at 660 and 725 nm, while absorbance difference peaks are at 368 nm, 414 nm, 656 nm, and 727 nm and isosbestic points are at 390 nm and 681 nm.

2. Native versus denatured phytochrome

As shown originally by Butler *et al.* (1964b), the extinction of the phytochrome chromophore is highly dependent upon the nature of its association with the apoprotein. Thus, mild denaturation or other minor modification of the protein often leads to a small blue shift in the absorbance maximum for Pr, which is evident in both absolute absorbance spectra as well as in Pr − Pfr difference spectra (e.g. Smith and Cyr, 1988). More severe denaturation leads to essentially complete bleaching of chromophore extinction.

3. Full size versus degraded phytochrome

Modification of phytochrome by proteolysis can also lead to small changes in its absorbance spectra. In particular, cleavage of *c.* 6 kDa from its *N*-terminus results in a small, *c.* 8 nm, blue shift in the absorbance maximum for Pfr. Since this *N*-terminal cleavage is the most rapid proteolytic cleavage, at least when phytochrome is in its Pr form, careful spectral analysis of a preparation can be helpful in assessing its state of degradation.

C. Quantitation

Two approaches are available to quantitation of phytochrome: spectrophotometric and immunochemical. While spectrophotometric assays are generally more convenient, they suffer from being (a) relatively insensitive, (b) vulnerable to chlorophyll-related artifacts, (c) dependent upon the presence of phytochrome chromophore, and (d) unable to discriminate readily among different populations of phytochrome (see Pratt, 1983, for extended discussion). Both because an immunochemical assay can overcome these limitations, and because spectrophotometric assays have already been reviewed extensively elsewhere (Pratt, 1983), the latter will not be discussed here.

Of available immunochemical assays, an enzyme-linked immunosorbent assay (ELISA) represents the best compromise among sensitivity, versatility, speed, and resolution. The protocol described here, which is a positive sandwich assay (Fig. 7.4), was developed for quantitation of phytochrome in completely crude, clarified extracts of plant tissue, whether etiolated or light-grown. The use of antigen-specific antibodies to coat the wells of an ELISA plate has the distinct advantage that the accuracy and sensitivity of the assay become essentially independent of the purity of the sample being analysed. Thus, a given amount of highly purified phytochrome will give the same activity as an identical quantity of phytochrome in a crude plant extract (Shimazaki *et al.*, 1983; Fig. 7.5). Moreover, this assay protocol yields a linear standard curve, which simplifies data analysis. Finally, by selecting appropriate MAbs, it is possible to perform an ELISA that is specific for only that phytochrome recognised by the MAbs being utilised.

1. ELISA for etiolated-oat phytochrome

If necessary to improve coating uniformity, prepare ELISA plates (either vinyl or polystyrene) in advance by soaking for at least 2 h in 95–100% ethanol and then rinsing with a stream of ethanol. While vinyl Costar 2596 plates have proved satisfactory in the past when washed in this way, it has been impossible to obtain uniform coating of recently obtained plates, even after washing with ethanol. It is therefore important to evaluate plates on a continuing basis, most easily by observing whether replicate wells yield reproducible data. Immulon 2 polystyrene plates (Dynatech Laboratories, No. 011-010-3450) have proven to be a useful alternative and do not require prior washing with ethanol.

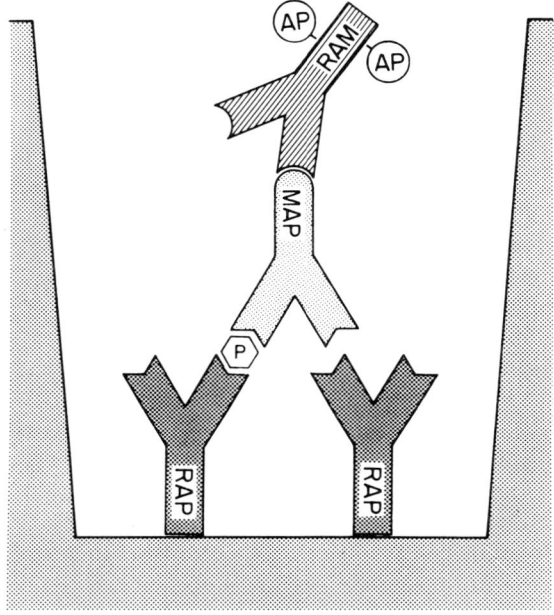

FIG. 7.4. Scheme illustrating a direct positive sandwich immunoassay that utilises antigen-specific antibodies from two animals. A well of an ELISA plate is coated with rabbit antibodies to phytochrome (RAP), to which phytochrome (P) is immunospecifically adsorbed, even from a completely crude plant extract. Phytochrome is then detected by sequential application of mouse monoclonal antibodies to phytochrome (MAP) followed by alkaline phosphatase (AP)-conjugated rabbit antibodies to mouse immunoglobulins (RAM).

On the day before use, pre-coat wells with 50 µl of immunopurified rabbit antibodies to etiolated-oat phytochrome (Fig. 7.4, RAP) at a concentration of 5 µg ml^{-1} in borate-saline (0.2 M Na borate, 75 mM NaCl, pH 8.5). Incubate overnight at 4°C.

Prepare crude extracts for phytochrome assay early the next morning. Pulverise frozen tissue under liquid nitrogen in a mortar and pestle. Homogenise the resultant powder in ice-cold extraction buffer (50 mM Tris-Cl, 0.2 M β-mercaptoethanol, pH 8.5 at 25°C) at a ratio of 2 ml buffer to 1 g tissue, centrifuge the homogenate at 40 000 × g for 15 min, and save the supernatant for immediate use.

Empty the pre-coated ELISA plate by shaking it vigorously. Wash three times by filling the wells with a stream of wash buffer (10 mM Tris-Cl, 0.05% Tween 20, 0.02% NaN$_3$, pH 8.0 at 25°C) from a squeeze bottle. Empty the plate completely between each wash. Saturate non-specific binding of protein to the plate by filling the wells with blocking solution (10 mM Na-phosphate, 140 mM NaCl, 1% bovine serum albumin, pH 7.4). Incubate for 30 min at room temperature. Wash the plate as above one or more times with wash solution and empty by shaking vigorously.

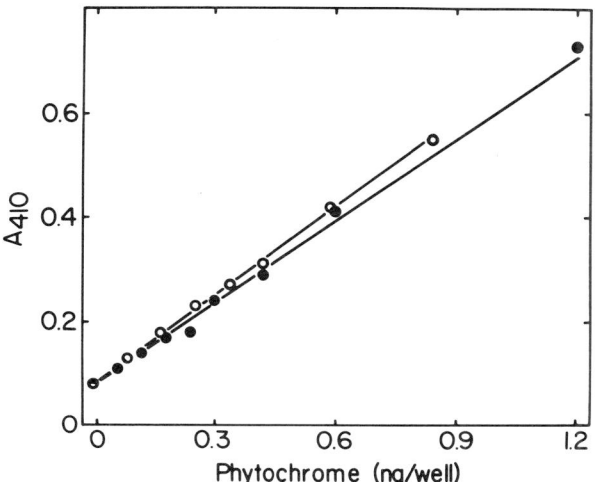

FIG. 7.5. Linear portion of a standard curve for quantitation of phytochrome by ELISA (●). Phytochrome quantities were also determined at various dilutions of a crude extract of etiolated oat shoots (○). In both cases, phytochrome amounts were determined independently by spectrophotometric assay (from Shimazaki *et al.*, 1983).

Fill wells either with 50 μl aliquots containing known quantities of phytochrome diluted in Diluent (10 mM Na-phosphate, 140 mM NaCl, 0.05% Tween 20, 0.02% NaN$_3$, 1% bovine serum albumin, pH 7.4), in order to prepare a standard curve (Fig. 7.5), or with 50 μl aliquots of unknown samples diluted as needed in Diluent (Fig. 7.4, P). For example, it is necessary to dilute extracts of etiolated oat shoots by 100- to 500-fold to bring them into the range of a typical standard curve. Incubate for 3 h at 4°C. One or more protease inhibitors can be added to the phytochrome samples, if desired.

Wash the plate three times as above. Add 50 μl of MAb in Diluent to each well. A mixture of three or four antibodies, each at 10 μg ml^{-1} and each detecting a different epitope on etiolated-oat phytochrome, will give the best results. Incubate for 2 h either at room temperature or at 37°C.

Again wash the plate three times as above. Add 50 μl of alkaline-phosphatase-conjugated rabbit antibody to mouse immunoglobulins (Fig. 7.4, AP-RAM). A typical commercial preparation (e.g. Sigma A-1902) works well diluted 500-fold in Diluent. Incubate for 2 h at room temperature or at 37°C.

After washing the plate three times as before, add 50 μl of *p*-nitrophenylphosphate solution (0.6 mg ml^{-1} in 9.6% (v/v) diethanolamine, 0.5 mM MgCl$_2$, pH adjusted to 9.8 with HCl) and incubate until colour development is sufficient, generally 30 min. Stop the reaction by addition of 50 μl of 3 N NaOH to each well. Measure absorbance at *c.* 400 nm, with reference to *c.* 500 nm if a two-wavelength measurement is possible. Figure 7.5 illustrates a typical standard curve, which provides sensitivity to the subfemtomol range.

2. Alternative, amplified ELISA

Amplification of alkaline phosphatase activity can greatly enhance sensitivity. The following protocol is adapted from Johannsson et al. (1986), who report that as little as 0.01 attomol (10^{-20} mol) of alkaline phosphatase can be detected. The ELISA protocol is as described above, with two modifications.

(1) The enhanced sensitivity leads to increased background. Consequently, the number of washes between each step is increased from three to five or more.
(2) Alkaline phosphatase-conjugated antibody is used at a dilution of 1000-fold rather than 500-fold.

The principle of the amplification is as follows. The initial substrate solution contains $NADP^+$ at pH 9.5, which is converted to NAD^+ by the alkaline phosphatase. After a suitable incubation period, a second substrate solution is added. This second solution is at pH 7.2, which effectively eliminates further activity of the alkaline phosphatase. The second substrate solution contains ethanol and alcohol dehydrogenase. NAD^+ is thereby reduced to NADH as ethanol is oxidized to acetaldehyde. The second substrate solution also contains a tetrazolium salt (INT) and a non-specific diaphorase. The diaphorase catalyses the oxidation of NADH back to NAD^+ with the concomitant reduction of INT, which yields an intensely coloured, violet reaction product. These two enzymes thus utilise NAD^+ in cyclic fashion, oxidising ethanol to reduce INT at a rate that is proportional to the amount of NAD^+ generated by alkaline phosphatase in the first step.

The necessary stock solutions are as follows. (1) Prepare 20 mM $NADP^+$ (Sigma N-0505) in water and store in 30 µl aliquots at $-20°C$. (2) Prepare 2.75 mM INT (Sigma I-8377) in 25 mM Na-phosphate, 20% (v/v) ethanol, pH 7.2, and store in 1.2 ml aliquots at $-20°C$. (3) Prepare a 5 mg ml^{-1} solution of diaphorase (Sigma D-2381) in 25 mM Na-phosphate, pH 7.2, and store in 150 µl aliquots at $-20°C$. (4) Prepare a 5 mg ml^{-1} solution of alcohol dehydrogenase (Sigma A-3263) in 50 mM tris-Cl, 40% (v/v) glycerol, 0.1% NaN_3, 10 mg ml^{-1} BSA, pH 8.0 at $25°C$, and store at $4°C$. Except for the alcohol dehydrogenase, reagents are stored in pre-measured aliquots so that they will not be repeatedly frozen and thawed. It is important that the alcohol dehydrogenase is not frozen as it will lose activity as a consequence.

Following incubation with alkaline phosphatase-conjugated antibody, wash the ELISA plate at least five times. Add to each well 50 µl of 100 µM $NADP^+$ (a 30 µl stock aliquot diluted to 6 ml with 50 mM diethanolamine, 1 mM $MgCl_2$, pH adjusted to 9.5 with HCl). Incubate at room temperature for an appropriate period, determined by trial and error, but usually in the range of 10–30 min. Without emptying the wells, add to each well 100 µl of second substrate solution prepared by mixing together 4.8 ml of 25 mM Na-phosphate, pH 7.2, 1.2 ml of INT stock solution, 150 µl of diaphorase stock solution, and 15 µl of alcohol dehydrogenase stock solution. Incubate for 30 min, or until colour development is sufficient. Stop the reaction by addition to each well of 50 µl of 0.4 N HCl. Measure absorbance at c. 500 nm, with reference to c. 400 nm if a two-wavelength measurement is possible. Because the violet reaction product precipitates with time, it is important to measure the plate as soon as feasible after stopping the reaction.

3. Comments

Controls are essential to test for possible non-specific activity in the assay, which might arise from non-specific adherence of any assay component to the well of the ELISA plate, and for accuracy. The most conservative control for non-specific activity is to coat the plate with non-immune immunoglobulins at the same concentration and of the same type as the immune immunoglobulins with which the plate is normally coated. A convenient method for testing the accuracy of the assay is to add a known amount of highly purified phytochrome to an unknown sample and verify that the expected increment in activity is attained (Shimazaki *et al.*, 1983).

An additional problem that we have encountered in immunoassays with plant extracts, whether a positive ELISA as described here (Shimazaki *et al.*, 1983) or a competitive radioimmunoassay (Hunt and Pratt, 1979a), is non-specific interference with antibody–antigen binding. In the former case, the result yields an underestimate in the amount of antigen in the plant extract, while in the latter, it yields an overestimate. Efforts to eliminate this artifact by additional preparation of crude plant extracts, for example by addition of EDTA or protease inhibitors, by $(NH_4)_2SO_4$ fractionation, or by filtration through Sephadex G-25, have been fruitless. Instead, we found it necessary to estimate the magnitude of the artifact by utilisation of a second immunoassay performed in parallel with the experimental immunoassay. This second immunoassay must be one that *a priori* should be totally non-responsive to the crude plant extracts being tested. In the case of the radioimmunoassay, this second assay was for either ovalbumin or apoferritin (Hunt and Pratt, 1979a), while in the case of the ELISA described here, it was for etiolated-pea phytochrome (Shimazaki *et al.*, 1983). Since the MAbs to etiolated-oat phytochrome that are used in the ELISA described here do not detect etiolated-pea phytochrome, any decrease in activity seen in an ELISA for etiolated-pea phytochrome, as a consequence of the addition of crude oat extract, must be ascribed to non-specific interference. Given that the extent of this non-specific interference is determined, correction of the experimental values is relatively straightforward (Shimazaki *et al.*, 1983). Fortunately, this problem does not arise when the crude extracts being assayed can be diluted 50- to 100-fold or more, as is the case when quantitating phytochrome from etiolated oat shoots.

While the protocol described here is tailored to the quantitation of etiolated-oat phytochrome, the specificity of the ELISA can be changed readily by substitution for the antigen-specific antibodies. For example, use of a MAb such as Oat-25, which is specific for an N-terminal epitope (Cordonnier *et al.*, 1985; see also Fig. 7.2), will permit detection of only that phytochrome which still has a relatively intact N-terminus. Alternatively, use of a MAb with sufficient specificity for the Pr form of phytochrome, such as LAS 41 (Holdsworth and Whitelam, 1987), will permit independent quantitation of this form. Similarly, while the ELISA described here is specific for etiolated-, as opposed to green-, oat phytochrome, the recent production of both MAbs and polyclonal rabbit antibodies (Pratt *et al.*, unpubl. res.) directed to and specific for green-oat phytochrome will facilitate independent quantitation of this pool of phytochrome. Finally, it should be emphasised that this sandwich ELISA does not require MAbs. It can, for example, be performed with polyclonal antibodies from two sources, such as rabbit and guinea pig, or with specific polyclonal antibodies from a single source. In this latter instance, one of the antibody preparations must be labelled, for example with alkaline phosphatase or with biotin, such that the assay will not detect the antibodies with which the ELISA plate was coated.

VI. PHYTOCHROME: PHOTOTRANSFORMATIONS

Phytochrome phototransformations can be considered from two perspectives: statistical and mechanistic. By statistical is meant the average behaviour of a population of molecules. By mechanistic is meant the behaviour of individual molecules. Thus, a consideration of phototransformation from a mechanistic perspective would include cryogenic and flash activation analyses in order to probe the pathways of phototransformation. Because such methodologies are beyond the scope of this chapter, they will not be discussed. Instead, the reader is referred to relatively recent summaries of work from three different laboratories (Rüdiger et al., 1985; Schaffner et al., 1985; Inoue, 1987) and to an older, but more comprehensive review (Kendrick and Spruit, 1977). Here, the focus will be on the statistical aspects of phototransformation, which can be elucidated with more conventional instrumentation. Of primary concern is the determination of (a) Pr → Pfr and Pfr → Pr phototransformation quantum yields (Φ_{Pr} and Φ_{Pfr}, respectively; see Table 7.4 for summary of terms and definitions), (b) absolute extinction spectra for Pr and Pfr, and (c) the proportion of phytochrome present as Pfr at photoequilibrium as a function of wavelength ($[Pfr]_\infty^\lambda$).

A. Quantum Yields and Photoequilibrium Values for Pfr

Two strategies exist for determining these photochemical parameters. The original approach was introduced by Butler et al. (1964a) and subsequently has been used most often (e.g. Pratt, 1975; Yamamoto and Smith, 1981; Vierstra and Quail, 1983b). This approach requires the measurement of (a) absorbance spectra for phytochrome after saturating irradiation with R and FR, (b) effective fluence rates for these actinic R and FR sources, and (c) initial phototransformation rates produced by these sources. While the relative simplicity of this initial-rates method makes it readily applicable, it is not as rigorous as the approach-to-equilibrium method that Kelly and Lagarias (1985) more recently introduced. Only this latter method will be outlined here.

Satisfactory measurement of phytochrome photochemical parameters requires that a number of conditions be met.

(1) The phytochrome in question must be homogeneous, in the sense that the opposing photoreactions represent a two-component system (i.e. Pr and Pfr, see Butler et al., 1964a; Yamamoto and Smith, 1981).
(2) Deriving from condition (1) is the necessity that actinic fluence rates be sufficiently low that photochemically active intermediates of phototransformation not be excited (Kelly and Lagarias, 1985).
(3) The phytochrome preparation being examined must be sufficiently pure that no other pigment absorbs light at the wavelengths of interest (Butler et al., 1964a).
(4) The spectrophotometer measuring beam must itself have no actinic effect.
(5) Measurements must be made rapidly enough such that any contribution to them by thermal reversion of Pfr to Pr is insignificant.

TABLE 7.4. Selected definitions and units for terms used to determine photochemical parameters of phytochrome phototransformations.

Term	Units	Definition
Φ_{Pr}	mole fraction	Pr → Pfr phototransformation quantum yield
Φ_{Pfr}	mole fraction	Pfr → Pr phototransformation quantum yield
$[Pfr]_\infty^\lambda$		Proportion of phytochrome as Pfr at photoequilibrium at wavelength λ
ε_{Pr}^λ	$1\ mol^{-1}\ cm^{-1}$	Molar extinction coefficient for Pr at wavelength λ
$\varepsilon_{Pfr}^\lambda$	$1\ mol^{-1}\ cm^{-1}$	Molar extinction coefficient for Pfr at wavelength λ
ε_{Pr}^r	$1\ mol^{-1}\ cm^{-1}$	Molar extinction coefficient for Pr at the red actinic wavelength
b_o	cm	Optical pathlength of actinic beam
E	$mol\ cm^{-2}\ s^{-1}$	Effective photon fluence rate
E_o	$mol\ cm^{-2}\ s^{-1}$	Incident photon fluence rate
A_r^λ		Sample absorbance at wavelength λ after saturating far-red irradiation
A_{fr}^λ		Sample absorbance at wavelength λ after saturating red irradiation
A_{max}^r		Sample absorbance at red actinic wavelength after saturating far-red irradiation
A_{min}^r		Sample absorbance at red actinic wavelength after saturating red irradiation
A_{max}^{fr}		Sample absorbance at far-red actinic wavelength after saturating red irradiation
σ^r	$cm^2\ mol^{-1}$	Photochemical cross-section for phototransformation, beginning with far-red-irradiated phytochrome, at the red actinic wavelength
σ^{fr}	$cm^2\ mol^{-1}$	Photochemical cross-section for phototransformation, beginning with red-irradiated phytochrome, at the far-red actinic wavelength
L	$mol\ cm^{-2}$	Absorbed photon fluence
α_{Pfr}^{fr}	$cm^2\ mol^{-1}$	Absorption cross-section for Pfr at the far-red actinic wavelength

Given that these conditions are met, three sets of data must be obtained: (1) effective photon fluence rates (E) of monochromatic R and FR actinic sources must be measured with precision; (2) absorbance spectra for the phytochrome preparation being examined must be recorded after saturating irradiation with these two actinic sources; and (3) photoconversion time courses as a function of absorbed fluence must be determined over a wide range. The derivation of equations that are then used to calculate phototransformation quantum yields and $[Pfr]_\infty$ for a reference red wavelength are already available (Butler et al., 1964a; Johns, 1969; Butler, 1972; Kelly and Lagarias, 1985). Only their use will be described here.

Accurate effective fluence rates are the most difficult data to obtain. When sample absorbance in the actinic path is low (<0.2) and the actinic beam is parallel and normal to the sample, it may be sufficient to measure the photon fluence rate with a well-calibrated thermopile, photocell, or other device, in which case a relatively simple mathematical correction for absorbance of the actinic beam as it passes through the sample can be made:

$$E = \frac{E_0(1 - 10^{-A})}{2.3A} \tag{7.3}$$

where E is the corrected, effective actinic photon fluence rate, E_0 is the incident actinic photon fluence rate, and A is the absorbance of the sample through the actinic light path and at the actinic wavelength. If desired, or if sample absorbance is high or if the geometry of the actinic beam is not ideal, effective actinic fluence rates may be measured by actinometry (Johns, 1969; Kelly and Lagarias, 1985).

Measure absorbance spectra following saturating irradiation with the actinic R and FR sources used for these photochemical measurements (see, e.g. Fig. 7.3(a), spectra for etiolated-oat phytochrome).

Determine phototransformation time courses by recording absorbance values, or preferably entire absorbance spectra, following different periods of irradiation with either actinic R or FR, beginning with a sample that had been pre-irradiated to saturation with either FR or R, respectively.

Derive photochemical cross-sections for phototransformation beginning with FR- and R-irradiated phytochrome at red and far-red actinic wavelengths, respectively (σ^r and σ^{fr}, respectively), from the phototransformation time courses as follows. Plot the logarithm of the percentage of reaction remaining to be completed ($\log[100(A_t - A_\infty)/(A_0 - A_\infty)]$) as a function of absorbed photon fluence (L), where A is measured at a wavelength at which a substantial change occurs during the phototransformation, A_t is the absorbance at a given absorbed photon fluence (i.e. at time $= t$), A_0 is the absorbance at zero time, and A_∞ is the absorbance at saturating photon fluence (i.e. at time $=$ infinity). A straight line should be obtained. The photochemical cross-section is then the reciprocal of the absorbed photon fluence when 36.8% (100%/e) of the reaction remains to be completed:

$$\sigma = \frac{1}{L_{36.8}} \quad \text{(Johns, 1969)}. \tag{7.4}$$

When beginning with a FR-irradiated sample, $\sigma = \sigma^r$ for the red actinic wavelength, and when beginning with a R-irradiated sample, $\sigma = \sigma^{fr}$ for the far-red actinic wavelength.

Determine the absorbed photon fluence, L, for use in constructing the previously described graph, by summing increments of absorbed photon fluence over the time intervals, Δt, between measurements of the extent of phototransformation:

$$L = \sum \left\{\frac{E\Delta t b_0}{V}\right\} \cdot \left\{\frac{1 - e^{-2.3A}}{2.3A}\right\} \tag{7.5}$$

where b_0 is the optical pathlength of the actinic beam through the sample, V is the volume of the sample in the cuvette, and A is the average absorbance of the sample through the actinic path and at the actinic wavelength during the increment of exposure (Eq. 24 in Johns, 1969; Eq. 3 in Kelly and Lagarias, 1985; note that the equation as given in Kelly and Lagarias contains a typographical error). If the geometry for actinic irradiation deviates significantly from the ideal (e.g. if the actinic beam is not parallel and is not normal to an optically flat surface of the cuvette containing the sample, or if there are internal reflections of the actinic beam within the sample), then correction of the pathlength for these non-idealities must be made. Procedures for performing this correction are given by Johns (1969) and by Kelly and Lagarias (1985). When the corrected pathlength is determined, it should be used in a modified form of Eq. 7.5:

$$L = \sum \left\{ \frac{E \Delta t b_0}{V} \right\} \cdot \left\{ \frac{1 - e^{-2.3Ab/b_0}}{2.3A} \right\} \tag{7.5a}$$

where b is the corrected pathlength.

Given σ^r and σ^{fr} (Eq. 7.4), calculate the ratio of quantum yields for the Pr → Pfr (Φ_{Pr}) and Pfr → Pr (Φ_{Pfr}) phototransformations from the following relationship:

$$\frac{\Phi_{Pr}}{\Phi_{Pfr}} = \frac{\sigma^r \cdot A^{fr}_{max}}{\sigma^{fr} \cdot A^{r}_{max}} \tag{7.6}$$

where A^{fr}_{max} is the absorbance of the sample at the far-red actinic wavelength following saturating irradiation with R and A^{r}_{max} is the absorbance of the sample at the red actinic wavelength following saturating irradiation with FR (Eq. 4 in Kelly and Lagarias, 1985). Calculate $[Pfr]^r_\infty$ as follows:

$$[Pfr]^r_\infty = 1 - [Pr]^r_\infty = 1 - \left\{ \left(\frac{A^r_{min}}{A^r_{max}} \right) \left(\frac{1}{1 + (\Phi_{Pr}/\Phi_{Pfr})} \right) \right\} \tag{7.7}$$

where $[Pfr]^r_\infty$ and $[Pr]^r_\infty$ are photoequilibrium values at the red actinic wavelength for Pfr and Pr, respectively, and A^r_{min} and A^r_{max} are the absorbance of the sample at the red actinic wavelength following saturating irradiation with R and FR, respectively (Eq. 3 in Butler et al., 1964a).

Given $[Pfr]^r_\infty$, it is possible to determine Φ_{Pr} and Φ_{Pfr} as follows. Because Pr does not absorb significantly in the far-red, it is simplest to calcuate first Φ_{Pfr}:

$$\Phi_{Pfr} = \frac{\sigma^{fr}}{\alpha^{fr}_{Pfr}} \tag{7.8}$$

where α^{fr}_{Pfr} is the absorption cross-section for Pfr at the far-red actinic wavelength (Kelly and Lagarias, 1985). Obtain α^{fr}_{Pfr} from the relationship:

$$\alpha^{fr}_{Pfr} = \frac{A^{fr}_{max} \cdot 2.3 \varepsilon^r_{Pr}}{A^r_{max} \cdot [Pfr]^r_\infty} \tag{7.9}$$

where ε^r_{Pr} is the molar extinction coefficient for Pr at the red actinic wavelength (Eq. 6 in Kelly and Lagarias, 1985). Given an absorbance spectrum for the sample after saturating irradation with FR (e.g. Fig. 7.3(a)) and a molar extinction coefficient for Pr at a given wavelength (e.g. 132 000 l mol^{-1} cm^{-1} for etiolated-oat phytochrome at maximum Pr absorbance; Kelly and Lagarias, 1985), calculation of ε^r_{Pr} is straightforward.

Given Φ_{Pfr}, determine Φ_{Pr} from Eq. 7.6. For Φ_{Pr}, Lagarias et al. (1987) report values of 0.152–0.154 and 0.172–0.174 for etiolated-oat and etiolated-rye phytochrome, respectively, while for Φ_{Pfr} they report values of 0.060–0.065 and 0.074–0.078.

B. Molar Extinction Spectra for Pr and Pfr

Because the absorbance spectra of Pr and Pfr overlap, it is impossible to obtain solutions of Pr that are not contaminated by Pfr, and vice versa. Since absorbance by Pr at far-red wavelengths is virtually nil (Fig. 7.3(a)), it is nevertheless possible to produce to a first approximation pure solutions of Pr by irradiation with FR. The inverse, however, is not possible. Thus, the absorbance spectrum of phytochrome after saturating irradiation with R reflects significant contributions from both Pfr and Pr. Fortunately, given $[Pfr]_\infty^r$, it is possible to calculate an absolute extinction spectrum for Pfr from an absorbance spectrum of a R-irradiated sample. Moreover, given an estimate for $[Pfr]_\infty^{fr}$ which is the proportion of phytochrome present as Pfr at photoequilibrium under FR, it is also possible to make the minor correction needed to account for residual Pfr in a FR-irradiated sample. It is assumed here that $[Pfr]_\infty^{fr} = 0.03$.

Calculate relative extinction spectra for Pr and Pfr from the following relationships:

$$\varepsilon_{Pr}^\lambda(\text{rel}) = \frac{A_r^\lambda - [Pfr]_\infty^{fr} \cdot \varepsilon_{Pfr}^\lambda(\text{rel})}{[Pr]_\infty^{fr}} \quad (7.10)$$

$$\varepsilon_{Pfr}^\lambda(\text{rel}) = \frac{A_{fr}^\lambda - [Pr]_\infty^r \cdot \varepsilon_{Pr}^\lambda(\text{rel})}{[Pfr]_\infty^r} \quad (7.11)$$

where $\varepsilon_{Pr}^\lambda(\text{rel})$ [$\varepsilon_{Pfr}^\lambda(\text{rel})$] is the relative extinction coefficient of Pr (Pfr) at wavelength λ, and A_r^λ (A_{fr}^λ) is the absorbance of the sample at wavelength λ after saturating far-red (red) irradiation. Because $\varepsilon_{Pr}^\lambda(\text{rel})$ is approximately equal to A_r^λ, it is best to approximate first $\varepsilon_{Pfr}^\lambda(\text{rel})$ by substituting A_r^λ for $\varepsilon_{Pr}^\lambda(\text{rel})$ in Eq. 7.11. With this first approximation of $\varepsilon_{Pfr}^\lambda(\text{rel})$, calculate $\varepsilon_{Pr}^\lambda(\text{rel})$ from Eq. 7.10. The difference between $\varepsilon_{Pr}^\lambda(\text{rel})$ and A_r^λ should be quite small (compare spectrum for etiolated-oat phytochrome as Pr in Fig. 7.3(a) to that for Pr in Fig. 7.6). Using this calculated value for $\varepsilon_{Pr}^\lambda(\text{rel})$, calculate $\varepsilon_{Pfr}^\lambda(\text{rel})$ with Eq. 7.11. To convert these relative values to molar extinction coefficients as a function of wavelength (ε_{Pr}^λ and $\varepsilon_{Pfr}^\lambda$), normalise the spectra to a value of 132 000 l mol^{-1} cm^{-1} at the wavelength of maximum absorbance for Pr, which is the molar extinction coefficient for etiolated-oat Pr at this wavelength (Kelly and Lagarias, 1985).

C. $[Pfr]_\infty$ as a Function of Wavelength

It is important to know the value of $[Pfr]_\infty$ at a given experimental wavelength in order to interpret physiological data obtained as a consequence of a given saturating irradiation at a given wavelength. With extinction spectra for Pr and Pfr (Fig. 7.6), and the ratio of Φ_R/Φ_{FR} (taken here to be 2.51, Lagarias et al., 1987), it is possible to calculate $[Pfr]_\infty$ as a function of wavelength (Eq. 1 in Lagarias et al., 1987):

$$[Pfr]_\infty^\lambda = \varepsilon_{Pr}^\lambda \left\{ \frac{\varepsilon_{Pr}^\lambda + \varepsilon_{Pfr}^\lambda}{(\Phi_R/\Phi_{FR})} \right\} \quad (7.12)$$

The results of this calculation are shown in Fig. 7.6.

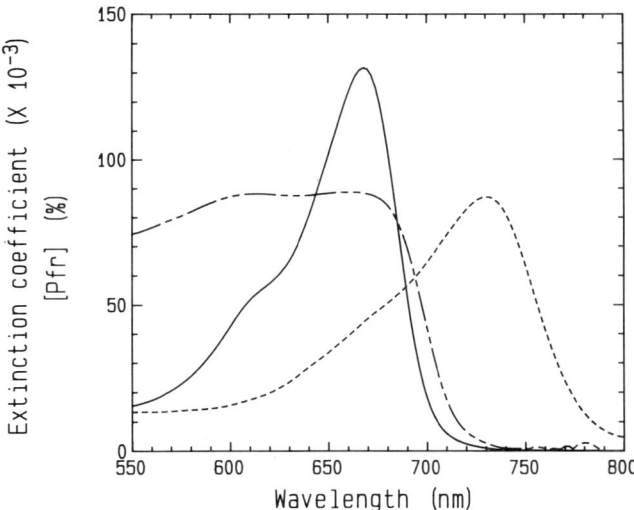

FIG. 7.6. Molar extinction spectra (1 mol^{-1} cm^{-1}) for etiolated-oat phytochrome as Pr (———) and Pfr (- - -), and the percentage of phytochrome present as Pfr at photoequilibrium as a function of wavelength ([Pfr], — - - —). These data are derived from the spectra for etiolated-oat phytochrome in Fig. 7.3(a), from the [Pfr]$_x$ value for phytochrome in red light determined by Kelly and Lagarias (1985), and from extinction coefficients determined by Lagarias *et al.* (1987). Absorbance maxima are at 670 nm and 731 nm for Pr and Pfr, respectively, while the isosbestic point is at 689 nm.

VII. PHYTOCHROME: MOLECULAR BIOLOGY

The tools of recombinant DNA technology can be used to contribute not only to a better understanding of phytochrome itself, but to learn more about how phytochrome functions at a molecular level. In the former case, cDNA clones can be used to predict primary amino acid sequences for phytochrome (Hershey *et al.*, 1985, 1987; Lissemore *et al.*, 1987; Sato, 1988) and, as described below, they can be used to map with precision amino acid sequences recognised by MAbs directed to this chromoprotein (Thompson *et al.*, 1989). In the latter case, information can be obtained about steps in the transduction chain between absorbance of light by phytochrome and biological responses, at least at transcriptional and translational levels (Schäfer and Briggs, 1986). Moreover, because phytochrome-mediated changes in gene expression are induced by brief, otherwise non-invasive pulses of R and FR, phytochrome provides an excellent model system for investigating control of gene expression at the molecular level.

Application of recombinant DNA technology to phytochrome is no different than to any other protein. Because generally applicable methods are already available in a number of handbooks (e.g. Maniatis *et al.*, 1989; Davis *et al.*, 1986; Berger and Kimmel, 1987), they will not be repeated here. Instead, one example of how phytochrome involvement in the control of gene transcription is documented will be presented. In addition, since phytochrome provides the best example from a plant system of the use of MAbs to elucidate the structure–function relationships of a protein (Cordonnier, 1989), the use of fusion peptides to map epitopes along the primary sequence of phytochrome will also be presented.

A. Phytochrome Control of Transcription of its Own Genes

Phytochrome regulates the expression of numerous genes, in both positive and negative fashion (Tobin and Silverthorne, 1985; Kuhlemeier et al., 1987; Silverthorne and Tobin, 1987; Colbert, 1988; Jenkins, 1988; Thompson, 1988). An example of how one documents this regulation could be selected from the many that have been described. Given, however, that phytochrome regulates the expression of its own genes at the mRNA level (Gottmann and Schäfer, 1982; Colbert et al., 1983, 1985), it seems most appropriate to select this phytochrome-mediated response for presentation here.

Phytochrome-modulated gene expression can result from brief very low fluence (VLF) or low fluence (LF) irradiations (Schäfer and Briggs, 1986), as well as presumably from prolonged irradiations at a relatively high fluence rate, as in the case of other phytochrome-mediated responses (Pratt and Cordonnier, 1989). Only VLF and LF responses have so far been well documented at the mRNA level. Moreover, LF responses are the easiest to assign unequivocally to phytochrome. Because phytochrome control of transcription of its own genes appears to occur in response to both VLF and LF irradiations in etiolated oat (*Avena sativa* L., cv. Garry) shoots, it serves as an ideal, as well as appropriate, example (Colbert et al., 1985; Colbert, 1988).

Begin with 4-day-old oat seedlings grown in total darkness and at 25°C. Perform all work under dim green light (Pratt, 1984), minimising exposure even to this safelight since VLF responses can be sensitive to even the most carefully prepared green safelight. Because it is easiest to prove phytochrome involvement by demonstrating FR reversibility of a R-induced effect, and because VLF responses can result from FR alone, an experimental protocol should include at a minimum the following: non-irradiated control, separate R and FR treatments, and a R exposure followed by FR to assess photoreversibility.

Ideally, the briefest practical pulses of R and FR should be used. Their durations can be determined from preliminary trial-and-error experiments. Given a sufficiently powerful irradiation system, such as a 500 W quartz halogen lamp, as little as 1–2 s may be required to saturate the responses to R and FR (McCurdy and Pratt, 1986). Suitable filtration of the output of the lamp can be obtained with interference filters at about 660 and 730 nm. Heat-absorbing glass, which is commonly used in commercial projection systems, usually absorbs FR strongly, and should therefore be avoided when working with FR.

Assessment of phytochrome mRNA levels by hybridisation assay with radiolabelled phytochrome cDNA is conveniently done after 3 h of dark incubation following the R and FR treatments (Colbert et al., 1985). It is important to note, however, that the time course of phytochrome-mediated responses, even at the mRNA level, varies considerably from gene to gene (Kaufman et al., 1986). Consequently, it is important to determine empirically the appropriate time to assess the abundance of a given mRNA. mRNA levels are quantitated in standard fashion via dot-blot, slot-blot or Northern-blot analyses. Phytochrome mRNA amounts are expressed either relative to an RNA species whose abundance can be assumed to remain unchanged during the experimental period, or, as in the case of Colbert et al. (1985), relative to a standard curve prepared with pure phytochrome mRNA that is transcribed *in vitro* from a suitable phytochrome cDNA clone.

Colbert et al. (1985) reported that a 5 s FR exposure resulted in a 46% decrease in phytochrome mRNA relative to the dark control. Because such a FR treatment is expected to produce a low but significant level of Pfr, it is likely that this FR effect results from a phytochrome-mediated, VLF response (Schäfer and Briggs, 1986). A 5 s R pulse induced a 92% decline in phytochrome mRNA level, indicative of strong, phytochrome-mediated repression of transcription of its own gene, an interpretation supported by subsequent *in vitro* nuclear run-on transcription experiments (Lissemore and Quail, 1988). Partial, but substantial, reversal of this R-induced response by 5 s of FR verified that phytochrome is responsible for repressing transcription of its own genes. Even though this reversal was to only 34% of the dark control level, it represented 58% reversal to the phytochrome mRNA level obtained with FR alone (Colbert *et al.*, 1985). The possibility that a phytochrome-mediated response might arise from FR alone, as documented in this example, emphasises the need to evaluate a R/FR reversibility assay with reference to what is observed with FR alone, rather than to what is observed for a dark control.

B. Epitope Mapping with Fusion Proteins

Investigators in several laboratories have obtained MAbs that interact with, and thereby identify, functionally important domains on phytochrome (Cordonnier, 1989). To maximise the information gained about phytochrome structure–function relationships as identified by MAbs, the precise amino acids to which a given MAb binds must be determined. While a variety of strategies exist for mapping epitopes recognised by MAbs, the use of fusion proteins is perhaps the most powerful. Previous uses of fusion proteins for epitope mapping either involved their generation at random (e.g. Mehra *et al.*, 1986; McGraw *et al.*, 1986), or relied upon the inability of a given MAb to detect a given fusion protein in making an epitope assignment (e.g. Doorbar *et al.*, 1988). In contrast, the method described here (1) does not rely upon the chance production of suitable fusion proteins, (2) yields wholly unambiguous data, with the final elucidation of an epitope based solely upon positive information (i.e. the ability of a MAb to bind to a given fusion protein rather than its inability to bind, which could arise from factors other than the absence of its epitope), and (3) is systematic, permitting one to narrow the definition of an epitope in organised fashion (Thompson *et al.*, 1989). Assignment of the epitope for MAb Pea-25 to a sequence of only seven amino acids will be presented here as an example of the method. Pea-25 is of special interest because it recognises the most highly conserved antigenic domain on phytochrome so far described, being found on a polypeptide the size of phytochrome from angiosperms through algae (Cordonnier *et al.*, 1986a).

Begin by obtaining the requisite cDNA sequence by any of a large number of presently available strategies. The clone utilised here, designated *9A*, was identified by immunochemical screening of a λgt11 expression library made with size-enriched, poly(A)$^+$-RNA from 3-day-old, etiolated oat shoots (Thompson *et al.*, 1989). Sequence analysis of the 5′ and 3′ ends of *9A* cDNA indicated that it probably derives from a type 3 gene as classified by Hershey *et al.* (1985), who also published the complete sequence of type 3 cDNA. This 2.2 kb *9A* sequence encodes amino acids 464 through 1129, which is the C-terminus of phytochrome, and extends through to an *Eco*RI site in the 3′, non-coding region (Fig. 7.7, the 3′ non-coding region is not shown). The *9A* insert was isolated from λgt11 by digestion with *Eco*RI, *Eco*RI to *Bam*HI linkers were attached and then digested with *Bam*HI. The isolated *9A* sequence with *Bam*HI ends was inserted into the *Bam*HI site in the polylinker region of pUC18, in frame with the *lacZa* sequence. The resultant construct is designated *pCIB315*. The subsequent use of this piece of cDNA to elucidate the epitope for MAb Pea-25 involves three steps.

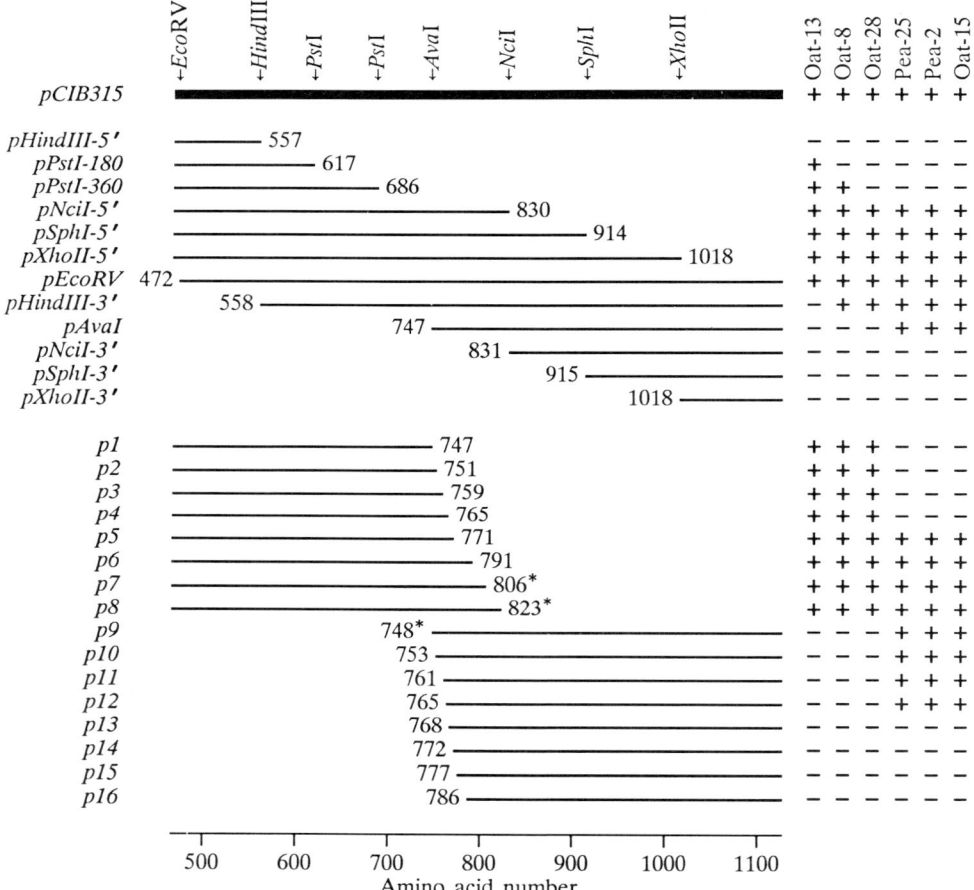

FIG. 7.7. Phytochrome cDNA subclones derived from clone *9A*, originally isolated in λgt11 and recloned in pUC18 as *pCIB315*. These subclones are also in pUC18, in frame with the *lacZa* sequence. Resultant fusion proteins were screened for antigenic activity by immunoblotting (see Fig. 7.8). Restriction sites are indicated at the top, amino acid residues beginning from the *N*-terminus of phytochrome at the bottom. Designation of each subclone is given to the left, its antigenic activity (+ or −) with respect to several MAbs to the right. Terminating or initiating amino acids are as indicated (those marked with an asterisk were not verified by sequencing) (from Thompson *et al.*, 1989).

In the first step, use restriction endonucleases to prepare two overlapping sets of nested DNA subclones, with each member of a set being about 300 bp shorter than the next longest member (Fig. 7.7, upper). The pUC18 clones containing these inserts are referred to here by the name of the enzymes used to generate them, followed if necessary by an indication of whether they are the 5' or 3' end of the original *9A* cDNA sequence (e.g. *pHindIII-5'*). Use the indicated endonuclease, together with a second endonuclease that would cut at only one end of the original pUC18 polylinker region, to excise the DNA that is discarded during the preparation of each of these subclones. Religate the open ends. In the case of the subclones that have had part of their 5' end deleted, take care to ensure that the religation is done in such a way that the phytochrome-coding sequence remains in frame with the *lacZa* sequence.

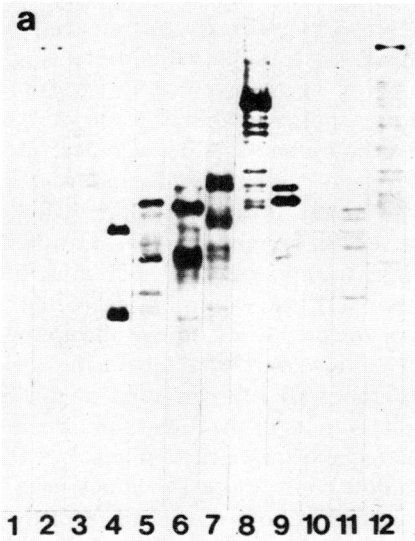

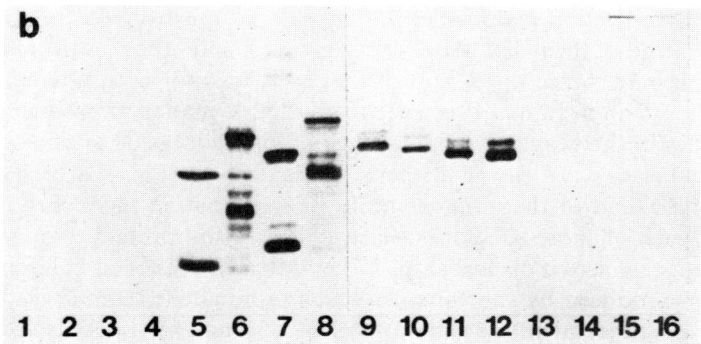

FIG. 7.8. Immunoblot analysis with MAb Pea-25 of fusion proteins derived from the cDNA subclones defined in Fig. 7.7. (a) Blot of fusion proteins derived from subclones generated by restriction endonuclease cleavage (upper part of Fig. 7.7). Note that each lane is derived from a different 10 to 20% linear-gradient polyacrylamide gel, such that comparison of sizes among lanes is not possible in this figure. (b) Blot of fusion proteins derived by Bal31 exonuclease digestion in the 5' direction from the *Sph*I site (subclones *p1–p8* in lanes 1–8, respectively), or in the 3' direction from a *Ppu*MI site, which is on the 5' side of the *Ava*I site (subclones *p9–p16* in lanes 9–16, respectively) (from Thompson *et al.*, 1989).

Transform host *Escherichia coli* (strain JM109) with the pUC18 subclones. Grow the transformed bacteria at 37°C overnight in 50 ml of LB-Amp medium (Maniatis *et al.*, 1982). In our experience, sufficient fusion protein is produced spontaneously so that addition of the inducer isopropyl-β-D-thiogalactopyranoside (IPTG) has been unnecessary. Harvest cells by centrifugation for 10 min at 1800 × g. Resuspend the cells in 1 ml of H_2O. Add 9 ml of 75 mM Tris-Cl, pH 6.8, 1.5% (w/v) SDS to the resuspended cells and lyse by boiling for 1 min. Store at or below $-20°C$. Immediately prior to use, thaw and dilute the lysate with SDS sample buffer by an amount that is determined by trial and error (3- to 10-fold dilution usually works well).

Electrophorese the diluted samples in an SDS polyacrylamide gel of suitable porosity (10–20% gradient generally works well), electrotransfer the protein to nitrocellulose, and immunostain. In addition to staining with MAbs (Fig. 7.8(a)), use non-immune mouse immunoglobulins as a negative control. Use rabbit polyclonal antibodies directed to phytochrome as a positive control to ensure that an antigenically active fusion protein has, in fact, been obtained. A detailed description of the immunoblotting protocol that is used by one of us is available elsewhere (Pratt et al., 1986). The approximate location of the epitope is defined by the region of overlap for the shortest subclone in each of the two sets that is immunostained by the MAb in question (*pNciI-5′* and *pAvaI* for Pea-25). Thus, the epitope for Pea-25 must be in the region encoded by DNA between the *Ava*I and *Nci*I sites (amino acids 747 and 830).

Tentative identification of the precise location of the epitope is achieved in the second step. Cleave *pCIB315* at a unique restriction site on the 3′ side of the region encoding the epitope being defined (the *Sph*I site was used in this instance). Digest with the exonuclease Bal31. Harvest aliquots of the digest mixture at appropriate intervals. The goal is to obtain a set of nested subclones, each of which is about 15 bp shorter than the previous one (Fig. 7.7, middle, only those subclones near the region of interest are shown). Although Bal31 digests in both directions from the site at which the plasmid was cleaved, the DNA lost in the 3′ direction is irrelevant as it is excised subsequently with a second restriction endonuclease that cuts uniquely in the original polylinker region at the 3′ end of the insert. After excising this 3′ end of the insert, religate the open ends of the plasmid. Since the 5′ end of the insert has not been altered, the resultant constructs all remain in frame. It is a relatively simple matter to produce a set of 50 or more subclones in this way, covering a range of about 150–200 amino acids.

Express and assay the fusion proteins as in step 1 (Fig. 7.8(b), lanes 1–8). By sequencing the 3′ end of the insert, identify the terminating phytochrome amino acid produced by each of those subclones that yields a fusion protein terminating near the site where antigenic activity is lost (Fig. 7.7, middle). Because Pea-25 immunostains the fusion protein produced by subclone *p5*, which terminates at amino acid 771, but not subclone *p4*, which terminates at amino acid 765, it may be concluded tentatively that the epitope is just to the *N*-terminal side of residue 771. This conclusion must be considered tentative because the absence of recognition may relate to considerations other than the absence of the epitope *per se*. For example, amino acids 765 through 771 may be required to induce recognisable structure at another position in the fusion protein, in which case these amino acids may not be part of the epitope itself. Alternatively, the predicted full length fusion protein may not accumulate to detectable levels, possibly because of rapid proteolytic cleavage within the host bacterium.

Conclusive identification of the epitope is achieved in the following third step. Cleave *pCIB315* at a unique restriction site on the 5′ side of the region known to contain the epitope, as determined in the first step above. In this case, the sequence was cleaved at a *Ppu*MI site just to the 5′ side of the *Ava*I site (Fig. 7.7). As in the second step, digest with Ba131, again harvesting aliquots at time points designed to give a nested set of subclones, each of which is about 15 bp shorter than before. In this case, however, cleave the harvested subclones with a second restriction endonuclease that is specific to a site in the pUC18 polylinker region that is at the 5' end of the insert. Religate the cut ends after this upstream fragment has been excised. Since the 5' end of the insert has been modified, not all constructs will be in frame. A larger number of subclones should therefore be prepared. Because those that are in frame will produce fusion proteins that are detectable by polyclonal antibodies to phytochrome, they are screened with such antibodies to select only in-frame constructs.

Assay in-frame constructs as above for the production of fusion protein to which Pea-25 will bind (Fig. 7.8(b), lanes 9–16). As in step 2, the sizes of the fusion proteins may be assessed by their mobility, as evidenced by position in the immunoblot. Sequence the 5' end of each insert that produces a fusion protein beginning near the location of the epitope (Fig. 7.7, lower, only selected constructs are shown). In this way, the N-terminal amino acid of each phytochrome sequence can be determined unambiguously. Because the subclone expressing phytochrome protein beginning with amino acid 765 contains the epitope, it is evident that the epitope must be wholly contained within structure from amino acid 765 towards the C-terminus. Because in the second step it was possible to conclude that the epitope must be wholly contained within structure from amino acid 771 towards the N-terminus, it is evident that the epitope is contained within the sequence from 765 through 771 (—pro—ile—phe—gly—ala—asp—glu—).

Consistent with this conclusion is the observation that a synthetic peptide containing this sequence competes with intact phytochrome for binding by Pea-25 in an ELISA (Pratt *et al.*, 1990). Moreover, the identical sequence is found in rice phytochrome (Kay *et al.*, 1989b) and, with the exception that alanine is replaced by serine or threonine, this sequence is also found at approximately the same location in the only other phytochrome sequences currently available, which are from *Cucurbita* (Sharrock *et al.*, 1986) and *Pisum* (Sato, 1988). While it is possible to define an epitope with comparable resolution with synthetic peptides (e.g. Geysen *et al.*, 1987), this alternative is considerably more laborious and expensive, especially as the size of the protein in question increases and the resolution desired becomes greater.

VIII. UV/BLUE-LIGHT PHOTORECEPTORS

A. Problems of UV/Blue-light Photoreceptor Identification

With the exception of some DNA photolyases (see below) no UV/blue-light photoreceptor has yet been isolated. The reason for this lies in the fact that no clear-cut photoreceptor assay is presently available. Investigations of the blue-light photoreceptors still depend, therefore, on action spectroscopy. Action spectra, such as those shown in Fig. 7.9, indicate that flavins are most likely involved in blue-light reception. There is a great need for a specific assay that would make it possible to distinguish the flavin receptor(s) from the bulk flavoproteins in the cell. The development of blue-light specific LIACs (light-induced absorbance changes) represent such an attempt (see below).

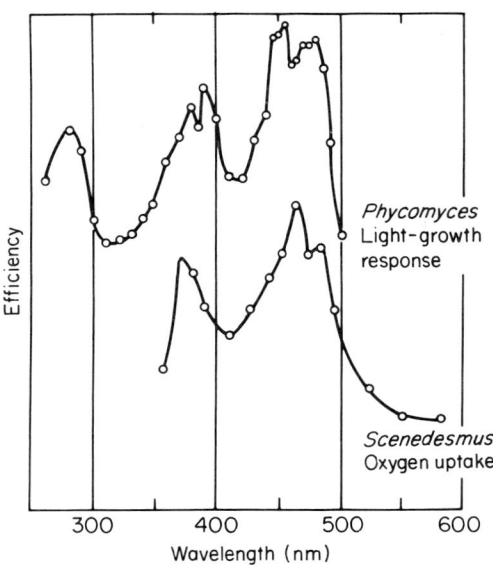

FIG. 7.9. Blue-light action spectra. Top: light-growth response of the *Phycomyces* sporangiophore (after Delbrück and Shropshire, 1960). Bottom: light-induced oxygen uptake of the single-celled, green alga *Scenedesmus obliquus* (after Wellburn et al., 1980).

B. Possible Methods for Photoreceptor Identification

1. Fluorescence spectroscopy

Most flavoproteins are non-fluorescent or only weakly fluorescent because of the strong interactions of the isoalloxazine ring with the aromatic amino acid residues (McCormick, 1977). Such strong interactions would make a flavoenzyme unsuitable as a photoreceptor, because the excited states would be immediately quenched and could not be used for primary photochemical reactions. It was, therefore, reasoned that a flavoprotein that functions as photoreceptor should bind only loosely to the apoprotein and thus allow efficient primary reactions to take place (Song, 1980). *In situ* the fluorescence should be quenched, just because of the expected efficient primary reactions. *In vitro*, however, where the tight coupling between photoreception and photochemistry is partially disrupted, some fluorescence with a short lifetime should occur. The fluorescence lifetime of the putative blue-light photoreceptor should be significantly shorter than that of free riboflavin (5.65 ns) or of weakly fluorescent enzymes such as D-amino acid oxidase (2.1 ns), because the primary photochemical reaction would efficiently depopulate the excited singlet state and thus shorten the fluorescence lifetime. The results of several investigators seem to bear out this reasoning. Flavoprotein preparations from corn coleoptiles contained a component with short fluorescence lifetime of <1 ns (Song, 1980). Similar preparations with short fluorescence lifetime were obtained in corn coleoptiles for a flavin bound to a *b*-type cytochrome (Sarkar et al., 1982). Pollock and Lipson (1985) found a flavoprotein in a membrane preparation of *Phycomyces* mycelia, which had a lifetime of <1 ns. Whether or not these flavoproteins are real blue-light photoreceptors *in vivo* is still unproven.

Plant photoreceptors are usually not located in special organelles, a fact which renders isolation attempts more difficult. The phytoflagellate *Euglena* is an exception in that light sensitivity is restricted to the paraflagellar body, which is a small paracrystalline swelling at the base of the flagellum. With the aid of a microspectrofluorometer it was possible to find flavin-like fluorescence emission near 525 nm (excitation with blue light), which was localised at the paraflagellar body (Benedetti and Lenci, 1977). Isolated flagella show pterin-like (emission near 450 nm) as well as flavin-like fluorescence (Galland *et al.*, 1989).

2. Light-induced absorbance changes (LIACs)

To distinguish the flavoprotein that functions as a photoreceptor from the bulk flavoproteins in the cell a specific assay is required. This problem was approached by analysis of light-induced absorbance changes (LIACs) that can possibly indicate a photochemical change of the photoreceptor. The types of LIAC that were most frequently studied indicate the blue-light-induced reduction of a *b*-type cytochrome (Fig. 7.10). These LIACs typically show absorption increase at 427 nm and absorption decrease near 450 nm. It is generally assumed that the flavin photoreceptor in its oxidised state is photochemically reduced by an unknown electron donor (for example NAD(P)H), thus creating an absorbance decrease at 450 nm. The reduced flavin would in turn transfer an electron to the *b*-type cytochrome and, therefore, increase the absorbance at 427 nm, i.e. the Soret peak of the cytochrome. LIACs of this type were found in mycelia of *Phycomyces*, *Neurospora*, in cell suspensions and cell homogenates of the slime mould *Dictyostelium* as well as in plants (for review see: Schmidt, 1980; Widell, 1987). The action spectra for the production of these LIACs indicate that flavoproteins are functioning as photoreceptors (inset Fig. 7.10). Below we describe an example of a typical LIAC-experiment.

(a) *LIAC in corn coleoptiles and* Neurospora *(Muñoz and Butler, 1975)*. Cell extracts or the whole organism are prepared under red safelight. Light-induced absorbance changes (LIAC) can be measured with a dual-wavelength spectrophotometer. The actinic light is usually very high (of the order of 20 W m^{-2}) and thus far above the thresholds of blue-light responses. Such intensities can be obtained from a xenon or a tungsten–halogen lamp (150–500 W) in appropriate combination with heat-absorbing glass (or a 15% solution of copper sulphate), interference filters and cutoff filters. The difference spectrum of the sample is obtained in the single beam-measuring mode by scanning the absorption spectrum before and after the actinic irradiation. The spectra are stored in an interfaced computer that also calculates the difference in the O.D. A reduced-oxidised difference absorption spectrum serving as a control is obtained by adding a few grains of dithionite to the sample.

The extent and the kinetics of the absorbance changes are measured in the double-beam mode setting the two wavelengths at positions that are indicative for cytochrome reduction—for example ΔA (424–466) or ΔA (560–578). Continuous measurements can be made before, during, and after the actinic irradiation, as long as the actinic light is prevented by cutoff filters from reaching the photomultiplier.

Absolute absorption spectra of *in vivo* material (for example mycelia of *Phycomyces*, *Trichoderma* or *Neurospora*) need to be corrected for wavelength-dependent light scattering. This can be done by measuring the spectrum of several layers of white tissue paper and subtracting it from the spectrum of the respective organism.

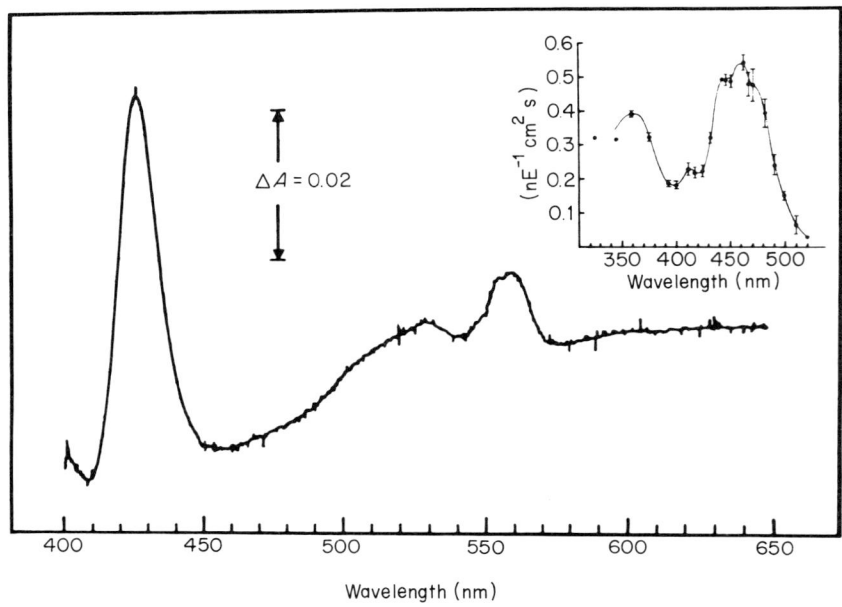

FIG. 7.10. Light-induced absorbance changes in mycelium of *Neurospora crassa*. Light-minus-dark difference spectrum after irradiation at 540 nm. Inset: action spectrum to produce the absorbance increase at 560 nm (after Muñoz and Butler, 1975).

3. Analogue-substitution experiments

A strong argument for the action of flavoproteins as blue-light photoreceptors came from substitution experiments with flavin analogues. Roseoflavin (8-dimethylamino-riboflavin) has a bathochromic shift (i.e. to longer wavelengths) compared with riboflavin. Sporangiophores of *Phycomyces* that were grown on a roseoflavin-containing medium showed a bathochromic shift of the action spectrum (Otto *et al.*, 1981). Roseoflavin-treated colonies of *Trichoderma* required six times more light to saturate conidiation (Horwitz *et al.*, 1984). However, no shift of the action spectrum to longer wavelengths was observed. Two photoresponses of *Neurospora*, namely photosuppression and phase shifting of circadian rhythm of conidiation, could be elicited at 540 nm by supplementing the growth medium with roseoflavin or the analogue 1-deazariboflavin, which also has a bathochromic shift (Paietta and Sargent, 1983). Using retinal analogues with either hypsochromic or bathochromic shifts it could be shown for *Chlamydomonas* that phototaxis is mediated by a rhodopsin photoreceptor (Foster *et al.*, 1983).

C. Photoactivation of Enzymes: Nitrate Reductase

Many enzymes can be activated by blue light on the transcriptional level and post-transcriptional level (for review see Ruyters, 1984; Horwitz and Gressel, 1986). A good example for a blue light-controlled enzyme is nitrate reductase, which is ubiquitously present among plants and microorganisms. Because its activity can be photostimulated *in vivo* and *in vitro* and because its activity was found to correlate with photostimulation of conidiophore formation in *Neurospora*, it can serve as a model system for blue-light photoreceptors (for review see Ninnemann, 1984, 1987). Very likely the following sequence of the two-electron transfer is valid for most nitrate reductases:

$$NAD(P)H \to FAD \to cyt\text{-}b_{557} \to \text{pterin-Mo-complex} \to NO_3.$$

In many plants nitrate reductase activity was found only in light but not in darkness (for review see Srivastava, 1980). The nitrate reductase of some algae and higher plants can exist *in vitro* and/or *in vivo* in an active and an inactive state, depending on their redox state. Inactivation can be brought about *in vitro* by NAD(P)H and subsequently reactivated with ferricyanide, suggesting that the enzyme is active in the oxidised and inactive in the reduced state. Reactivation of inactivated spinach nitrate reductase was also obtained by addition of FMN or FAD in the presence of white or specifically blue light (Aparicio *et al.*, 1976). In *Chlamydomonas* the nitrate reductase remains inactive in darkness and can be reactivated in the presence of nitrate by blue light, but not by red light (Azuara and Aparicio, 1983). Also in *Neurospora* blue light could reactivate purified enzyme that was inactivated by cyanide reduction. The action spectrum for reactivation indicated a flavin as the blue-light photoreceptor (Roldán and Butler, 1980).

As an example for such reactivation experiments we give the protocol used by Aparicio *et al.* (1976). For a general review about this topic the reader is referred to recent reviews (Ninnemann, 1984, 1987).

(a) *Photoreactivation of spinach nitrate reductase (Aparicio et al., 1976)*. Partially purified nitrate reductase is equilibrated in Tris buffer, pH 8.2, in the presence of FAD. The preparation is left with NADH and KCN for 20 min in darkness. Excess NADH and KCN is removed by passing the preparation over a Sephadex G-25 column. The inactivated nitrate reductase thus obtained is stable in darkness for several days and can be used for light experiments. The enzyme preparation is irradiated in glass cuvettes kept at 4°C in ice water. The enzyme activity during or following irradiation can be measured spectrophotometrically by following the nitrate-dependent NADH oxidation at 340 nm. Under these conditions half-maximal photoreactivation by blue light occurs at $3 \times 10^4 \, \text{J m}^{-2}$, which is extremely high compared to *in vivo* blue-light responses. Removal of the FAD or FMN results in a 6–8 times higher fluence requirement for photoreactivation.

IX. DNA PHOTOLYASES

Photoreactivation or photorepair of UV-induced DNA damage is a vital function ubiquitously present in the plant and animal kingdoms (Rupert, 1975). Irradiation of DNA with far-UV light induces the formation of pyrimidine dimers, mainly *cis–syn* thymine–thymine dimers, that obstruct further replication and would be lethal if they remained in the cell. Besides general dark-repair mechanisms that are non-specific for dimer repair, the pyrimidine dimers can be specifically removed by the photoreactivating enzymes, which can be classified as deoxyribonucleate cyclobutane dipyrimidine photolyases, or for short, DNA photolyases (Minato and Werbin, 1972). Upon absorption of light between 300–600 nm, DNA photolyases split the cyclobutane ring that was formed by cycloaddition of adjacent pyrimidines; the two original pyrimidines are then regenerated.

The only UV/blue-light photoreceptors that have been isolated to date are the DNA photolyases of *Escherichia coli*, *Saccharomyces cerevisiae* and *Streptomyces griseus*. The reason for this one has to seek in the fact that an assay for these enzymes—the monomerisation of thymine dimers—has been available for a long time while it has been lacking for other UV/blue-light photoreceptors. The DNA photolyase of *E. coli* was purified to homogeneity. It is a monomer with an apparent M_r of 49 000 in agreement with the M_r of 53 994 that was calculated from the gene sequence (Sancar et al., 1984). The chromophores were identified as $FADH_2$ (Sancar and Sancar, 1984; Jorns et al., 1984) and 7,8-dihydropterin with substitutions at positions 5 and 6 (Wang et al., 1988). Both chromophores seem to be involved in the light reception and contribute to the *in vivo* and also the *in vitro* action spectra peaks near 390 nm (Sancar et al., 1987). The DNA photolyase of *Streptomyces griseus* has an oxidised deazaflavin as a chromophore (7,8-didemethyl-8-hydroxy-5-deazaflavin derivative) (Eker et al., 1981). There are no indications for a second chromophore. The precise action mechanism of the enzyme is unknown. Because thymine dimers can be monomerised not only by electron donation, but also by electron abstraction, it seems likely that the deazaflavin chromophore acts from the oxidised state.

A. Action Spectra for *In vitro* Photoreactivation

DNA containing pyrimidine dimers is mixed with an excess of photolyase such that all dimers are complexed and an intense light flash (flash-photolysis 1 ms) (Harm et al., 1971) or a light pulse of several seconds or minutes duration is applied to the mixture. This treatment suffices to completely remove the dimers. Under these conditions the enzymatic activity can be described by the following sequence of reactions (Sancar et al., 1987):

$$E + S \xrightarrow{k_1} ES \xrightarrow{k_3} E + P \tag{7.13}$$

$$\frac{-d[ES]}{dt} = k_3[ES] \tag{7.14}$$

$$\frac{[ES]}{[ES]_0} = \exp(-k_3 t) \tag{7.15}$$

The first-order rate constant k_3 has the dimension of s^{-1} and is thus equal to $k_3 =$

ε_λ(m² mol⁻¹)$\Phi_\lambda N_\lambda$ (mol$_{(photons)}$ m⁻² s⁻¹), where ε_λ is the molar extinction coefficient, Φ_λ the quantum effectiveness and N_λ the photon fluence rate. Equation 7.15 can thus be rewritten in the form:

$$\frac{[ES]}{[ES]_0} = \exp(-\varepsilon_\lambda \Phi_\lambda F_\lambda) \qquad (7.16)$$

where F_λ is the total photon fluence applied during the duration Δt of the flash ($F_\lambda = N_\lambda \Delta t$). At a photon-fluence at which e^{-1} dimers remain the photoreactivation cross-section ($\varepsilon_\lambda \Phi_\lambda$) equals F^{-1}. Because ε_λ is known for the purified enzyme, Φ_λ can be obtained. For the purified photolyase of *E. coli*, which exists in the blue form, i.e. the flavin-semiquinone radical, the value was rather low (0.062) at 384 nm (Sancar *et al.*, 1987). Reduction of the enzyme with dithionite caused a drastic increase to a value of $\Phi_\lambda = 1$, similar to the values obtained for *in vivo* experiments. From these results one can conclude that the flavin chromophore exists *in vivo* as $FADH_2$ and not as the semiquinone radical (Sancar *et al.*, 1987).

B. Assays for DNA Photolyases

The activity of photolyases can be tested *in vitro* by a number of different methods. All assays begin with complexing the photolyase preparations with UV-irradiated DNA containing a known amount of pyrimidine dimers and subsequent irradiation with long-wavelength light ($\approx$ 320–500 nm). All manipulations are done under green, yellow or red safelight. Also gold-fluorescent light has been used.

1. Spectroscopic method

The spectroscopic method takes advantage of the fact that two monomeric thymines undergo a hypsochromic shift to 240 nm in the dimeric form (Jorns *et al.*, 1985). Dimer formation thus results in a reduction of the O.D. of the total DNA at 260 nm. To detect the small changes of the optical density one needs to work with oligo$(dT)_n$ as substrate to avoid interference caused by purines. Treatment of dimer-containing oligonucleotides (minimal length $\approx$ 4 nucleotides) with photolyase reverses the hypsochromic shift and the O.D. at 260 nm can be quantitatively restored. The detection limit of approximately 150 pmol of repaired dimer does not compare favourably with the 100 times more sensitive assays that have been developed with phage and plasmid DNA (see below). This disadvantage is, however, fully made up by the extreme speed of this assay, which greatly exceeds the speed of the assays described below.

2. Chromatographic detection

DNA labelled with [³H]thymine is irradiated with far-UV light, complexed to photolyase and irradiated above 340 nm. The mixture is then hydrolysed with 97% formic acid and chromatographed on Whatman No. 1 paper with *n*-butanol–water (86:14). The thymine and the thymine dimer-containing regions of the chromatogram are excised, the material is eluted with water, and the radioactivity is measured. The ratio of the radioactivity contained in the monomers and the dimers is a measure for the photolyase activity (Terry and Setlow, 1967). ¹⁴C-labelled *cis–syn* thymine dimers required as reference markers for the chromatography can be prepared by irradiating a solution of [methyl-^{14}C]thymine in water with about 230 J m^{-2} at 254 nm (Beukers and Berends, 1950).

3. DEAE disc assay

This assay takes advantage of the fact that dimers are resistant to nuclease treatment as detailed below (Farland and Sutherland, 1979). After photoreactivation of ^{3}H-labelled, dimer-containing DNA the preparation is incubated during one hour with deoxyribonuclease I, snake venom phosphodiesterase and bacterial alkaline phosphatase type III-S. This treatment causes complete digestion of DNA that is lacking dimers, while dimer-containing DNA retains sequences that are resistant to nuclease treatment and which bond to DEAE-substituted paper. Two aliquots of the DNA-enzyme mixture are spotted on two DE-81 (Whatman) discs (21 mm). One disc is washed twice in 1 mM ammonium formate and once in water to remove [^{3}H]nucleosides and inorganic phosphate, while another disc remains unwashed. Washed and unwashed discs are transferred to individual scintillation vials and agitated for 15 min in 1 mM NaCl to elute [^{3}H]oligonucleotides. Aquasol is added and the resulting clear solution is counted in a scintillation counter. The percentage, N, of nuclease-resistant DNA is obtained in the following way: N = (c.p.m. of washed disc/c.p.m. unwashed disc) × 100. Control experiments involving formic acid hydrolysis with subsequent thin layer chromatography show that N is proportional to the amount of pyrimidine dimers in the UV-irradiated DNA.

4. Alkaline agarose gel electrophoresis method

After termination of photolyase activity the DNA-enzyme mixture is treated with the UV-endonuclease from *Micrococcus luteus*, which cleaves specifically behind a pyrimidine dimer (Freeman et al., 1986). The sample is then denatured by alkaline treatment (0.5 N NaOH) and run on a 0.4% alkaline agarose gel in 50 mM NaCl and 4 mM EDTA. DNA bands are stained with ethidium bromide. The number of endonuclease-sensitive sites per kilobase (ESS kb^{-1}) is given by the formula:

$$\frac{\text{ESS}}{\text{kb}} = \frac{1}{L_{n(+\text{endo})}} - \frac{1}{L_{n(-\text{endo})}}$$

where L_n is the average length of the DNA bands on the gel with and without endonuclease treatment.

C. *In vivo* Transformation Assay

If the DNA containing the thymine dimers carries a suitable marker, it can be used for transformation of *E. coli* or possibly another host organism missing that marker gene. A widely used method employs DNA bearing a streptomycin resistance marker gene to transform *Hemophilus influenzae*. Another useful substrate for transformation is the *E. coli* plasmid pBR322 bearing genes for ampicillin and tetracycline resistance (Sancar et al., 1987). Because a single thymine dimer is sufficient to inhibit plasmid (or other DNA) replication completely the successful transformation of the recipient cell indicates complete removal of the dimers. To achieve a high sensitivity for the assay care should be taken that the recipient strain lacks repair enzymes that could possibly remove the thymine dimers of the transforming DNA. One uses, therefore, strains that are $recA^-$ (general recombinational and postreplicative repair), $uvrA^-$ (general excision repair of UV damage) and also pho^- (lack of DNA photolyase itself) (Sancar et al., 1987). In the actual assay a known amount of plasmids with a known amount of dimers are complexed with the photolyase in darkness. After irradiation (340–460 nm) the DNA is added to the recipient cells under conditions that are appropriate for DNA uptake (high Ca^{2+} in E. coli). The number of extransformants is proportional to the photolyase activity.

REFERENCES

Aparicio, P. J., Roldán, J. M. and Calero, F. (1976). *Biochem. Biophys. Res. Commun.* **70**, 1071–1077.
Azuara, A. P. and Aparicio, J. P. (1983). *Plant Physiol.* **71**, 286–290.
Benedetti, P. A. and Lenci, F. (1977). *Photochem. Photobiol.* **26**, 315–318.
Berkelman, T. R. and Lagarias, J. C. (1986). *Anal. Biochem.* **156**, 194–201.
Berger, S. L. and Kimmel, A. R. (eds) (1987). *Methods Enzymol.* **152**, 812 pp.
Beukers, R. and Berends, W. (1950). *Biochim. Biophys. Acta* **41**, 550–551.
Björn, L. O. (1971). *Physiol. Plant.* **25**, 300–307.
Bradford, M. M. (1976). *Anal. Biochem.* **72**, 248–254.
Butler, W. L. (1972). In "Phytochrome" (K. Mitrakos and W. Shropshire, Jr, eds), pp. 185–192. Academic Press, New York.
Butler, W. L., Norris, K. H., Siegelman, H. W. and Hendricks, S. B. (1959). *Proc. Natl. Acad. Sci. USA* **45**, 1703–1708.
Butler, W. L., Hendricks, S. B. and Siegelman, H. W. (1964a). *Photochem. Photobiol.* **3**, 521–528.
Butler, W. L., Siegelman, H. W. and Miller, C. O. (1964b). *Biochemistry* **3**, 851–857.
Chai, Y. G., Singh, B. R., Song, P.-S., Lee, J. and Robinson, G. W. (1987). *Anal. Biochem.* **163**, 322–330.
Colbert, J. T. (1988). *Plant Cell Environ.* **11**, 305–318.
Colbert, J. T., Hershey, H. P. and Quail, P. H. (1983). *Proc. Natl. Acad. Sci. USA* **80**, 2248–2252.
Colbert, J. T., Hershey, H. P. and Quail, P. H. (1985). *Plant Mol. Biol.* **5**, 91–101.
Cordonnier, M.-M. (1989). *Photochem. Photobiol.* **49**, 821–831.
Cordonnier, M.-M. and Pratt, L. H. (1982). *Plant Physiol.* **69**, 360–365.
Cordonnier, M.-M. and Pratt, L. H. (1990). In "Photobiology" (E. Riklis, ed.), Plenum Press, New York, in press.
Cordonnier, M.-M., Greppin, H. and Pratt, L. H. (1985). *Biochemistry* **24**, 3246–3253.
Cordonnier, M.-M., Greppin, H. and Pratt, L. H. (1986a). *Plant Physiol.* **80**, 982–987.
Cordonnier, M.-M., Greppin, H. and Pratt, L. H. (1986b). *Biochemistry* **25**, 7657–7666.
Correll, D. L., Edwards, J. L., Klein, W. H. and Shropshire, W., Jr (1968). *Biochim. Biophys. Acta* **168**, 36–45.
Datta, N. and Roux, S. J. (1985). *Photochem. Photobiol.* **41**, 229–232.
Davis, L. G., Dibner, M. D. and Battey, J. F. (1986). "Basic Methods in Molecular Biology". Elsevier, New York, 388 pp.
Delbrück, M. and Shropshire, W., Jr (1960). *Plant Physiol.* **35**, 194–204.
Doorbar, J., Evans, H. S., Coneron, I., Crawford, L. V. and Gallimore, P. H. (1988). *EMBO J.* **7**, 825–833.
Eker, A. P. M., Dekker, R. H. and Berends, W. (1981). *Photochem. Photobiol.* **33**, 65–72.
Farland, W. H. and Sutherland, B. M. (1979). *Anal. Biochem.* **97**, 376–381.
Foster, K. W., Saranak, J., Patel, N., Zarilli, G., Okabe, M., Kline, T. and Nakanishi, N. (1984). *Nature* **311**, 756–760.

Freeman, S. E., Blackett, A. D., Monteleone, D. C., Setlow, R. B., Sutherland, B. M. and Sutherland, J. C. (1986). *Anal. Biochem.* **158**, 119–129.
Galland, P. and Senger, H. (1988a). *J. Photochem. Photobiol. B* **1**, 277–294.
Galland, P. and Senger, H. (1988b). *Photochem. Photobiol.* **48**, 811–820.
Galland, P., Keiner, P., Dörnemann, D., Senger, H., Brodhun, B. and Häder, D.-P. (1990). *Photochem. Photobiol.* (submitted).
Gardner, G., Pike, C. S., Rice, H. V. and Briggs, W. R. (1971). *Plant Physiol.* **48**, 686–693.
Geysen, H. M., Tainer, J. A., Rodda, S. J., Mason, T. J., Alexander, H., Getzoff, E. D. and Lerner, R. A. (1987). *Science* **235**, 1184–1190.
Gottmann, K. and Schäfer, E. (1982). *Photochem. Photobiol.* **35**, 521–525.
Grimm, R. and Rüdiger, W. (1986). *Z. Naturforsch.* **41c**, 988–992.
Grimm, R., Lottspeich, F. and Rüdiger, W. (1987). *FEBS Lett.* **225**, 215–217.
Harm, W., Rupert, C. S. and Harm, H. (1971). *In* "Photophysiology" (A. C. Giese, ed.), Vol. 6, pp. 279–324. Academic Press, New York.
Hartmann, K. M. (1982). *In* "Biophysik—Ein Lehrbuch" (W. Hoppe, W. Lohmann, H. Markl and H. Ziegler, eds), pp. 122–152. Springer-Verlag, Berlin, Heidelberg and New York. English edition (1983), pp. 115. Springer-Verlag, Berlin.
Hartmann, K. and Cohnen-Unser, I. (1972). *Ber. Deutsch. Bot. Ges.* **85**, 481–551.
Hershey, H. P., Barker, R. F., Idler, K. B., Lissemore, J. L. and Quail, P. H. (1985). *Nucl. Acids Res.* **13**, 8543–8559.
Hershey, H. P., Barker, R. F., Idler, K. B., Murray, M. G. and Quail, P. H. (1987). *Gene* **61**, 339–348.
Holdsworth, M. L. and Whitelam, G. C. (1987). *Planta* **172**, 539–547.
Holmes, M. G. (1984a). *In* "Techniques in Photomorphogenesis" (H. Smith and M. G. Holmes, eds), pp. 43–79. Academic Press, London.
Holmes, M. G. (1984b). *In* "Techniques in Photomorphogenesis" (H. Smith and M. G. Holmes, eds), pp. 81–107. Academic Press, London.
Horwitz, B. A. and Gressel, J. (1986). *In* "Photomorphogenesis in Plants" (R. E. Kendrick and G. H. M. Kronenberg, eds), pp. 159–186. Martinus Nijhoff/Dr. W. Junk, Dordrecht, The Netherlands.
Horwitz, B. A., Malkin, S. and Gressel, J. (1984). *Photochem. Photobiol.* **40**, 763–769.
Hunt, R. E. and Pratt, L. H. (1979a). *Plant Physiol.* **64**, 327–331.
Hunt, R. E. and Pratt, L. H. (1979b). *Plant Physiol.* **64**, 332–336.
Hunt, R. E. and Pratt, L. H. (1980). *Biochemistry* **19**, 390–394.
Inoue, Y. (1987). *In* "Phytochrome and Photoregulation in Plants" (M. Furuya, ed.), pp. 117–126. Academic Press, Tokyo.
Jagger, J. (1967). "Introduction to Research in Ultraviolet Photobiology". Prentice-Hall, Englewood Cliffs, NJ.
Jagger, J. (1977). *In* "The Science of Photobiology", 1st edn (K. C. Smith, ed.), pp. 1–26. Plenum Press, New York.
Jenkins, G. I. (1988). *Photochem. Photobiol.* **48**, 821–832.
Johannsson, A., Ellis, D. H., Bates, D. L., Plumb, A. M. and Stanley, C. J. (1986). *J. Immunol. Meth.* **87**, 7–11.
Johns, H. E. (1969). *Methods Enzymol.* **16**, 253–316.
Jones, A. M. and Quail, P. H. (1986). *Biochemistry* **25**, 2987–2995.
Jorns, M. S., Sancar, G. B. and Sancar, A. (1984). *Biochemistry* **23**, 2673–2679.
Jung, J. and Song, P.-S. (1979). *Photochem. Photobiol.* **29**, 419–421.
Kaufman, L. S., Roberts, L. L., Briggs, W. R. and Thompson, W. F. (1986). *Plant Physiol.* **81**, 1033–1038.
Kay, S. A., Keith, B., Shinozaki, K., Chye, M.-L. and Chua, N.-H. (1989a). *Plant Cell* **1**, 351–360.
Kay, S. A., Keith, B., Shinozaki, K. and Chua, N.-H. (1989b). *Nucleic Acids Res.* **17**, 2865–2866.
Kelly, J. M. and Lagarias, J. C. (1985). *Biochemistry* **24**, 6003–6010.
Kendrick, R. E. and Kronenberg, G. H. M. (eds) (1986). "Photomorphogenesis in Plants". Martinus Nijhoff/Dr. W. Junk, Dordrecht, The Netherlands.
Kendrick, R. E. and Spruit, C. J. P. (1977). *Photochem. Photobiol.* **26**, 201–214.

Kerscher, L. and Nowitzki, S. (1982). *FEBS Lett.* **146**, 173–176.
Kidd, D. G. and Lagarias, J. C. (1990). *J. Biol. Chem.* **265**, 7029–7035.
Kuhlemeier, C., Green, P. J. and Chua, N.-H. (1987). *Ann. Rev. Plant Physiol.* **38**, 221–257.
Lagarias, J. C., Kelly, J. M., Cyr, K. L. and Smith, W. O. Jr (1987). *Photochem. Photobiol.* **46**, 5–13.
Laemmli, U. K. (1970). *Nature (Lond.)* **227**, 680–685.
Lissemore, J. L. and Quail, P. H. (1988). *Mol. Cell. Biol.* **8**, 4840–4850.
Lissemore, J. L., Colbert, J. T. and Quail, P. H. (1987). *Plant Mol. Biol.* **8**, 485–496.
Litts, J. C., Kelly, J. M. and Lagarias, J. C. (1983). *J. Biol. Chem.* **258**, 11 025–11 031.
Maniatis, T., Fritsch, E. F. and Sambrook, J. (1989). "Molecular Cloning: A Laboratory Manual", Cold Spring Harbor Laboratory, Cold Spring Harbor, New York, in 3 vols.
McCurdy, D. W. and Pratt, L. H. (1986). *Planta* **167**, 330–336.
McGraw, R., Frazier, D., de Serres, M., Reisner, H. and Stafford, D. (1986). *Blood* **67**, 1344–1348.
Mehra, V., Sweetser, D. and Young, R. A. (1986). *Proc. Natl. Acad. Sci. USA* **83**, 7013–7017.
Minato, S. and Werbin, H. (1972). *Photochem. Photobiol.* **15**, 97–100.
McCormick, D. B. (1977). *Photochem. Photobiol.* **26**, 169–182.
Montagnoli, G. (1980). In "Photoreception and Sensory Transduction in Aneural Organisms" (F. Lenci and G. Colombetti, eds), pp. 23–44. Plenum Press, New York.
Mumford, F. E. and Jenner, E. L. (1966). *Biochemistry* **5**, 3657–3662.
Muñoz, V. and Butler, W. L. (1975). *Plant Physiol.* **55**, 421–426.
Ninnemann, H. (1984). In "Blue Light Effects in Biological Systems" (H. Senger, ed.), pp. 95–109. Springer-Verlag, Berlin, Heidelberg and New York.
Ninnemann, H. (1987). In "Blue Light Responses: Phenomena and Occurrence in Plants and Microorganisms" (H. Senger, ed.), Vol. I, pp. 17–30. CRC Press, Baca Raton, FL.
Otto, M.-K., Jayaram, M., Hamilton, R. H. and Delbrück, M. (1981). *Proc. Natl. Acad. Sci. USA* **78**, 266–269.
Paietta, J. and Sargent, M. L. (1983). *Plant Physiol.* **72**, 764–766.
Parks, B. M., Jones, A. M., Adamse, P., Koornneef, M., Kendrick, R. E. and Quail, P. H. (1987). *Plant Mol. Biol.* **9**, 97–107.
Poff, K. L., Burkhart, U., Häder, D.-P. and Vierstra, R. (1984). *Photochem. Photobiol.* **39**, 119–122.
Pollock, J. A. and Lipson, E. D. (1985). *Photochem. Photobiol.* **41**, 351–354.
Pratt, L. H. (1973). *Plant Physiol.* **51**, 203–209.
Pratt, L. H. (1975). *Photochem. Photobiol.* **22**, 33–36.
Pratt, L. H. (1980). In "Photoreceptors and Plant Development" (J. DeGreef, ed.), pp. 103–119. Antwerpen University Press, Antwerpen, Belgium.
Pratt, L. H. (1983). *Encycl. Plant Physiol. New Ser.* **16**, 152–177.
Pratt, L. H. (1984). In "Techniques in Photomorphogenesis" (H. Smith and M. G. Holmes, eds), pp. 175–200. Academic Press, London.
Pratt, L. H. and Cordonnier, M.-M. (1987). In "Phytochrome and Photoregulation in Plants" (M. Furuya, ed.), pp. 83–94. Academic Press, Tokyo.
Pratt, L. H. and Cordonnier, M.-M. (1989). In "The Science of Photobiology", Second Edition (K. C. Smith, ed.), pp. 273–304. Plenum Press, New York.
Pratt, L. H., McCurdy, D. W., Shimazaki, Y. and Cordonnier, M.-M. (1986). In "Modern Methods of Plant Analysis, New Series" (H. F. Linskens and J. F. Jackson, eds), Vol. 4, pp. 50–74. Springer-Verlag, Berlin.
Pratt, L. H., Cordonnier, M.-M. and Lagarias, J. C. (1988). *Arch. Biochem. Biophys.* **267**, 723–735.
Pratt, L. H., Cordonnier, M.-M. and Crossland, L. (1990). In "Photobiology", (E. Riklis, ed.) Plenum Press, New York, in press.
Rice, H. V., Briggs, W. R. and Jackson-White, C. J. (1973). *Plant Physiol.* **51**, 917–926.
Roldán, J. M. and Butler, W. L. (1980). *Photochem. Photobiol.* **32**, 375–381.
Rüdiger, W., Eilfeld, P. and Thümmler, F. (1985). In "Optical Properties and Structure of Tetrapyrroles" (G. Blauer and H. Sund, eds), pp. 349–366. Walter de Gruyter & Co., Berlin.

Rupert, C. S. (1975). In "Molecular Mechanisms for Repair of DNA", Part A (P. C. Hanawalt and R. B. Setlow, eds), pp. 73–87. Plenum Press, New York.
Ruyters, G. (1984). In "Blue Light Effects in Biological Systems" (H. Senger, ed.), pp. 283–301. Springer-Verlag, Berlin, Heidelberg, New York and Tokyo.
Sancar, A. and Sancar, G. B. (1984). *J. Mol. Biol.* **172**, 223–227.
Sancar, A., Smith, F. W. and Sancar, G. B. (1984). *J. Biol. Chem.* **259**, 6028–6032.
Sancar, G. B., Jorns, M. S., Payne, G., Fluke, D. J., Rupert, C. and Sancar, A. (1987). *J. Biol. Chem.* **262**, 492–498.
Sarkar, H. K., Song., P.-S. Leong, T.-Y. and Briggs, W. R. (1982). *Photochem. Photobiol.* **35**, 593–595.
Sato, N. (1988). *Plant Mol. Biol.* **11**, 697–710.
Schäfer, E. and Briggs, W. R. (1986). *Photobiochem. Photobiophys.* **12**, 305–320.
Schaffner, K., Braslavsky, S. E. and Holzworth, A. R. (1985). In "Optical Properties and Structure of Tetrapyrroles" (G. Blauer and H. Sund, eds), pp. 367–382. Walter de Gruyter & Co., Berlin.
Schmidt, W. (1980). *Struct. Bond.* **41**, 1–44.
Schuman Jorns, M. (1985). *Biochemistry* **24**, 1856–1861.
Senger, H. (ed.) (1980). "The Blue Light Syndrome". Springer-Verlag, Berlin, Heidelberg and New York.
Senger, H. (ed.) (1984). "Blue Light Effects in Biological Systems". Springer-Verlag, Berlin, Heidelberg, New York and Tokyo.
Senger, H. (ed.) (1987). "Blue Light Responses: Phenomena and Occurrence in Plants and Microorganisms," Vols. 1 & 2. CRC Press, Boca Raton, FL.
Sharrock, R. A. and Quail, P. H. (1989). *Genes Develop.* **3**, 1745–1757.
Sharrock, R. A., Lissemore, J. L. and Quail, P. H. (1986). *Gene* **47**, 287–295.
Shimazaki, Y. and Pratt, L. H. (1985). *Planta* **164**, 333–344.
Shimazaki, Y., Moriyasu, Y., Pratt, L. H. and Furuya, M. (1981). *Plant Cell Physiol.* **22**, 1165–1173.
Shimazaki, Y., Cordonnier, M.-M. and Pratt, L. H. (1983). *Planta* **159**, 534–544.
Siegelman, H. W. and Firer, E. M. (1964). *Biochemistry* **3**, 418–423.
Silverthorne, J. and Tobin, E. M. (1987). *BioEssays* **7**, 18–23.
Smith, W. O. Jr and Cyr. K. L. (1988). *Plant Physiol.* **87**, 195–200.
Snapka, R. M. and Fuselier, C. O. (1977). *Photochem. Photobiol.* **25**, 415–420.
Srivastava, H. S. (1980). *Phytochemistry* **19**, 725–733.
Terry, C. E. and Setlow, R. B. (1967). *Photochem. Photobiol.* **6**, 799–803.
Thompson, L. K., Pratt, L. H., Cordonnier, M.-M., Kadwell, S., Darlix, J.-L. and Crossland, L. (1989). *J. Biol. Chem.* **264**, 12 426–12 431.
Thompson, W. F. (1988). *Plant Cell Environ.* **11**, 319–328.
Tobin, E. M. and Silverthorne, J. (1985). *Ann. Rev. Plant Physiol.* **36**, 569–593.
Tokuhisa, J. G. and Quail, P. H. (1989). *Photochem. Photobiol.* **50**, 143–152.
Tokuhisa, J. G., Daniels, S. M. and Quail, P. H. (1985). *Planta* **164**, 321–332.
Towbin, H., Staehelin, T. and Gordon, J. (1979). *Proc. Natl. Acad. Sci. USA* **76**, 4350–4354.
Vierstra, R. D. and Quail, P. H. (1982). *Proc. Natl. Acad. Sci. USA* **79**, 5272–5276.
Vierstra, R. D. and Quail, P. H. (1983a). *Biochemistry* **22**, 2498–2505.
Vierstra, R. D. and Quail, P. H. (1983b). *Plant Physiol.* **72**, 264–267.
Vierstra, R. D. and Quail, P. H. (1986). In "Photomorphogenesis in Plants" (R. E. Kendrick and G. H. M. Kronenberg, eds), pp. 35–60. Martinus Nijhoff/Dr. W. Junk, Dordrecht, The Netherlands.
Vierstra, R. D., Cordonnier, M.-M., Pratt, L. H. and Quail, P. H. (1984). *Planta* **160**, 521–528.
Wang, B., Jordan, S. P. and Schuman Jorns, M. (1988). *Biochemistry* **27**, 4222–4226.
Wang, Y.-C., Pratt, L. H., Cordonnier, M.-M. and Stewart, S. J. (1989). In "Annual Symposium on Photomorphogenesis in Plants" (book of abstracts), Freiburg, FRG, Abstract 25.
Wellburn, F. A. M., Wellburn, A. R. and Senger, H. (1980). *Protoplasma* **103**, 35–54.
Widell, S. (1987). In "Blue Light Responses: Phenomena and Occurrence in Plants and Microorganisms" (H. Senger, ed.), Vol. II, pp. 89–98. CRC Press, Boca Raton, FL.
Yamamoto, K. T. and Smith, W. O. Jr (1981). *Plant Cell Physiol.* **22**, 1159–1164.

8 Optical Techniques in the Study of Photosynthesis

PAUL MATHIS

Service de Biophysique, Département de Biologie, Centre d'Etudes Nucléaires de Saclay, 91191 Gif-sur-Yvette Cedex, France

I.	Introduction	231
II.	Light absorption: electronic transitions	233
	A. General considerations	233
	B. Steady-state absorption spectroscopy	234
	C. Kinetic absorption spectroscopy	235
	D. Linear dichroism	247
	E. Special techniques	248
III.	Light emission	249
	A. Fluorescence lifetime measurements	250
	B. Phosphorescence	250
	C. Delayed light emission	250
	D. Thermoluminescence	251
	E. ODMR: FMDR	251
IV.	Vibrational spectroscopy	251
	A. General considerations	251
	B. Infrared absorption	252
	C. Resonance Raman spectroscopy	253
	Acknowledgements	254
	References	254

I. INTRODUCTION

Of all biological processes, photosynthesis is probably the one most related to light, and it is thus not surprising that optical techniques play a key role in experimental studies of this process (Clayton, 1970). These techniques are all based on the spectroscopic properties of molecules, which exist in a number of discrete states, each with its own energy. The transition between these states is associated with the absorption or the emission of a quantum of radiation, with conservation of the energy: $E_2 - E_1 = h\nu$, where h is Planck's constant, and ν is the frequency of the light. Transitions between electronic states involve radiation ranging from the ultraviolet to the near infrared, whereas transitions between vibrational states involve medium infrared. This review will not cover the theoretical aspects of molecular spectroscopy, which are dealt with in the textbooks by Steinfeld (1985), Birks (1970) and Salem (1966). Rather it will stay on the technical side and try to cover the gap between biological problems and the practical aspects of their investigation by optical techniques: electronic spectroscopy (absorption, fluorescence, delayed emission) and vibrational spectroscopy (infrared absorption, resonance Raman scattering). Most techniques will be introduced, but only kinetic flash absorption will be covered in detail. For detailed information on experimental methods, one can refer to Rabek (1982). Commercial sources for pieces of equipment are not given here. They can be found in the many cited original references.

Optical techniques offer three major advantages: their great sensitivity, their great potential time resolution, and their non-destructive character. In the case of photosynthesis, they facilitate the identification of the molecules present in a sample, enable many details on the state of a given molecule (redox state, excitation state, orientation in the membrane, interaction with other molecules, modes of binding, etc.) to be obtained, and also enable reaction kinetics to be studied. It is well known that photosynthetic structures involve a great variety of pigments and redox active molecules, and that several of the chemical species, such as chlorophyll and plastoquinone, include several functional and structural subsets: optical methods contribute decisively to unravelling these complexities within relatively unperturbed biological structures (see for example many applications in Scheer and Schneider, 1988).

Two properties of the photosynthetic apparatus in plants render optical measurements often rather difficult, however. First, the photoactive reaction centres are associated with a large number of pigments, 50–1000 molecules typically, which function as a light-harvesting antenna. This tends to dilute the features of interest within a high background of light absorption, and also to confer an actinic character to any kind of measuring light which will be mainly absorbed by the antenna and its excitation energy transferred to the reaction centres. A second property is the light scattering by the photosynthetic apparatus, which is made of bilayer membranes located within organelles with their own envelopes. Isolated reaction centres are devoid of these two complications since they have only a small antenna, and are practically optically clear; however, they do not constitute the complete photosynthetic apparatus.

Although not the main objective of this review, it is worth mentioning that optical technology is also exploited for appropriate excitation of the photosynthetic apparatus. In particular, single flashes can be provided, especially laser flashes, which ensure a single turnover of all reaction centres in a studied sample, and their effect analysed by multiple optical and non-optical techniques (for example, oxygen evolution, electron paramagnetic resonance spectroscopy (EPR)). This potentiality is a great advantage which is not offered in other enzymatic systems.

II. LIGHT ABSORPTION: ELECTRONIC TRANSITIONS

A. General Considerations

Transitions between electronic states of molecules give rise to broad absorption bands, from the UV to about 1.5 µm, in the near infrared. In general, these bands are rather well understood in terms of molecular electronic wave functions. In the case of pigments and electron carriers involved in photosynthesis, the most prominent features are displayed by tetrapyrrolic molecules (chlorophylls, phycobiliproteins, cytochromes) and by polyenes (carotenoids), the structures of which include a large array of delocalised π electrons. Quinones display less intense absorption bands in the UV, and metal ions give rise to rather weak but characteristic absorption: iron in Fe–S centres, copper in plastocyanin, or manganese in the water-oxidising enzyme (San Pietro, 1971, 1980; Vernon and Seely, 1986; Lichtenthaler, 1987). A few spectra are presented in Fig. 8.1 for illustration.

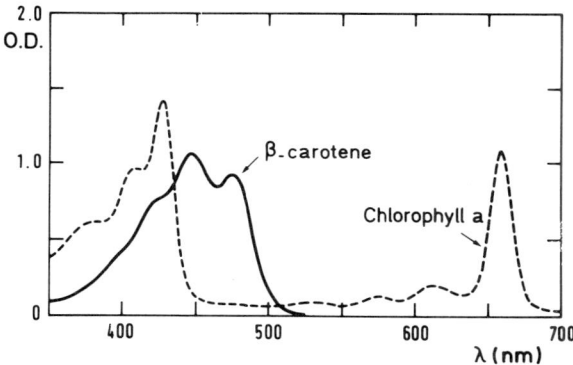

FIG. 8.1. Absorption spectra of pure chlorophyll a (11 µM, in diethylether) and β-carotene (8.6 µM, in petroleum ether). Cuvette path: 1.0 cm (data kindly provided by J. Kleo).

These absorption properties have a rather rich information content. Without being exhaustive, four aspects can be mentioned:

(1) The shape of a spectrum is rather characteristic of a given molecule and can be used for an approximate chemical identification. Molecular concentrations can also be measured by application of Beer–Lambert's Law: at a given wavelength, the absorbance A is given by $A = \log_{10}(I_0/I) = \varepsilon\, cl$, where I_0 and I are the incident and exiting light beam intensities, ε is the molar extinction coefficient, c is the molar concentration and l is the optical pathlength.

(2) For a given molecule, the shape of the absorption spectrum is distorted by interaction with the solvent and, more importantly for photosynthetic pigments and proteins, by interaction with neighbouring pigments. In particular, chlorophyll–chlorophyll interactions induce strong absorption shifts which can be interpreted within the exciton theory (see, e.g. Warshel and Parson, 1987).

(3) Transient states of molecules, such as excited states within the singlet or triplet manifold or radical anions or cations, have spectra different from the parent molecule. This allows identification and kinetic studies.

(4) In chiral molecules, the absorption is slightly different for left or right circularly polarised light. This property is known as circular dichroism. It is strongly influenced by the interactions experienced by the molecule under study and, for that reason, is often used to understand chlorophyll–protein or chlorophyll–chlorophyll interactions.

B. Steady-state Absorption Spectroscopy

Recording of the absorption spectrum is now a simple operation, and many commercial spectrophotometers are available for this purpose (although most of them do not scan the near infrared, from 800 to 1400 nm, which is mandatory in studies of photosynthetic bacteria). However, several properties of photosynthetic systems deserve special attention and discussion.

1. Optical perturbations

Several optical perturbations can severely distort the spectra. The first, most obvious one, is the actinic effect of the measuring light, which can induce a photochemical reaction and change the state of the photosynthetic material. Some states are rather long-lived, such as the redox state of the manganese complex in Photosystem II (PSII), the state of the bound quinone Q_B in PSII and bacteria, and P_{700}^+, the primary donor of Photosystem I (PSI), in the absence of reductant. This problem is also important at low temperatures. The answer is obviously to always use monochromatic measuring light of low intensity on the sample, and to check for eventual actinic effects by recording the spectrum with different measuring beam intensities.

Photosynthetic preparations usually have a particulate nature inducing a λ (wavelength)-dependent light scattering which distorts the recorded spectra. Outside the absorption bands, the scattered intensity follows a λ^{-4} law and can thus be approximately corrected. Also the contribution of scattering can be decreased by collecting light under a wide solid angle after the cuvette (Latimer, 1959; Shibata, 1973). In some cases, intense light scattering can result in an effective light path much greater than the sample thickness, due to multiple back-scattering. In that case the absorption bands are enhanced (Butler, 1962, 1972). This effect can take place in a microcrystallised sample at low temperature.

With large, strongly absorbing particles such as chloroplasts, absorption spectra are also distorted by particle flattening (the so-called sieve effect; light hitting a chloroplast will be nearly totally absorbed whereas light passing outside will not be absorbed at all). This effect can be approximately corrected, by making some assumptions about the optical properties of a single particle. The correction factor can have a value of about two in the main chlorophyll absorption band of whole chloroplasts, and so it is safer to work as much as possible with smaller particles (Duysens, 1956; Pulles et al., 1976), such as reaction centre complexes, in which the flattening effect is negligible.

2. Differential absorption spectroscopy

In photosynthesis research, it is often important to record a difference spectrum, i.e. the absorption difference between two steady-state conditions of a sample. The transition from one state to another is usually caused by light excitation which populates an excited state or changes the redox equilibrium in the sample. When the created state is stable, such as after illumination at low temperature (Butler, 1972) or after redox poising (Chance, 1972), there are well-known techniques for recording the difference spectrum. For low temperature differential spectroscopy, good commercial spectrophotometers can be used, after coupling with a microcomputer. Three specifications of the instrument are of special interest: the sensitivity (to measure small variations on top of a large background of absorption), the wavelength reproducibility (to measure narrow absorption shifts which often take place at low temperature), and the solid angle of light collection in the case of measurements on scattering samples. Of recent application in photosynthesis is the highly specialised technique known as 'photochemical hole burning' where a difference spectrum is recorded at very low temperature (~ 1.5 K), with extremely high spectral resolution, in response to the actinic excitation of a sample with monochromatic light. The photochemical hole is persistent over many minutes, although it gives information on the initial short-lived excited states (Boxer et al., 1986; Meech et al., 1986).

For unstable light-induced states a good sensitivity is provided by modulated techniques: the measuring light is modulated at a frequency compatible with the lifetime of the unstable state, and the detecting device includes a lock-in amplifier with frequency and phase matching. On this basis, several types of instruments give useful information. The actinic light source can be turned on as a long square pulse and its beam allowed to fall on the sample through a chopper which is out of phase with the chopper on the measuring beam. Its effect is recorded separately at each wavelength of the measuring light. This technique is well adapted to slowly reversible states (several seconds) and to states which accumulate progressively with a low quantum efficiency. It has been applied very successfully to study the primary partners in reaction centres (see Klevanik et al., 1977 for example). Out-of-phase synchronisation of the choppers avoids saturation of the photodetector by the actinic beam and allows measurements at wavelengths where chlorophyll fluoresces (675–750 nm in plant photosystems). Many variations of the modulated method can be set up, when the spectral data are of prime interest, but in practice, this can lead to a loss of kinetic information.

C. Kinetic Absorption Spectroscopy

The methodologies of kinetic absorption spectroscopy derive from the flash photolysis technique invented by Norrish and Porter. Basically it consists of exciting a sample under study with a short flash of light, and in monitoring the effect of that flash with another light beam. The measuring capability is based on the difference in the absorption spectra of the initial molecules and of the photochemical products. The interest of the method in photosynthesis research can be illustrated by the reaction scheme of Photosystem II (Fig. 8.2). Excitation with a short light pulse excites antenna pigments, which transfer energy to the primary electron donor P_{680}. When promoted to its singlet excited state, P_{680} transfers an electron to the primary electron acceptor (a pheophytin molecule named I), and the formation of the primary radical pair (P_{680}^+, I^-) is followed by secondary steps of electron transfer leading to reduction of the primary quinone Q_A and to the oxidation of the tyrosine Z. There is some probability that the radical pair will recombine, forming the triplet state of P_{680}, which can decay by two mechanisms: de-excitation to the ground state or energy transfer to a carotene molecule which becomes excited to its triplet state ^{3}Car. Kinetic absorption spectroscopy is the method of choice to identify all these steps, in terms of kinetics and of chemical identification, since each species displays its own characteristic absorption spectrum.

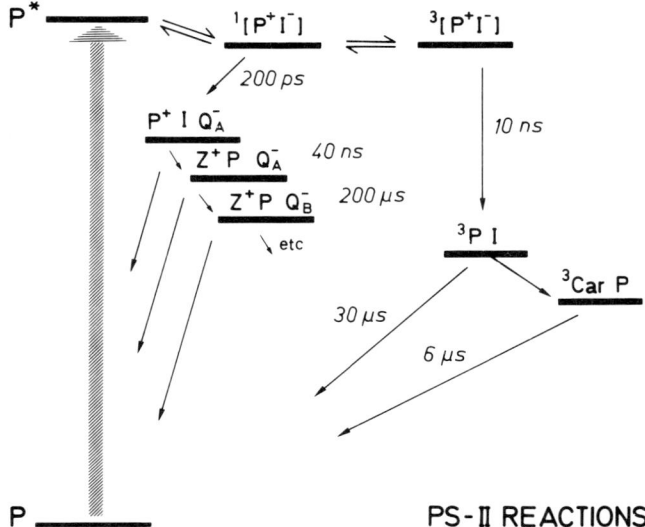

FIG. 8.2. Reaction scheme for excited states and electron transfer steps in the Photosystem II reaction centre. The scheme is limited to a few basic reactions and is far from being complete (for details, see Mathis and Rutherford, 1987). Indicated reaction times are very approximate.

1. Methodology: general considerations

Apparatus for flash absorption measurements are laboratory-built according to two schemes, which differ in the means of obtaining the time coordinate. Absorbance can either be measured directly versus time at a single wavelength with a constant monitoring light, or the absorption can be sampled by short measuring flashes fired with variable delays with respect to the actinic flash (detecting flash method). Set-ups are organised around five elements: cuvette, excitation source, measuring light, photo-detector and signal recording. The technique, as applied to photosynthesis research, was reviewed in detail a few years ago (Mathis, 1977). Thus, this presentation will mainly deal with the new technical developments. It should be clear that there is no universal push-button apparatus for kinetic absorption and that each set-up has to be built according to specific objectives.

(a) *Cuvette and cuvette housing.* The cuvette shape must be optimised according to the absorbance at the exciting and measuring wavelength, to provide a rather homogeneous excitation and a maximum signal. For example, a long path cuvette will be appropriate for clear samples, whereas a short path has to be used for scattering materials. Small cuvettes (1 mm or less) are required in picosecond measurements. Recent progress has been made in temperature regulation, which is now rather practical down to 4.2 K, and feasible to below 1.5 K, for specific applications. Several manufacturers offer optical cryostats cooled with a flow of helium gas which is controllably warmed above the liquefaction temperature.

(b) *Excitation.* Ideally, the excitation light pulse should be short in comparison to the kinetics of interest. It should also be strong, in order to saturate all the reaction centres in the cuvette, but not too strong, to avoid superfluous excitation, which results in an excess of excited states in the antenna and in biphotonic processes. The excitation should also be as selective as possible, especially in fast kinetics studies, but spectral regions of high absorbance by the sample should not be used, to avoid inhomogeneous excitation. Fantastic advances in pulsed laser technology give the user an extended choice in terms of pulse duration, energy, wavelength (including bandwidth) and repetition rate: nearly everything is possible, but most pulsed lasers are expensive. Xenon flashes are much cheaper, and they remain of great interest in all cases where neither very short nor wavelength-selective excitation is required.

(c) *Measuring light.* This is often a key factor. Let us recall the absorbance change ΔA per cm, measured at the wavelength where the molecule under study (the concentration of which changes by Δc) absorbs with a molar absorptivity (extinction coefficient) ε:

$$\Delta A = \varepsilon \Delta c = \log_{10} \frac{I}{I + \Delta I} \simeq -\frac{1}{2.3} \cdot \frac{\Delta I}{I} \quad \text{for small } \Delta A \tag{8.1}$$

where I is the measuring light intensity at the cuvette exit. In practical cases in photosynthesis research, ΔA will be of the order of 10^{-3} and the accuracy in relative I measurement should thus be at least of that order. Various sources of noise go against that accuracy: fluctuations in the emission of the measuring light source (this has to be analysed and solved according to the source and to the studied time domain), turbulence in the sample and mechanical vibrations of the optical components, electrical pick-up (especially 50 Hz ripple and also nanosecond electromagnetic radiation deriving from triggering devices such as Pockels cells or xenon flash ignition) which can usually be avoided with good electrical connections. All these problems being solved, one is left with the shot noise from the detector: for a photocurrent i, the noise is Δi, so that the signal-to-noise ratio S/N is given by:

$$\frac{S}{N} = \frac{i}{\Delta i} = \left(\frac{i}{2e\Delta f}\right)^{\frac{1}{2}} \tag{8.2}$$

where i and Δi are in amps, Δf is the high-pass electrical bandwidth in Hz, and $e = 1.6 \times 10^{-19}$ C. The formula shows that, for a given time response or Δf, i should be as large as possible. In practice we have $i \propto IQTr$, where Q is the detector quantum efficiency at the measuring wavelength, and Tr is the transmission factor of the optical components between the cuvette output and the detector surface. It thus means that I should be as large as possible, especially when a short time response is desired. This requirement is limited by light source brightness, and also by the need to avoid an actinic effect of the measuring light. This last source of artifact is limited by always using a monochromatic monitoring light, by using measuring light of a wavelength which is weakly absorbed, for kinetic studies, and when necessary by opening the measuring light only shortly before the actinic flash.

Recent progress has been made in CW (continuous wave) lasers which can be used as measuring light (argon-pumped dye lasers, and laser diodes for the near infrared), and in the generation of short pulses which can be used in the detecting flash method.

(d) *Photodetectors.* For efficient kinetic absorption spectroscopy, the photodetector should satisfy four mandatory requirements: (1) a linear measurement of I, at a current level defined by Eq. 8.2; (2) a quantum efficiency Q as high as possible, to improve the S/N ratio; (3) an appropriate time response; (4) a rapid recovery of linear response after a saturating over-illumination by the actinic flash (the direct light of which is mostly blocked by filters, but the detector is very often perturbed by leaky direct light and by fluorescence). Photomultipliers are still of interest, mostly in the UV where their quantum yield is better than photodiodes. In most instances, however, solid-state detectors are preferable, after coupling with a low-noise amplifier: silicon photodiodes below 1.0 µm, and germanium photodiodes between 0.9 and 1.5 µm. Optical multichannel analysers can be used in those limited cases where a non-monochromatic measuring light can be sent onto the cuvette.

(e) *Recording devices.* For nanoseconds or slower time-scales, practically all measurements involve an analog-to-digital conversion to facilitate handling of data. This often includes averaging to improve the S/N ratio, which increases as the square root of the number of averaged experiments. Transient digitisers are now common for microsecond resolution. They are becoming increasingly commercially available for the nanosecond time domain and one apparatus extends into the subnanosecond region.

2. *Picosecond techniques*

Picosecond spectroscopy has been the field in which the most important advances have been made during the last few years. This is partly due to a biological incentive: the search for a better understanding of the primary steps of energy transfer in the antenna and of electron transfer in reaction centres. The main reason, however, is technological: picosecond lasers are now fairly common, and subpicosecond lasers, together with the associated optics, are progessing quickly toward 'femtosecond' spectroscopy (1 fs = 10^{-15} s), although they remain very expensive and complex to run. Picosecond pulses are obtained by a process called 'mode-locking': a saturable dye inserted in a laser cavity fractionates the light into a train of picosecond pulses separated by an equal time interval. One pulse is usually isolated and amplified. A basic diagram of picosecond absorption measurements is given in Fig. 8.3. All set-ups are built according to the pump-probe structure which is similar to the original apparatus of Norrish and Porter (see e.g. Rockley *et al.*, 1975; Rentzepis, 1978; Shuvalov *et al.*, 1979a,b; Kirmaier *et al.*, 1983; Shank, 1983; Gillbro *et al.*, 1985a; Migus *et al.*, 1985; Nuijs *et al.*, 1985; Nuss *et al.*, 1985; Woodbury *et al.*, 1985; Gore *et al.*, 1986; Chekalin *et al.*, 1987; Schatz *et al.*, 1987; Wasielewski *et al.*, 1987). The extremely good synchronisation between pump and probe pulses is obtained by starting from a unique laser pulse: part of the pulse is used as excitation source (pump pulse), and part of it is focused on a water cuvette where it creates a continuum of white light of short duration which is used as measuring light (probe pulse). The time coordinate is provided by placing a variable optical delay on the measuring beam. Measurements are performed for various values of the delay. Fluctuations of the measuring pulse intensity are accounted for by use of a reference beam, i.e. a fraction of the probe pulse which does not go through an excited part of the cuvette. The signal-to-noise is increased by averaging the effect of many flashes. The large amount of information is handled by a computer. Several parameters define the specification of the equipment: (1) The pulse duration which is around 20 ps in many cases and can be as short as 100 fs in some apparatus. This duration limits the time resolution. (2) The wavelength of the pump pulse (530 nm obtained by frequency-doubling a YAG laser, or longer wavelengths obtained by pumping a dye) and its energy, which depends on the number of amplifying stages. (3) The repetition rate which is often 10 Hz, but which can be as high as 400 kHz under conditions where the pulses are not amplified.

Three areas of application illustrate the usefulness of picosecond absorption in photosynthesis research:

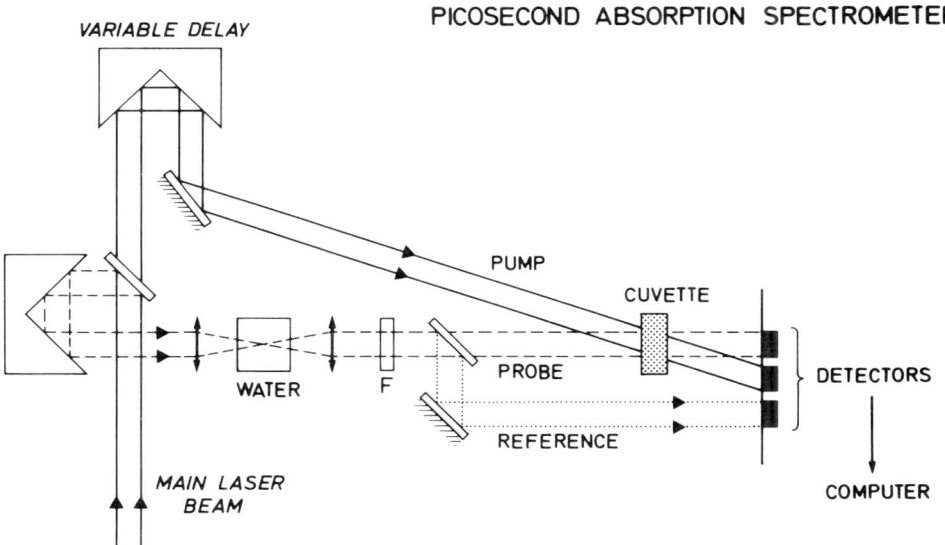

FIG. 8.3. Simplified scheme of a picosecond absorption spectrometer (adapted from Migus *et al.*, 1985). F is a filter which selects the desired wavelength band from the continuum generated in the water cuvette.

(a) *Measurements of rates of energy transfer*. In the photosynthetic antenna, energy transfer can be considered as a succession of steps, from one pigment molecule to another, and from one pigment subset to another. All these steps are very fast, but they can be resolved by (sub)picosecond spectroscopy when the pigments have a different absorption spectrum. Excitation should of course be selective. Without being exhaustive, let us mention a few studies which have permitted the path of energy transfer in the photosynthetic apparatus to be partially unravelled: energy transfer between different classes of phycobiliproteins in phycobilisomes (Wendler *et al.*, 1984; Gillbro *et al.*,

1985a), energy transfer between (bacterio)chlorophylls absorbing at different wavelengths in isolated pigment–protein complexes from bacteria and plants (Gillbro et al., 1985b; Nuijs et al., 1985; Bergström et al., 1986), and ultrafast (100 fs or less) transfer within the reaction centre (Breton et al., 1986). Energy transfer from carotenoids to chlorophylls has been studied *in vitro* in a synthetic model system (Wasielewski et al., 1986).

(b) *Elucidation of the primary photochemical acts.* The chemical nature of the partners involved in the very primary electron transfer in bacterial reaction centres has been a field of intense research, especially since X-ray crystallography revealed that a bacteriochlorophyll is located on the potential path of electrons from the special pair of bacteriochlorophylls (the primary donor) to bacteriopheophytin. A related question is the physical nature of the short-lived reactive excited state: is it localised on the special pair or does it extend to neighbouring bacteriochlorophylls (Creighton et al., 1988)? After some controversy, resulting probably from excessively long excitation pulses or from non-selective excitation, recent work showed that bacteriochlorophyll cannot be resolved as an intermediate electron acceptor (Kirmaier et al., 1985a; Breton et al., 1986; Martin et al., 1986; Wasielewski and Tiede, 1986).

(c) *Electron acceptors in reaction centres: identification and kinetics.* Early picosecond studies revealed that, in purple bacterial reaction centres, a bacteriopheophytin is reduced in less than 10 ps, and then reduces the quinone Q_A in about 200 ps. Similar studies were conducted more recently on other classes of reaction centres; this has been facilitated by their purification and the increased sensitivity of the apparatus. In plant PSII the results are not yet fully conclusive, but they tend to show that pheophytin is reduced in less than 100 ps in PSII membranes (less than 10 ps in reaction centres), and re-oxidised in 200–500 ps (Danielius et al., 1987; Schatz et al., 1987). In PSI the specialised chlorophyll *a* named A_0 is reduced in the picosecond range, and A_0^- is re-oxidised with a half-time of 32 ps, presumably by electron transfer to the secondary acceptor A_1 (Shuvalov et al., 1986; Wasielewski et al., 1987). These studies are rendered difficult by the rather poor enrichment of active PSI or PSII reaction centres, and by the less favourable optical characteristics of their active constituents as compared with purple bacteria.

Still in the picosecond time domain, the newly developed technique of accumulated photon echo gives access to the lifetime of an excited state which precedes a slowly decaying photochemical product. The technique has been applied to *Rhodobacter sphaeroides* and *Rhodopseudomonas viridis* reaction centres (Meech et al., 1986) and to PSII (Vink et al., 1987).

Before leaving this field it is worth mentioning that, in the (sub)picosecond time domain, absorption techniques provide a much better time resolution than fluorescence techniques: they are essentially limited by the duration of the laser pulses, which is always decreasing thanks to time compression tricks, whereas fluorescence measurements are limited by the time response of photodetectors and amplifiers. On time-scales slower than about 1 ns, fluorescence will usually provide better sensitivity and time resolution than absorption, but the former technique is of course limited to the study of fluorescing pigments.

3. Nanosecond measurements

Absorption measurements in the nanosecond time range remain relatively difficult to perform. To obtain the time coordinate, the detecting flash method cannot be used: optical delays used in picosecond measurements are not practical (a 100 ns delay would require a 30 m optical path) and electronic time delays introduce jitters of the order of 1 μs for firing a conventional flash. Absorption measurements have thus to be performed with a quasi-continuous light source and a detector with nanosecond time resolution (Mathis, 1977; de Vault, 1978). Nanosecond absorption set-ups have the configuration of classical kinetic spectrophotometry (Fig. 8.4): a cuvette is crossed by a continuous measuring beam and excited by an actinic laser pulse. As a matter of fact, the 'nanosecond barrier' is not an absolute one: picosecond technology can extend to about 20 ns (obtained with a 6 m optical delay), and the nanosecond techniques can now provide a time resolution below one nanosecond (about 0.5 ns).

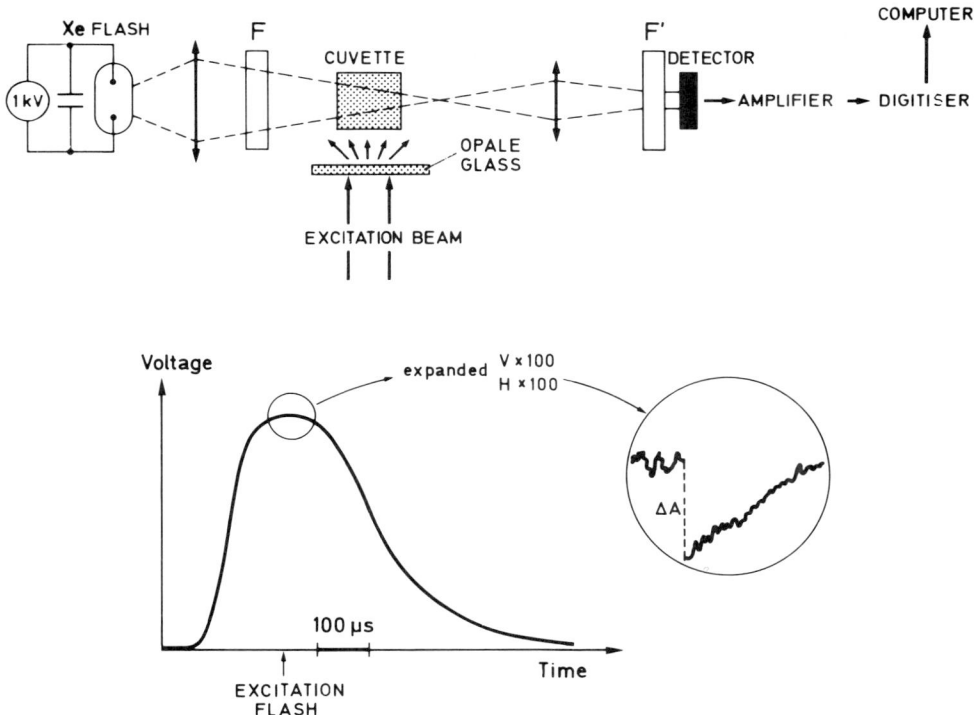

FIG. 8.4. Simplified scheme of a nanosecond absorption spectrometer with a xenon flash as a source of measuring light. F,F': complementary optical filters (F can be a monochromator). The time profile of the voltage at the amplified output is shown on a microsecond time and, after an horizontal (time) and vertical (voltage) expansion, on a nanosecond time-scale, showing the small ΔA induced by the excitation flash.

Excitation flashes are provided by various types of lasers. In their so-called Q-switch mode the pulse duration is 10–20 ns. Shorter pulses can be obtained either by slicing the pulse (down to about 1 ns) with a Pockels cell or by using a mode-locking cavity and a single pulse selector. The YAG laser is now the most commonly used type of laser. Its fundamental wavelength (1060 nm) is not convenient, but frequency doubling is efficient and brings the wavelength to 530 nm, which is often directly useful or can be used to pump a dye laser. The user can choose between various dyes, giving wavelengths between 540 nm and the near infrared, either as a broad band or as a tunable narrow band. YAG lasers operate at up to 50 pulses per second (pps). The ruby laser is less versatile. Its fundamental wavelength (694 nm) is quite efficient to excite photosynthetic pigments (especially PSI), but its rate of operation is limited at about 0.2 pps, and its capabilities for wavelength change are rather limited. Excimer lasers emit in the near UV (248–351 nm). They are efficient in dye pumping and can be triggered at a high frequency around 100 Hz.

For nanosecond kinetics, the measuring light should have a rather high intensity (see Eq. 8.2), which usually cannot be delivered by conventional light sources. Xenon flashes can be used; a choke is inserted in the discharge circuit to smooth the emission time profile, so that the light intensity can be considered as nearly constant on a 1 µs timescale. Recent studies have taken advantage of CW lasers as measuring light sources (van Best and Mathis, 1978a). The main advantage of these lasers is that they deliver a nearly parallel beam which can be weakly focused through the cuvette. The detector can thus be placed at a rather long distance from the cuvette, a configuration which greatly reduces the detector perturbation by the excitation flash and especially by the fluorescence from the biological material. The beam is focused on a small spot, which permits its measurement by small fast-response photodiodes. Two types of CW lasers have been used: dye lasers pumped by an argon laser, and laser diodes. These solid state emitters are very easy to operate, in spite of their fragility. Laser diodes emit commonly in the following spectral ranges: 740–870 nm (GaAlAs), around 1200 nm and around 1300 nm (InGaAsP). Shorter wavelengths (670–740 nm) have recently been made available.

The photodetector will usually be a silicon photodiode (with a UV-enhanced response if required from the manufacturer), or a germanium photodiode above 1000 nm. Photomultipliers have a rather large sensitive surface and they are preferred when the measuring beam cannot be easily focused, or below 500 nm where their quantum efficiency is better than that of photodiodes. They should be selected according to their ability to measure linearly a rather strong light and be wired accordingly, with a low gain (less than 10^4) and a high voltage. Photodiodes are much easier to set up, provided they are well coupled to a low-noise amplifier. They should be carefully protected from electromagnetic perturbations associated with pulsed lasers, especially from Pockels cells.

The application of nanosecond measurements will be illustrated by three examples:

(a) *Properties of the primary biradical in photosynthetic reaction centres.* Excitation of reaction centres leads to the rapid formation of the biradical state P^+I^- (P, primary donor; I, primary acceptor). Normally this state decays very quickly, by subnanosecond electron transfer from I^- to the secondary acceptor. When this acceptor is artificially reduced or removed, the biradical P^+I^- decays much more slowly. Nanosecond absorption measurements gave a $t_{\frac{1}{2}}$ of the order of 10 ns in all types of reaction centres (Parson et al., 1975; Sétif et al., 1985; Hansson et al., 1988) and permitted the identification of I: bacteriopheophytin in purple bacteria, pheophytin in PSII and chlorophyll a (A_0) in PSI. The properties of the radical pair which have been revealed are of considerable interest. The pair arises in a singlet state; it then oscillates between that state and an isoenergetic triplet state, and eventually recombines into several possible states: ground state, singlet excited state of P, triplet excited state of P. These various paths are influenced by external magnetic fields and by temperature. Their kinetic properties have been studied in great detail, mostly in purple bacteria, by nanosecond spectroscopy. These studies permit a rather accurate physical description of the radical pair (Blankenship *et al.*, 1977; Shuvalov and Parson, 1981; Norris *et al.*, 1982, 1987; Schenck *et al.*, 1982; Wasielewski *et al.*, 1983; Chidsey *et al.*, 1984; Ogrodnik *et al.*, 1987). The question is also of great interest in the case of PSII. The lifetime of the radical pair seems to be influenced by the antenna size. This has been interpreted in terms of an equilibrium between the radical pair and the excited state of antenna chlorophyll, a property which can explain the increased fluorescence yield of chlorophyll in PSII when the secondary quinone Q_A is reduced ('variable fluorescence': Schatz *et al.*, 1987; Hansson *et al.*, 1988; Schlodder and Brettel, 1988).

(b) *Secondary electron donation in PSII*. In the history of research on PSII, it has been difficult to analyse electron transfer from the oxygen-evolving enzyme to photo-oxidised P_{680}. A good tool for these studies came from the observation that P_{680}^+ displays a weak absorption band ($\varepsilon = 7000 \, M^{-1} \, cm^{-1}$) around 820 nm, like the chlorophyll *a* radical cation in solution. This spectral region is technically very interesting for kinetic studies (it should be noted that equivalent bands at 820 nm, 1200 nm and 1300 nm in the oxidised primary donor of PSI, *Rh. sphaeroides* and *Rps. viridis*, respectively, have similar advantages): the thylakoid has negligible absorption in that region, so that the measuring light can be very strong without having any actinic effect. Also this is a region where solid-state photodiodes have a maximum quantum efficiency and laser diodes (as sources of measuring light) are commonly available. A specialised set-up was built, with which we showed that P_{680}^+ is normally re-reduced in the nanosecond range, a reaction which is slowed down to the microsecond domain when oxygen evolution is inhibited (van Best and Mathis, 1978a,b; Conjeaud *et al.*, 1979). These studies were extended to show that the rate of reduction varies according to the oxidation state (S-state) of the manganese cluster (Brettel *et al.*, 1984; Schlodder *et al.*, 1984). These kinetic properties are still not understood. In particular it is not at all clear why the kinetics of P_{680}^+ reduction are influenced by the oxygen-evolving complex. Nanosecond techniques at 820 nm remain very useful in the study of PSII because of their relative simplicity. Also, with a CW laser diode as measuring light source, the same set-up can be used to cover the nano- and microsecond time ranges, i.e. the range of P_{680}^+ reduction kinetics, with practically no modification. It should be noted, however, that the measurements are especially easy with clear suspensions of small particles: with scattering suspensions it is not possible simultaneously to focus the measuring light on a small photodiode and to get rid of the chlorophyll fluorescence.

(c) *Triplet–triplet energy transfer*. When chlorophyll is excited by light it is initially converted directly into a singlet excited state, from which processes of singlet energy transfer (in the antenna) and ultimately of electron transfer (in reaction centres) take place. Conversion from the singlet excited to the triplet excited state, a process known as intersystem crossing, can also take place, albeit at a rate (around $10^8 \, s^{-1}$) which is slow compared to other *in vivo* processes (note that for chlorophyll in organic solvents, intersystem crossing to the triplet state is the major fate from the singlet excited state). A small amount of triplet chlorophyll (^{3}Chl) is thus populated, a process which is very important because of its potential danger for the cell. Triplet states of phorphyrins can efficiently react with oxygen, forming highly reactive singlet oxygen which is known to destroy cellular components. Flash absorption measurements showed that ^{3}Chl is dissipated in the nanosecond domain by energy transfer to carotenoids which are present in the same pigment–protein complexes, populating their triplet state ^{3}Car (Kramer and Mathis, 1980). *In vitro* studies have shown that ^{3}Car has an intense and characteristic absorption band around 510–530 nm. This property facilitates nanosecond studies using a xenon flash as measuring light source. The kinetics of this energy transfer have been studied in various organelles and isolated pigment–protein complexes. Synthetic models have been studied *in vitro*, reproducing nanosecond energy transfer under certain structural conditions (Gust *et al.*, 1985). A similar energy transfer takes place in bacterial reaction centres from the primary donor triplet state, which is populated by recombination in the primary radical pair. ^{3}Car is effectively populated in the submicrosecond time range (Cogdell *et al.*, 1975; Schenck *et al.*, 1982; Sebban and Lindqvist, 1987), but a good kinetic parallelism between the decay of ^{3}P and the rise of ^{3}Car has not yet been observed. In PSI and PSII it has not yet been shown convincingly that ^{3}Car could be formed from the primary radical pair.

4. Slower measurements with DC measuring light

Slower measurements are in principle technically easier to perform. Excitation flashes are easier to handle and their perturbation (optical: direct scattered light or chlorophyll fluorescence; or electronic) easier to eliminate. The intensity of the measuring light can be weaker, for an equivalent signal-to-noise ratio. Finally, detectors are very classical, as are also the electronics and the recording components such as transient averagers. In fact, much information on the photosynthetic apparatus was gained during the period 1960–1980, thanks to microsecond and millisecond absorption techniques (Witt, 1971; Ke, 1972, 1973; Mathis, 1977): identification of primary donors, the role of cytochromes and quinones, the use of the carotenoid electrochromic shift as a criterion of transmembrane electron transfer, population and decay of ^{3}Car, the triplet state of the primary donors, etc. This field now evolves in two directions:

(a) The use of well-defined absorption kinetics for more detailed biochemical properties of reaction centre proteins. These approaches require accurate and easy-to-handle apparatus. To cite just but a few of these applications:

(1) studies of the stoichiometries of PSI and PSII in plant photosynthetic membranes and of their variations with growth conditions, by means of measurement of P_{700}, of Q_A and of the carotenoid electrochromic shift (e.g. Thielen and van Gorkom, 1981);
(2) systematic check of the activity of suborganelle preparations in the course of fractionation studies;
(3) interaction of extrinsic proteins (plastocyanin, soluble cytochromes) with reaction centres: effect of substitution, of protein modification and of the ionic composition of the suspending medium (Overfield *et al.*, 1979; Bottin and Mathis, 1985; Hall *et al.*, 1987; Tiede, 1987; van der Wal *et al.*, 1987);

(4) detailed study of the kinetics of the back-reaction (P^+, Q_A^-) in bacterial reaction centres, including the effect of temperature, heterogeneity, relation with the protonation of as yet unknown sites, effect of isotopic substitution, and influence of Q_A replacement by other quinones of selected electrochemical and structural properties (see, e.g. Kleinfield et al., 1985; Shopes and Wraight, 1985, 1987; Gunner et al., 1986; Okamura and Feher, 1986; Parot et al., 1987);
(5) studies with reaction centres incorporated in artificial membranes (e.g. Packham et al., 1980);
(6) investigation of the effect of mutations leading to resistance against herbicide-type inhibitors, in PSII and in purple bacteria: change in the rate of electron transfer from Q_A to Q_B and of (P^+, Q_A^-) recombination;
(7) more generally, structure–function studies on the electron transfer proteins can be largely based on kinetic flash absorption for the functional part of the work, which complements structural measurements and increasingly site-directed protein engineering.

(b) The other direction of present research is to develop the methodology of flash absorption to find out more about molecular species which were previously undetected because of their weak absorption, especially in those spectral regions more difficult to access such as the UV or the blue and red parts of the spectrum (where antenna pigments absorb maximally). It is also hoped to discover more from *in vivo* experiments, where there are the associated problems of intense light scattering and weakness of the signals. These studies do not require major technical developments, but in each case a good adaptation is needed to optimise the apparatus in view of the particular study. A few examples can be cited, related to plant thylakoids:

(1) the identification of the secondary acceptor A_1 in PSI through microsecond measurements in the UV at low temperature (Brettel et al., 1986). The difference spectrum fits approximately that of the formation of the radical anion of phylloquinone;
(2) measurement of the difference spectrum due to the oxidation of the secondary donor Z in PSII. Weak bands in the UV were observed, together with electrochromic shifts in the main chlorophyll absorption bands (e.g. Schatz and van Gorkom, 1985);
(3) measurement of the difference spectra due to reduction of the quinone acceptors Q_A and Q_B in PSII (Schatz and van Gorkom, 1985). Both spectra can be attributed to plastoquinone radical anions because of their positive band around 320 nm, but they are rather different in the visible where the features are due to electrochromic shifts of pigments;
(4) difference spectra due to the transitions between the various S-states of the oxygen-evolving apparatus in PSII. The direct results are contaminated by contributions from Q_B and perhaps from Fe^{2+} and from changes in light scattering. This important field is still the object of active research and controversy (see, e.g. Lavergne, 1987; Kretschmann et al., 1988);
(5) observation of the formation of the carotenoid radical cation Car^+ in PSII, under special circumstances (Schenck et al., 1982). This carotenoid oxidation is greatly favoured by some classes of uncouplers;

(6) measurement of proton flows across the thylakoid associated with light-induced electron transfer (Junge, 1987). pH changes are measured as absorption changes of pH-indicating dyes (neutral red, cresol purple). These measurements require great accuracy and good control of the biological material.

5. *Slower measurements with detecting flashes*

In the microsecond and slower time ranges, absorption changes induced by an actinic flash can be monitored by weak detecting flashes with a variable delay with respect to the actinic flash. This is similar to the picosecond techniques, and to the initial method of Norrish and Porter, except that wavelengths are scanned one at a time. Such an apparatus has been developed by Joliot *et al.* (1980), and shown to provide an extremely good accuracy in ΔA measurements. The device has been used to study the carotenoid electrochromic shifts, the redox reactions of cytochromes in the cytochrome b/f complex, and PSII electron carriers (see, e.g. Joliot and Joliot, 1986, 1988; Lavergne, 1987).

The basic organisation of a detecting flash apparatus is given in Fig. 8.5. The detecting flash should be short (about 1 μs) and very stable in energy. It is necessary to split the light from the flash in two beams: a measuring beam which goes through the excited cuvette, and the reference beam which crosses a reference cuvette. The flash energy on each beam is measured with a photodiode. The sensitivity of the apparatus relies largely on the equilibration of these two beams and on the quality of the optical construction. The kinetics are constructed point by point, about 10 points being usually sufficient.

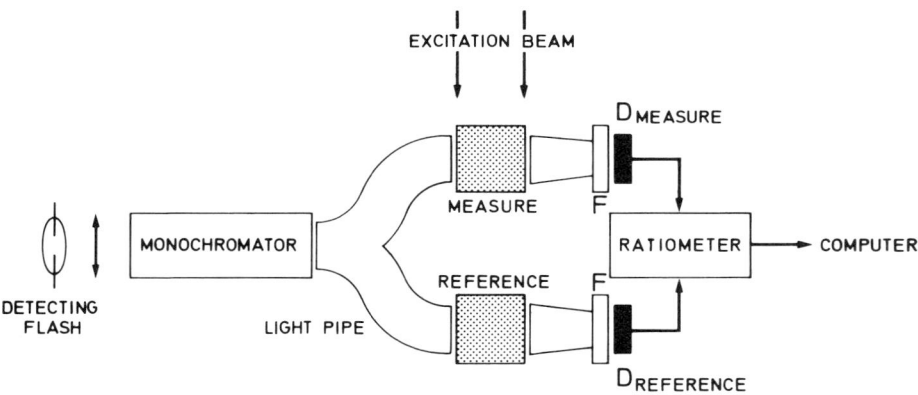

FIG. 8.5. Schematic organisation of a detecting flash absorption spectrometer, according to Joliot *et al.* (1980). There are two identical cuvettes containing the biological material, only one of them is excited by the excitation flash. D is a detector (photodiode) and F an optical filter. The delay between the detecting and actinic flashes is varied stepwise by an electronic generator.

It might be of interest to attempt a comparison of the detecting flash set-up with a set-up using continuous measuring light, being clear that in both cases the spectra are reconstituted wavelength by wavelength. In the set-up developed by Joliot *et al.* (1980), a large part of the accuracy certainly resides in the good equilibration of the measuring and of the reference beams: this can also be realised in a classical set-up. Part of the optical quality also resides in the large solid angle of collection of light at the exit of the cuvettes, which ensures minimum spectral distortions. This also can be incorporated in classical set-ups. This optical arrangement, however, has a drawback when carrying out measurements in the red and near infrared parts of the spectrum: the detector is disturbed by the actinic flash fluorescence artifact, and the measurements are distorted by the eventual variations in the yield of fluorescence excited by the measuring beam. A major difference between the two types of set-up, of course, resides in the mode of time delivery of the measuring light: a large intensity during short time intervals or a more permanent weaker light. If the measuring light intensity, in both cases, is adjusted to the same energy, time-integrated over the time interval of interest (in order to have nearly the same acceptable level of actinic effect), then the flash technique will provide very good accuracy at a few time points, whereas the classical technique will give a noisier curve but with a full time course. The information content is probably equivalent. Clearly the detecting flash technique will be better when measurements are needed over a large time range (for example one point at 10^{-n} s after the actinic flash, with $n = 5, 4, 3$, etc.); the equivalent in the classical technique would be to decrease the measuring light intensity continuously with time, an operation which is certainly not practical.

D. Linear Dichroism

A sample displays linear dichroism when it differentially absorbs a linearly polarised light beam with two perpendicular directions (named H, V, horizontal, vertical). This observation requires both that the studied molecules are oriented in the sample, and that the studied electronic transition is polarised in the molecules (Michl and Thulstrup, 1986). In photosynthetic membranes, four categories of molecules have a sufficiently dissymetric cloud of π electrons to give linearly polarised transitions: chlorophylls, carotenoids, phycobilins and cytochromes. Their orientation in photosynthetic structures can thus be studied, provided these structures can be oriented (Breton and Vermeglio, 1982).

Several methods of orientation can be used, which are more or less suited to the properties of the studied structures: magnetic field (mostly for organelles), deposition on a plane support (for membranes), and uniaxial compression (for pigment–protein complexes in polyacrylamide gels). Orientation can also be provided by an hydrodynamic flow or by stretching a polymer film in which the material is included, or by the action of an electric field. The sample can also remain isotropic on a macroscopic scale and be rendered anisotropic by the action of the excitation light: this is the method of photoselection. The dichroism is measured as a steady-state spectrum, in a spectrophotometer modified so as to pass through the sample a measuring light beam which is alternately H- or V-polarised, and to measure the difference in their intensity after passing through the sample. The dichroism can be measured in flash kinetic experiments, for transient species (Kirmaier et al., 1983, 1985b).

Linear dichroism (LD) has been very useful in determining the orientation of pigments and electron carriers in photosynthetic structures. Without trying to be exhaustive, a few examples can be cited:

(1) in bacterial reaction centres the respective orientations of practically all the Q_X and Q_Y transitions of the bacteriochlorophylls and bacteriopheophytins have

been measured, showing a well-defined positioning of these species. From similar studies with chromatophore membranes it is possible to predict the orientation of the reaction centres in the membrane (Breton, 1985);

(2) light-harvesting complexes have also been studied, either *in situ* in the membrane or after isolation (Breton, 1986). The molecular orientations are usually lower than in reaction centres: this might be due to some disorder, or to the greater complexity of antenna systems in terms of number of constituent molecules;

(3) chlorophyll analogues have been incorporated in apomyoglobin crystals. LD measurements were performed on these crystals, the atomic structure of which has been determined by X-ray crystallography. This type of study has enabled the orientations of optical transition moments with respect to the molecular framework to be elucidated (Boxer *et al.*, 1982; Moog *et al.*, 1984);

(4) in PSI particles, photoselection studies gave the relative orientation of P_{700} and of neighbouring antenna molecules of chlorophyll *a* and of β-carotene (Junge *et al.*, 1977).

E. Special Techniques

Many methods take advantage of the absorption of light to obtain information on molecules, molecular assemblies or large structures, or on reaction mechanisms.

1. Circular dichroism (CD)

This property reports the capacity for absorbing differently left- or right-circularly polarised light. It requires the presence of asymmetric centres in the molecule. For isolated chlorophyll molecules, the CD amplitude (or rotational strength) is weak. In several pigment–protein complexes, however, a rather large rotational strength is observed (Sauer, 1975; Pearlstein *et al.*, 1982). It is attributed primarily to chlorophyll–chlorophyll exciton interaction, and partly to chlorophyll–protein interaction. CD spectra are one of the optical properties which have to be quantitatively accounted for in structural models of pigment–protein complexes (Scherz and Parson, 1984; Parson and Warshel, 1987; Scherer and Fischer, 1987). The enhancement of rotational power by magnetic field is named magnetic circular dichroism (MCD). This technique is especially useful for studying spin states in iron complexes (Vickery, 1978; Gadsby and Thomson, 1986) but it has not been much utilised in photosynthesis research, although MCD is also observed in chlorophyll (Houssier and Sauer, 1970).

2. Stark effect

This effect is a shift in the energies of electronic states upon application of an electric field. The shift is different for the ground state and for the singlet excited states when the molecular electric dipole is different in the two states. In that case the electric field disturbs the absorption spectrum. This technique has been helpful in understanding the physical nature of the excited state of the primary donor in bacterial reaction centres (Braun *et al.*, 1987; Lockhart and Boxer, 1987; Lösche *et al.*, 1987).

3. ADMR: absorption detected magnetic resonance

This technique concomitantly provides electron paramagnetic resonance data and absorption difference spectra. The absorption level of a sample is detected under a constant illumination which excites the reaction centres. In published applications, normal electron transfer is blocked and the light thus creates a steady-state population of the primary donor triplet state ^3P. This steady-state level is perturbed by an oscillating microwave field, in the absence of a strong external magnetic field, and the resulting oscillating absorption variation is measured versus wavelength, giving a (^3P minus P) difference spectrum. This measurement is extremely sensitive and selective, since the microwave frequency is adjusted to the resonance between triplet sublevels. Its application is limited to very low temperatures, between 1.5 and 10 K. Difference spectra (^3P minus P) have been obtained in bacterial reaction centres, in PSI and in PSII (Den Blanken and Hoff, 1983; Den Blanken et al., 1983; Lous and Hoff, 1987). With polarised measuring light, it is possible to measure accurately the linear dichroism of triplet spectra (Hoff et al., 1985a). The ADMR method can also be extended to measure spin–spin relaxation times of triplet donors by means of spin echo (Lous and Hoff, 1987). The (^3P minus P) difference spectrum can be measured at higher temperature (110–280 K) by a technique based on the effect of a weak a.c. magnetic field, which modulates the steady-state concentration of ^3P under continuous illumination (Hoff et al., 1985b). A time-resolved version of this technique, known as RYDMR (reaction yield detected magnetic resonance) is useful for studying reaction centre primary reactions (Bowman et al., 1981; Norris et al., 1982).

4. Microspectrophotometry

This technique has enjoyed renewed interest now that it is possible to obtain small crystals of photosynthetic membrane proteins (Wacker et al., 1986; Kühlbrandt et al., 1988). It can be extended so as to permit linear dichroism or absorption kinetic studies.

5. Reflectance spectroscopy, photoacoustic spectroscopy

These techniques give an indirect access to absorption spectra, since they rely on losses by reflection or by thermal dissipation (Moore, 1983; Braslavsky, 1986). They are of special interest for photosynthetic objects which have a poorly defined geometry or are optically very dense, such as lichens, leaves and fruits.

III. LIGHT EMISSION

Emission of a photon by a molecule is coupled to its return to the ground state from the lowest excited singlet state (fluorescence) or triplet state (phosphorescence). In photosynthetic structures, emitting molecules are chlorophylls or analogous pigments and also phycobiliproteins. Fluorescence is usually preceded by energy transfer steps. In some instances, it is preceded by charge separation and electron transfer, followed by charge recombination: in these latter cases the terminology 'delayed light' or 'delayed fluorescence' is used. Thermoluminescence is a special case of delayed light. Light emission will be covered very briefly in this chapter, focusing mainly on recent developments. Major applications of fluorescence to photosynthesis research are presented in Chapter 9 (this volume) and have been reviewed recently (Govindjee et al., 1986).

A. Fluorescence Lifetime Measurements

These measurements are a major tool in the study of energy transfer processes. The phase shift method has led to important progress in the study of proteins (Alcala et al., 1987) and is still used in photosynthesis research, but most recent measurements use the pulse method. Whereas flash absorption requires strong flashes which nearly saturate reaction centres in a cuvette, only weak flashes are needed for fluorescence lifetime measurements. They are produced by mode-locked CW lasers (duration around 15 ps) or by synchrotron radiation sources. The time profile of light emission is reconstructed by photon counting (single photon technique), which provides a direct time resolution of 100–300 ps. The effective time resolution can be as good as 20 ps after correction for the apparatus response time. Problems and results have been adequately reviewed (Karukstis and Sauer, 1983; Holzwarth, 1986; for more recent work, see Hodges and Moya, 1987; Owens et al., 1987; Schatz et al., 1987). In all systems the data are very complex and do not yet permit unambiguous interpretations. When stronger flashes are used to excite the fluorescence, processes of exciton annihilation take place, which also can be interpreted in terms of energy transfer (Geacintov and Breton, 1986).

B. Phosphorescence

Emission from the chlorophyll triplet state occurs with a low yield, at wavelengths significantly longer than fluorescence, because the triplet energy level is lower than that of the singlet. Phosphorescence has been observed with chlorophyll *in vitro*. In most photosynthetic structures, however, phosphorescence is not observed, probably because the chlorophyll triplet state is rapidly quenched by carotenoids (see Section II.C.3(c)) which do not phosphoresce. Phosphorescence from the triplet primary donor has been observed in PSI (the attribution is not fully documented; see Shuvalov, 1976) and more recently in bacterial reaction centres (Takis and Boxer, 1988).

C. Delayed Light Emission

Delayed fluorescence accompanies, with a weak yield, the various kinetic phases of charge recombination. For that reason it has been largely used in early studies on electron transfer kinetics. These researches have been recently reviewed (Govindjee et al., 1986). It is worth mentioning the recent development of a single apparatus for delayed light measurements with intact leaves (Schreiber and Schliwa, 1987).

Nanosecond delayed fluorescence has been clearly identified in bacterial reaction centres (Sebban and Barbet, 1984; Woodbury and Parson, 1986). Fluorescence kinetics are more complex than the kinetics of radical pair recombination, as studied by flash absorption. This different behaviour was interpreted as an indication of relaxation phenomena in the radical pair. In PSII, it has been proposed that nanosecond radical-pair recombination is the origin of the variable fluorescence F_v (Klimov and Allakhverdiev, 1978). This question is of great conceptual interest to understand the relationships between the PSII reaction centre and its antenna. In fact, it seems that the radical pair is in equilibrium with the excited state of P_{680} and through it with the excited state of antenna chlorophyll.

D. Thermoluminescence

Thermoluminescence (TL) is a technique of very general interest. In photosynthesis research, a sample is illuminated under controlled conditions and then cooled to 77 K. During warming, successive thermal barriers for charge recombination are crossed. They give rise to successive peaks of chlorophyll delayed fluorescence which can be recorded with a sensitive photon-counting apparatus. The method has been reviewed in detail (Sane and Rutherford, 1986). In practice, TL is especially useful for probing charge stabilisation in PSII at the acceptor side (Q_A/Q_B equilibrium, protonation, effect of herbicides) and donor side (S-state turnover, protonation, effect of various treatments which inhibit oxygen evolution). TL has the inconvenience of being an indirect method. Its main interests reside in the need for only a small amount of material and the lack of stringent optical requirements, permitting work with intact leaves or cells.

E. ODMR: FDMR

Optically detected magnetic resonance (ODMR) has mainly relied on fluorescence measurements (whence the name FDMR, which is the fluorescence equivalent of ADMR, see Section II.E.3). The method has been applied to the triplet state of the primary donor P of reaction centres, providing the triplet sublevel decay rates and the zero-field splitting parameters D and E (Clarke, 1982; Levanon and Norris, 1982).

IV. VIBRATIONAL SPECTROSCOPY

A. General Considerations

Absorption by molecules of mid-wavelength infrared light (2–20 µm) is due to transitions between vibrational states. In large molecules, spectroscopic analysis can practically divide the vibrational transitions into two classes of vibrational modes associated either with localised functional groups (such as C=O, C—O, O—H, N—H, etc. which experience stretching and bending motions) or with collective movements of a large molecular skeleton. Group vibrations are sensitive to local interactions such as hydrogen bonding. They are thus very important for understanding how molecules interact with each other in large molecular assemblies such as pigment–protein complexes, in which pigments and redox centres are bound to the polypeptides by non-covalent interactions. Vibrational spectroscopy is unique in providing information at the atomic level in systems when NMR cannot be utilised. It is a good complement to X-ray crystallography.

Vibrational transitions can be studied by two methods (Parker, 1983) (Fig. 8.6): (1) infrared absorption directly records the spectrum of light absorption by vibrating modes (in terms of wavelength or frequency, and of intensity); and (2) resonance

Raman spectroscopy, RRS, basically provides the same information, but with specific selection rules which limit its use to those vibrational modes coupled to an electronic absorption transition. This gives it a useful selectivity. The application of both methods in photosynthesis research will be briefly examined.

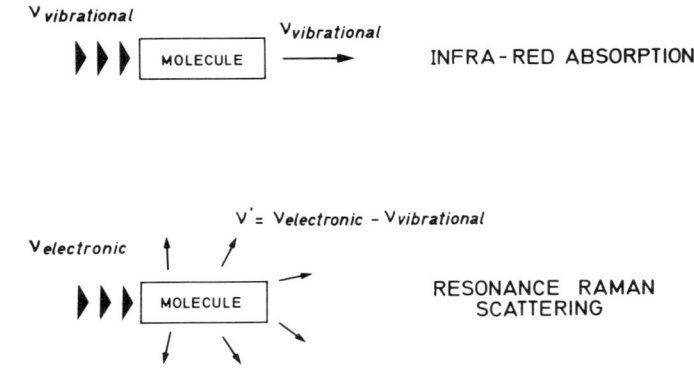

FIG. 8.6. Vibrational frequencies of a molecule can be measured by a direct absorption method (infrared absorption) or by the frequency shift induced in the light of a frequency corresponding to an electronic transition and which is scattered by the molecule (resonance Raman scattering).

B. Infrared Absorption

Infrared absorption by a biological sample is very complex, due to the overlapping contribution of many vibrators, and especially of water. In partially dehydrated films of pigment–protein complexes, several amide peaks can be resolved, corresponding to N—H stretching (around $3300\,cm^{-1}$) or bending (around $1550\,cm^{-1}$) and to C=O stretching ($1650\,cm^{-1}$). β-Sheet and α-helical structures have different vibration frequencies. This property enables the α-helix content to be evaluated, and also the transmembrane orientation of the helices to be determined (Breton and Nabedryk, 1984; Nabedryk et al., 1984).

The study of individual vibrations is now possible by differential Fourier transform infrared spectroscopy (FTIR) which provides high resolution and accuracy. The sample is a thin film of concentrated biological material, which is partly dehydrated. The spectrum is recorded with the sample as is (reference spectrum) and also under or after actinic illumination. The sample should be highly stable. The difference spectrum shows the effect of light, which is prominent in the region of carbonyl frequencies (Nabedryk et al., 1986; Tavitian et al., 1986; Gerwert et al., 1988). According to the experimental conditions, this spectrum can be related to predicted changes in the samples, such as oxidation of the primary donor P, or reduction of an acceptor. For confirmation, the spectrum can also be compared with the difference spectrum due to oxidation or reduction of isolated molecules in electrochemical cells. This method is quite new, and under development. It should permit the analysis of subtle local changes in the polypeptides, pigments and redox centres which presumably accompany photoinduced electron transfer. It has already been of great interest in the study of rhodopsin and bacteriorhodopsin.

C. Resonance Raman Spectroscopy

Raman scattering is a process by which a photon impinging on a molecule is scattered with an exchange of energy corresponding to one vibrational quantum. The probability of scattering is very weak in classical Raman scattering, but is greatly enhanced ($\times 10^5$ or so) by a resonance phenomenon when the wavelength of the incident light is close to that of an electronic transition in the studied molecule (Parker, 1983; Rousseau and Ondrias, 1984). The property is the reason for the selectivity of the method: choosing the wavelength of the probing light allows the Raman scattering to be enhanced and so enables vibrational frequencies in specific molecules to be measured. This is especially important for photosynthetic pigments, but also for electron carriers such as iron–sulphur proteins or cytochromes.

For technical reasons, the development of resonance Raman spectroscopy (RRS) is associated with that of CW lasers: the frequency shift is 200–1700 cm^{-1}, and so the probing light should have a corresponding very narrow bandwidth. The association of ion lasers with dye lasers now allows the near UV and visible ranges to be completely covered. The scattering efficiency is very low, a property which results in three technical constraints: (1) the Raman light is easily obscured by fluorescence. For that reason it is difficult to work in the Q_Y region of chlorophylls (see, however, the recent work by Bocian *et al.*, 1987); (2) the probe light needs to be rather strong, with the risk of photodegradation of the sample. The sample is often cooled to low temperature to slow down the degradation; (3) the probe light has to be focused on a small spot of the sample (increasing the potential degradations) to obtain a good spectral resolution of the scattered light. The entire optics need to be very good, and the detection system sensitive.

The attribution of the Raman data in terms of molecular vibrational modes raises the same problems as infrared absorption. It is based largely on parallel studies on molecules *in vitro* and on biological complexes, utilising the effects of chemical or isotopic substitution. To be active in RRS, a vibration should be coupled to π electrons involved in the probe electronic transition. This necessary condition is not sufficient and at present it is difficult to rationalise the intensity of Raman lines.

RRS has now provided a good deal of information on photosynthetic structures (Lutz and Robert, 1988). Two examples can be picked out. In *Rhodobacter sphaeroides*, the RR spectrum of spheroidene is strikingly different in the antenna and in the reaction centre, especially for a C=C stretching mode around 1540 cm^{-1} (Lutz *et al.*, 1976). On the basis of comparisons with RR spectra of carotenoids in solution, it could be shown that spheroidene is an all-*trans* isomer in the antenna, and a 15-*cis* isomer in the reaction centre (Lutz *et al.*, 1976, 1987; Koyama *et al.*, 1982). Time-resolved studies showed that the *cis* configuration is retained upon excitation of spheroidene to its triplet state. Another good example is the mode of binding of bacteriochlorophylls in the primary donor P of bacterial reaction centres. The (P minus P$^+$) and (P minus ^{3}P) difference spectra were obtained; they are nearly identical and can be attributed to the neutral ground state P. From the position of specific Raman lines, the molecular interactions experienced by the Mg atoms and by keto and acetyl carbonyls could be described. They result from binding to five protein ligands and probably not from bonding interactions within the special pair of bacteriochlorophylls. Interspecies differences have been observed (Robert and Lutz, 1986; Zhou *et al.*, 1987).

Time-resolved RRS is of increasing importance. The Raman process is essentially instantaneous (shorter than 10^{-13} s), and the time-resolved method is of the pump-probe type, so the time resolution can be in the (sub)picosecond domain. Such fast studies have been conducted on visual pigments and on bacteriorhodopsin. They could prove useful for probing ultra-fast changes in the binding of pigments and redox centres, following charge separation.

ACKNOWLEDGEMENTS

I would like to thank my colleagues in Saclay, and especially Drs J. Breton, J. M. Ducruet, M. Lutz, E. Nabedryk, B. Robert, A. W. Rutherford and P. Sétif, for providing an environment that promotes the use of optical methods to study photosynthesis, with the constant objectives of greater technical achievement and greater understanding of biological problems. Drs G. Paillotin and A. Vermeglio, as well as many foreign visitors, have also helped greatly.

REFERENCES

Alcala, J. R., Gratton, E. and Prendergast, G. (1987). *Biophys. J.* **51**, 925–936.
Bergström, H., Sundström, V., van Grondelle, R., Akesson, E. and Gillbro, T. (1986). *Biochim. Biophys. Acta* **852**, 279–287.
Birks, J. B. (1970). "Photophysics of Aromatic Molecules". Wiley-Interscience, London.
Blankenship, R. E., Schaafsma, T. J. and Parson, W. W. (1977). *Biochim. Biophys. Acta* **461**, 297–305.
Bocian, D. F., Boldt, N. J., Chadwick, B. W. and Frank, H. A. (1987). *FEBS Lett.* **214**, 92–96.
Bottin, H. and Mathis, P. (1985). *Biochemistry* **24**, 6453–6460.
Bowman, M. K., Budil, D. E., Closs, G. L., Kostka, A. G., Wraight, C. A. and Norris, J. R. (1981). *Proc. Natl. Acad. Sci. USA* **78**, 3305–3307.
Boxer, S. G., Kuki, A., Wright, K. A., Katz, B. A. and Xuong, N. H. (1982). *Proc. Natl. Acad. Sci. USA* **79**, 1121–1125.
Boxer, S. G., Lockhart, D. J. and Middendorf, T. R. (1986). *Chem. Phys. Lett.* **123**, 476–482.
Braslavsky, S. E. (1986). *Photochem. Photobiol.* **43**, 667–675.
Braun, H. P., Michel-Beyerle, M. E., Breton, J., Buchanan, S. and Michel, H. (1987). *FEBS Lett.* **221**, 221–225.
Breton, J. (1985). *Biochim. Biophys. Acta* **810**, 235–245.
Breton, J. (1986). In "Encyclopedia of Plant Physiology", New Series, (L. A. Staehelin and C. J. Arntzen, eds), Vol. 19, pp. 319–326. Springer-Verlag, Berlin.
Breton, J. and Nabedryk, E. (1984). *FEBS Lett.* **176**, 355–359.
Breton, J. and Verméglio, A. (1982). In "Photosynthesis: Energy Conversion by Plants and Bacteria" (Govindjee, ed.), pp. 153–194. Academic Press, New York.
Breton, J., Martin, J.-L., Migus, A., Antonetti, A. and Orszag, A. (1986). *Proc. Natl. Acad. Sci. USA* **83**, 5121–5125.
Brettel, K., Schlodder, E. and Witt, H. T. (1984). *Biochim. Biophys. Acta* **766**, 403–415.
Brettel, K., Sétif, P. and Mathis, P. (1986). *FEBS Lett.* **203**, 220–224.
Butler, W. L. (1962). *J. Opt. Soc. Am.* **52**, 292–299.
Butler, W. L. (1972). *Proc. Natl. Acad. Sci. USA* **69**, 3420–3422.
Chance, B. (1972). *Methods Enzymol.* **24**, 322–335.
Chekalin, S. V., Matveets, Yu. A. and Yartsev, A. P. (1987). *Rev. Phys. Appl.* **22**, 1761–1771.
Chidsey, C. E. D., Kirmaier, C., Holten, D. and Boxer, S. G. (1984). *Biochim. Biophys. Acta* **766**, 424–437.

Clarke, R. H. (1982). In "Light Reaction Path of Photosynthesis" (F. K. Fong, ed.), pp. 196–233. Springer-Verlag, Berlin.
Clayton, R. K. (1970). "Light and Living Matter". McGraw-Hill, New York.
Cogdell, R. J., Monger, T. G. and Parson, W. W. (1975). Biochim. Biophys. Acta **408**, 189–199.
Conjeaud, H., Mathis, P. and Paillotin, G. (1979). Biochim. Biophys. Acta **546**, 280–291.
Creighton, S., Hwang, J.-K., Warshel, A., Parson, W. W. and Norris, J. (1988). Biochemistry **27**, 774–781.
Danielius, R. V., Satoh, K., Van Kan P. J. M., Plijter, J. J., Nuijs, A. M. and Van Gorkom, H. J. (1987). FEBS Lett. **213**, 241–244.
De Vault, D. (1978). Methods Enzymol. **54**, 32–46.
den Blanken, H. J. and Hoff, A. J. (1983). Biochim. Biophys. Acta **724**, 52–61.
den Blanken, H. J., Hoff, A. J., Jongenelis, A. P. J. M. and Diner, B. (1983). FEBS Lett. **157**, 21–27.
Duysens, L. N. M. (1956). Biochim. Biophys. Acta **19**, 1–12.
Gadsby, P. M. A. and Thomson, A. J. (1986). FEBS Lett. **197**, 253–257.
Geacintov, N. E. and Breton, J. (1986). In "Encyclopedia of Plant Physiology: Photosynthetic Membranes" (A. Staehelin and C. Arntzen, eds), Vol. 19, pp. 310–318. Springer-Verlag, Berlin.
Gerwert, K., Hess, B., Michel, H. and Buchanan, S. (1988). FEBS Lett. **232**, 303–307.
Gillbro, T., Sandström, A., Sundström, V., Wendler, J. and Holzwarth, A. R. (1985a). Biochim. Biophys. Acta **808**, 52–65.
Gillbro, T., Sundström, V., Sandström, A., Spangfort, M. and Andersson, B. (1985b). FEBS Lett. **193**, 267–270.
Gore, B. L., Doust, T. A. M., Giorgi, L. B., Klug, D. R., Ide, J. P., Crystall, B. and Porter, G. (1986). J. Chem. Soc. Faraday Trans. 2 **82**, 2111–2115.
Govindjee, Amesz, J. and Fork, D. C. (1986). "Light Emission by Plants and Bacteria". Academic Press, New York.
Gunner, M. R., Robertson, D. E. and Dutton, P. L. (1986). J. Phys. Chem. **90**, 3783–3795.
Gust, D., Moore, T. A., Bensasson, R. V., Mathis, P., Land, E. J., Chachaty, C., Moore, A. L., Liddell, P. A. and Nemeth, G. A. (1985). J. Am. Chem. Soc. **107**, 3631–3640.
Hall, J., Zha, X., Durham, B., O'Brien, P., Vieira, B., Davis, D., Okamura, M. and Millett, F. (1987). Biochemistry **26**, 4494–4500.
Hansson, O., Duranton, J. and Mathis, P. (1988). Biochim. Biophys. Acta **932**, 91–96.
Hodges, M. and Moya, I. (1987). Biochim. Biophys. Acta **892**, 42–47.
Hoff, A. J., den Blanken, H. J., Vasmel, H. and Meiburg, R. F. (1985a). Biochim. Biophys. Acta **806**, 389–397.
Hoff, A. J., Lous, E. J., Moehl, K. W. and Dijkman, J. A. (1985b). Chem. Phys. Lett. **114**, 39–43.
Holzwarth, A. R. (1986). Photochem. Photobiol. **43**, 707–725.
Houssier, C. and Sauer, K. (1970). J. Am. Chem. Soc. **92**, 779–791.
Joliot, P. and Joliot, A. (1986). Biochim. Biophys. Acta **849**, 211–222.
Joliot, P. and Joliot, A. (1988). Biochim. Biophys. Acta **933**, 319–333.
Joliot, P., Beal, D. and Frilley, B. (1980). J. Chim. Phys. **77**, 209–216.
Junge, W. (1987). Proc. Natl. Acad. Sci. USA **84**, 7084–7088.
Junge, W., Schafferricht, H. and Nelson, N. (1977). Biochim. Biophys. Acta **462**, 73–85.
Karukstis, K. K. and Sauer, K. (1983). J. Cell. Biochem. **23**, 131–158.
Ke, B. (1973). Biochim. Biophys. Acta **301**, 1–33.
Ke, B. (1972). Methods Enzymol. **24**, 25–53.
Kirmaier, C., Holten, D. and Parson, W. W. (1983). Biochim. Biophys. Acta **725**, 190–202.
Kirmaier, C., Holten, D. and Parson, W. W. (1985a). FEBS Lett. **185**, 76–82.
Kirmaier, C., Holten, D. and Parson, W. W. (1985b). Biochim. Biophys. Acta **810**, 49–61.
Kleinfeld, D., Okamura, M. Y. and Feher, G. (1985). Biophys. J. **48**, 849–852.
Klevanik, A. V., Klimov, V. V., Shuvalov, A. A. and Krasnovskii, A. A. (1977). Dokl. Akad. Nauk SSSR **236**, 241–244.
Klimov, V. V. and Allakhverdiev, S. I. (1978). Dokl. Akad. Nauk SSSR **242**, 1204–1207.
Koyama, Y., Kito, M., Takii, T., Saiki, K., Tsukida, K. and Yamashita, J. (1982). Biochim. Biophys. Acta **680**, 109–118.

Kramer, H. and Mathis, P. (1980). *Biochim. Biophys. Acta* **593**, 319–329.
Kretschmann, H., Dekker, J. P., Saygin, Ö. and Witt, H. T. (1988). *Biochim. Biophys. Acta* **932**, 358–361.
Kühlbrandt, W., Becker, A. and Mäntele, W. (1988). *FEBS Lett.* **226**, 275–279.
Latimer, P. (1959). *Plant Physiol.* **34**, 193–199.
Lavergne, J. (1987). *Biochim. Biophys. Acta* **894**, 91–107.
Levanon, H. and Norris, J. R. (1982). *In* "Light Reaction Path of Photosynthesis" (F. K. Fong, ed.), pp. 152–185. Springer-Verlag, Berlin.
Lichtenthaler, H. K. (1987). *Methods Enzymol.* **148**, 350–382.
Lockhart, D. J. and Boxer, S. G. (1987). *Biochemistry* **26**, 664–668.
Lösche, M., Feher, G. and Okamura, M. Y. (1987). *Proc. Natl. Acad. Sci. USA* **84**, 7537–7541.
Lous, E. J. and Hoff, A. J. (1987a). *Proc. Natl. Acad. Sci. USA* **84**, 6147–6151.
Lous, E. J. and Hoff, A. J. (1987b). *Chem. Phys. Lett.* **140**, 620–625.
Lutz, M. and Robert, B. (1988). *In* "Biological Applications of Raman Spectroscopy" (T. G. Spiro, ed.), Vol. III, Ch. 8, pp. 347–411. John Wiley & Sons, New York.
Lutz, M., Kléo, J. and Reiss-Husson, F. (1976). *Biochim. Biophys. Res. Commun.* **69**, 711–717.
Lutz, M., Szponarski, W., Berger, G., Robert, B. and Neumann, J.-M. (1987). *Biochim. Biophys. Acta* **894**, 423–433.
Martin, J.-L., Breton, J., Hoff, A. J., Migus, A. and Antonetti, A. (1986). *Proc. Natl. Acad. Sci. USA* **83**, 957–961.
Mathis, P. (1977). *In* "Topics in Photosynthesis", Vol. 2, "Primary Processes of Photosynthesis", Ch. 7 (J. Barber, ed.), pp. 270–302, Elsevier, Amsterdam.
Mathis, P. and Rutherford, A. W. (1987). *In* 'Photosynthesis, New Comprehensive Biochemistry" (J. Amesz, ed.), Vol. 15, pp. 63–96. Elsevier, Amsterdam.
Meech, S. R., Hoff, A. J. and Wiersma, D. A. (1986). *Proc. Natl. Acad. Sci. USA* **83**, 9464–9468.
Michl, J. and Thulstrup, E. W. (1986). *In* "Spectroscopy with Polarized Light" VCH Publications, New York.
Migus, A., Antonetti, A., Etchepare, J., Hulin, D. and Orszag, A. (1985). *J. Opt. Soc. Am.* **B2**, 584–594.
Moog, R. S., Kuki, A., Fayer, M. D. and Boxer, S. G. (1984). *Biochemistry* **23**, 1564–1571.
Moore, T. A. (1983). *Photochem. Photobiol. Rev.* **7**, 187–221.
Nabedryk, E., Biaudet, P., Darr, S., Arntzen, C. J. and Breton, J. (1984). *Biochim. Biophys. Acta* **767**, 640–647.
Nabedryk, E., Mäntele, W., Tavitian, B. A. and Breton, J. (1986). *Photochem. Photobiol.* **43**, 461–465.
Norris, J. R., Bowman, M. K., Budil, D. E., Tang, J., Wraight, C. A. and Closs, G. L. (1982). *Proc. Natl. Acad. Sci. USA* **79**, 5532–5536.
Norris, J. R., Budil, D. E., Tiede, D. M., Tang, J., Kolaczkowski, S. V., Chang, C. H. and Schiffer, M. (1987). *In* "Progress in Photosynthesis Research" (J. Biggins, ed.), Vol. 1, pp. 1.4.363–1.4.369. Martinus Nijhoff Publishers, Dordrecht, Netherlands.
Nuijs, A. M., van Grondelle, R., Joppe, H. L. P., van Bochove, A. C. and Duysens, L. N. M. (1985). *Biochim. Biophys. Acta* **810**, 94–105.
Nuss, M. C., Zinth, W., Kaiser, W., Kölling, E. and Oesterhelt, D. (1985). *Chem. Phys. Lett.* **117**, 1–7.
Ogrodnik, A., Remy-Richter, N. and Michel-Beyerle, M. E. (1987). *Chem. Phys. Lett.* **135**, 576–581.
Okamura, M. Y. and Feher, G. (1986). *Proc. Natl. Acad. Sci. USA* **83**, 8152–8156.
Overfield, R. E., Wraight, C. A. and Devault, D. (1979). *FEBS Lett.* **105**, 137–142.
Owens, T. G., Webb, S. P., Mets, L., Alberte, R. S. and Fleming, G. R. (1987). *Proc. Natl. Acad. Sci. USA* **84**, 1532–1536.
Packham, N. K., Tiede, D. M., Mueller, P. and Dutton, P. L. (1980). *Proc. Natl. Acad. Sci. USA* **77**, 6339–6343.
Parker, F. S. (1983). "Applications of Infra-red, Raman and Resonance Raman Spectroscopy in Biochemistry". Plenum Press, New York, 550 pp.
Parot, P., Thiery, J. and Verméglio, A. (1987). *Biochim. Biophys. Acta* **893**, 534–543.
Parson, W. W. and Warshel, A. (1987). *J. Am. Chem. Soc.* **109**, 6152–6163.

Parson, W. W., Clayton, R. K. and Cogdell, R. J. (1975). *Biochim. Biophys. Acta* **387**, 265–278.
Pearlstein, R. M., Davis, R. C. and Ditson, S. L. (1982). *Proc. Natl. Acad. Sci. USA* **79**, 400–402.
Pulles, M. P. J., Van Gorkom, H. J. and Verschoor, G. A. M. (1976). *Biochim. Biophys. Acta* **440**, 98–106.
Rabek, J. F. (1982). "Experimental Methods in Photochemistry and Photophysics", Vol. 2. John Wiley, New York.
Rentzepis, P. M. (1978). *Methods Enzymol.* **54**, 3–32.
Robert, B. and Lutz, M. (1986). *Biochemistry* **25**, 2303–2309.
Rockley, M. G., Windsor, M. W., Cogdell, R. J. and Parson, W. W. (1975). *Proc. Natl. Acad. Sci. USA* **72**, 2251–2255.
Rousseau, D. L. and Ondrias, M. R. (1984). *Optical Techniques in Biological Research* (D. L. Rousseau, ed.), pp. 65–132. Academic Press, New York.
Salem, L. (1966). "The Molecular Orbital Theory of Conjugated Systems". W. A. Benjamin Inc, New York.
San Pietro, A. (1971). "Methods in Enzymology", Vol. 23, Part A. Academic Press, New York.
San Pietro, A. (1980). "Methods in Enzymology", Vol. 69, Part C. Academic Press, New York.
Sane, P. V. and Rutherford, A.-W. (1986). *In* "Light Emission by Plants and Bacteria" (Govindjee, J. Amesz and D. C. Fork, eds), pp. 329–360. Academic Press, New York.
Sauer, K. (1975). *In* "Bioenergetics of Photosynthesis" (Govindjee, ed.), pp. 115–181. Academic Press, New York.
Schatz, G. H. and van Gorkom, H. J. (1985). *Biochim. Biophys. Acta* **810**, 283–294.
Schatz, G. H., Brock, H. and Holzwarth, A. R. (1987). *Proc. Natl. Acad. Sci. USA* **84**, 8414–8418.
Scheer, H. and Schneider, S. (1988). "Photosynthetic Light-Harvesting Systems". Walter de Gruyter, Berlin.
Schenck, C. C., Blankenship, R. E. and Parson, W.W. (1982a). *Biochim. Biophys. Acta* **680**, 44–59.
Schenck, C. C., Diner, B., Mathis, P. and Satoh, K. (1982b). *Biochim. Biophys. Acta* **680**, 216–227.
Scherer, P. O. J. and Fischer, S. F. (1987). *Chem. Phys. Lett.* **137**, 32–36.
Scherz, A. and Parson, W. W. (1984). *Biochim. Biophys. Acta* **766**, 666–678.
Schlodder, E. and Brettel, K. (1988). *Biochim. Biophys. Acta* **933**, 22–34.
Schlodder, E., Brettel, K., Schatz, G. H. and Witt, H. T. (1984). *Biochim. Biophys. Acta* **765**, 178–185.
Schreiber, U. and Schliwa, U. (1987). *Photosynth. Res.* **11**, 173–182.
Sebban, P. and Barbet, J.-C. (1984). *FEBS Lett.* **165**, 107–110.
Sebban, P. and Lindqvist, L. (1987). *Photosynth. Res.* **13**, 57–67.
Sétif, P., Bottin, H. and Mathis, P. (1985). *Biochim. Biophys. Acta* **808**, 112–122.
Shank, C. V. (1983). *Science* **219**, 1027–1031.
Shibata, K. (1973). *Biochim. Biophys. Acta* **304**, 249–259.
Shopes, R. J. and Wraight, C. A. (1985). *Biochim. Biophys. Acta* **806**, 348–356.
Shopes, R. J. and Wraight, C. A. (1987). *Biochim. Biophys. Acta* **893**, 409–425.
Shuvalov, V. A. (1976). *Biochim. Biophys. Acta* **430**, 113–121.
Shuvalov, V. A. and Parson, W. W. (1981). *Proc. Natl. Acad. Sci. USA* **78**, 957–961.
Shuvalov, V. A., Ke, B. and Dolan, E. (1979a). *FEBS Lett.* **100**, 5–8.
Shuvalov, V. A., Klevanik, A. V., Sharkov, A. V., Kryukov, P. G. and Ke, B. (1979b). *FEBS Lett.* **107**, 313–316.
Shuvalov, V. A., Nuijs, A. M., van Gorkom, H. J., Smit, H. W. J. and Duysens, L. N. M. (1986). *Biochim. Biophys. Acta* **850**, 319–323.
Steinfeld, J. I. (1985). "Molecules and Radiation: An Introduction to Modern Molecular Spectroscopy". MIT Press, Cambridge, MA.
Takiff, L. and Boxer, S. G. (1988). *Biochim. Biophys. Acta* **932**, 325–334.
Tavitian, B. A., Nabedryk, E., Mäntele, W. and Breton, J. (1986). *FEBS Lett.* **201**, 151–157.
Thielen, A. P. G. M. and van Gorkom, H. J. (1981). *Biochim. Biophys. Acta* **635**, 111–120.
Tiede, D. M. (1987). *Biochemistry* **26**, 397–410.
Van Best, J. and Mathis, P. (1978a). *Biochim. Biophys. Acta* **503**, 178–188.
Van Best, J. and Mathis, P. (1978b). *Rev. Sci. Instrum.* **49**, 1332–1335.

Van der Wal, H. N., Van Grondelle, R., Millett, F. and Knaff, D. B. (1987). *Biochim. Biophys. Acta* **893**, 490–498.
Vernon, L. P. and Seely, G. R. (1966). *In* "The Chlorophylls". Academic Press, New York.
Vickery, L. E. (1978). *Methods Enzymol.* **54**, 284–302.
Vink, K. J., De Boer, S., Plijter, J. J., Hoff, A. J. and Wiersma, D. A. (1987). *Chem. Phys. Lett.* **142**, 433–438.
Wacker, T., Gad'on, N., Becker, A., Mäntele, W., Kreutz, W., Drews, G. and Welte, W. (1986). *FEBS Lett.* **17**, 267–273.
Warshel, A. and Parson, W. W. (1987). *J. Am. Chem. Soc.* **109**, 6143–6152.
Wasielewski, M. R. and Tiede, D. M. (1986). *FEBS Lett.* **204**, 368–372.
Wasielewski, M. R., Bock, C. H., Bowman, M. K. and Norris, J. R. (1983). *Nature* **303**, 520–522.
Wasielewski, M. R., Liddell P. A., Barrett, D., Moore, T. A. and Gust, D. (1986). *Nature* **322**, 570–572.
Wasielewski, M. R., Fenton, J. M. and Govindjee (1987). *Photosynth. Res.* **12**, 181–190.
Wendler, J., Holzwarth, A. R., and Wehrmeyer, W. (1984). *Biochim. Biophys. Acta* **765**, 58–67.
Witt, H. T. (1971). *Q. Rev. Biophys.* **4**, 365–477.
Woodbury, N. W. and Parson, W. W. (1986). *Biochim. Biophys. Acta* **850**, 197–210.
Woodbury, N. W., Becker, M., Middendorf, D. and Parson, W. W. (1985). *Biochemistry* **24**, 7516–7521.
Zhou, Q., Robert, B. and Lutz, M. (1987). *Biochim. Biophys. Acta* **890**, 368–376.

9 Chlorophyll Fluorescence Transients

PETER HORTON[1] and JOHN R. BOWYER[2]

[1]*Robert Hill Institute, Department of Molecular Biology and Biotechnology, University of Sheffield, Western Bank, Sheffield S10 2TN, UK*
[2]*Department of Biochemistry, Royal Holloway and Bedford New College, University of London, Egham Hill, Egham, Surrey TW20 0EX, UK*

I.	Origins of chlorophyll fluorescence	260
	A. Chlorophyll proteins of the thylakoid membrane	260
	B. Fluorescence emissions from Photosystems II and I	260
II.	Factors affecting chlorophyll fluorescence yield	261
	A. Fluorescence and photochemistry	261
	B. Other photochemical quenchers of fluorescence	263
	C. Non-photochemical quenching of chlorophyll fluorescence	263
	D. The importance of membrane organisation in determining the characteristics of chlorophyll fluorescence	266
III.	Measurement of fluorescence changes in the picosecond to nanosecond time domain	267
IV.	Measurement of fluorescence changes in the microsecond to millisecond time domain	268
	A. Flash excitation fluorescence apparatus	268
	B. Kinetics of Q_A^- oxidation	271
	C. Binding of inhibitors to the Q_B site	272
V.	Measurement of fluorescence changes in the millisecond to second time domain	274
	A. Apparatus	274
	B. Kinetics in the absence of DCMU	275
	C. Kinetics in the presence of inhibitors of Q_A^- oxidation	276
	D. Measurements using the flash excitation fluorescence apparatus	279
VI.	Measurement of long-term quenching of chlorophyll fluorescence	282
	A. Apparatus	282

	B.	Separation of photochemical and non-photochemical quenching components using modulated fluorescence	283
	C.	Experimental problems and precautions	285
	D.	Identification of non-photochemical quenching components	287
	E.	Use of q_N and q_P as probes of metabolic activity	288
VII.		Use of quenching analysis to determine quantum efficiency	289
		Acknowledgements	292
		References	293
		Abbreviations	296

I. ORIGINS OF CHLOROPHYLL FLUORESCENCE

A. Chlorophyll Proteins of the Thylakoid Membrane

A detailed discussion of the organisation of the pigment–protein complexes in the thylakoid membrane is beyond the scope of this chapter. However, a few general statements to indicate the origins of chlorophyll fluorescence are appropriate. In common with all photosynthetic systems, the pigment–protein complexes of the plant thylakoid membrane are organised into reaction centre and antenna complexes. For Photosystem II (PSII) the reaction centre itself probably contains just four chlorophylls, including the primary donor P_{680} (Nanba and Satoh, 1987). Closely associated with this, to form the PSII-core, are the chlorophyll proteins CP47 and CP43 (Akabori *et al.*, 1988). The PSII core has ~ 35 chlorophyll *a* per reaction centre (Glick and Melis, 1988). The peripheral antenna system is both complex and variable (Morrissey *et al.*, 1989). The major light-harvesting chlorophyll *a/b* complex (LHC-II) is heterogeneous and appears to be in two pools. A peripheral pool is the major phosphorylated population which (a) can migrate away from PSII in state 2 and (b) can vary in amount dependent on growth conditions. In addition there is a more tightly-bound LHC-II population that is probably linked to the PSII core via specialised chlorophyll proteins CP29 and CP24. For Photosystem I, the core antenna and reaction centre are associated with the 62 and 58 kDa polypeptides with ~ 100 chlorophylls per reaction centre (Malkin, 1986). The LHC-I complex accounts for the peripheral antenna and there are no reports of variable LHC-I:PSI ratios.

B. Fluorescence Emissions from Photosystems II and I

At room temperature most of the measured fluorescence emission comes from PSII and has a maximum at 685 nm. This emission arises from the antenna chlorophyll, although it is not established whether one particular complex is involved. Emission from PSI has maxima at 690 and 720 but is normally quenched by the PSI reaction centre. At liquid nitrogen temperature the PSII emission band is split, with bands at ~ 685 and 695 nm. These have been assigned to the CP47 and 43, respectively (Govindjee and Wasielewski, 1989). Emission from LHC-II is observed as a weak band at 680 nm. Fluorescence from PSI is enhanced by a factor of about 20 at 77 K, with a major emission at 735 nm which emanates from the peripheral antenna, and a minor peak at around 715–720 nm (Sauer, 1989).

II. FACTORS AFFECTING CHLOROPHYLL FLUORESCENCE YIELD

The numerous factors which determine the yield of fluorescence have been the subject of many reviews (Krause and Weis, 1984; Briantais *et al.*, 1986; Baker and Horton, 1987) and a recent compilation (Walker and Osmond, 1989) contains a number of articles on new techniques in the measurement and interpretation of chlorophyll fluorescence. Therefore only a brief summary will be given here. In all of the subsequent discussion only fluorescence from pigments associated with PSII will be considered. Earlier detailed accounts of the use of chlorophyll fluorescence to study PSII electron transport are given by Briantais *et al.* (1986), van Gorkom (1986) and Lavorel *et al.* (1986).

The fluorescence yield of PSII, Φ_f, has been described in terms of the competition between fluorescence and other processes which bring about the relaxation of the excited state to the ground state:

$$\text{if} \quad \Phi_f = \frac{F}{I} \tag{9.1}$$

$$\text{then} \quad \Phi_f = \frac{k_f}{k_f + k'} \tag{9.2}$$

where F is the intensity of fluorescence, I is the incident light intensity, k_f is the rate constant for fluorescence and k' the sum of the constants for other dissipative processes which include k_d (thermal dissipation), k_p (utilisation for photochemistry) and k_t, for energy transfer from PSII to PSI. This competitive relationship between fluorescence, photochemistry and non-radiative dissipation has been confirmed experimentally (Briantais *et al.*, 1972), and forms the basis for all subsequent treatments that are mentioned below. The yield of chlorophyll fluorescence is therefore affected by two types of processes—firstly photochemical events (i.e. k_p); secondly, non-photochemical processes ($k_d + k_t$).

A. Fluorescence and Photochemistry

The relationship between chlorophyll fluorescence and photochemistry was first described by Duysens and Sweers (1963), who introduced the term Q. Q was defined as a quencher when oxidised and was originally considered to be the primary electron acceptor of PSII. Thus, with the primary acceptor reduced the reaction centre cannot carry out photochemistry and is therefore closed. With it oxidised, the reaction centre is open. Thus two levels of fluorescence can be defined: F_o, the level of chlorophyll fluorescence when all PSII centres are open, and F_m, the maximum level of chlorophyll fluorescence when all PSII centres are closed.

From Eq. 9.2 it is predicted that Φ'_p, the yield of photochemistry at any state of the reaction centre population, will depend on fluorescence according to Eq. 9.3:

$$\Phi'_p = 1 - \frac{F}{F_m} \tag{9.3}$$

Hence, the maximum yield of photochemistry, when all the centres are open and the fluorescence is at F_o, is given by:

$$\Phi_p = \frac{F_m - F_o}{F_m} = \frac{F_v}{F_m} \qquad (9.4)$$

The potential yield of photochemistry, Φ_p, is hence given by the ratio F_v/F_m, which has been shown to be approximately 0.83 for a wide variety of plants (Bjorkman and Demmig, 1987). Moreover, the ratio F_v/F_m was shown to be directly proportional to the measured quantum yield of photosynthetic O_2 evolution measured at saturating CO_2 (Demmig and Bjorkman, 1987). It should be noted that this upper limit for the F_v/F_m is less than the measured quantum yield of photochemistry, which approaches 1; one explanation of this discrepancy is given below.

A number of subsequent experiments have shown that Q is identical to Q_A, the primary bound quinone in the reaction centre. Photoreduction of Q_A gives rise to an absorption change at 320 nm indicative of reduction to a semiquinone. However, the identity of Q as the *primary* electron acceptor of PSII was disproved by the finding that a transient electron acceptor 'I' existed prior to Q_A; I was later shown to be a molecule of phaeophytin. Thus we can represent PSII photochemistry by the following series of reactions:

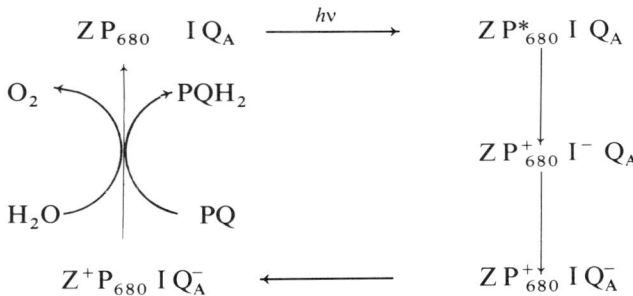

The reaction centre in state P_{680} I Q_A^- is the highly fluorescent form of the closed centre that gives rise to the F_m level of fluorescence. However, this state is still capable of charge separation to $P_{680}^+ I^-$ and the relaxation of this state back to P_{680} I may result in luminescence. It has been suggested that all F_v derives from this reaction and is, in fact, luminescence, providing an explanation as to why maximum emission is obtained from a reaction centre that can still carry out primary charge separation. This hypothesis appears to be inconsistent with measurements of fluorescence lifetimes and it has been suggested instead that with Q_A in its reduced state, the rate of charge separation is decreased, possibly due to an electrostatic effect of the presence of Q_A^- (Schatz et al., 1988) or the transmembrane electric field (Keuper and Sauer, 1989). Furthermore, Schlodder and Brettel (1988) have suggested that the state $P_{680}^+ I^- Q_A^-$ decays to a lower energy state $(P_{680}^+ I^- Q_A^-)_{II}$ which is not at equilibrium with excitation in the antenna, and that this process is slower than the oxidation of I^- by Q_A. These different theories make little difference to the uses of chlorophyll fluorescence described below, but it is worth noting the absence of a simple relationship between Φ_f and the redox state of the reaction centre. It should be pointed out that the fact that F_v/F_m is less than the predicted value of 1 led to the suggestion that k_p may not be zero in Q_A^- centres, k_p in closed centres representing charge separation followed by recombination (Clayton, 1980).

Because of energy transfer between PSII units, the relationship between the redox state of Q_A and F is curvilinear (see below). Although the Q_A redox state is not a good indicator of photochemistry, nevertheless, the yield of photochemistry is still directly proportional to $F_m - F$. However, during steady-state photosynthesis, it is common practice now to use the term q_P as a parameter describing the photochemical utilisation of excitation energy. This has become necessary because other factors cause a decline in fluorescence so that a simple measurement of F cannot be used under most conditions. q_P is defined as $(F_m' - F_s)/(F_m' - F_0')$, where F_m' and F_0' are the maximum and minimum fluorescence under that particular condition and F_s is the observed fluorescence; in this way, q_P is a parameter that is normalised according to the presence of other quenching processes and is a measure of the proportion of available excitation energy that is being used photochemically. With all centres open q_P is 1 and it declines towards a value of zero as the centres become closed. q_P is most readily determined using modulated fluorescence (see Section VI.B).

B. Other Photochemical Quenchers of Fluorescence

In addition to Q_A, a number of different quenchers have been identified which are photochemical. A detailed discussion of these is beyond the scope of this chapter. In brief, quenching remains when all of Q_A is reduced, either upon a single flash or during redox titration, and this quenching can be removed by a succession of subsequent flashes or a further decrease in redox potential. The quenchers are referred to as Q_2 and Q_L; their chemical identities are unknown (for a review, see Black et al., 1986).

C. Non-photochemical Quenching of Chlorophyll Fluorescence

Illumination of plants induces changes in fluorescence that cannot be ascribed simply and directly to changes in the redox state of Q_A. This quenching can thus be observed by changes in the fluorescence yield when PSII centres are closed, and is therefore referred to as non-photochemical quenching of chlorophyll fluorescence. In the same way as q_P is used to quantify photochemical quenching, a coefficient q_N is a measure of non-photochemical quenching. q_N is defined as $(F_v - F_v')/F_v'$, which therefore has a value of zero in the fully relaxed state, approaching a maximum value of 1 when all absorbed excitation energy is being dissipated or transferred away from PSII.

It should be pointed out that this distinction is essentially an empirical one, quenching being defined as non-photochemical if it cannot be removed by illumination with saturating light or in the presence of DCMU (3-(3,4-dichlorophenyl)-1,1-dimethylurea) (which both result in closure of PSII centres, DCMU acting by blocking the oxidation of Q_A^- by Q_B, the secondary quinone electron acceptor of PSII). For each of the several processes that give rise to non-photochemical quenching, the molecular and physical basis of the dissipation of energy is not known. Thus, the distinction between photochemical and non-photochemical quenching in a mechanistic sense could become clouded if there were to be a rapid cycling of electrons in the PSII reaction centre that could prevent accumulation of Q_A^- even in saturating light (Schreiber and Neubauer, 1989). Further, it should be remembered that thermal dissipation of excitation energy can be brought about by quenching species that are generated directly by photochemical events—both P_{680}^+ (Butler, 1980a) and I^- (Klimov et al., 1977) are quenchers in the reaction centre and their presence could underlie non-photochemical quenching in some of the processes described below.

1. Quenching associated with the proton gradient across the thylakoid membrane

Quenching of chlorophyll fluorescence driven by the 'high energy state' of the thylakoid membrane was first reported by Murata and Sugahara (1969). This quenching was later shown to depend on the light-induced formation of a ΔpH across the membrane and was suggested to be due to protonation of the thylakoid interior (Wraight and Crofts, 1970; Krause, 1973). This quenching, which became known as q_E, can be induced in the presence of DCMU (which results in all PSII traps being closed) by conditions in which cyclic electron transport around PSI are promoted. q_E develops with a half-time of 10–30 s and at steady-state can be shown to correlate with other indicators of the ΔpH, such as the quenching of 9-aminoacridine fluorescence (Briantais et al., 1979; Krause et al., 1988; Oxborough and Horton, 1988). The molecular mechanism of q_E is not known and appears not to be simply a direct consequence of a protonation reaction at the thylakoid lumenal surface. It was shown in early work that q_E could be inhibited by glutaraldehyde fixation (Mohanty et al., 1972). There are often significant time delays between ΔpH formation and the development of quenching (e.g. Rees et al., 1989). Moreover it is possible to eliminate q_E with antimycin A without altering ΔpH (Oxborough and Horton, 1987). It has been suggested that q_E results from alteration within the PSII reaction centre (Weis and Berry, 1988), perhaps facilitating the formation of P_{680}^+ or I^-. However, there are alternative suggestions that q_E is a result of energy dissipation in one of the antenna complexes. One theory is that the quenching requires formation of zeaxanthin from violaxanthin via the xanthophyll cycle (Demmig-Adams et al., 1989). It is possible that more than one process may contribute to q_E. q_E is thought to be responsible for the major part of the non-photochemical quenching observed *in vivo*. This is based on the generally similar kinetic features of q_N when measurements made with intact chloroplasts and protoplasts (in which q_E can be measured) are compared with leaves (Horton, 1985).

2. State transitions

Bonaventura and Myers (1969) first showed that thylakoid membranes possess a mechanism that enables corrections to be made for imbalance in the relative rates of excitation of PSII and PSI. Because room temperature fluorescence reflects the excitation of PSII, these changes are detectable as alterations in the yield of fluorescence. In state 1, a maximum proportion of excitation is delivered to PSII and fluorescence is high, whereas in state 2 the proportion of energy reaching PSII is decreased and hence the fluorescence is lowered. The state change results predominantly from a change in the relative absorption cross-sections or antennae sizes of PSII and PSI, but there can also be an increase in spillover of excitation from PSII to PSI. The former would result in an unchanged F_v/F_m because the F_o level would be lowered by exactly the same degree as F_m, whereas the latter would cause F_v/F_m to fall. It should be pointed out that a fluorescence change due to an absorption cross-section change is not strictly *quenching* since it is really a change in I in Eq. 9.1 rather than in k'. In contrast, spillover can be described as a process that removes excitation away from PSII in competition with emission as fluorescence and is correctly defined as non-photochemical quenching. However, for practical reasons, the fluorescence decrease due to the transition to state 2 has been included under q_N and has been referred to as q_T (Horton and Hague, 1988).

A molecular mechanism of the state transition is suggested by the changes that can be observed *in vitro* upon phosphorylation of the light-harvesting complex of PSII (Horton and Black, 1981). A linear relationship between the level of phosphorylation and the fluorescence decrease was observed (Horton et al., 1981). Examination of fluorescence emission spectra recorded at $-196°C$ showed that phosphorylation leads to an increase in rate of excitation delivery to PSI (Krause and Behrend, 1983). There is now abundant evidence that the transition from state 1 to state 2 is associated with migration of the phosphorylated LHC-II from the appressed membrane region to the unappressed stromal lamellae (for review, see Horton, 1983a). The state transition is induced by changes in redox state of the intersystem electron transfer chain, but is also influenced by the ΔpH, the ionic composition of the aqueous phase in contact with the stromal thylakoid surface, the developmental state of the chloroplast and the temperature. A detailed discussion of these phenomena is beyond the scope of this chapter, but they are discussed in a number of papers elsewhere (Horton and Black, 1983; Black and Horton, 1984; Telfer et al., 1984; Weis, 1985). A final point, of considerable practical importance, is that the protein phosphatase activity that is necessary for reversal from state 2 to state 1 by bringing about dephosphorylation of LHC-II can be inhibited by NaF; thus addition of NaF can be used to stabilise state 2, allowing detailed analysis of state 2, or can be used to probe for the presence of the state transition in the presence of other non-photochemical quenching processes (see Section VI.D.1).

3. Quenching under conditions of light stress

Exposure of plants to a level of irradiance in excess of that which can be used for photosynthesis has been shown to cause photoinhibition. Photoinhibition has been defined as the decrease in quantum yield and photosynthetic capacity associated with *damage* to the photosynthetic apparatus and has been correlated with large changes in the yield of chlorophyll fluorescence (Baker and Horton, 1987). However, this definition of photoinhibition is complicated by the fact that the thylakoid membranes show an apparently *protective* response that is readily reversible but that has symptoms that resemble photoinhibitory damage. Bjorkman first suggested that the decline in F_v/F_m that is observed under light stress (q_1) could be caused by either a decrease in the probability for photochemistry (a rise in F_o) or an increase in thermal dissipation of excitation energy (a decrease in F_m). It can be shown theoretically that both these effects could produce the experimentally observed linear relationship between the decline in the quantum yield of O_2 evolution and the decrease in F_v/F_m. Measurement of fluorescence during photoinhibition of both leaves and chloroplasts showed that increases in F_o and decreases in F_m were occurring (Bradbury and Baker, 1986; Demmig and Bjorkman,

1987). The kinetic differences of these changes gave rise to the suggestion that photoinhibition was associated with two processes, one causing a rise in F_o and the other a decrease in F_v.

At least a part of the quenching observed under light stress appears to be due to a dissipative process in the antenna chlorophylls that has a protective function. This quenching has been correlated with the conversion of violaxanthin to zeaxanthin (Demmig et al., 1987; Demmig-Adams et al., 1989). This quenching is diagnosed by: (a) decreases in both F_m and F_o, (b) a decrease in quantum yield but not the light-saturated rate of electron transport, (c) a linear relationship between the decline in F_v/F_m and the decrease in quantum yield. However, it is clear that quenching can be seen in systems where no changes in the carotenoid population are occurring and that the damaged PSII reaction centre is also a quencher (Cleland et al., 1987). The mechanism of this quenching is unknown but could be due to either P_{680}^+, Chl^+ or I^-. Because at least two distinct processes can contribute to the observed quenching, use of chlorophyll fluorescence as an assay of photoinhibition has to be employed with great care.

4. Quenching by oxidised plastoquinone

It has been known for many years that oxidised quinones can quench the fluorescence of chlorophyll solutions. Vernotte et al. (1979) showed that non-photochemical quenching of fluorescence in thylakoids could also be brought about by oxidised plastoquinone. Whilst this is of little physiological importance, this phenomenon has to be taken into account when estimating quenching coefficients and may be especially important when comparing conditions where there may be differences in the redox state of plastoquinone.

5. The physiological significance of non-photochemical quenching

Whilst the possible physiological significance of the state transitions has been appreciated for many years, it is only recently that an important role has been put forward for q_E and the protective component of q_I. The importance of maintaining balance between the light-driven processes in the thylakoid and the reactions of carbon assimilation has been frequently pointed out (Horton, 1985, 1987) and it was realised that the thermal dissipation of excitation energy observed as q_N could provide a way of 'down regulating' the light-harvesting system under conditions of excess irradiance (Horton, 1987; Weis et al., 1987; Krause et al., 1988). q_N, particularly q_E, would provide a means of feedback control, via the ΔpH, which would sense the balance between irradiance and metabolic capacity (Weis et al., 1987; Horton, 1989). In Section VII the methods used to explore quantitative aspects of this feedback control are described. It has also been demonstrated that the dissipation of excitation energy observed as both q_E (Krause and Behrend, 1986) and q_I (Demmig et al., 1987) can protect against photoinhibitory damage to the reaction centre of PSII.

D. The Importance of Membrane Organisation in Determining the Characteristics of Chlorophyll Fluorescence

The high F_v/F_m ratio and the ability of a photosynthetic system to show light-induced quenching is dependent on the lateral segregation of the protein complexes in the thylakoid membrane. The importance of divalent cations in maintaining the high fluorescent state was first detected by Homann (1969) and Murata (1969). The formation of the appressed membranes, which allow separation of the LHC-II and PSII complexes from PSI and prevent spillover of excitation energy, is dependent on the screening effect of cations (Barber, 1982). Consequently, suspension of chloroplast thylakoids in a medium of low Mg^{2+} concentration (less than 1 mM in the presence of 10 mM Na^+) results in a decrease in fluorescence yield (due to spillover from PSII to PSI) and membrane unstacking. Addition of Mg^{2+} (approx. 5 mM) then causes a fluorescence increase and the re-establishment of lateral segregation.

Any treatment which disrupts this precise membrane organisation will modify fluorescence and this has allowed the study of the effects of environmentally induced perturbations of chloroplast membranes to be investigated. For example, an increase in temperature results in dissociation of the PSII reaction centre from its light-harvesting system, resulting in an increase in F_o (Schreiber and Berry, 1977). Significantly, plants which show different tolerances to survival at high temperatures show the increase in F_o at different temperatures (Bilger et al., 1990).

III. MEASUREMENT OF FLUORESCENCE CHANGES IN THE PICOSECOND TO NANOSECOND TIME DOMAIN

The absorption of light by chlorophyll results in the formation of higher excited singlet states which are rapidly converted by internal conversion to the lowest excited singlet state (Shipman, 1982). A fraction of the energy of this state is lost as fluorescence emission. Following an ultra-short excitation pulse, the decay profile of the fluorescence emission can yield information on the organisation of the light-harvesting pigments and the kinetics of energy conversion in the photochemical reaction centres. Fluorescence lifetime measurements require relatively complex and expensive apparatus which is unlikely to be available in most plant biochemistry laboratories. A very brief summary is given below, but the reader is referred to Karukstis and Sauer (1983), O'Connor and Phillips (1984), Holzwarth et al. (1985), Yamazaki et al. (1985), Holzwarth (1986), Moya et al. (1986) and Schatz et al. (1987) for detailed accounts. These measurements are generally made using single photon timing, in which the delays between the excitation laser pulse and the emission of single photons detected using a photomultiplier are monitored. The profile of photon counts as a function of time after the excitation pulse is deconvoluted into a number of different kinetic components. This deconvolution is improved by measurement of photon emission at a number of different wavelengths so that the emission spectrum of each kinetic component is obtained (Knorr and Harris, 1981; Wendler and Holzwarth, 1987). Because of the complex organisation of chlorophyll in intact thylakoid membranes, measurements are also made on isolated pigment–protein complexes and on reaction centres with their associated antenna proteins.

IV. MEASUREMENT OF FLUORESCENCE CHANGES IN THE MICROSECOND TO MILLISECOND TIME DOMAIN

A number of electron transfer reactions in PSII occurring in the microsecond to millisecond time domain can be monitored by measurement of chlorophyll fluorescence. Following a saturating actinic flash the chlorophyll fluorescence rises as a result of almost complete reduction of Q_A. The fluorescence decay kinetics in the microsecond to millisecond time domain primarily reflect electron transfer from Q_A^- to Q_B, or, under certain conditions, a back-reaction with the donor side of PSII. P_{680}^+ is also a fluorescence quencher (Butler, 1980a). Following a saturating flash, about 90% of P_{680}^+ is re-reduced in $<1\,\mu s$ by Tyr_z (Brettel et al., 1984), reflecting the equilibrium constant between P_{680}^+ and Tyr_z. (Tyr_z is tyrosine-161 of the D1 protein of PSII, the immediate electron donor Z to P_{680}^+.) The remainder is reduced with a $t_{1/2}$ of around 50 µs as Tyr_z^+ is re-reduced by the water-splitting complex (Ford and Evans, 1985; Schlodder et al., 1985). In the presence of DCMU, which blocks Q_A^- oxidation on a millisecond time-scale, the microsecond kinetics of P_{680}^+ re-reduction can be monitored by following the increase of chlorophyll fluorescence resulting from the disappearance of the P_{680}^+ quencher.

A. Flash Excitation Fluorescence Apparatus

The principle of this device is that a (normally) saturating actinic flash is used to reduce Q_A to Q_A^-, and a weak measuring flash, fired at specified times before and after the actinic flash, is used to sample the chlorophyll fluorescence yield as a probe for the redox state of Q_A^-. By varying the time interval between the actinic and measuring flashes, a kinetic profile of Q_A^- can be determined. Such a method was first described by Joliot (1974), and has since been developed by A. R. Crofts and H. Robinson, whom we would like to thank for providing detailed descriptions of their apparatus. The merits of using a detecting flash rather than continuous illumination have been discussed by Mathis (Chapter 8, this volume) in the context of spectrophotometry. The use of a detecting flash provides very good signal/noise at a few time points because a relatively high intensity pulse can be used to obtain data with minimum actinic effect, whereas a weaker exciting light must be used for continuous illumination to avoid this. The alternative is to use a pulsed excitation source with a high modulation frequency but at a sufficiently low intensity to cause negligible actinic effect. Such a device is described in Section VI.A. The detecting flash apparatus is particularly useful when measurements over extended time ranges are required. The measuring flash is designed to sample about 1% of the reaction centres. The time resolution of the device depends on the emission profiles of the flash sources. Using xenon flash bulbs, with pulse widths at half maximal intensity of 2–3 µs, the minimum practical time between the actinic and measuring flashes is $\sim 40\,\mu s$, owing to the tail of the actinic flash.

The minimum time interval between successive firing of the measuring flash is ~ 3 ms, which means that for rapid kinetic transients, each time point must be obtained from a fresh sample excited by the actinic flash. For slower kinetics, of course, all measurements can be made on the same sample. To facilitate the provision of fresh, dark-adapted samples, a large reservoir of chloroplasts is connected to a flow cuvette in such a way that material can be transferred to and from the cuvette by application of gas pressure under the control of solenoid valves.

1. Instrumental requirements

Instrument function, which is controlled by a microcomputer, involves precise timing of firing the flash lamps, acquisition of fluorescence intensities and conversion to digital form for storage in the computer's memory, transfer of fresh dark-adapted material from a reservoir to the flow cuvette, operation of a reservoir stirrer and electronic shutter, and a printer/plotter for data output. A block diagram of the apparatus is shown in Fig. 9.1.

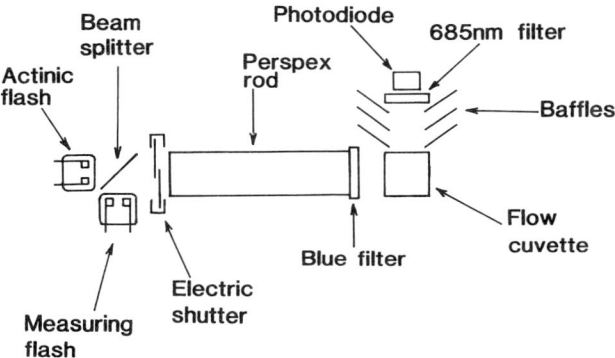

FIG. 9.1. Schematic representation of the flash excitation fluorimeter designed to measure changes in chlorophyll fluorescence yield over the 10^{-5} to $>10^2$ s time domain, based on the design of H. Robinson and A. R. Crofts. A more detailed account of the apparatus is given in the text.

2. Notes on components

A number of alternatives exist for most of the components in the fluorescence device. Specific components used by Robinson and Crofts are given here.

The main requirement of the computer is that it should be one to which lab-built boards can be added. The measuring and actinic flashes are provided by FX201 bulbs, recently superseded by FX802, with the appropriate trigger packs (EG & G, Boston, MA, USA) giving flashes with 2 µs width at half maximal intensity. As an alternative, the measuring flash can be provided by a Stroboslave 1539-A (GenRad, Concord, MA, USA). The beam-splitter (microscope slide) is positioned to give an identical illumination geometry from both flash lamps, while greatly attenuating the measuring flash. The position of the measuring flash is adjusted to excite $< 1\%$ of the PSII centres. The actinic flash energy is selected to excite $\sim 95\%$ of the centres. This avoids reduction of the Q_2 fluorescence quencher which does not appear to be on the normal path of electron transfer to Q_B (Joliot and Joliot, 1979). Fine-tuning of the illumination intensity of both lamps can be achieved by adjusting the voltage of the discharge capacitor using a voltage-controlled d.c./d.c. convertor (PS-350, EG & G, Boston, MA, USA). The fluorescence is detected by a photodiode or photomultiplier (EMI 9558 or EMI 9698) screened by a 685 nm interference filter (10 nm bandwidth). The actinic and measuring flashes are filtered by a blue filter (CS4-96, Corning) to eliminate wavelengths longer than 600 nm. The light is delivered to the cuvette by a polished perspex rod 1–3 ft

(30–90 cm) long, both to provide a more uniform light intensity at the sample and to physically remove the flashlamp trigger circuitry from the photodetector.

The accurate measurement of fluorescence yield requires uniform measuring flash intensity. To reduce variation in flash intensity, the lamp should ideally be fired at a fixed frequency, and should also be fired a few times before opening the electronic shutter (Vincent Associates, Rochester, NY, USA). Because firing the lamp at a fixed frequency is not always compatible with experimental design, an alternative procedure is to measure the intensity of each detecting flash using a second photodiode coupled to an analogue-to-digital (A/D) convertor. Each fluorescence yield measurement can then be normalised to the same excitation intensity.

The integrator should be capable of integrating the output from the detector over a 3 µs period, during the peak output from the detecting flash. The A/D should be 8 or 12 bit capable of conversion in 25 µs.

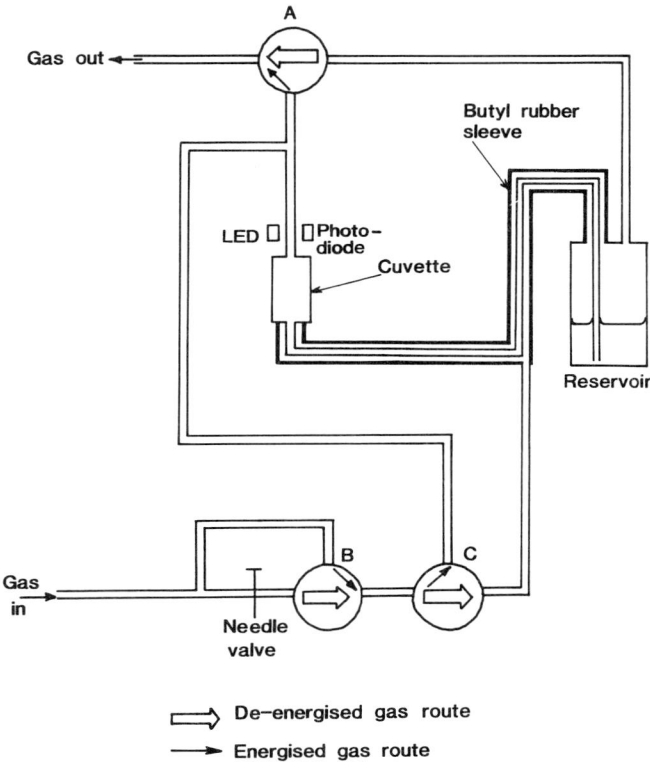

FIG. 9.2. Schematic representation of the gas-pressure driven system for pumping suspensions of photosynthetic material from a darkened reservoir into a flow cuvette for fluorescence measurements on dark-adapted material, based on the design of H. Robinson and A. R. Crofts. The gas used is normally high purity nitrogen or argon. A, B and C are 12 V three-way solenoid valves. All three valves are de-energised to flush nitrogen through the reservoir (particularly important when carrying out experiments at a controlled redox potential). To transfer suspension from the reservoir into the flow cuvette, valves A and B (by-passing the needle valve) are energised until the meniscus interrupts the light beam passing from the LED to the photodiode situated on opposite sides of the glass tube connected to the exit part of the flow cuvette. To empty the cuvette, valves B and C are energised for a specified period of time.

Dark-adapted material is transferred from a reservoir (100 ml) to the flow cuvette (0.6 ml) and back by alternate application of nitrogen or argon gas pressure to the reservoir and the gas space above the liquid in the flow cuvette (Fig. 9.2). The reservoir is stirred by a magnetic stirrer, but this is turned off during measurements.

This device was designed for measurements on chloroplast suspensions or intact algae, but a similar apparatus may be constructed for measurements on leaves (Robinson, 1986).

3. Preparation of chloroplasts

A more or less standard method for preparing chloroplasts suitable for chlorophyll fluorescence measurements to monitor Q_A^- kinetics is given below (modified from Robinson and Crofts, 1983), although many minor variations are reported in the literature.

50 g of washed, chilled leaf material is ground for 10 s at full speed in a blender in 100 ml of 0.4 M sucrose, 20 mM Hepes (pH 7.6), 15 mM NaCl, 5 mM $MgCl_2$, 2 mM EDTA and 2 mg ml^{-1} bovine serum albumin at 4°C. All subsequent operations are carried out at 4°C. The suspension is filtered through at least 12 layers of cheesecloth, centrifuged for 10 min at $2000 \times g$, and resuspended using a soft paint brush in 0.4 M sucrose, 20 mM Hepes (pH 7.6), 15 mM NaCl. To oxidise any Q_B^- which is present in freshly prepared dark-adapted chloroplasts, while avoiding oxidation of the Fe(II) in the reaction centre (Bowes et al., 1979), the suspension is diluted in 0.1 M sucrose, 20 mM Hepes (pH 7.6), 20 mM KCl, 5 mM $MgCl_2$ and 50 µM ferricyanide, to a final concentration of 0.15 mg chlorophyll ml^{-1}. Chlorophyll concentration is measured using the method of Arnon (1949). The suspension is then stirred slowly on ice in absolute darkness for 1 h, centrifuged for 10 min at $2000 \times g$, and resuspended in 0.4 M sucrose, 20 mM Hepes (pH 7.6), 15 mM NaCl at a chlorophyll concentration of 1–3 mg ml^{-1}. Typically, this would be diluted to 5 µg chlorophyll ml^{-1} in a dark stirred reservoir vessel at 22°C containing 100 ml of 0.1 M sucrose, 20 mM Hepes (pH 7.6), 10 mM KCl, 5 mM $MgCl_2$, 100 µM methyl viologen as electron acceptor, with 0.1 µM valinomycin and 0.1 µM nigericin to prevent any build-up of a protonmotive force across the thylakoid membrane, which itself quenches chlorophyll fluorescence (Briantais et al., 1980; Krause et al., 1982, and Section II.C.1). As an alternative to the use of ferricyanide to oxidise residual Q_B^-, 20 µM benzoquinone may be included which will oxidise Q_B^- over a seconds time-scale (Lavergne, 1982a). The material is allowed to dark adapt for at least 10 min in the reservoir before measurements are commenced.

B. Kinetics of Q_A^- oxidation

The simplest application of the device is to measure the kinetics of Q_A^- oxidation after the first, second and subsequent saturating actinic flashes from the dark-adapted state in which the chloroplasts have been pre-treated with ferricyanide or benzoquinone to oxidise the residual Q_B^- (Bowes and Crofts, 1980; Bowes et al., 1980; Robinson and Crofts, 1983, 1984; and Fig. 9.3). The reactions occurring are as follows:

$$Q_A + PQ \text{ pool} \rightleftharpoons Q_A Q_B \xrightarrow{h\nu_1} Q_A^- Q_B \rightleftharpoons Q_A Q_B^- \xrightarrow{h\nu_2} Q_A^- Q_B^- \rightleftharpoons$$

$$Q_A Q_B^{2-} + 2H^+ \rightleftharpoons Q_A PQH_2 \rightleftharpoons Q_A + PQH_2 \text{ pool}$$

After correction has been made for the non-linear relationship between $F(t) - F_o$ and $[Q_A^-]$ (see Section V.C.2), it is found that the kinetics of oxidation of Q_A^- after a single flash, in which any residual Q_B^- has been pre-oxidised, are biphasic. The slower component is thought to reflect the binding of plastoquinone to the Q_B site (Taoka, 1989).

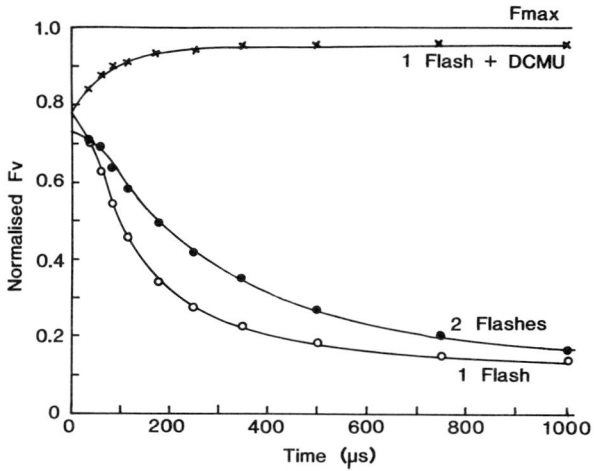

FIG. 9.3. Decay of variable fluorescence in pea thylakoids after one or two saturating flashes in the presence or absence of DCMU. Normalised F_v is $F - F_o/F_m - F_o$, where F_o is the fluorescence level from dark-adapted thylakoids and F_m is the fluorescence level after several actinic flashes in the presence of DCMU (10 μM). Reproduced with minor modification from Robinson and Crofts (1987), with permission.

C. Binding of Inhibitors to the Q_B Site

If the kinetics of chlorophyll fluorescence changes are monitored at saturating concentration of Q_B site inhibitor on a submillisecond time-scale, there is a rise in fluorescence, associated with a slow phase of P_{680}^+ reduction, followed by a steady fluorescence which decays on a seconds time-scale (Fig. 9.3). If, however, the kinetics of Q_A^- oxidation are measured in the presence of a subsaturating concentration of an inhibitor of Q_A^- oxidation, the kinetics are strongly biphasic. There is a rapid component showing a half-time similar to uninhibited centres ($t_{1/2}$ 150–200 μs) and a slow component ($t_{1/2} \sim 10$ ms) reflecting the fraction of inhibited centres. The displacement of inhibitor by plastoquinone to enable formation of $Q_A \, Q_B^-$ in a fraction of centres can be followed by observing the loss of fluorescence in the slow phase. In these experiments, hydroxylamine (30 μM) is added to rapidly and essentially irreversibly reduce components on the donor side of PSII (Bennoun and Joliot, 1969). This inhibits the back-reaction of electrons from Q_A^- to the donor side, which would otherwise compete with forward oxidation by Q_B. At higher concentrations (5–10 mM), NH_2OH irreversibly destroys the native donor complex. This effect of NH_2OH results in a quenching of chlorophyll fluorescence, and elimination of the modulation of the fluorescence level by the different 'S' states of the water-splitting complex (Delosme and Wurmser, 1971).

The degree of inhibition by a particular concentration of inhibitor may be measured by determining the chlorophyll fluorescence level at 200 μs after an actinic flash (from the dark-adapted state) in control thylakoids, in the presence of the inhibitor, and in the presence of 10 μM DCMU which would be expected to completely block Q_A^- oxidation on this time-scale. Assuming that the differences in amplitude of fluorescence between the inhibited and control states are proportional to the amount of Q_A^-, and that the intermediate fluorescence level F_i associated with blocked PSII centres is given by $F_i = F_{(200\,\mu s)} - F_o$, then the fraction of inhibited centres is given by

$$\frac{(F_i/F_o)_I - (F_i/F_o)_{CON}}{(F_i/F_o)_{DCMU} - (F_i/F_o)_{CON}} \tag{9.5}$$

(I, inhibitor present; CON, control; DCMU, 10 μM DCMU present). The I_{50} is then the concentration of inhibitor required to give a fractional inhibition of 0.5.

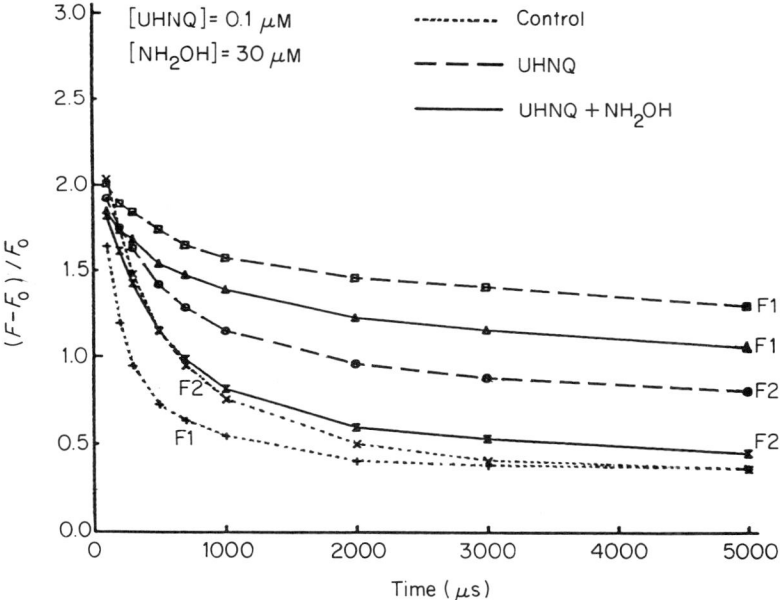

FIG. 9.4. Use of the flash fluorimeter to study the effect of 3-undecyl hydroxynaphthoquinone (UHNQ) on the kinetics of Q_A^- oxidation in the presence and absence of hydroxylamine, after one (F1) or two (F2) flashes (F2 given 1s after F1). The control rates show the normally observed faster rate of oxidation of Q_A^- formed on the second flash than on the first. In the presence of UHNQ, Q_A^- oxidation is markedly inhibited after one flash, but after two flashes, the degree of inhibition is diminished. This reflects displacement of UHNQ from its binding site by a molecule of plastoquinone from the pool, and the consequent oxidation of Q_A^-. Thus, by the time the second flash is delivered, the Q_B site is occupied by plastosemiquinone (Q_B^-), which binds tightly and prevents binding of UHNQ. The effect is seen more clearly in the presence of NH_2OH, which prevents oxidation of Q_A^- by the back reaction. Reproduced with minor modification from Taoka et al. (1983), with permission.

The flash fluorimeter may also be used to study the effect of the redox state of the Q_A Q_B acceptor complex on herbicide binding. This is illustrated in Fig. 9.4 which shows the effect of introduction of an electron into the Q_A Q_B complex on the binding of 3-undecyl-2-hydroxy-1,4-naphthoquinone (UHNQ). Note that quinones also quench chlorophyll fluorescence; this can be corrected by measuring F_{max} in the presence of DCMU and the quinone. The loss of inhibition of Q_A^- oxidation by UHNQ after two flashes is not observed with DCMU. This reflects the fact that the rate constant for dissociation of UHNQ from the Q_B site is much higher than that for strongly binding inhibitors such as DCMU. Such inhibitors only dissociate from the Q_B site on a minutes time-scale, and this can be followed by measuring the slow decline of fluorescence associated with Q_A^- oxidation as Q_B sites occupied by DCMU are transiently vacated to be occupied by plastoquinone (Taoka, 1989).

V. MEASUREMENT OF FLUORESCENCE CHANGES IN THE MILLISECOND TO SECOND TIME DOMAIN

Measurement of chlorophyll fluorescence has been particularly important in following and identifying changes in the photosynthetic system upon transition from dark to light on a millisecond time-scale and beyond. The major processes that give rise to non-photochemical quenching of fluorescence change with time constants between seconds and minutes. Thus, an extremely powerful and important measurement of fluorescence is that made in the ms–s time domain when non-photochemical processes are constant. Generally, samples are dark-adapted to ensure all centres are in the open state and then the light-induced rise in fluorescence is recorded as the centres are gradually converted to the closed state.

A. Apparatus

Chlorophyll fluorescence in the millisecond to second time domain can be measured very simply. For example, a d.c. light source (e.g. a tungsten halogen lamp of 100–250 W, driven by a regulated power supply), blocked by a blue-green filter (e.g. Corning 4-96 or Schott BG18) can be used to excite fluorescence which is detected through an appropriate red long-pass filter (e.g. Schott RG665 or Corning 2-64). Overlap between these filters is reduced by further defining the blue excitation and/or red emission using additional filters. Alternatively, a red excitation can be used with fluorescence detected in the far-red. Red excitation sources are light-emitting diodes (LEDs), or lamp sources filtered with broad band filters (e.g. Schott RG610 and 680 short-pass interference filters). Fluorescence can be detected with a red-sensitive photomultiplier tube (e.g. EMI9558) or a less expensive photodiode. Excitation and emission are measured at 90° in a 1 cm^2 cuvette or alternatively 'front face', most easily using a branched fibre optic. The latter arrangement is suitable for assay of leaves as well as liquid phase samples. The exciting light not only excites chlorophyll fluorescence, but also causes turnover of the photochemical reactions. In the milliseconds to seconds time domain, Q_A^- initially accumulates in the reduced state due to over-reduction of components in the electron transport chain between PSI and PSII. The resulting changes in fluorescence are usually recorded using a shutter to initiate illumination rapidly and a transient recorder or equivalent data storage device for capturing the data. The fluorescence versus time profiles in such an experiment are referred to as fluorescence induction transients. An apparatus to measure such transients is available commercially from Hansatech Ltd., and a lab-made device is described in Paterson and Arntzen (1982). This kind of

B. Kinetics in the Absence of DCMU

Using a photographic shutter to initiate illumination, the rise of fluorescence follows a distinctive pattern (Fig. 9.5). First the F_o level is observed, provided the ratio between the light intensity and the opening time of the shutter is appropriate (it is necessary to ensure that the fluorescence rise from F_o is slower than the shutter speed). Following this, in intact systems or *in vitro* in the absence of added electron acceptor, a rise to an F_i level is followed by a sigmoidal rise to a peak, F_p (Forbush and Kok, 1968). In strong light, in which Q_A becomes nearly completely reduced, F_p may be equivalent to F_m, but the final extent of Q_A reduction, and therefore F_p, is determined by the ratio of intercepted irradiance and electron flux out of the plastoquinone pool. *In vivo*, this depends on the capacity for $NADP^+$ reduction.

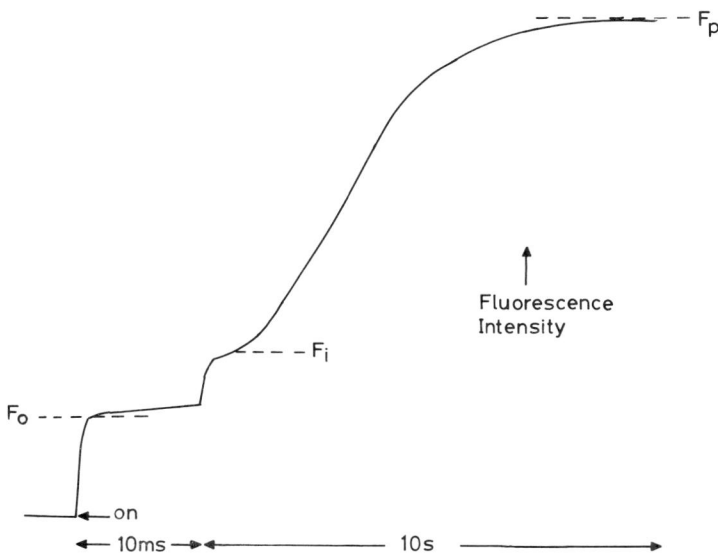

FIG. 9.5. Kinetics of fluorescence induction upon illumination of a dark-adapted barley leaf. A dark-adapted leaf segment was illuminated in air. The rise from F_o to F_p shows a characteristic plateau F_i, often followed by a dip before a sigmoidal rise to F_p. Note the use of a split time base to resolve F_o and F_p.

The rise from F_o to F_i is thought to represent the reduction of Q_A in a population of PSII centres not connected to the plastoquinone pool (Lavergne, 1974, 1982b; Melis, 1985). The F_i level can be maintained in low light intensity because these centres are slow to re-oxidise or, in the case of thylakoids, at higher light intensities if potassium ferricyanide is added to keep the plastoquinone pool oxidised (Melis, 1985). The rise to F_p reflects the reduction of the remaining Q_A, which in this kind of steady-state measurement at relatively low light does not occur until the plastoquinone pool is reduced. In fact the area above the induction curve can be shown to be a measure of the number of quanta used photochemically by the PSII centres (Malkin and Kok, 1966)—

hence the area above the curve from F_o to F_p has been used as a measure of the pool of electron acceptors available to Q_A.

The kinetics of rise from F_i to F_p have been used in various types of ecophysiological studies (Hetherington and Smillie, 1983). The rise time is determined by the efficiency of electron transfer by PSII and is particularly sensitive to inhibition of electron donation from the O_2-evolving complex. For example, the rise time is a very good indicator of stress-induced damage caused by chilling (Hetherington and Smillie, 1983). The technique has also been used to make comparative assays of sizes of the electron acceptor pools in sun and shade plants (Chow and Anderson, 1987; Schreiber et al., 1989).

It has been noted that the induction kinetics in strong light do not conform to the simple notion of Q_A reduction. Thus, as the light intensity is raised, a saturation pattern is reached (Schreiber and Neubauer, 1987). Whilst the rate of the rise from F_o to a level I_1 can be shown to be 'photochemical' in being determined by light intensity, two further phases from I_1 to I_2 and I_2 to F_p were 'thermal' (i.e. rate independent of light intensity). Explanation for this phenomenon presently remains obscure and explanations ranging from the saturation of a PSII cyclic pathway to the presence of PSI fluorescence have been implicated. Of importance, however, is that at I_1, Q_A is fully reduced (Schreiber et al., 1989). I_1 to I_2 correlates with the reduction of the plastoquinone pool and I_2 to F_p with the reduction of P_{700}, the primary electron donor of PSI.

C. Kinetics in the Presence of Inhibitors of Q_A^- Oxidation

The quantitative aspects of the fluorescence rise kinetics have been exploited extensively when electron transport is blocked with inhibitors which block Q_A^- oxidation by binding to the Q_B site (e.g. DCMU) and which therefore block electron transfer out of the reaction centre to the plastoquinone pool. In such conditions only a single electron transfer to Q_A is generally possible upon illumination. The rate of rise from F_o to F_m (since all traps will become closed) is now a measure only of the rate of capture of incident excitation. The area above the fluorescence induction curve is again equal to the number of quanta used photochemically in the closure of the PSII centres, but because only one charge separation is possible per centre, the area above the induction curve now becomes proportional to the number of reaction centres (Melis and Homann, 1976). The fluorescence induction kinetics are complex, however, and it was shown that a more rapid sigmoidal phase could be separated from a slower exponential phase (Fig. 9.6). It can be shown that, since the total area above the curve represents the number of quanta used to drive the fluorescence from F_o to F_m, that each area segment

$$A = \int [F_m - F_{(t)}] \, dt$$

is equal to the number of charge separations during time interval $t_2 - t_1$. Thus the kinetics of area growth were shown to depend on light intensity, i.e. it is a measure of the rate of light utilisation. The principal uses of this technique have been in the investigation of PSII heterogeneity and in describing the transfer of energy between PSII centres. Measurement of fluorescence induction transients on the millisecond timescale can also be used to study the potency of inhibitors of Q_A^- oxidation, both in intact leaves and in isolated chloroplasts. This is of particular interest because many inhibitors of Q_A^- oxidation are herbicidal through their effect at this site. Chlorophyll fluorescence measurements on intact leaves therefore provide a means of studying herbicidal effects at the primary target site non-invasively (Renger and Schreiber, 1986, and references therein).

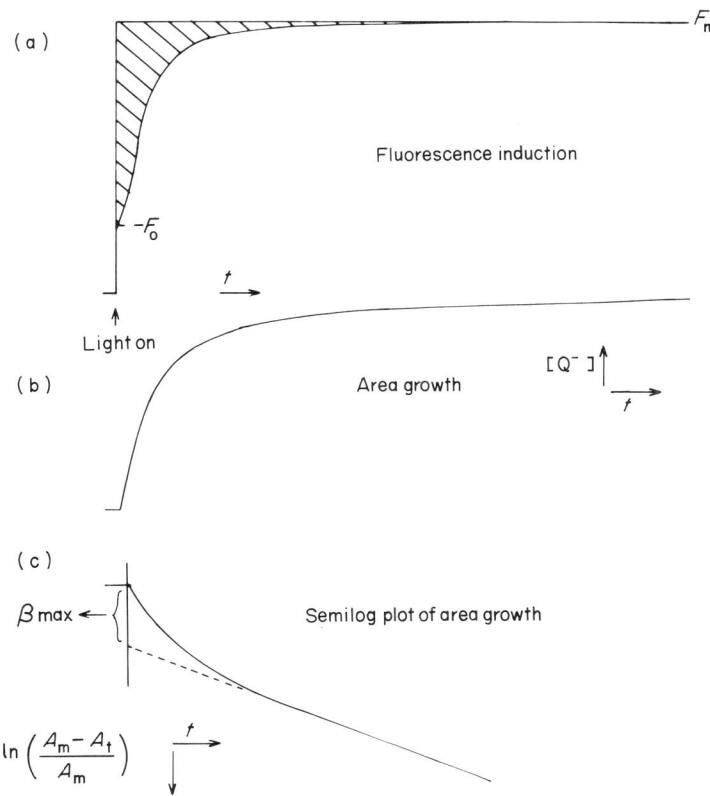

FIG. 9.6. Fluorescence induction in the presence of DCMU. Pea thylakoid membranes illuminated in the presence of 10 μM DCMU. (a) Kinetics of F_v rise. (b) Kinetics of area growth. (c) Plot of logarithm of area growth illustrating how the proportion of $PSII_\beta$ β_{max} is calculated.

1. *The detection of $PSII_\alpha/PSII_\beta$ heterogeneity*

The biphasic kinetics of fluorescence induction have been interpreted as indicating the presence of two types of PSII centres, differing primarily in their antennae sizes. $PSII_\alpha$ accounts for the sigmoidal rise in fluorescence and $PSII_\beta$ the exponential phase. The proportion of the area accounting for $PSII_\beta$ can be obtained by plotting the logarithm of the area growth and determining the *y*-axis intercept at time zero (Fig. 9.6). Assuming complete blockage by DCMU, routinely this gives a value of around 0.3 in chloroplasts and indicates that this proportion of PSII is in the β form. The rate constants K_α and K_β calculated in this way are measures of the relative antenna sizes of $PSII_\alpha$ and $PSII_\beta$ (Melis, 1989). The differences in the proportion of $PSII_\alpha$ and $PSII_\beta$ in plants grown under different light regimes, in mutants deficient in LHC-II and in different chloroplast types have been used to illustrate the flexibility in the composition and organisation of the thylakoid membrane.

2. Connectivity between PSII centres

The sigmoidal kinetics of the α phase have been explained by energy transfer between PSII reaction centres. The larger antenna size of $PSII_\alpha$ is thought to allow such transfer, whereas this is not possible for $PSII_\beta$ which therefore exist as separate 'packages'. Energy transfer between PSII centres was first examined by Joliot and Joliot (1964) and later by Paillotin (1976). The probability of energy transfer is most easily assessed by considering the relationship between F_v and the extent of closure of reaction centres as determined by the area growth. The plot of F_v against $1 - Q_A$ is hyperbolic if there is energy transfer. The analysis is complicated by the α/β heterogeneity described above and it is best to subtract the β phase before constructing the plot. According to Joliot and Joliot (1964) the relationship between fluorescence and Q_A is given by the equation:

$$\frac{F - F_o}{F_v} = \frac{(1-p)(1-A)}{1-p(1-A)} \tag{9.6}$$

where p is a parameter associated with the probability of energy transfer between PSII centres, A is the fraction of open centres and F_v is equal to $F_m - F_o$. In fact, p is related to the ratio F_v/F_m such that $p = k\,F_v/F_m$, where k is a connection parameter. Butler (1980b) approached this in a more satisfactory manner using a slightly more complex and probably more accurate model of energy transfer at PSII. Thus it was recognised that, for a separate package model

$$\frac{F - F_o}{F_v} = 1 - A \tag{9.7}$$

For a matrix model, in which all centres can be considered to share the same antennae:

$$\frac{F - F_o}{F_v} = \frac{1 - A}{1 + (F_v/F_o)\,A} \tag{9.8}$$

Butler suggested that the organisation of the thylakoid membrane was most likely to be explained by a connected package model, in which a relatively small number of reaction centres share the same antenna. For this:

$$\frac{F - F_o}{F_v} = \frac{1 - A}{1 + (F_v/F_o)\,\psi T_{(22)}\,A} \tag{9.9}$$

where $\psi T_{(22)}$ describes the probability of energy transfer between PSII units when all are open.

3. Experimental problems

The quantitative analysis of the induction curve in the presence of DCMU is only reliable if several important precautions are taken. First, the sample must be evenly illuminated, requiring dilute chlorophyll, a measuring beam of low absorbtivity (i.e.

green light) and a cuvette in which there are no obscured areas at the bottom or edges. It is possible that many slow phases of fluorescence induction reported in the literature are entirely artifactual. Second, determining the true F_m is particularly important if an accurate estimate of $PSII_\beta$ is to be obtained. Third, it is often the case that the fitting of just two components seems unsatisfactory and it has been suggested that three or more components are required. In summary, like many fluorescence measurements, the ease by which data can be collected should not disguise the need for very rigorous design of apparatus and experiments. It is also important, if possible, to check results obtained against other assays of the reaction centre of PSII; e.g. the α/β heterogeneity first described only in terms of fluorescence assay was given greater credibility when the rate of reduction of Q_A as measured from both the A_{320} (absorption change at 320 nm attributed directly to formation of Q_A^-) and C_{550} (absorption change at 550 nm attributed to an electrochromic bandshift in the absorption of a phaeophytin molecule located close to Q_A, in response to its reduction to Q_A^-) absorption signals were found to be identical to the rate of area growth (Melis and Schreiber, 1979).

4. Determination of the potency of inhibitors of Q_A^- oxidation

At subsaturating concentrations of inhibitor, when the sample is first illuminated, there is a rapid rise in fluorescence from F_o to an intermediate level F_i in centres blocked by the inhibitor, followed by a more gradual rise in non-blocked centres (Fig. 3 in Paterson and Arntzen, 1982). When all centres have been blocked, the fluorescence rises rapidly to the maximum level, F_m. The fraction of centres which are non-inhibited can be calculated using the formula

$$\frac{(F_m - F_o) - (F_i - F_o)}{F_m - F_o}$$

This value is plotted against herbicide concentration, and the I_{50} (concentration of herbicide required to give 50% inhibition) determined from the plot (Govindjee et al., 1973; Paterson and Arntzen, 1982).

This measurement can be refined to take into account the non-linear relationship between $F - F_o$ and the fraction of closed centres using Eq. 9.6. The value of p, normally around 0.55, is obtained by comparing the fluorescence induction curve with curves calculated using Eq. 9.6 with varying values of p, applied to a $[Q_A^-]$ versus time profile measured under the same conditions. This is normally obtained directly from the fluorescence induction curve because the area over the fluorescence induction curve at a particular time point is linearly related to $[Q_A^-]$ (Malkin and Kok, 1966; Murata et al., 1966; Lavorel et al., 1986; van Gorkom, 1986). It should also be borne in mind that Q_A^- oxidation is blocked in a fraction of PSII centres even in the absence of inhibitors (see Section V.B and references therein). These centres bind inhibitors (Graan and Ort, 1986) but will contribute to F_i irrespective of whether an inhibitor is bound.

D. Measurements using the Flash Excitation Fluorescence Apparatus

The apparatus described in Section IV.A is well-suited for certain types of measurements on the $Q_A Q_B$ acceptor complex in the millisecond to second time domain, as outlined below.

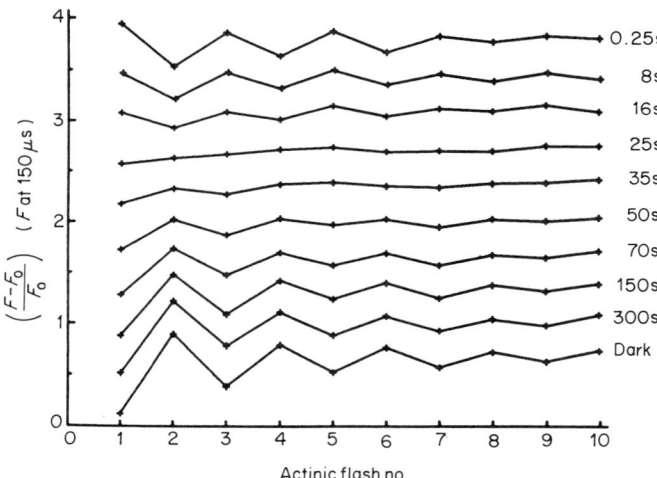

FIG. 9.7. Rephasing of the binary oscillation in chlorophyll fluorescence level after each of a series of saturating flashes (250 ms between flashes), as a function of the time in the dark after one saturating flash. The dark time intervals are listed on the right of the figure. The graphs have been offset for clearer presentation. Reproduced from Robinson and Crofts (1983), with permission.

1. Measurement of the equilibrium constant between Q_A and Q_B

Measurement of the pH dependence of the equilibrium constant for sharing a single electron between Q_A and Q_B provides a useful probe for the thermodynamics of reduction and protonation of plastoquinone at the Q_B site in wild-type chloroplasts and in mutants with alterations in the D1 protein. The method is described in Taoka (1989) and Robinson and Crofts (1983, 1984). It relies on measurement of the recombination rate of Q_A^- (in the presence of DCMU) and of Q_B^- with the S_2 state of the water-splitting complex. The recombination of Q_A^- with S_2 in the presence of DCMU is measured by monitoring the decay of the high fluorescence state (half-time around 1.5 s). This method cannot be applied to Q_B^- because its redox state does not directly affect the chlorophyll fluorescence yield. If a train of flashes is given to dark-adapted chloroplasts in the absence of DCMU at a frequency of 4 Hz, and the fluorescence level monitored 150 µs after each flash, the level shows a damped binary oscillation, low after the first flash. This results from the more rapid oxidation of Q_A^- by Q_B than by Q_B^-. If a single pre-flash is given, generating Q_B^-, followed by a train of flashes starting at varying times after the pre-flash, the concentration of Q_B^- at the start of the flash train will decrease with increasing time after the pre-flash as it recombines with S_2. This will result in a rephasing of the binary oscillation in fluorescence during the subsequent flash train (Fig. 9.7). Measurement of the decay of the oscillation as a function of dark time between the pre-flash and the flash train reveals a half-time for a first-order recombination process between Q_B^- and S_2 of 22 s for ~80% of the centres. (The extent of oscillation is measured from the sum of the differences of the measured fluorescence in each sequence, compared to a range established by the difference between similar sums for the Q_B fully reduced (0.25 s dark time) starting state and Q_B fully oxidised (300 s dark time).) Assuming that Q_B^- back-reacts through Q_A, and that DCMU does not modify the properties of Q_A^-, the ratio of the back-reaction half-times minus 1 gives an apparent equilibrium constant of 14–15 for sharing an electron between Q_A and Q_B (Robinson and Crofts, 1983; see also Wraight, 1979).

$$\frac{t_{1/2} \text{ (slow)}}{t_{1/2} \text{ (fast)}} = 1 + K_{1, \text{app}} \tag{9.10}$$

This apparent equilibrium constant includes the binding constant for plastoquinone to the Q_B site.

$$Q_A^- + \text{plastoquinone} \underset{}{\overset{K_o}{\rightleftharpoons}} Q_A^- Q_B \underset{}{\overset{K_1}{\rightleftharpoons}} Q_A Q_B^-$$

One way of estimating K_o is by measuring the fraction of centres in which Q_A^- is rapidly oxidised by Q_B in a first-order process ($t_{1/2}$ 150 μs). It is assumed that this reflects centres in which the Q_B site is occupied by plastoquinone when the flash is given, whereas the slower reaction is rate-limited by the binding of plastoquinone to the Q_B site. Between 60 and 70% of Q_A^- is oxidised rapidly, giving a K_o value of 500 M^{-1}.

$$K_1 = \frac{[Q_A Q_B^-]}{[Q_A^- Q_B]}$$

$$= \frac{[Q_A Q_B^-]}{([Q_A^-] [\text{plastoquinone}] . K_o)} \tag{9.11}$$

$$K_{1, \text{app}} = \frac{[Q_A Q_B^-]}{([Q_A^-] + [Q_A^- Q_B])}$$

$$\therefore \quad K_1 = \frac{K_{1, \text{app}} (1 + [\text{plastoquinone}] K_o)}{[\text{plastoquinone}] K_o} \tag{9.12}$$

The concentration of plastoquinone in the lipid bilayer has been estimated as 5 mM, giving, for a K_o of 500 M^{-1} and $K_{1, \text{app}}$ of 15 at pH 7.6, a value for K_1 of 21.

2. *Measurement of the rate of herbicide binding to the Q_B site*

A further development of the flash fluorimeter is to couple the apparatus to a stopped flow system to measure the kinetics of herbicide binding after rapid mixing of the herbicide with thylakoids. A complete description of the apparatus and protocols is given in Taoka (1989), but a brief description will be given here to illustrate the usefulness of the method.

A sample of chloroplast suspension is mixed rapidly with buffer containing the herbicide at the required concentration. The extent of herbicide binding to PSII at various times (t) after mixing is determined by firing an actinic flash at the sample at time (t), followed by a weak measuring flash given 10 ms later. It is assumed that after 10 ms, Q_A will be reduced only in centres with herbicide bound. Back-reaction of Q_A^- with the donor side is inhibited by including 5 mM NH_2OH. The time course of herbicide binding is obtained by repeating the experiment with varying times (t) between mixing and the actinic flash. The F_o level is measured by sampling the fluorescence level after rapid mixing using a weak measuring flash without the actinic flash. Fractional Q_A^- data is obtained by correcting for the non-linear relationship between $(F(t)-F_o)/F_o$ and $[Q_A^-]$ using a probability of intersystem crossing of 0.55 (Joliot and Joliot, 1964).

To measure the kinetics of herbicide binding to the Q_B^- state, the dark-adapted chloroplast suspension is subjected to a single saturating actinic flash before mixing with the herbicide solution. The fraction of centres with herbicide bound at time (t) after mixing is determined by measuring the fractional concentration of Q_A^- by sampling F using a weak measuring flash. It is assumed that Q_A^- will be present in centres in which herbicide has bound, arising from the back-reaction of Q_B^- with Q_A (Velthuys, 1982; Vermaas et al., 1984), which is not expected to limit the rate of herbicide binding.

VI. MEASUREMENT OF LONG-TERM QUENCHING OF CHLOROPHYLL FLUORESCENCE

When applied with longer time-scales the kinds of fluorescence measurement described in Sections V.A–C using d.c. measuring light become less satisfactory for several reasons. First, because the excitation light has to be filtered in d.c. measurement, it is often difficult to obtain high enough intensities to drive electron transport fast enough for physiological work. Second, by trying to use as high an intensity as possible it becomes difficult to resolve the F_o level because significant rise to F_i and F_p occurs within the shutter opening time. Third, and most important, the fluorescence signal obtained is ambiguous since photochemical and non-photochemical parameters are now affecting the yield.

A. Apparatus

For these reasons, most measurements are now made using a modulated fluorimeter. In this technique, the excitation source is modulated (either mechanically or electronically) and the modulated fluorescence measured using an amplifier designed to reject signals which do not have that modulation frequency. This technique was used, for example, in the classic work of Bonaventura and Myers (1969) during the discovery of state transitions. It was first used to tackle the problem of separating q_P and q_N by Quick and Horton (1984). Since then, two commercially available devices have been marketed which measure modulated fluorescence: the Hansatech Modulated Fluorimeter (Ogren and Baker, 1985) and the Walz Chlorophyll Fluorimeter (Schreiber et al., 1986). The former uses an array of yellow LEDs (modulated at 100 Hz) to excite fluorescence which is detected by a photodiode in conjunction with a two-stage lock-in amplifier. The Walz fluorimeter uses 1 μs light pulses from a single red LED, at a frequency of either 1.6 kHz or 100 kHz, and fluorescence is again detected by a photodiode whose output is transmitted to a Pulse Amplitude Modulation (PAM) amplifier for signal discrimination and measurement. The high modulation frequency of this device means that measurements can also be made in the submillisecond time domain, providing an alternative to the flash fluorescence device (Section IV) for measuring Q_A^- oxidation kinetics in the absence and presence of inhibitors (Schreiber, 1986). A particular advantage of the modulation technique is that the excitation or 'measuring beam' intensity can be set to very low levels (i.e. $<1\ \mu E\ m^{-2}\ s^{-1}$) such that PSII centres remain open and the F_o level is detected and maintained—this level has sometimes been called the 'dark' level of fluorescence. The Walz Fluorimeter is particularly useful in this regard since it can automatically switch between 1.6 kHz modulation for F_o measurement to 100 kHz during actinic illumination to achieve a higher signal-to-noise ratio. The Walz Fluorimeter has found wide application in the measurement of fluorescence from whole leaves or leaf discs. Its use in quenching analysis (Section VI.B) is facilitated by collecting and manipulating data in a microcomputer. Software written for this purpose is now available (e.g. Walz DA100; McAuley and Scholes, University of Sheffield).

Important in the investigation of the response of the photosynthetic system to changing conditions and in understanding the processes of fluorescence quenching has been the development of apparatus that allows simultaneous assays of fluorescence, gas exchange and other optical signals that probe the function of the thylakoid membrane (Fig. 9.8). Apparatus for the assay of fluorescence, O_2 evolution and 9-aminoacridine fluorescence has been described (Horton, 1983b; Oxborough and Horton, 1988). A cuvette with vertical optics designed by Schreiber and available from Walz can be used to make simultaneous assay of O_2 evolution, P_{700} (the primary electron donor of PSI) and chlorophyll fluorescence. Active preparations of intact chloroplasts and protoplasts (Leegood and Walker, 1985) have been particularly useful in exploring the basis for the chlorophyll fluorescence signals *in vivo*. For leaves, simultaneous measurement of CO_2 uptake, chlorophyll fluorescence and P_{700} has been described (Harbinson *et al.*, 1989). The leaf disc O_2 electrode designed by Delieu and Walker (1983) can be readily adapted to make fluorescence measurements with the Walz Fluorimeter and by using a six branch fibre optic it is possible to add the A_{820} measurement of P_{700}. With these systems it is now possible to obtain a detailed description of thylakoid function both *in vitro* and *in vivo*. A particularly important use of these simultaneous measurements is given in Section VII.

B. Separation of Photochemical and Non-photochemical Quenching Components using Modulated Fluorescence

The origin of the technique of 'quenching analysis' was the work of Bradbury and Baker (1981), who showed that if a leaf illuminated to generate a quenched state is given a pulse of strong light, any PSII centres open prior to the pulse will become closed. The normalised increase in yield could then be used to indicate the amount of fluorescence being quenched photochemically, the difference between the initial F_m and the pulse-induced F_m' representing non-photochemical quenching. This early work used two d.c. sources, the pulse-induced fluorescence being recorded on a transient recorder; this

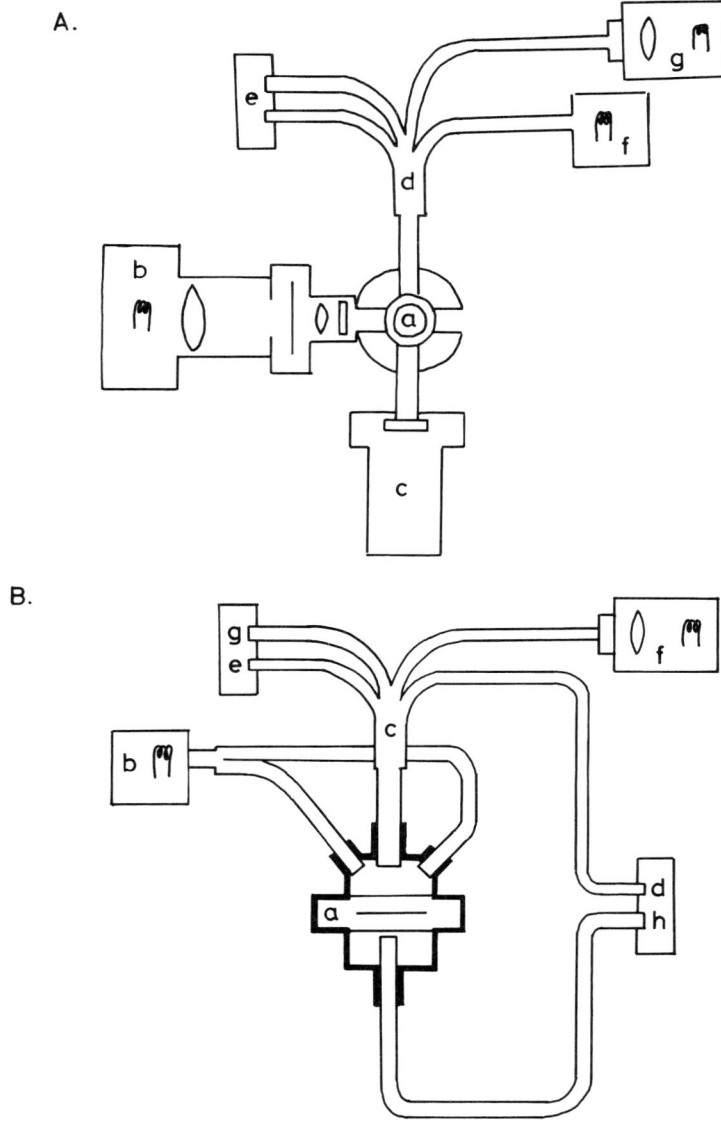

FIG. 9.8. (A) Schematic of apparatus for simultaneous assay of modulated chlorophyll fluorescence, 9-aminoacridine fluorescence and O_2 concentration. (a) Hansatech liquid phase O_2 electrode chamber. (b) Excitation source for 9-aminoacridine fluorescence comprising d.c. tungsten–halogen source, light chopper and filters (Ealing 420 nm short-pass interference filter, Corning 5-58 and 7-39 glass filters). (c) Photomultiplier tube for detection of 9-aminoacridine fluorescence blocked by 460 nm narrow band, 450 nm wide band and 480 nm short-pass interference filters, the signal being detected by a lockin amplifier. (d) Four-branched Walz fibre optic leading to emitter–detector unit (e) for chlorophyll fluorescence and tungsten–halogen sources for actinic (f) and shuttered saturation pulse illumination (g). The vacant port can be used to provide a saturating pulse of higher intensity or a supplementary far-red light to promote Q_A^- oxidation in the dark.

(B) Apparatus for simultaneous assay of A_{820}, modulated chlorophyll fluorescence and IRGA (infrared gas analysis). (a) ADC gas exchange chamber, modified to accept fibre optic illumination, (b) actinic light from a branched fibre from a Schott light source, (c) Four-branched Walz fibre optic with inputs from (d) 820 nm emitter, (e) fluorescence excitation emitter and (f) light saturation pulse; (g) fluorescence detector and (h) fibre optic collecting signal for 820 nm detector. This apparatus requires two Walz PAM 101 Units and a Balzers Calflex X Filter in front of the fluorescence detector to prevent detection of 820 nm light (Schreiber *et al.*, 1988).

technique therefore still suffered from the rest of the problems described above for d.c. measurement. A more convenient and powerful approach to separating q_P and q_N arose from the use of modulated fluorescence (Quick and Horton, 1984)—here the pulse can be of extremely high intensity (up to 10 000 µE m^{-2} s^{-1}), it need not be chromatically defined, and there is the convenience of the fluorescence rise immediately being seen as a 'spike' on the fluorimeter readout essentially already normalised without any need for computation or for a separate data recorder. The validation of the technique was made possible by an earlier discovery by Krause and co-workers (1982) that DCMU, rapidly introduced into a sample *in vitro*, causes PSII trap closure within 1 s and a consequent removal of photochemical quenching. It was demonstrated in experiments using protoplasts that q_P measured by DCMU addition and by light pulsing were very similar (Quick and Horton, 1984).

Using the Walz or Hansatech Fluorimeter it is now possible to obtain a rigorous quantitation of fluorescence quenching. The terms q_P and q_N have clear numerical definitions defined above and are calculated from the fluorescence trace as shown in Fig. 9.9.

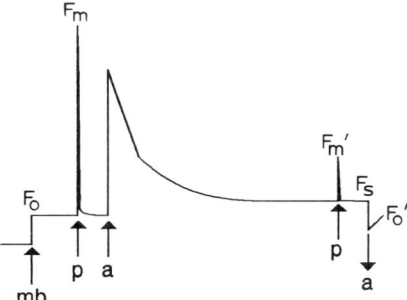

FIG. 9.9. Procedure and parameters of importance in analyses of quenching. mb, measuring beam; a, actinic light; p, saturation pulse. This fluorescence trace was obtained from a dark-adapted barley leaf strip using a Walz chlorophyll fluorimeter in combination with a Hansatech LD2 leaf disc oxygen electrode.

C. Experimental Problems and Precautions

1. Characteristics of the pulse

(a) *Pulse duration.* This should be as short as possible, within the limit determined by the longest time constant in the measuring system. Obviously the pulse has to be long enough to deliver sufficient energy to close all the PSII traps. Typical pulse characteristics are 1 s duration with an intensity of 5000 µE m^{-2} s^{-1}. The pulse can be defined by the automatic switching given by the Walz PAM103 unit together with the FL103 light source. Alternatively an electronic shutter can be used with any appropriate timing device. Recently, Schreiber *et al.* (1989) have described the advantages of using a long xenon flash of 50 ms duration (Walz XST 103).

(b) *Pulse intensity.* This has to be high enough for saturation of PSII traps under conditions of maximum electron transport. The latter point is particularly important since, for example, a more intense pulse is required to saturate Q_A in uncoupled thylakoid electron transfer than in intact chloroplasts reducing CO_2. Similarly, the pulse intensity required to saturate a given sample may change with time; a dark-adapted sample with no q_N is easier to saturate than an illuminated sample with high q_N. Conversely, it is inadvisable to use a vastly excessive pulse intensity since the pulse itself is not non-intrusive (see Section (c) below). Unusual fluorescence responses can also be induced when excessively high pulse intensities are used which may result from changes in non-photochemical quenching during the pulse. There is still considerable debate about how to determine whether the pulse is saturating and how best to ensure maximum accuracy of q_P determination. It is best to adopt a purely empirical approach of testing whether intensity changes give significant differences in the ΔF observed and how possible errors in q_P would alter the conclusions being made from the measurement.

(c) *Pulse interval.* To determine quenching parameters at steady-state only a single pulse is necessary. To obtain kinetic information about changes in photosynthetic parameters during transients, pulses need to be applied repetitively. However, within a 1 s pulse significant amounts of electron transfer occur. This has two important consequences. First, the observed steady-state electron transfer rate will be increased, perhaps significantly if the pulse interval is too short. Second, the pulse gives rise to Q_A reduction, H^+ translocation, ΔpH build-up and q_E quenching, which need a finite length of time to relax. The pulse interval, again, has to be chosen according to conditions and requirements. In the original work of Quick and Horton (1984), the criterion used was to obtain identical fluorescence kinetics with and without pulses, but obviously this is not always possible. The same considerations are as important here as in Section (b) above in that an interval appropriate for one condition may not be applicable to another.

2. *Measuring F_o in the quenched state*

It was first observed by Bilger and Schreiber (1987) that q_N was associated with quenching of F_o; apart from the intrinsic interest in terms of what this might say about the mechanism of quenching, this of course presents a problem in the calculation of q_P and q_N. In a highly quenched state in high light q_P calculated on the basis of the initial F_o will be nonsensical. There is a problem, however, in determining the F_o in the quenched state (F_o'). Upon turning off the actinic light, Q_A^- oxidation seems to take at least several seconds. During this time q_N relaxes, and the 'dark' fluorescence can show complex kinetics. Imposition of far-red light can speed up the Q_A^- oxidation. However, it should also be stated that complete accuracy of F_o' is probably impossible, but it is also probable that the 85–90% re-oxidation of Q_A^- that occurs within 1 s after darkening is sufficiently accurate for most purposes. Indeed the slowly relaxing component of Q_A^- is likely to belong to the non-Q_B centres that in any case do not participate in electron transfer (Section V.B). Under stress or unusual experimental conditions, extra care has to be taken since Q_A^- re-oxidation may be significantly slowed.

3. Measurement of F_m and F_o

A second problem of standardisation of parameters is how to define the F_m and F_o values. Since q_N affects both parameters, F_m and F_o should be determined when q_N is completely relaxed. However, it is not possible to know when q_N is zero, particularly *in vivo*. Residual thylakoid energisation may persist in darkness and it is also likely that the dark state is somewhere between state 1 and 2. *In vitro*, F_m and F_o can be reliably measured, but since F_v/F_m ratios of chloroplasts are invariably closer to 0.75 than the *in vivo* norm of 0.83, there is perhaps some persistent q_N developed during isolation. There can be no absolute condition or experimental protocol here. A standard period of dark adaptation is obviously a necessity, but its length has to be determined according to the particular kind of experiment being done.

D. Identification of Non-photochemical Quenching Components

It was shown in experiments on chloroplasts that q_N was heterogeneous; DCMU and uncouplers do not relax all q_N. The fact that the kinetics of relaxation of fluorescence in darkness following illumination are multiphasic also suggested that several of the processes described in Section II.C could be contributing to the observed non-photochemical quenching. It therefore became important to find methods for separating q_N into its component parts.

1. Use of specific inhibitors to resolve non-photochemical quenching

In isolated chloroplasts and protoplasts addition of inhibitors can be used to induce relaxation of q_N. Thus, when DCMU is added, the inhibition of electron flow leads to collapse of ΔpH and q_E (Krause *et al.*, 1982). In isolated chloroplasts simultaneous measurement of quenching of chlorophyll and 9-aminoacridine fluorescence is possible, allowing the behaviour of q_E and ΔpH to be compared. Uncouplers such as nigericin or NH_4Cl can also be added to collapse ΔpH and determine the extent of q_E. It was observed in chloroplasts (Horton, 1983b) and protoplasts (Quick and Horton, 1984) that dissipation of ΔpH does not lead to complete relaxation of q_N. This 'residual' quenching was examined in detail in isolated protoplasts and could be shown to comprise two components (Horton and Hague, 1988). One component, formed at low light intensity, relaxed with a $t_{1/2}$ of ~ 5 min after DCMU addition. This relaxation could be completely blocked by addition of NaF, a well-known inhibitor of protein phosphatases. It was suggested that this NaF-sensitive slow relaxation of quenching was a state 2 to state 1 transition, the reversal of the transition that had occurred during illumination.

It is important to use uncouplers to test that q_E has relaxed—this may be particularly important in algae, which appear to be able to maintain q_E in darkness for extended periods or after DCMU addition (Lee *et al.*, 1990). With careful use of DCMU, NaF and nigericin it is possible to resolve q_N into q_E, q_T and a third component that relaxes only partially and very slowly—this has been designated q_I. As discussed above, this quenching is difficult to resolve into that due to photodamage and that due to protective quenching in the antenna.

2. Use of dark relaxation kinetics to resolve non-photochemical quenching

It has been observed in protoplasts (Horton and Hague, 1988; Horton *et al.*, 1989) and leaves (Demmig and Winter, 1988; Quick and Stitt, 1989; Walters and Horton, 1990) that relaxation of q_N upon darkening is multiphasic. A three phase decay is normally observed and it is probable that *in some cases* these represent q_E, q_T and q_I. Thus, in protoplasts and leaves NaF inhibits the middle phase, whereas far-red light brought about stimulation of the rate of relaxation. Similarly, low intensity blue-green light given to preferentially excite PSII in order to induce a state 1 to state 2 transition gave a quenching that relaxed monophasically and had the same kinetics as the middle-phase observed after high light. Infiltration of leaves with low concentrations of nigericin (an uncoupler to reduce the ΔpH) or tentoxin (an ATP synthase inhibitor to increase the ΔpH) respectively caused a diminution or an enhancement of the amplitude of the fast phase (Quick and Stitt, 1989).

However, there are a number of problems with using this technique to resolve q_N. First, it is possible that ΔpH can be maintained in darkness following an extended period of illumination. Metabolite pools take some time to relax back to dark levels and during this period the chloroplast could stay partially energised. A clear example has been found in the green alga *Dunaliella*, which has a q_E component that persists for several minutes in darkness (Lee *et al.*, 1990). In leaves, the magnitude of the middle phase, which is significantly larger than previously measured amplitudes of fluorescence change due to a state transition *in vitro*, suggests that it is unlikely to be totally due to q_T.

Second, relaxation kinetics can only be measured if saturating pulses are given repetitively over a period of time. As discussed above, the pulses are perturbing the photosynthetic system, an effect which is particularly evident in the absence of actinic light. The dependency of relaxation kinetics upon the pulse frequency has been investigated by Quick and Stitt (1989). In leaves, a pulse frequency any greater than 1 pulse per 100 s causes substantial quenching. This seems to be because the middle q_T component of q_N saturates at low light intensities. Thus, too high a pulse frequency will perhaps prevent relaxation of all but the q_E component. Of even greater danger is that the pulse regime employed during relaxation may cause or re-establish a q_N component not present during steady-state illumination.

E. Use of q_N and q_P as Probes of Metabolic Activity

The problems of identifying q_E components mean that q_N *in vivo*, at present, cannot be conclusively split into the known processes that will cause quenching. This is particularly a problem in work which attempts to use q_N as an indicator of the 'energy status' of the leaf, as a probe of metabolic processes. Moreover, even if q_E could be 'extracted' from q_N it is clear, as discussed in Section II.C.1, that q_E is not a simple indicator of ΔpH. Nevertheless, observation of changes in q_N *in vivo* are often strongly suggestive of changes in the ATP/ADP ratio in the chloroplast (e.g. Walker *et al.*, 1983) and direct assay of metabolites, whenever made, has confirmed these predictions (Furbank and Foyer, 1986). Similarly, experiments performed on isolated chloroplasts and protoplasts (Horton, 1983b,c; Quick and Horton, 1984; Horton *et al.*, 1984; Furbank *et al.*, 1987) have provided a substantial underpinning of the interpretation of the behaviour of q_N *in vivo* with respect to induction (Walker, 1981), oscillatory behaviour (Walker *et al.*, 1983) and the effects of phosphate limitation (Sivak and Walker, 1986).

The interpretation of the complex kinetics of fluorescence induction observed in intact systems has been discussed in detail previously (Horton, 1985; Sivak and Walker, 1985; Baker and Horton, 1987). Considerable confusion can arise from a need to find a system of terminology to describe the induction kinetics. The PSMT nomenclature of earlier work (Papageorgiou, 1975) breaks down when there are multiple transients or when superficially similar transients arise from fundamentally different causes. Quenching analysis employed together with simultaneous assay of O_2 evolution or CO_2 uptake can distinguish between different kinds of transients. Illumination of most intact systems results in the classical PSM transient that is associated with a burst in the rate of O_2 evolution; PS involves mainly q_N, whereas SM is due to re-reduction of oxidised Q_A^-. The PSM transient is usually complete within 30 s, before there has been much q_N formation. The quenching from M to T is associated with the acceleration of CO_2-dependent O_2 evolution and comprises increases in both q_N and q_P; under some circumstances q_N can develop early in the induction period and then decrease as photosynthesis increases, perhaps causing a discontinuity in the MT quenching. The MT phase can be interrupted by dramatic transients in fluorescence yield which can take the form of dampening oscillations (Walker *et al.*, 1983). In dark-adapted leaves or protoplasts there is a single transient, with q_N, q_P and the rate of photosynthesis changing in parallel with a small phase shift between each (Horton and Quick, 1984). In the past, confusion has arisen from the assumption that q_P is a consistent measure of the rate of electron transfer; recent work has identified control mechanisms in the thylakoid which have the effect of stabilising the redox state of PSII against changes due either to increased light intensity or to decreased turnover of NADPH and ATP (see Section VII). A recent treatment of oscillatory behaviour by Peterson *et al.* (1988) took into account this regulation.

VII. USE OF QUENCHING ANALYSIS TO DETERMINE QUANTUM EFFICIENCY

One of the developing uses of fluorescence measurement was initiated by Weis and Berry (1988). This work has its origins in the early photochemical work that lead to the definition of Q and the description of the relationship between F_v and photochemical rate. The important parameter here is q_P. Theory predicts that q_P is a linear indicator of Φ_s, the rate of electron transport through PSII divided by the light intensity. This relationship has been demonstrated very clearly experimentally and in Fig. 9.10 data obtained using isolated PSII particles are shown. *In vivo*, however, q_P and Φ_s are not linearly related. The extent of this deviation from linearity is variable and is observed whether O_2 evolution or CO_2 uptake is measured. *In vivo* either high CO_2 or low O_2 should be used since photorespiration would inevitably lead to a breakdown of the q_P versus Φ_s relationship.

The decline in Φ_s as the light intensity increases has of course been interpreted in terms of closing of PSII centres. The non-linear relationship between q_P and Φ_s means that the decline in Φ_s is *not* only due to the decline in q_P—in other words even the *open*

PSII centres are becoming *intrinsically* less efficient. Weis and Berry introduced the term Φ_p, the ratio Φ_s/q_P, to measure the intrinsic yield of PSII. Essentially this term corrects for the proportion of closed centres and at its limit with $q_P = 1$, $\Phi_p = \Phi_s$, equal to what we define as the quantum yield of photosynthetic electron transport.

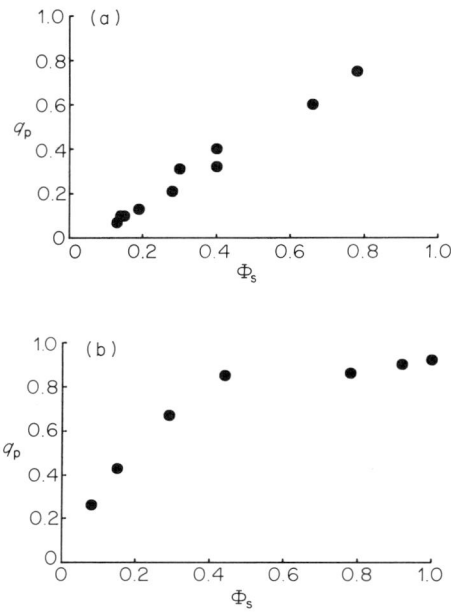

FIG. 9.10. Relationship between q_p and Φ_s in (a) isolated PSII particles and (b) cells of the green alga *Dunaliella* at a range of light intensities. Measurements were made in an apparatus similar to that described in Fig. 9.8(A), Φ_s being the rate of net oxygen evolution divided by incident light intensity.

The intrinsic yield Φ_p declines at high light or upon CO_2 decrease or temperature decrease. Attempts have been made to try to uncover the mechanism of Φ_p decline using fluorescence analysis. Weis and Berry (1988) noticed a linear relationship between Φ_p and q_N that led them to propose a mechanism for PSII regulation involving interconversion between an active unquenched state to a quenched inactive state of the reaction centre. The linearity of this relationship has been questioned and more recent experimentation has suggested a hyperbolic relationship with $\Phi_p = 0$ at $q_N = 1$ (Weis and Lechtenberg, 1989). A typical set of data obtained from a barley leaf is shown in Fig. 9.11. The data suggest that when $q_N > 0.8$, Φ_p declines more strongly, as predicted from a model in which q_N is due to antenna quenching (Horton et al., 1990). Nevertheless, between q_N of 0.3 and 0.7, linearity is observed and could be used as a way of employing q_P and q_N values to measure electron transport rate *in vivo* without recourse to conventional gas exchange. Thus, using a calibration curve of the type shown in Fig. 9.11 a number of studies have used fluorescence to study electron transport under different gas phase compositions and light intensity (e.g. Sharkey et al., 1989). These give useful insights but are based on the assumptions that the Φ_p–q_N relationship is fixed under all conditions. This assumption is flawed, firstly because Φ_p can decrease as a result of mechanisms other than q_N; any electron transfer pathway that leads to oxidation of Q_A but not net O_2 evolution will give rise to Φ_p change. O_2 consumption by a Mehler reaction or a cyclic flow of electrons in PSII would cause Φ_p to decrease. Secondly, as discussed above, q_N is heterogeneous and these components would affect Φ_p in different ways.

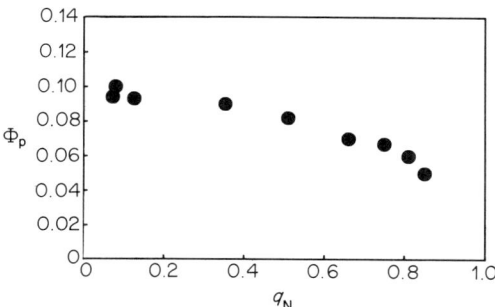

FIG. 9.11. Relationship between the intrinsic yield of PSII (Φ_p) and the non-photochemical quenching of chlorophyll fluorescence (q_N) obtained from barley leaves illuminated with a range of light intensities in a leaf disc electrode.

Whilst the use of Φ_p has been a useful one, a study by Genty et al. (1989) approached the discrepancy between q_P and Φ_s from first principles, suggesting that the theoretical relationship between Φ_p and F_v/F_m should persist even under conditions when q_N causes a F_v/F_m decline. This was convincingly confirmed under several experimental conditions. A plot of F_v'/F_m' against measured Φ_p is shown in Fig. 9.12. A linear relationship extrapolating to the origin was observed. This relationship has several important implications. First, linearity between F_v'/F_m' and Φ_p only occurs if there is a decline in the rate constant for photochemistry (k_p) giving a F_o rise or, for quenching, if there is dissipation in the antenna chlorophyll; this suggests a mechanism for the increase in q_N and for the control of PSII efficiency which is fundamentally different from that put forward by Weis and Berry (1988). Second, this model suggests that the ratio between the amplitude of the fluorescence rise induced by the pulse (ΔF) and the F_m' is directly proportional to Φ_s, a prediction confirmed by experiment (see Fig. 9.13). This has great practical importance. To predict the electron flow rate from the Φ_p/q_N relationship requires measurement of q_N and q_P, whereas the measurement of $\Delta F/F_m'$ is a parameter simply obtained from amplitudes of fluorescence before and after a pulse—thus neither *the original F_m* (for which the dark-adapted state is required) nor *the steady state F_o* (which requires darkening following illumination) is required to use fluorescence to measure Φ_s. Thus Φ_s can be determined for any leaf under any conditions within a few seconds. However, again, this is based on the assumption that Φ_s and $\Delta F/F_m$ are always related in the same way. This has to be demonstrated rather than assumed.

In conclusion, if either the analysis of Weis and Berry or that of Genty et al. is to be used to predict electron flow *in vivo*, then it first has to be substantiated more than either has been so far. Nevertheless, the development of these analytical procedures has given rise to powerful approaches to the study of the regulation of photosynthesis. The ratio

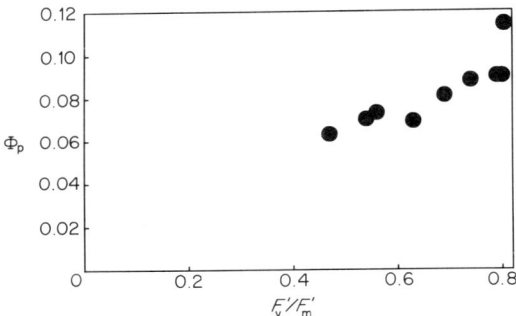

FIG. 9.12. Relationship between Φ_p and the ratio F_v'/F_m' recorded as in Fig. 9.11.

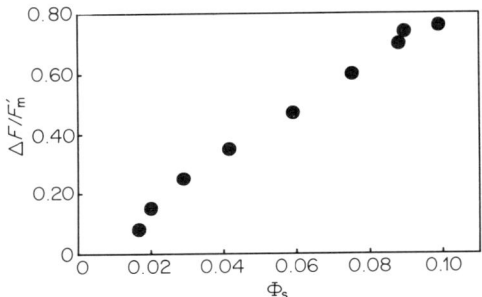

FIG. 9.13. Linear relationship between the ratio $\Delta F/F_m'$ and Φ_s for the data shown in Figs 9.11 and 9.12.

of Φ_p to F_v'/F_m' is particularly significant. Non-photochemical dissipation of excitation energy will affect Φ_p and F_v'/F_m' equally. However, a photochemical dissipation will lower Φ_p and not F_v'/F_m'. The features of this analysis have been developed using *in vitro* systems and have allowed clear demonstration of the presence of photochemical and non-photochemical dissipation bringing about the decline in Φ_p (Rees and Horton, 1990). It is possible, therefore, to compare predicted electron fluxes from either model with those measured by gas exchange, differences being a quantitative estimate of dissipative electron transfer. Moreover, simultaneous assay of P_{700} can then distinguish between a Mehler reaction and a dissipation in PSII; a constant ratio $\Delta F/F_m':P_{700}^+$ would indicate that electron flux through PSII and PSI is equal and that any deviation between Φ_p and F_v'/F_m' is then due to a Mehler reaction.

ACKNOWLEDGEMENTS

We wish to thank Debbie Rees for many helpful discussions and for critically reading the manuscript and Julie Scholes for giving us permission to use the data presented in

Figs 9.10 to 9.13. We are also indebted to Howard Robinson and Tony Crofts for providing unpublished data on their flash fluorescence device, and to Shinichi Taoka for providing a draft copy of his PhD Thesis (University of Illinois at Urbana-Champaign). This work was supported in part by grants to PH from the AFRC and SERC, and to JRB from the SERC.

REFERENCES

Akabori, K., Tsukamoto, H., Tsukihara, J., Nagatsuka, T., Motokawa, O. and Toyoshima, Y. (1988). *Biochim. Biophys. Acta* **932**, 345–357.
Arnon, D. I. (1949). *Plant Physiol.* **24**, 1–15.
Baker, N. R. and Horton, P. (1987). *In* "Photoinhibition" (D. J. Kyle, C. B. Osmond and C. J. Arntzen, eds), pp. 145–168. Elsevier, Amsterdam.
Barber, J. (1982). *Ann. Rev. Plant Physiol.* **33**, 261–295.
Bennoun, P. and Joliot, P. (1969). *Biochim. Biophys. Acta* **189**, 85–94.
Bilger, W. and Schreiber, U. (1987). *Photosynth. Res.* **10**, 303–308.
Bilger, W., Lange, O. L. and Schreiber, U. (1990). *In* "Plant Responses to Stress" (O. L. Lange and J. Tenhunen, eds), p. 18, Springer-Verlag, Berlin.
Bjorkman, O. and Demmig, B. (1987). *Planta* **170**, 489–504.
Black, M. T. and Horton, P. (1984). *Biochim. Biophys. Acta* **767**, 568–573.
Black, M. T., Brearley, T. H. and Horton, P. (1986). *Photosynth. Res.* **8**, 193–207.
Bonaventura, C. and Myers, J. (1969). *Biochim. Biophys. Acta* **189**, 366–383.
Bowes, J. and Crofts, A. R. (1980). *Biochim. Biophys. Acta* **590**, 373–384.
Bowes, J. M., Crofts, A. R. and Itoh, S. (1979). *Biochim. Biophys. Acta* **547**, 320–335.
Bowes, J., Crofts, A. R. and Arntzen, C. J. (1980). *Arch. Biochem. Biophys.* **200**, 303–308.
Bradbury, M. and Baker, N. R. (1981). *Biochim. Biophys. Acta* **635**, 542–551.
Bradbury, M. and Baker, N. R. (1986). *Plant Cell Environ.* **9**, 289–297.
Brettel, K., Schlodder, E. and Witt, H. T. (1984). *Biochim. Biophys. Acta* **766**, 403–415.
Briantais, J.-M., Vernotte, C. and Moya, I. (1972). *Biochim. Biophys. Acta* **325**, 530–538.
Briantais, J.-M., Vernotte, C., Picaud, M. and Krause, G. H. (1979). *Biochim. Biophys. Acta* **548**, 128–138.
Briantais, J.-M., Vernotte, C., Picaud, M. and Krause, G. H. (1980). *Biochim. Biophys. Acta* **591**, 198–202.
Briantais, J.-M., Vernotte, C., Krause, G. H. and Weis, E. (1986). *In* "Light Emission by Plants and Bacteria" (Govindjee, J. Amesz and D. C. Fork, eds), pp. 539–583, Academic Press, New York.
Butler, W. L. (1980a). *Proc. Natl. Acad. Sci. USA* **69**, 3420–3422.
Butler, W. L. (1980b). *Proc. Natl. Acad. Sci. USA* **77**, 4697–4701.
Chow, W.S. and Anderson, J. M. (1987). *Aust. J. Plant Physiol.* **14**, 9–19.
Clayton, R. K. (1980). "Photosynthesis: Physical Mechanisms and Chemical Patterns". Cambridge University Press, Cambridge.
Cleland, R. E., Melis, A. and Neale, P. J. (1987). *Photosynth. Res.* **9**, 79–88.
Delieu, T. and Walker, D. A. (1983). *Plant Physiol.* **73**, 534–541.
Delosme, R. and Wurmser, R. (1971). *C.R. Acad. Sci. Paris* **t272**, (Serie D), 2828–2831.
Demmig, B. and Bjorkman, O. (1987). *Planta* **171**, 171–184.
Demmig, B. and Winter, K. (1988). *Aust. J. Plant Physiol.* **15**, 151–162.
Demmig, B., Winter, K., Kruger, A. and Czygan, F. C. (1987). *Plant Physiol.* **84**, 218–224.
Demmig-Adams, B., Winter, K., Kruger, A. and Dzygan, F. C. (1989). *Plant Physiol.* **90**, 887–893.
Duysens, L. N. M. and Sweers, H. E. (1963). *In* "Studies on Microalgae and Photosynthetic Bacteria" (Jpn. Soc. Plant Physiol., eds), pp. 353–372. University of Tokyo Press, Tokyo.
Forbush, B. and Kok, B. (1968). *Biochim. Biophys. Acta* **162**, 243–253.
Ford, R. C. and Evans, M. C. W. (1985). *Biochim. Biophys. Acta* **807**, 1–9.

Furbank, R. T. and Foyer, C. H. (1986). *Arch. Biochem. Biophys.* **246**, 240–244.
Furbank, R. T., Foyer, C. H. and Walker, D. A. (1987). *Biochim. Biophys. Acta* **894**, 165–173.
Genty, B., Briantais, J.-M. and Baker, N. R. (1989). *Biochim. Biophys. Acta* **990**, 87–92.
Glick, R. E. and Melis, A. (1988). *Biochim. Biophys. Acta* **934**, 151–155.
Govindjee, and Wasielewski, M. R. (1989). *In* "Photosynthesis" (W. R. Briggs, ed.), pp. 71–103. Alan R. Liss Inc., New York.
Govindjee, Papageorgiou, G. and Rabinowitch, E. (1973). *In* "Practical Fluorescence" (G. G. Guilbault, ed.), pp. 543–575. Marcel Dekker, New York.
Graan, T. and Ort, D. R. (1986). *Biochim. Biophys. Acta* **852**, 320–330.
Harbinson, J., Genty, B. and Baker, N. R. (1989). *Plant Physiol.* **90**, 1029–1034.
Hetherington, S. E. and Smillie, R. M. (1983). *Plant Physiol.* **72**, 1043–1050.
Holzwarth, A. R. (1986). *Photochem. Photobiol.* **43**, 707–725.
Holzwarth, A. R., Wendler, J. and Haehnel, W. (1985). *Biochim. Biophys. Acta* **807**, 155–167.
Homann, P. (1969). *Plant Physiol.* **44**, 932–936.
Horton, P. (1983a). *FEBS Lett.* **152**, 47–52.
Horton, P. (1983b). *Proc. Roy. Soc. Lond. B* **217**, 405–416.
Horton, P. (1983c) *Biochim. Biophys. Acta* **724**, 404–410.
Horton, P. (1985). *In* "Photosynthetic Mechanisms and the Environment" (J. Barber and N. R. Baker, eds), pp. 135–187. Elsevier, Amsterdam.
Horton, P. (1987). *In* "Progress in Photosynthesis Research" (J. Biggins, ed.), Vol. 2, pp. 681–688. Martinus-Nijhoff, Dordrecht.
Horton, P. (1989). *In* "Photosynthesis" (W. R. Briggs, ed.), pp. 393–406. Alan R. Liss Inc., New York.
Horton, P. and Black, M. T. (1981). *Biochim. Biophys. Acta* **635**, 53–62.
Horton, P. and Black, M. T. (1983). *Biochim. Biophys. Acta* **722**, 47–52.
Horton, P. and Hague, A. (1988). *Biochim. Biophys. Acta* **896**, 332–338.
Horton, P., Allen, J. F., Black, M. T. and Bennett, J. (1981). *FEBS Lett.* **125**, 193–196.
Horton, P., Lee, P. and Anderson, S. (1984). *In* "Advances in Photosynthesis Research" (C. Sybesma, ed.), Vol. III, pp. 657–660. Dr. W. Junk, The Hague.
Horton, P., Crofts, J., Gordon, S., Oxborough, K., Rees, D. and Scholes, J. D. (1989). *Phil. Trans. Roy. Soc. Lond. B* **323**, 269–279.
Horton, P., Noctor, G. and Rees, D. (1990). *In* "Perspectives in Biochemical and Genetic Regulation of Photosynthesis" (I. Zelitch, ed.), pp. 145–158. Alan R. Liss Inc., New York.
Joliot, A. (1974). *In* "Proceedings of the Third International Congress on Photosynthesis" (M. Avron, ed.), Vol 1, pp. 315–322. Elsevier, Amsterdam.
Joliot, A. and Joliot, P. (1964). *C.R. Acad. Sci. Paris* **258**, 4622–4625.
Joliot, P. and Joliot, A. (1979). *Biochim. Biophys. Acta* **546**, 93–105.
Karukstis, K. K. and Sauer, K. (1983). *J. Cell. Biochem.* **23**, 131–158.
Keuper, H. J. K. and Sauer, K. (1989). *Photosynth. Res.* **20**, 85–103.
Klimov, V. V., Klevanik, A. V., Shuvalov, V. A. and Krasnovsky, A. A. (1977). *FEBS Lett.* **82**, 183–186.
Knorr, F. J. and Harris, J. M. (1981). *Anal. Chem.* **53**, 272–276.
Krause, G. H. (1973). *Biochim. Biophys. Acta* **292**, 715–728.
Krause, G. H. and Behrend, U. (1983). *Biochim. Biophys. Acta* **723**, 176–181.
Krause, G. H. and Behrend, U. (1986). *FEBS Lett.* **200**, 298–302.
Krause, G. H. and Weis, E. (1984). *Photosynth. Res.* **5**, 139–157.
Krause, G. H., Vernotte, C. and Briantais, J.-M. (1982). *Biochim. Biophys. Acta* **679**, 116–124.
Krause, G. H., Laasch, H. and Weis, E. (1988). *Plant Physiol. Biochem.* **26**, 445–452.
Lavergne, J. (1974). *Photochem. Photobiol.* **20**, 377–386.
Lavergne, J. (1982a). *Biochim. Biophys. Acta* **679**, 12–18.
Lavergne, J. (1982b). *Photobiochem. Photobiophys.* **3**, 273–285.
Lavorel, J., Breton, J. and Lutz, M. (1986). *In* "Light Emisssion by Plants and Bacteria" (Govindjee, J. Amesz, and D. C. Fork, eds), pp. 57–98. Academic Press, New York.
Lee, C. B., Rees, D. and Horton, P. (1990). *Photosynth. Res.* **24**, 167–174.
Leegood, R. C. and Walker, D. A. (1985). *In* "Techniques in Bioproductivity and Photosynthesis" (J. Coombs, D. O. Hall, S. P. Long and J. M. O. Scurlock, eds), 2nd edn, pp. 118–132. Pergamon Press, Oxford.

Malkin, R. (1986). *Photosynth. Res.* **10**, 197–201.
Malkin S. and Kok, B. (1966). *Biochim. Biophys. Acta* **126**, 413–432.
Melis, A. (1985). *Biochim. Biophys. Acta* **808**, 334–342.
Melis, A. (1989). *Phil. Trans. Roy. Soc. Lond. B* **323**, 397–409.
Melis, A. and Homann, P. (1976). *Photochem. Photobiol.* **23**, 343–350.
Melis, A. and Schreiber, U. (1979). *Biochim. Biophys. Acta* **547**, 47–57.
Mohanty, P. K., Zilinskas-Braun, B. and Govindjee (1972). *FEBS Lett.* **20**, 273–276.
Morrissey, P. J., Glick, R. E. and Melis, A. (1989). *Plant Cell Physiol.* **30**, 335–344.
Moya, I., Sebban, P. and Haehnel, W. (1986). *In* "Light Emission by Plants and Bacteria" (Govindjee, J. Amesz and D. C. Fork, eds), pp. 161–190. Academic Press, New York.
Murata, N. (1969). *Biochim. Biophys. Acta* **172**, 242–251.
Murata, N. and Sugahara, K. (1969). *Biochim. Biophys. Acta* **189**, 182–189.
Murata, N., Nishimura, M. and Takamiya, A. (1966). *Biochim. Biophys. Acta* **120**, 23–33.
Nanba, O. and Satoh, K. (1987). *Proc. Natl. Acad. Sci. USA* **84**, 109–112.
O'Connor, D. V. and Phillips, D. (1984). "Time-correlated Single Photon Counting", pp. 37–54. Academic Press, London.
Ogren, E. and Baker, N. R. (1985). *Plant Cell Environ.* **8**, 539–597.
Oxborough, K. and Horton, P. (1987). *Photosynth. Res.* **12**, 119–128.
Oxborough, K. and Horton, P. (1988). *Biochim. Biophys. Acta* **934**, 135–143.
Paillotin, G. (1976). *J. Theoret. Biol.* **58**, 237–252.
Papageorgiou, G. (1975). *In* "Bioenergetics of Photosynthesis" (Govindjee, ed.), pp. 320–371. Academic Press, New York.
Paterson, D. R. and Arntzen, C. J. (1982). *In* "Methods in Chloroplast Molecular Biology" (M. Edelman, R. B. Hallick and N.-H. Chua, eds), pp. 109–110. Elsevier Biomedical Press, Amsterdam.
Peterson, R. B., Sivak, M. N. and Walker, D. A. (1988). *Plant Physiol.* **88**, 158–163.
Quick, W. P. and Horton, P. (1984). *Proc. Roy. Soc. Lond. B* **220**, 371–382.
Quick, W. P. and Stitt, M. (1989). *Biochim. Biophys. Acta* **977**, 287–296.
Rees, D. and Horton, P. (1990). *Biochim. Biophys. Acta* **1016**, 219–227.
Rees, D., Young, A., Noctor, G., Britton, G. and Horton, P. (1989). *FEBS Lett.* **256**, 85–90.
Renger, G. and Schreiber, U. (1986). *In* "Light Emission by Plants and Bacteria" (Govindjee, J. Amesz and D. C. Fork, eds), pp. 587–619, Academic Press, New York.
Robinson, H. H. (1986). *In* "Advanced Agricultural Instrumentation: Design and Use" (W. G. Gensler, ed.), pp. 92–106, Martinus Nijhoff, Dordrecht.
Robinson, H. and Crofts, A. R. (1983). *FEBS Lett.* **153**, 221–226.
Robinson, H. H. and Crofts, A. R. (1984). *In* "Advances in Photosynthesis Research" (C. Sybesma, ed.), Vol. 1, pp. 477–480. Martinus Nijhoff/Dr. W. Junk, The Hague.
Robinson, H. H. and Crofts, A. R. (1987). *In* "Progress in Photosynthesis Research" (J. Biggins, ed.), Vol. II, pp. 429–432. Martinus Nijhoff, Dordrecht.
Sauer, K. (1989). *In* "Photosynthesis" (W. R. Briggs, ed.), pp. 105–122. Alan R. Liss Inc., New York.
Schatz, G. H., Brock, H. and Holzwarth, A. R. (1987). *Proc. Natl. Acad. Sci. USA* **84**, 8414–8418.
Schatz, G. H., Brock, H. and Holzwarth, A. R. (1988). *Biophys. J.* **54**, 397–405.
Schlodder, E. and Brettel, K. (1988). *Biochim. Biophys. Acta* **933**, 22–34.
Schlodder, E., Brettel, K. and Witt, H. T. (1985). *Biochim. Biophys. Acta* **808**, 123–131.
Schreiber, U. (1986). *Photosynth. Res.* **9**, 261–272.
Schreiber, U. and Berry, J. A. (1977). *Planta* **136**, 233–238.
Schreiber, U. and Neubauer, C. (1987). *Z. Naturforsch.* **42c**, 1255–1264.
Schreiber, U. and Neubauer, C. (1989). *FEBS Lett.* **258**, 339–342.
Schreiber, U., Schliwa, U. and Bilger, W. (1986). *Photosynth. Res.* **10**, 51–62.
Schreiber, U., Klughammer, C. and Neubauer, C. (1988). *Z. Naturforsch.* **43c**, 686–694.
Schreiber, U., Neubauer, C. and Klughammer, C. (1989). *Phil. Trans. Roy. Soc. Lond. B* **323**, 241–251.
Sharkey, T. D., Berry, J. A. and Sage, R. F. (1989). *Planta* **176**, 415–424.
Shipman, L. L. (1982). *In* "Photosynthesis" (Govindjee, ed.), Vol. 1, pp. 275–291. Academic Press, New York.
Sivak, M. N. and Walker, D. A. (1985). *Plant Cell Environ.* **6**, 439–488.

Sivak, M. N. and Walker, D. A. (1986). *New Phytol.* **102**, 499–512.
Taoka, S. (1989). In "Kinetics of Electron Transfer and Binding of Inhibitors in the Two Electron Gate of Chloroplasts". PhD Thesis, University of Illinois at Urbana-Champaign.
Taoka, S., Robinson, H. H. and Crofts, A. R. (1983). In "The Oxygen Evolving System of Photosynthesis" (Y. Inoue, A. R. Crofts, Govindjee, N. Murata, G. Renger and K. Satoh, eds), pp. 369–381. Academic Press, New York.
Telfer, A., Hodges, M., Millner, P. A. and Barber, J. (1984). *Biochim. Biophys. Acta* **766**, 554–562.
van Gorkom, H. (1986). In "Light Emission by Plants and Bacteria" (Govindjee, J. Amesz and D. C. Fork, eds), pp. 267–289. Academic Press, New York.
Velthuys, B. R. (1982). In "Function of Quinones in Energy Conserving Systems" (B. L. Trumpower, ed.), pp. 401–408. Academic Press, New York.
Vermaas, W. F. J., Dohnt, G. and Renger, G. (1984). *Biochim. Biophys. Acta* **765**, 74–83.
Vernotte, C., Briantais, J.-M., Armond, P. and Arntzen, C. J. (1975). *Plant Sci. Lett.* **4**, 115–123.
Vernotte, C., Etienne, A. and Briantais, J.-M. (1979). *Biochim. Biophys. Acta* **545**, 519–527.
Walker, D. A. (1981). *Planta* **153**, 273–278.
Walker, D. A. and Osmond, C. B. (1989). "New Vistas in Measurement of Photosynthesis". Cambridge University Press, Cambridge.
Walker, D. A., Sivak, M. N., Prinsley, R. T. and Cheesbrough, J. K. (1983). *Plant Physiol.* **73**, 542–549.
Walters, R. and Horton P. (1990). In "Proceedings of the 8th International Congress on Photosynthesis" (M. Baltscheffsky, ed.), Kluwer, Dordrecht (in press).
Weis, E. (1985). *Biochim. Biophys. Acta* **807**, 118–126.
Weis, E. and Berry, J. A. (1988). *Biochim. Biophys. Acta* **894**, 198–208.
Weis, E. and Lechtenberg, D. (1989). *Phil. Trans. Roy. Soc. Lond. B* **323**, 253–268.
Weis, E., Berry, J. A. and Ball, T. (1987). In "Progress in Photosynthesis Research" (J. Biggins, ed.), pp. 553–556. Martinus Nijhoff, Dordrecht.
Wendler, J. and Holzwarth, A. R. (1987). *Biophys. J.* **52**, 717–728.
Wraight, C. A. (1979). *Biochim. Biophys. Acta* **548**, 309–327.
Wraight, C. A. and Crofts, A. F. (1970). *Eur. J. Biochem.* **17**, 319–323.
Yamazaki, I., Mimuro, M., Tamai, N., Yamazaki, T. and Fujita, Y. (1985). *FEBS Lett.* **179**, 65–68.

ABBREVIATIONS

A_{820}, absorption at 820 nm; DCMU, 3-(3,4-dichlorophenyl)-1,1-dimethylurea; F, fluorescence intensity; F_i, intermediate fluorescence level in fluorescence induction curve associated with blocked PSII centres; F_m, maximum level of chlorophyll fluorescence when all PSII centres are closed; F_o, level of chlorophyll fluorescence when all PSII centres are open; F_p, maximum level of fluorescence in fluorescence induction curve in the absence of inhibitors of PSII electron transfer; F_v, difference between F_m and F_o; Φ_f, fluorescence yield of PSII; Φ_s, rate of electron transport through PSII divided by light intensity; I, incident light intensity *or* intermediate phaeophytin acceptor of PSII; LED, light-emitting diode; LHC-I, light-harvesting chlorophyll a/b protein complex serving PSI; LHC-II, light-harvesting chlorophyll a/b protein complex serving PSII; PSI, Photosystem I; PSII; Photosystem II; P_{680}, primary electron donor of PSII; P_{700}, primary electron donor of PSI; Q_A, primary quinone electron acceptor of PSII; Q_B, secondary quinone electron acceptor of PSII; q_E, coefficient for quenching of fluorescence by the "high energy state"—a component of q_N; q_I, coefficient for fluorescence quenching associated with light stress; q_N, coefficient for non-photochemical quenching of chlorophyll fluorescence; q_P, coefficient for quenching of chlorophyll fluorescence due to photochemical utilisation by PSII; q_T, coefficient for fluorescence quenching due to the state 1 to state 2 transition—a component of q_N; tyr_z, tyrosine-161 of the D1 protein of PSII, the immediate electron donor Z to P^+_{680}; UHNQ, 3-undecyl-2-hydroxy-1,4-naphthoquinone.

10 Structure and Dynamics of Plant Membranes

P. J. QUINN and W. P. WILLIAMS

Department of Biochemistry, King's College London, Campden Hill, London W8 7AH, UK

I.	Introduction	297
II.	Isolation of plant cell membranes	298
III.	Structure of membrane constituents	300
	A. Phase behaviour of plant membrane lipids	301
	B. Structure of membrane proteins	316
IV.	Membrane dynamics	320
	A. Membrane fluidity	320
	B. Phase separations	326
	C. Lipid–protein interactions	331
V.	Conclusions	334
VI.	References	334

I. INTRODUCTION

Membranes are cytological features observed in all living cells. They play an essential role in preserving a distinct identity of the cell compared with its surroundings. Plant cells are distinguished from animal cells in that they are bounded by a cell wall which constrains the boundary of the living cell represented by the plasma membrane. In addition to the plasma membrane, plant cells contain a system of intracellular membranes of varying complexity. In common with all eukaryotes, plant cells are characterized by the presence of a nucleus surrounded by a double-membrane nuclear envelope. Other subcellular membranes create a system of cavities and compartments within the cell which serve to regulate and organise the various functions which it performs. Membranes also represent the site of numerous enzyme reactions and the special environment created in and around the membranes is believed to be important in regulating activity of these reactions. Most of the proteins which form integral components of membranes appear to have a role in transporting solutes and ions across membranes or act as redox components in photosynthetic and oxidative processes. The role of the membrane in organising the multicomponent complexes concerned in these reactions is recognised but not well understood.

The current dogma concerning the structure of biological membranes is embodied in the fluid mosaic model of Singer and Nicolson (1972). According to this model, the membrane lipids are arranged in a bilayer configuration which serves as a matrix for the insertion and/or attachment of membrane proteins. The membranes are said to be fluid in the sense that the components are able to diffuse readily in the plane of the membrane but are constrained in their motion from one side of the membrane to the other. This model, since its formulation, has proved fairly durable but it is obvious that no single model is likely to encompass the structure of all cell membranes which probably differ, at least in detail, one from another. Some of the major questions that still remain to be resolved include the nature of the interaction between membrane lipids and intrinsic proteins and whether there is any functional significance in such interactions. The factors responsible for directing newly synthesised components of membranes to their ultimate location is just one of the unknown factors in membrane biogenesis and differentiation. Studies of plant membrane systems, particularly the chloroplast, have been helpful in understanding these questions. The dynamic nature of plant cell membranes has also to be understood in terms of the traffic of membrane vesicles within the cell. These events involve molecular segregation within the structure and fusion between the different membranes in a manner which appears to be highly specific and directed. Finally, the mechanisms which preserve membrane homeostasis appear to be intimately linked to the metabolic turnover of individual membrane constituents and seem to involve the need to have a bewilderingly complex assortment of molecular species of lipids within each of the major lipid classes. The way these mechanisms are controlled and even their objectives are, as yet, largely unexplained.

The application of biophysical methods has proved to be useful in approaching many of these questions. The isolation and characterisation of different plant cell membranes is naturally an essential prerequisite for such studies. Once isolated, membrane preparations or their constituents can then be subjected to detailed biochemical and biophysical studies.

II. ISOLATION OF PLANT CELL MEMBRANES

As a first step in the characterisation of plant cell membranes, it is necessary to free the structures from other cellular constituents. In the case of plant cells, the cell wall represents a major obstacle in preparing a so-called homogenate of the plant tissue. Other complications arise because plant organs and tissues often contain more than one cell type, each performing particular functions, and this is presumably reflected to some extent in the type of constituents present in their respective membranes. In order to study the properties of the different cell membranes it is necessary to free them from within the cellular structures in a condition that hopefully preserves their *in vivo* composition and functional integrity.

There are a number of methods commonly employed to destroy the integrity of plant cells and tissues. The choice of shearing, grinding, cutting or pressure devices depends upon the resilience of the tissue, and this can often be modified by prior treatment of the tissue with enzymes, particularly those that will digest the cell wall such as cellulase and pectinase. The preferred method for particular tissues has usually been developed on the basis of experience; the objective in all instances is to produce a suspension of membranous particles with common subcellular origins. In the preparation of homogenates it is important to recognise that subcellular compartments are destroyed and the medium surrounding the membranes in their cellular locations is replaced by a buffer of very different composition. One of the consequences of tissue destruction is that constraints on enzymes capable of modifying membrane composition such as proteases, lipases, oxidases, etc. are often removed and unless appropriate precautions are taken, extensive modification of membrane composition may result. The common strategies adopted to prevent the actions of these enzymes are to perform the procedures at low temperatures and as quickly as possible and to include protease inhibitors such as phenylmethylsulphonyl halides in the homogenising medium. Another factor concerned with changing the environment of membranes by placing them in dilute buffer solutions is that proteins that are normally soluble in the cytoplasm may adsorb to membranes in an adventitious manner. Conversely, the forces holding some of the extrinsic membrane proteins may alter so that they dissociate from their normal binding sites. It must be emphasised that membrane proteins are generally considered as all those proteins, including those trapped inside vesicles during homogenisation, that are associated with an isolated membrane fraction. It is clear that certain assumptions are required in the light of the aforementioned possibilities.

Two strategies for the homogenisation of plant tissues have been developed, one in which no attempt is made to digest the cell wall and the other in which cell membranes and organelles are isolated from protoplasts. In the case of chloroplast isolation, direct homogenisation of dicotyledon leaf tissue as described by Stokes and Walker (1971) gives an adequate yield of largely intact chloroplasts. The preferred method of isolation of chloroplasts from leaves of monocotyledons, on the other hand, is from protoplasts prepared by cellulase and pectinase digestion of leaf tissue (Edwards *et al.*, 1978) and isolated by density gradient centrifugation (Evans *et al.*, 1972). Disruption of protoplasts to yield subcellular organelles can then be achieved by relatively gentle mechanical shear forces (Rathman and Edwards, 1976).

Isolation of other subcellular membranes from plant homogenates often requires different methods from those developed for animal cell membranes. The endoplasmic reticulum, when isolated as a microsomal fraction by sedimentation methods, is sometimes heavily contaminated with other subcellular membranes (Lord *et al.*, 1973; Quail, 1979). Experiments performed on etiolated hypocotyls of soybean (Williamson *et al.*, 1975) and cultured carrot cells (Gripsover *et al.*, 1984) showed that conventional centrifugation methods could only achieve fractions enriched to less than 70% with rough surface endoplasmic reticulum. The contaminating membranes, based on analysis of specific enzyme activity of marker enzymes, were almost entirely from mitochondria. Improved methods have been reported (Morre *et al.*, 1984) in which crude microsomal fractions obtained from sucrose density gradients are subjected to free-flow electrophoretic separation.

Other membranes that have represented a challenge in their separation have been the membranes of the chloroplast envelope. The first attempt to isolate these membranes was published by Douce and his collaborators (Douce et al., 1973; Douce, 1974) and refined subsequently by Cline (1981). The envelope which consists of two aligned membranes with close proximity between the membranes confined to specific points of contact (Knoll and Brdiczka, 1983; Cline et al., 1985) can only be separated with difficulty. These contact points are believed to be involved in protein translocation into the chloroplast (Schmidt and Mishkind, 1986; Weisbeek et al., 1986). Their detailed structure has recently been studied by freeze-fracture and freeze-substitution followed by ultrathin sectioning (Cremers et al., 1988). It was found that a proportion of the envelope membranes of pea chloroplasts suspended in isotonic medium were separated in blister-like structures that came into close contact in small regions or contact sites and which could be separated on exposure of the chloroplast to hypertonic conditions. In freeze-substituted preparations small vesicles were shown to adhere to the inner surface of the outer envelope membrane and their origin as inner membrane was inferred on the basis of immunocytochemical methods employing colloidal gold conjugated antibodies against ribulose-*bis*-phosphate carboxylase and protein A.

Successful methods for separation of appressed membranes of grana stacks from membrane derived from non-appressed and stromal lamellae have also been developed (Andersson and Anderson, 1980). These methods, which are based on aqueous polymer two-phase partition methods, have been exploited to characterise differences in lipid composition between these membranes (Gounaris et al., 1983a; Murphy and Woodrow, 1983). Some caution is necessary in the interpretation of these findings because of the possibility that the separation method itself may alter the composition of the membrane (Gounaris et al., 1986).

III. STRUCTURE OF MEMBRANE CONSTITUENTS

Once membranes have been isolated, preferably in an undegraded state, an analysis of the structure and properties of the constituents can be undertaken. In the case of the lipid components, a great deal of useful information about the phase behaviour of these molecules can be obtained from an examination of pure molecular species of lipids prepared either synthetically or isolated laboriously from plant membranes.

The ultimate structure of cellular macromolecules and macromolecular assemblies is conventionally established on the basis of diffraction methods. Because the wavelength of X-rays approximates the length of a covalent bond and hence the distance separating the scattering centres represented by the electron clouds surrounding the constituent atoms, this is the most common form of diffraction method. Other forms of diffraction, such as the scattering of electrons by heavy metals and neutron scattering from the nuclei, particularly of deuterons, also serve usefully in the study of molecular architecture.

One of the essential prerequisites for resolution of the three-dimensional structure of any macromolecular entity by X-ray diffraction methods is that the diffracting lattice is sufficiently extensive so as to be capable of detection within a reasonable time. While only one photon in about 10 000 that is incident on a specimen is scattered by entry into regions of high electron density, the total number of scattered X-rays per unit time is a function of both the intensity of the incident X-ray beam and the physical size of the diffracting lattice. In general, the lattice size can be increased by producing an ordered array of macromolecules such as a crystal and this is conventionally how the X-ray diffraction studies of soluble proteins have been resolved. In the case of membranes, however, the existence of naturally occurring arrays suitable for X-ray analysis is restricted to structures like nerve myelin and disc membranes of the outer segment of visual retinal rods (Blaurock, 1982). Attempts to stack other membranes into orderly arrays have been made with varying degrees of success and often at the expense of membrane structure, particularly when the water activity is reduced.

Diffraction methods have been employed to examine the phase behaviour of membrane lipids and their synthetic analogues when these are dispersed in aqueous systems. The arrangement of lipid molecules in these phases turns out to be highly ordered both in the short (intermolecular) and long range (phase structures) (Shipley, 1973). Membrane proteins, on the other hand, are not as amenable to crystallisation as their water-soluble counterparts, although some success has been achieved in producing two-dimensional arrays of some intrinsic membrane proteins and these have provided considerable insight into the three-dimensional structure.

A. Phase Behaviour of Plant Membrane Lipids

As will be seen from subsequent sections, plant membranes comprise a rich variety of membrane lipids. Each of the major lipid classes, distinguished by the type of polar residue, consists of a whole range of molecular species that differ in the length and extent of *cis* unsaturation of the hydrocarbon chains and their position of attachment to the glycerol backbone. While there is no obvious pattern in the molecular species composition of a given membrane, the relative proportions of the different lipid constituents is preserved within remarkably narrow limits when the environmental conditions are similar (Quinn and Williams, 1985). Very little is known about the biochemical mechanisms that are responsible for regulating the lipid composition of membranes or maintaining the ratio of membrane lipids to proteins, but it is presumably the function of lipase enzymes that are capable of sensing the physical state of the lipids and hence the susceptibility of the enzymic substrates to hydrolysis.

Despite the apparent lack of consistency in the chemistry of the lipid components, each molecular species can be categorised on the basis of the phase formed when the lipid is dispersed in aqueous systems. These phases include bilayer arrangements, with gel, crystal, or liquid-crystal configurations of the hydrocarbon component, an hexagonal-II structure, and a number of cubic phases. The polymorphic behaviour of membrane polar lipids is, in general, largely dependent on the activity of water in the dispersion and the temperature. A number of reviews have been published dealing with lipid polymorphism of biological membranes (Luzzati, 1968, 1974; Shipley, 1973; Verkleij, 1985). Table 10.1 shows the phase behaviour of the molecular species normally found in cell membranes of each of the major lipid classes when dispersed in aqueous systems under physiological conditions. It can be seen that the lipids fall into two categories: a liquid-crystal bilayer configuration and an hexagonal-II structure.

TABLE 10.1. Structural characteristics of native membrane lipids under physiological conditions.

Bilayer-forming	Non-bilayer-forming
Phosphatidylcholines	Phosphatidylethanolamines
Phosphatidylinositols	Monohexosyldiacylglycerols
Phosphatidylglycerols	Cardiolipins/Calcium
Phosphatidylserines	
Dihexosyldiacylglycerols	
Sulpholipids	

1. Monolayer properties

The surface-active properties of polar lipids from plant membranes have been examined by monomolecular film techniques. Monogalactosyldiacylglycerols isolated from a number of different plants and microorganisms have been reported by Bishop *et al.* (1980). Surface pressure–area isotherms of lipids containing different numbers of *cis* unsaturated double bonds per molecule showed that as the double-bond index varied between 0.6 and 3.9 there were no dramatic changes in the liquid-expanded type of monolayer formed. This contrasted with the fully saturated lipid which behaved as a typically condensed film. It was noteworthy that the presence of polyunsaturated fatty acyl residues does not appreciably expand monolayers beyond the extent observed for lipid containing a single monoenoic fatty acyl residue. The lift-off areas of these films, i.e. the area occupied by each molecule of the film at a point where the surface tension decrease is first detected, was found to vary from 1.10 to 1.25 nm^2 molecule^{-1}. These values are somewhat less than corresponding values for phospholipids such as phosphatidylcholine. Values of approximately 1.30, 1.70 and 2.90 nm^2 molecule^{-1} have been reported for dioleoyl-, dilinoleoyl- and dilinolenoyl-phosphatidylcholines, respectively (Demel *et al.*, 1972). The smaller lift-off areas for galactolipid could possibly reflect an increased interaction between neighbouring headgroups that results from hydrogen bonding between adjacent galactosyl residues. A number of studies have been reported of the surface-active properties of 1,2-distearoyl derivatives of mono- and digalactosyl lipids (Sen *et al.*, 1981b; Tomoaia-Cotisel *et al.*, 1983a,b, 1988). Two particular points of interest in the behaviour of these monolayers are the mechanism of collapse of the film and the hysteretic behaviour from which the structures of the film can be deduced at low surface pressures. The mechanism of monolayer collapse at the air–water interface has been studied by measuring the time dependence of surface pressure changes while the monolayer is maintained at constant area (Sims and Zografi, 1972; Gabrielli *et al.*, 1976, 1979; Gabrielli and Guarini, 1978; Baglioni *et al.*, 1980) or determining the rate of decrease in film area at constant surface pressure (Gaines, 1966; Neuman, 1976; Smith and Berg, 1980). It was found that the collapse process in monolayers of the distearoyl derivative of monogalactosyldiacylglycerol, when compressed to pressures exceeding 50 mN m^{-1}, exhibited a sharp loss of film stability consistent with an abrupt rupture or fracture process. High pressure (<40 mN. m^{-1}) monolayers of the digalactosyl lipid, on the other hand, showed time-dependent changes consistent with a more gradual change in the structure of the surface phase akin to a nucleation and growth process of a collapsed structure. This is illustrated in Fig. 10.1 which shows collapse structures of

1,2-distearoyl- (a) monogalactosylglycerol and (b) digalactosyldiacylglycerol monolayers compressed above collapse pressure. In the case of the monogalactolipid, regions of multilamellar structure in the surface layer contrast with collapsed digalactosyl monolayer where multilayer structures interspersed with areas indistinguishable from liquid-expanded monolayers exist together, indicating collapse by a nucleation and growth process rather than by a fracture mechanism.

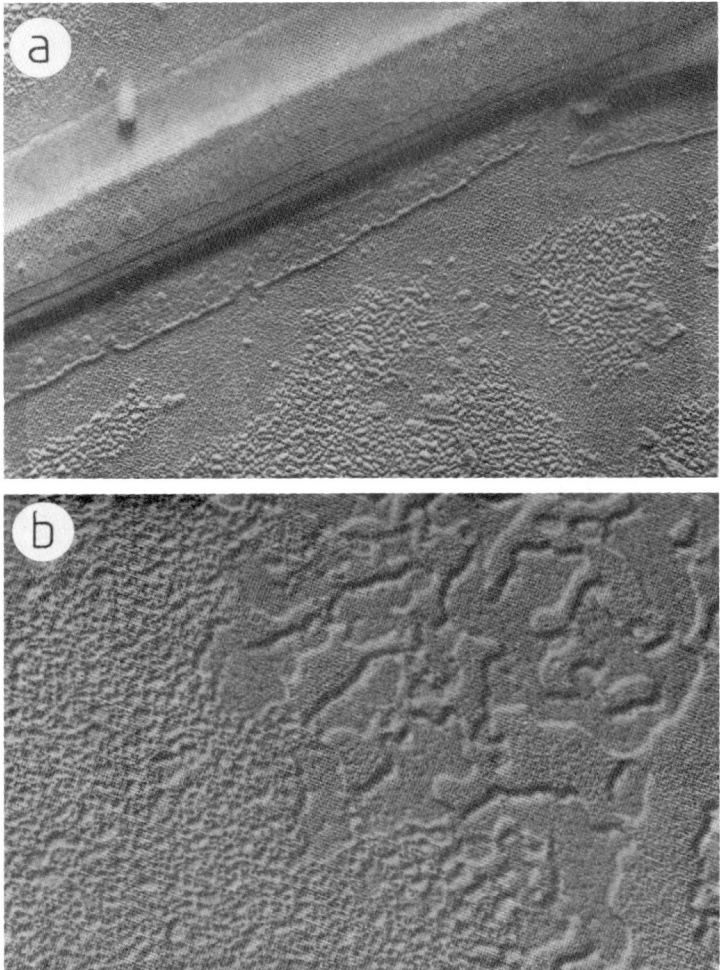

FIG. 10.1. Shadow-cast electron micrographs of monomolecular films of distearoyl derivatives of (a) monogalactosyl- and (b) digalactosyldiacylglycerols. Data from Tomoaia-Cotisel *et al.* (1983a).

The stability of monomolecular films of distearoylmonogalactosylglycerol at low surface pressures has been examined in compression and expansion cycles. The high-pressure portion of the isotherm in these films ($>19\,\mathrm{mN\,m^{-1}}$) is identical in compression and expansion modes but there is a marked and reproducible hysteresis in the pressure–area relationship in the region of low molecular densities ($a < 45\,\mathrm{nm^2}$

molecule^{-1}). Figure 10.2 shows the relationship between the two-dimensional compressibility of monolayers of distearoylmonogalactosylglycerol and the surface pressure when monolayers are spread at the air–water interface at 20°C. The discontinuity observed in this relationship at a pressure of 11.5 mN. m^{-1} was interpreted according to the Eherenfest (1933) theory of transition order as a second-order transition between liquid-expanded and liquid-condensed states.

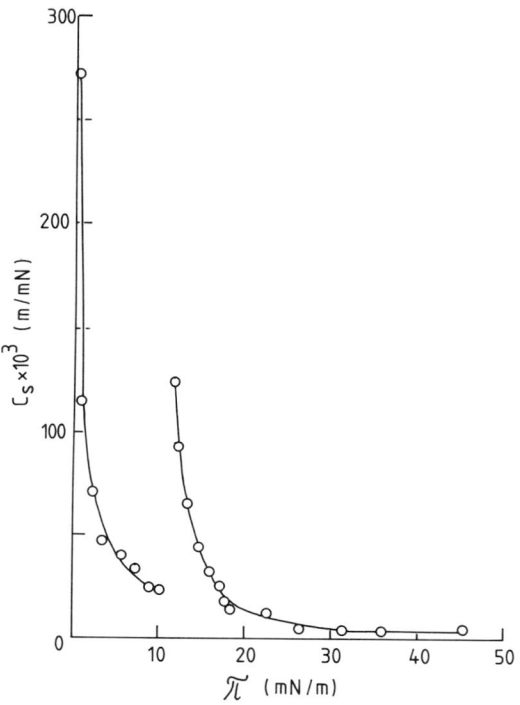

FIG. 10.2. Relationship between two-dimensional compressibility (C_s) and surface pressure (π) of monolayers of distearoylmonogalactosylglycerol spread on an air–water interface at 20°C. Data from Tomoaia-Cotisel et al. (1988)

2. Diffraction methods

The structure of plant lipids has been examined extensively by X-ray diffraction techniques. One of the earliest reported studies was by Rivas and Luzzati (1969), who constructed lipid–water phase diagrams of polar lipids isolated from maize chloroplasts. These preparations, however, were mixtures of different lipid classes as well as mixed fatty acyl substituents and therefore gave little detailed information about the phase behaviour of single molecular species of lipid. Data on mixed acyl chain monogalactosyl and digalactosyl lipids extracted from chloroplast thylakoid membranes were reported by Kreutz (1970); however, these lipids were not hydrated and provided little information about their likely behaviour in aqueous systems. Phase diagrams of plant lipids dispersed in aqueous systems have been reported by Shipley et al. (1973). A phase diagram of monogalactosyldiacylglycerol in water measured over the concentration range $c = 1.0$–0.5, where $c =$ mass of lipid (g)/mass (lipid + water) (g), and the temperature range $-15°C$ to $80°C$ was constructed. An hexagonal phase, showing maximum hydration at $c = 0.78$, predominates under these conditions. At higher water concentrations, excess water co-exists with the hydrated hexagonal phase. Calculations of the surface areas per molecule for the two possible hexagonal-type structures, i.e. hexagonal-I or hexagonal-II, as a function of lipid concentration, suggested that the structure formed was an hexagonal-II type. The distance between the cylinder axes increases with temperature to a limiting value at 6.25 nm at 0°C when the lipid is fully hydrated. The area occupied by each molecule in the lipid–water interface was calculated to be 0.47 nm^2. X-ray diffraction studies (Sen *et al.*, 1981b, 1983) on hydrated samples of dilinolenoyl and distearoyl derivatives of monogalactosyldiacylglycerol have indicated the dominant effect of the acyl chain on phase behaviour of this lipid. Thus the wide-angle X-ray diffraction patterns obtained from the dilinolenoyl derivative contained only a single diffuse reflection centred at 0.46 nm, consistent with a disordered configuration. By contrast, the pattern of reflections obtained from the distearoyl derivative was found to be strongly dependent on the thermal history of the sample but indicated the existence of gel phase structures as well as crystal phases. Low-angle diffraction patterns indicated that both the gel and crystal phases were bilayer arrangements.

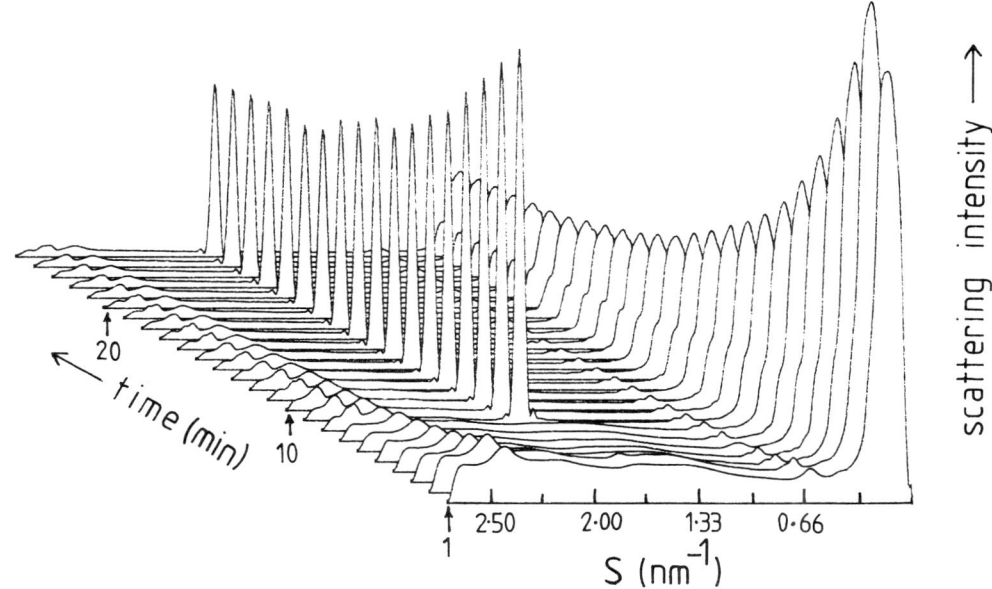

FIG. 10.3. Synchrotron X-ray scattering intensity as a function of reciprocal spacing for hydrated distearoylmonogalactosylglycerol during storage at 20°C after cooling from the liquid-crystal phase. Data from Lis and Quinn (1986).

Conventional X-ray diffraction techniques are limited primarily to the characterisation of the static structures of lipid dispersions. Dynamic changes and studies of metastable states of lipid dispersions can, however, be examined by exploiting the high brilliance of synchrotron X-rays (Caffrey, 1985; Tenchov *et al.*, 1987). A time-resolved synchrotron X-ray study of the structural changes associated with the phase transition between lamellar gel and crystalline bilayer phase of monogalactosyldiacylglycerol has been reported by Lis and Quinn (1986). The phase change was found to be co-operative and to take place isothermally after the lamellar gel phase had been held for 8 min at 20°C. This can be seen from the X-ray scattering patterns recorded sequentially during storage and shown in Fig. 10.3. It can be seen that the transition is relatively fast and is completed within 8 min. Secondly, the transition appears to be highly co-operative with simultaneous changes taking place in the packing of the different regions of the molecule oriented in the bilayer configuration. Thus, there is a clear transition from a single diffraction maximum centred at a spacing of 0.41 nm typical of the hexagonally packed chains in the gel phase to two spacings at 0.38 and 0.40 nm, respectively, characteristic of the crystal phase. Simultaneously, there is a transition in scattering angles in the intermediate range, where the gel phase spacings located at 0.61 nm and 0.69 nm are interpreted as a rectangular packing of the galactose residues at the aqueous interface. These spacings are transformed to a single intense diffraction maxima centred at 0.60 nm, suggesting hexagonal packing of the sugar residues. The conclusions from this study were that the rearrangement of the acyl chains into a more closely packed subcell requires the headgroups to reorient to reduce the steric hindrance between the bulky galactose residues. Examination of this lipid using time-resolved X-ray diffraction techniques has enabled the mechanism of phase transitions to be identified (Lis and Quinn, 1987) and the existence of intermediate phase states between transitions to be identified (Quinn and Lis, 1987).

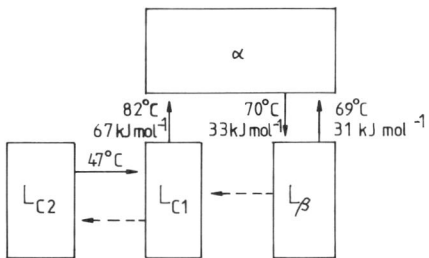

FIG. 10.4. Thermotropic phases of 1,2-distearoylmonogalactosyldiacylglycerol in excess water showing their interrelationships. The continuous arrows correspond to co-operative phase transitions whereas the dashed arrows show the gradual relaxation processes of metastable into more stable phases at 20°C.

Based on X-ray diffraction studies and other methods, a block phase diagram of distearoylmonogalactodiacylglycerol has been constructed and is illustrated in Fig. 10.4. This shows that some transitions between phases are co-operative and take place over a narrow range of temperature. Other transitions are non-co-operative relaxation processes and require some incubation time at defined temperatures to observe. It is noteworthy that the crystal phase characterised in Fig. 10.3 slowly relaxes into a second crystal phase in which the packing of the acyl chains of the lipid undergoes a gradual change. The high-temperature stable phase of this saturated lipid is a non-bilayer phase but the characteristics have yet to be determined. The structure of the phases is determined from analysis of the relationship between diffraction orders of X-rays scattered at low angles from specimens that possess long-range order. Thus in multi-bilayer systems the spacings of the first- and higher-order Bragg reflections are in the ratio 1:1/2:1/3, in good agreement with theoretical predictions (Reiss-Husson, 1967; Luzzati, 1968; Luzzati et al., 1968; Rand et al., 1971; Shipley, 1973; Janiak et al., 1976; Harlos and Eibl, 1980). In practice, the relative intensity of the various Bragg reflections is highly variable and depends to a large extent on the nature of the sample. In some multilamellar systems, for example, only the first order of reflection can be observed (Harlos and Eibl, 1980; Marsh and Seddon, 1982), in which case the unambiguous assignment of the phase cannot be made. The first-order low-angle reflection represents the lamellar repeat distance in bilayer structures in which the dimension of the bilayer thickness plus the intervening water layer between successive lipid bilayer sheets represents the d-spacing.

Other types of ordered long-range structure can be recognised from the characteristic ratio of their higher-order Bragg reflections. Sharp reflections in the ratio $1:1/\sqrt{3}:1/\sqrt{4}:1/\sqrt{7}$ are another commonly observed combination in lipid dispersions (Luzzati et al., 1968; Marsh and Seddon, 1982; Seddon et al., 1984). Again applying a model approach, this order is consistent with a two-dimensional periodicity of hexagonally packed cylinders. The orientation of the molecules within the cylinders for naturally occurring polar lipids is the hexagonal-II arrangement. The arrangement of the molecules in hexagonal-II and lamellar structure is illustrated in Fig. 10.5. This shows that the polar headgroups of the molecules in the hexagonal-II arrangement are oriented towards the aqueous core of the cylinder with the hydrocarbon chains radiating out into the interior of the structure. Arrangements of the hexagonal-I type in which the polar headgroups reside at the periphery of the cylinders and the hydrocarbon chains project into the cores are not normally found amongst membrane lipids, but are formed by certain detergents and membrane destabilising agents such as the lysophosphatides. The highest-order reflection of the hexagonal phase, d, is related to the diameter of the cylinders, a, by the equation $d = \sqrt{3}/2a$.

Other three-dimensional phases to which structures can be assigned include rectangular, rhombohedric and cubic phases. Provided there are sufficient orders of reflection, structures can be readily deduced (Luzzati and Tardieu, 1974). Cubic phases are fairly common amongst membrane lipids (Lindblom et al., 1979; Larsson et al., 1980). They can take up a number of forms, one of which consists of a structure based upon deformed bilayer units that are attached to create two different continuous aqueous networks.

3. Electron microscopy

A direct method for the examination of lipid polymorphism is the use of electron microscopic techniques. Possibly the most useful of these techniques has been freeze-fracture electron microscopy. The development of reliable thermal quenching procedures has helped to overcome relaxation processes, which in the case of lipid phases occur on a relatively fast time-scale (seconds). Replicas prepared from the deposition of platinum on the fracture surface are relatively durable in the electron beam and the phase can be examined directly from the surface features at a resolution of around 3 nm (Zingstein, 1972; Verkleij and Ververgaert, 1978). The planes of weakness through which the fracture plane passes are generally believed to be the hydrocarbon interior of the structures where the terminal ends of the hydrocarbon chains are located. Direct observation of the structure, which provides localised information, contrasts with

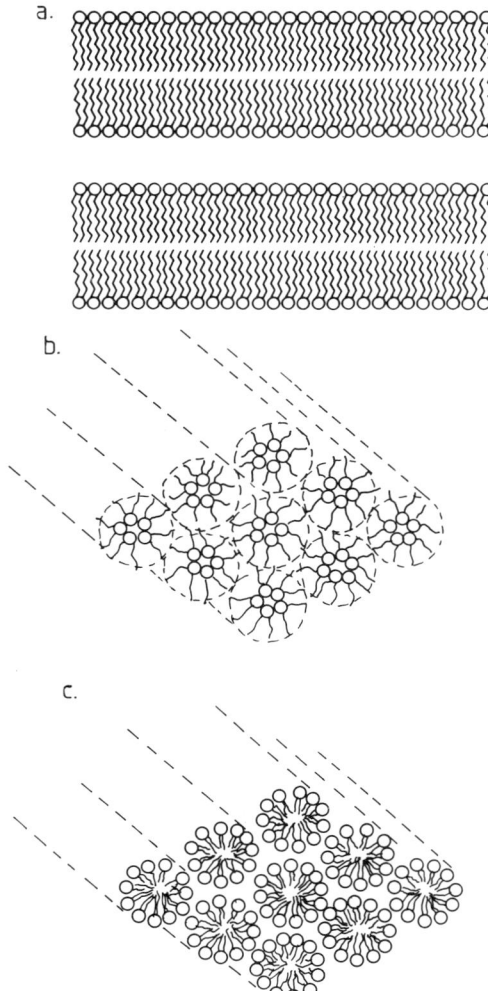

FIG. 10.5. Molecular arrangement of membrane polar lipids in (a) bilayer, (b) hexagonal-II and (c) hexagonal-I structure. The circles represent the polar group and the lines represent the hydrocarbon chains.

spectroscopic and diffraction methods, which produce only average information of structure throughout the specimen under examination. Thus, the freeze-fracture technique has led to the characterisation of phase separations such as lipidic particles, interpreted as inverted spherical micelles sandwiched within a bilayer structure, which are difficult to determine unambiguously by other methods (Sen et al., 1981a). Larger scale phase separations such as those that occur between saturated and unsaturated mixtures of monogalactosyldiacylglycerols (Mansourian and Quinn, 1986) are also easily distinguished by freeze-fracture microscopy (Fig. 10.6) as well as by the other techniques. Figure 10.6 shows electron micrographs of phase separations observed in mixed lipid systems of plant lipids dispersed in aqueous systems. One of the difficulties in applying the freeze-fracture technique to the examination of lipid phases is the difficulty of fixing the specimen in phases that exist at temperatures above ambient. Relaxation phenomena in mixed lipid phases have been observed in large lipid aggregates, presumably due to the poor thermal conductivity and slower rates of thermal quenching (Sen *et al.*, 1982a). Current methods of thermal quenching including jet-freezing (Moor *et al.*, 1976) permit sample preparation without using cryopreservatives, which are known to alter lipid polymorphic behaviour (Sen *et al.*, 1982b).

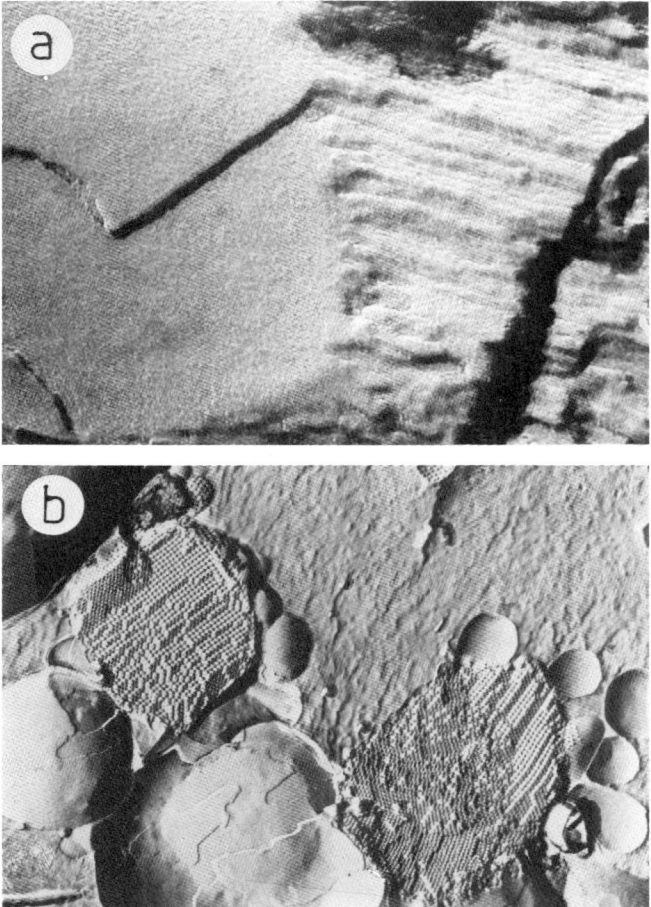

FIG. 10.6. Phase separations observed in mixtures of plant lipids dispersed in aqueous systems by freeze-fracture electron microscopy (a) dioleoyl- : distearoylmonogalactosylglycerols, 50 : 50 mole ratio; (b) mono- and digalactosyldiacylglycerols from bean chloroplasts, 2 : 1 mole ratio.

4. Calorimetry

A number of methods are available for characterising the transition between phases formed by polar lipids dispersed in aqueous systems. The most common method is to measure enthalpy changes by some form of calorimetry. Differential scanning calorimetry (Silvius, 1982; McElhaney, 1986; Small, 1986) is one method which accurately measures the heat required to sustain a constant change in temperature within a lipid dispersion compared with a reference sample of similar heat capacity. Although the method provides an accurate temperature at which the conversion from one phase to another takes place, as well as a measure of the enthalpy of the particular transition, it does not provide any structural information about the transition. However, useful information on phase mixing and compound formation can be derived from the method. Moreover, the relative co-operativity of a phase transition can be evaluated directly from the thermogram since it is related to the sharpness of the excess specific heat absorption curve of an endothermic transition. This is generally expressed as the temperature width at half height of the heat absorption peak or as the temperature difference between the onset or low temperature boundary of the phase transition and the completion of the upper limit of the transition temperature. Using this criterion very pure single component lipids exhibit highly co-operative chain melting endotherms with temperature widths at half-height of less than 0.1°C, but this may extend to 10–15°C for complex mixtures and biological membranes.

Another application of differential scanning calorimetry is to investigate the perturbing effects of membrane proteins on lipid phase transitions and phase separation processes that take place within biological membranes. As emphasised above, however, independent methods such as X-ray diffraction, freeze-fracture electron microscopy, etc., must be used to verify the precise structural transition.

5. Dilatometry

Phase transitions of the type that can be conveniently examined by calorimetric methods also involve changes in specific volume. The specific volume of each particular mesophase is characteristic of that phase and transition between phases results in a change in specific volume. This change, as a function of temperature, can be monitored accurately by direct forms of dilatometry (Wilkinson and Nagle, 1978; Tristram-Nagle *et al.*, 1987). The method can be adapted to measure changes in specific volume during temperature scans of lipid dispersions or in a thermal quench mode in which the temperature is lowered as rapidly as possible to a designated quenching temperature, where it is subsequently held while changes in volume are recorded as a function of time. The relationship between absolute specific volume of dipalmitoylphosphatidylcholine dispersed in excess water as a function of temperature is shown in Fig. 10.7. All of the transitions characterised in this phospholipid can be observed in the figure and the temperatures at which these phase changes occur agree precisely with data derived from calorimetric studies. Like calorimetry, the structural transitions cannot be determined from measurements of specific volume changes, and it is necessary to assign these using direct methods such as X-ray diffraction.

6. Magnetic resonance spectroscopy

A variety of spectroscopic methods can be used to examine lipid polymorphism. Nuclear magnetic resonance (NMR) spectroscopy, for example, can be used to provide both static and dynamic information about lipid phases. It is possible, therefore, to derive rotational and translational diffusion coefficients and correlation times for molecular motions of lipid molecules within phase structures. In respect of the structural properties of these systems, information on average conformation of the lipid acyl chains and polar headgroups can be obtained as well as an indication of specific and non-specific interactions between molecules in mixed lipid systems. Another important parameter that can also be derived is the order parameter, which is a measure of the angular distribution of molecules about a preferred molecular orientation. In bilayer arrays, for example, there is effective axial symmetry about the normal to the bilayer surface referred to as the director (Seelig and Neiderberger, 1974; Stockton et al., 1974; McLauchlan et al., 1975).

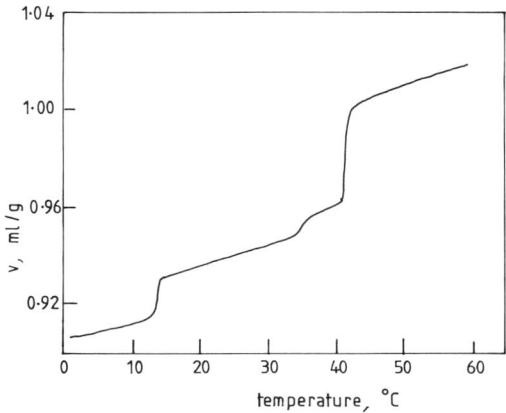

FIG. 10.7. Relationship between absolute specific volume, v, of aqueous dispersions of 1,2-dipalmitoylphosphatidylcholine and temperature.

Translation diffusion of lipids in lipid–water systems can be measured by pulsed field gradient NMR techniques (Charvolin and Rigny, 1971; Bull and Lindman, 1974; Roeder et al., 1976; Lindblom and Wennerstrom, 1977; Kuo and Wade, 1979; Stilbs et al., 1984; Stilbs, 1986). One of the advantages of NMR methods is that a direct measurement of diffusion coefficients can be obtained without probes or model-dependent assumptions. Second, the time during which measurements can be made may be varied according to the displacement distances the particular molecules experience within the structure. It is therefore possible to measure molecular displacements over distances considerably greater than the dimensions of the components of the macro-molecular structure, for example an inverted micelle.

Of particular importance in NMR methods in the examination of lipid polymorphism is the fact the quadrupolar splitting of deuterium nuclei can be examined in suitably synthesised analogues of membrane lipids. This method has wide and very specific applications in that the deuterium can be substituted for protons in all domains of the lipid without significantly perturbing the physical properties of the molecules or the aggregates which they form. It is also possible to substitute deuterated water for protonated water and selective replacement techniques such as those described above have been the subject of a number of reviews (Seelig, 1977; Seelig and Seelig, 1980). The deuterium quadrupolar resonance in anisotropic arrangements is split to an extent which depends on the size of the lipid structure and decreases as motional averaging tends to an isotropic type. An order parameter can be derived from the extent of splitting of the deuterium resonances for particular residues of lipids in the different polymorphic phases (Gally *et al.*, 1980). The spectral features of phospholipid with deuterium substituted in the hydrocarbon chains in bilayer, hexagonal-II and isotropic phases are presented in Fig. 10.8.

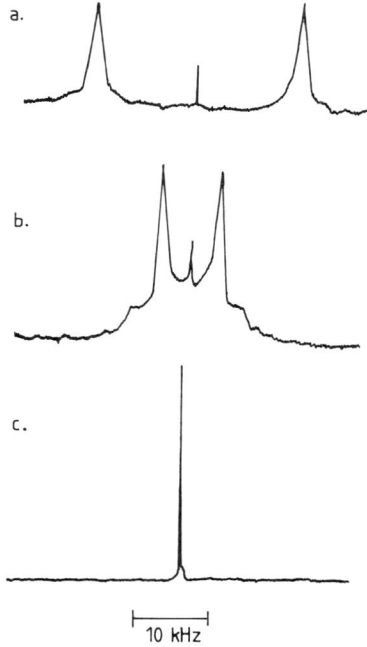

FIG. 10.8. Typical ^{2}H-NMR spectra of aqueous dispersions of phospholipid containing deuterium substitution at the [9,10] position of the acyl chains in (a) bilayer (b) hexagonal-II and (c) micellar configurations.

Lipid translational diffusion coefficients in cubic and lamellar liquid crystal phases and diffusion coefficients of water in cubic phases and in pure water can be used to distinguish different types of cubic phases (Lindblom and Wennerstrom, 1977). Structures which have continuous regions of both hydrocarbon chains and water (Scriven, 1976; Lindblom *et al.*, 1979), and either discontinuous hydrocarbon regions with continuous water domains (Eriksson *et al.*, 1982), or with discontinuous water domains but with continuous hydrocarbon regions (water-in-oil emulsions), can all be distinguished by deuterium NMR methods. If diffusion of lipid molecules can take place over macroscopic distances without polar groups entering the hydrocarbon phase or hydrocarbons being exposed to water, the diffusion coefficient, in the case of bicontinuous cubic systems, is found to be of the same order as in liquid-crystal bilayers. Furthermore, if water diffusion is of the same order as in bulk water, then the cubic phase is of an oil-in-water emulsion type (Rilfors *et al.*, 1986).

Another NMR method that has been used to characterise phospholipid polymorphism is ^{31}P-NMR (Seelig, 1978; Cullis and De Kruijff, 1978a,b; De Kruijff *et al.*, 1985; Cullis *et al.*, 1985; Tilcock, *et al.*, 1986). The particular feature of this method is that it provides a convenient and quantitative estimate of macromolecular structures formed by the major lipid–water systems. The information is derived from spectral lineshapes of proton-decoupled spectra, which give an indication of the chemical shielding anisotropy of the lipid phosphate group. When the phosphate segment assumes different orientations, characteristic resonances are observed at different frequencies. The ^{31}P-NMR spectrum of phospholipids in bilayers is typical of a shielding tensor that is axially symmetric around a director axis consisting of a low-field shoulder and a high-field peak. This spectrum, together with ^{31}P-NMR spectra obtained from phospholipids in other phases, is shown in Fig. 10.9. When phospholipids in the bilayer configuration are oriented, individual narrow resonances with an angular dependent chemical shift are observed (Hemminga and Cullis, 1982). Studies of the orientation dependence of ^{31}P-NMR spectra indicate that molecules with their long axes perpendicular to the magnetic field contribute to the high-field peak, while molecules oriented parallel to the magnetic field correspond to resonances at a position within the shoulder region. The director axis coincides with the bilayer normal about which the phospholipid segment rotates.

The lineshape of the ^{31}P-NMR spectrum of molecules in a bilayer configuration can be used to determine the rate of reorientation of the molecules resulting from Brownian motion (Burnell et al., 1980). Thus, when the rate of reorientation of the phosphate segment is fast the lineshape progressively narrows until eventually, when motion is effectively isotropic on the time scale of the measurement (< 10 μs), motional averaging leads to a narrow and symmetric resonance line (Fig. 10.9(b)). Motion of this type is characteristic of phospholipid molecules in small, rapidly tumbling bilayer vesicles, micelles, inverted micelles and a variety of cubic phases. Phospholipid molecules in the hexagonal-II structure can also be identified from the ^{31}P NMR spectrum, despite the fact there is virtually no difference in the motion experienced by the phosphate segment in bilayer and hexagonal-II structures. Nevertheless, the macroscopic orientation of the lipid cylinders within the spectrometer magnetic field gives rise to additional averaging of the chemical shift anisotropy. The resulting spectrum of molecules in an hexagonal-II arrangement is illustrated in Fig. 10.9(c).

7. Infrared spectroscopy

Other forms of spectroscopy used to characterise lipid polymorphism include vibrational spectroscopy. Amongst the earliest reports of spectroscopic studies of hydrated lipids was that of Chapman et al. (1967), who examined lipids at lower water contents by conventional infrared spectroscopy. One of the major problems with the method, however, is that the absorption of water tends to dominate the spectrum. Raman spectroscopy is one way that this problem can be overcome and many studies have been undertaken using this technique (Lord and Mendelsohn, 1981; Cary, 1982; Wong, 1984). Recent developments such as the use of attenuated total reflection methods (Fringeli and Gunthard, 1981) and interferometric Fourier transform infrared spectroscopy, together with computer data analysis, have been able to overcome the problem of water absorption and to provide useful information to characterise lipid polymorphism (Casal and Mantsch, 1984). Absorption bands corresponding to —CH_2— antisymmetric and symmetric stretching modes can be used to determine *trans/gauche* isomerisations of the acyl chains. Absorption bands associated with —CH_2— bending or scissoring modes can also be distinguished which reflect acyl chain packing and conformation. C—C stretching and low-frequency vibrational modes detected by Raman spectroscopy have also been used to assign phases dependent on the

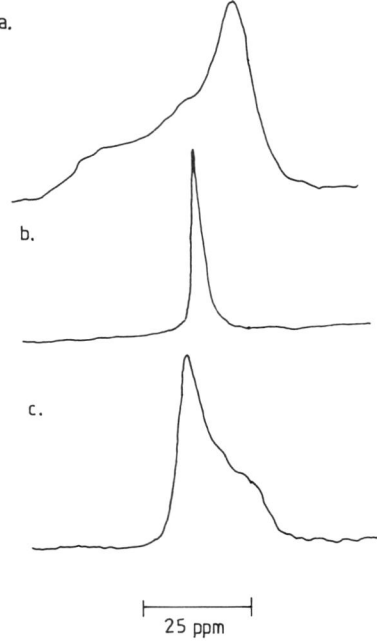

FIG 10.9. Typical ^{31}P-NMR spectra of aqueous dispersions of phospholipids in (a) bilayer, (b) micellar and (c) hexagonal-II arrangements.

conformation of acyl chains. The infrared spectrum is altered in characteristic ways by the substitution of deuterons for protons. Specific substitutions in lipid molecules, as in the case with deuterium NMR spectroscopy, provide useful information on motion and structure of lipids (Smith and Mantsch, 1979; Mendelsohn and Koch, 1980) as well as biomembrane systems (Sunder et al., 1978; Casal et al., 1979, 1980, 1982, 1983; Cameron et al., 1982). Arrondo et al. (1984) have also identified absorption bands corresponding to residues of the polar headgroup of lipid molecules.

8. Birefringence

Information about the relative orientation of molecules in an assembly such as a lipid phase can be obtained using an intrinsic optical property of the material referred to as birefringence. If the chemical bonds of molecules of a particular material have a preferred orientation, they transmit light polarised in one direction differently compared to that polarised at another. This means for any given frequency of vibration there are two different indices of refraction for light polarised perpendicularly and parallel to the respective axes of the oriented bonds of the sample. This difference in refractive index is referred to as birefringence. Birefringence can be expressed in terms of the difference between the refractive index of the extraordinary wave and the refractive index of the ordinary wave, $n_e - n_o$, where subscripts refer to the extraordinary and ordinary waves, respectively. Experimentally, $n_e - n_o$ is measured and is equated with the dielectric constant, K, by the expression:

$$n_e - n_o = \left(\frac{1}{2\tilde{n}}\right)(K_\| - K_\perp) \tag{10.1}$$

where $K_\|$ and $K_\perp$ are the dielectric constants for light vibrating along and at right-angles to the axes of the chemical bonds in the ensemble and $\tilde{n}$ is the average refractive index of the material. The measured birefringence is a composite of two types of birefringence known as form and intrinsic, or crystalline, birefringence. In crystals, birefringence can be explained solely in terms of the intrinsic and isotropic properties of the molecules comprising the crystal, and may be positive or negative. Birefringence, however, may arise from anisotropy on a much larger scale than that of the molecular dimensions. This occurs when particles of isotropic material are arranged into an ordered structure whose size is large relative to the dimensions of the molecules, but small compared with the wavelength of light. This is referred to as form birefringence because it depends on the form of the presentation of the molecules in the system and is negative if this is a series of parallel sheets. Form and intrinsic contributions to birefringence can in practice be separated if the refractive index of the surrounding medium is varied independently of the refractive index of the membrane, or lipid phase, until the form component is eliminated. Replacing the suspending medium with glycerol or sucrose buffers is the usual method employed, but in all cases it is essential that there are no structural alterations in the phase or the membrane resulting from media substitution and that the thickness and distance apart of the molecules within the phase remains unaltered.

The observation of birefringence in a polarising microscope can be of two types. Plain polarised light focused through the sample enters an analyser placed above the objective lens. The analyser is oriented with its axis of activity at right-angles to that of the polariser so that unless some optically active material is inserted into the optical path between the polariser and analyser no light passes through to the observer and the field remains dark. Observations using the polarising microscope have shown that dry amorphous lipid extracts prepared from a variety of cell membranes are not birefringent. When the lipid is warmed in dilute salt solution, however, the molecules become hydrated, and so-called myelin figures are observed which show positive birefringence. The addition of aqueous medium causes an ordering of the molecules into bilayers or other ordered structures and it is this molecular order which is responsible for the birefringent properties. When the membrane lipids carry no net charge, such as zwitterionic phosphatidylcholines, bilayers that form in water are separated by relatively thin layers of water and they give rise to a strong positive birefringence. Positive birefringence is to be expected when the hydrocarbon chains are oriented predominantly perpendicular to the plane of the bilayers. This may alter, however, depending on the thickness of the aqueous layers. Inclusion of lipids which carry a net charge, irrespective of whether this is positive or negative, causes swelling of the structure and a separation of the adjacent bilayers resulting in an increase in the intensity of the negative form birefringence. This can lead to an overall loss in sign, or even reversal in sign, of the observed birefringence.

Values for birefringence of lipid dispersions are similar to those of biological membranes which have orientations suitable for observations. Birefringence measurements of multibilayer lecithin sheets and single planar bilayers give values of about $+0.017$ and $+0.022$, respectively. This compares with values reported for the myelin sheath of large single fibres of the frog sciatic nerve of about +0.011 and intact bovine retinal rods of +0.02. It is noteworthy that birefringence is greater for crystalline or gel arrangements of the hydrocarbon chains of the lipids compared with fluid, disordered chains where the chain orientation is less well defined.

B. Structure of Membrane Proteins

With few exceptions (one being cytochrome c) membrane proteins cannot be separated from lipids in a water-soluble form and subsequently crystallised into a three-dimensional lattice suitable for conventional X-ray diffraction analysis. One of the reasons for this is that membrane proteins depend on their interaction with the lipids or other membrane proteins to preserve their native configuration. There is, however, a range of relatively mild detergents available which is able to replace membrane lipids and prevent loss of biochemical function. Detergent extraction of biological membranes allows the solubilisation of membrane proteins, which can then be isolated into purified fractions and reconstituted with defined lipids. Subsequent removal of the detergent produces membrane vesicles containing a single membrane protein, sometimes exhibiting a preferred orientation with respect to the sidedness of the vesicle, and providing useful systems with which to examine biochemical functions.

Another feature of detergent complexes of some membrane proteins is that they can be aggregated into close-packed two-dimensional arrays which have sufficient order for analysis by diffraction methods. Since the specimen is only one layer of molecules thick, it is not possible to use the usual X-ray methods described above because of radiation damage to the sample. To avoid this problem transmission electron microscopic methods have been modified to enable the detection of electron scattering intensities and their subsequent conversion into structural co-ordinates of the two-dimensional protein–detergent crystals. The intensity with which electrons are scattered varies according to the electrical potential within the structure which, for pratical purposes, is roughly proportional to electron density and, in turn, to atomic number. In this sense the electron diffraction method is analogous to X-ray diffraction, since the Fourier transform of electron scattering yields the profile of atomic density through the structure.

Apart from the usual problems of phase determination, the advantage of the electron scattering method is that very low doses of electrons can be employed (< 1 electron per unit cell) and, in order to reconstruct the image, information is combined from a large number of unit cells. This avoids radiation damage to the sample which would otherwise destroy the structure. The strategy for constructing a three-dimensional image of a two-dimensional array of molecules is achieved by analysing specimens tilted with respect to the angle of incidence of the electron beam. This method relies on the fact that the Fourier transform of a transmission electron micrograph of a two-dimensional crystal is a central section through the three-dimensional Fourier transform of the crystal, that is, through a lattice of lines in reciprocal space perpendicular to the plane of the crystal. Thus the amplitudes and phases of Fourier terms along each lattice line can be obtained by combining the Fourier transforms from an appropriate number of different, tilted views of the two-dimensional array. As in X-ray diffraction analysis, an inverse Fourier transform produces a three-dimensional map of the density distribution within the monomolecular lattice.

1. Cytochrome b–c₁

Some examples of membrane protein complexes that have been examined by electron scattering methods are presented in Fig. 10.10 Two-dimensional crystals of the cytochrome b–c_1 complex from mitochondria have been formed using a mixture of phosphatidylcholine and phosphatidylserine with the detergent, Triton, to solubilise the protein (Karlsson *et al.*, 1983). Upon removal of the detergent, the protein–lipid complexes aggregate into two-dimensional crystals. The protein complexes are then negatively stained with uranyl ions to enhance the contrast in the specimen. Examination of the electron-density distribution within the array shows that the protein complexes are associated into dimers with the dimers alternating in their orientation across the two-dimensional array. The complex has a molecular weight of approximately 550 kDa and the monomeric units consist of at least eight separate subunits. The long axes of the dimer are vertically oriented with respect to the membrane plane and are about 15 nm in length. The monomers are in contact with a 5 nm band corresponding to the membrane lipid domain and the distribution of protein on either side of the membrane is asymmetric. The cytochrome c_1 components are located on the cytoplasmic side of the membrane in a domain that contains 20% of the protein mass and extends 3 nm into the aqueous phase. The cytochrome b, on the other hand, protrudes about 7 nm into the aqueous phase on the opposite side of the membrane (matrix surface) and occupies a volume equivalent to 50% of the total protein mass. The remaining 30% of the protein within the membrane contains the iron–sulphur proteins in addition to some components of cytochrome b. Structural studies of the subunits present in these domains have also been undertaken using electron diffraction methods and are consistent with their location in the intact cytochrome b–c_1 complex. Preliminary studies of the structure of NADH-ubiquinone reductase (complex I) have also been reported by this group (Leonard *et al.*, 1987).

2. Cytochrome oxidase

Cytochrome oxidase crystals have been prepared using both Triton and deoxycholate detergents (Fuller *et al.*, 1979; Costello and Frey, 1982; Frey *et al.*, 1982). The arrangement of molecules is different in each case; the crystals formed in Triton consisting of two membrane layers formed by collapse of a vesicle whereas those in deoxycholate consist of a single two-dimensional array. A three-dimensional reconstruction of uranyl acetate-stained specimens formed in deoxycholate show that cytochrome oxidase exists as monomers arranged roughly in the shape of a Y. The overall length is about 11 nm with the arms of the Y extending 5 nm in length. The centre-to-centre spacing of the ends of the arms is about 4 nm and they protrude into the aqueous phase on the matrix surface of the membrane. The stem of the Y represents a single domain of the molecule extending into the aqueous phase on the cytoplasmic surface of the inner mitochondrial membrane.

3. Light-harvesting chlorophyll–protein complex

The last example shown in Fig. 10.10 is the light-harvesting chlorophyll a/b–protein from chloroplasts which plays an essential role in converting light energy into a form that can be utilised by the photosystem reaction centre to separate charge. Two-dimensional crystals have been formed from Triton-solubilised complexes, stained with uranyl acetate and examined by electron diffraction methods (Kuhlbrandt *et al.*, 1983). The protein complexes are aggregated into trimers composed of three structurally equivalent subunits and arranged in symmetrically related positions (Kuhlbrandt, 1984; Li, 1985). The protein complex is asymmetric in shape and spans the membrane. The long axis has a length of about 6 nm so that, assuming a membrane thickness of about 4.5–5 nm, the protein can protrude only about 1–1.5 nm into the aqueous phase on either side of the membrane. One side of the complex is composed of three hook-like lobes extending 2 nm into the complex. Each lobe consists of a vertically oriented cylindrical domain about 1.5–2.0 nm in diameter extending across the membrane and protruding as three separate arms on the opposite side of the membrane. This produces an arrangement with a significant difference in the amount of surface area exposed on either side of the membrane. This may play a role in stabilising membrane–membrane interactions involved in the membrane stacking of the grana thylakoids and in displaying the antennae chlorophylls in an advantageous manner to trap the available light. Even greater resolution of the structure of the complex may be expected from studies of the recently reported three-dimensional crystals (Kuhlbrandt, 1987).

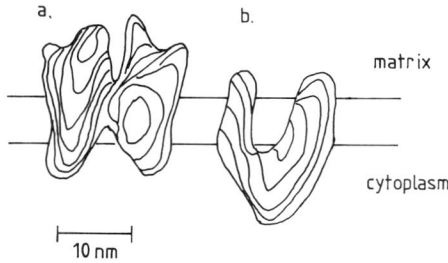

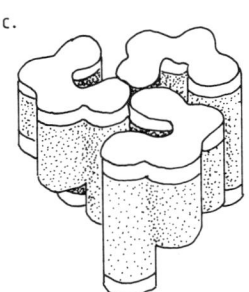

FIG. 10.10. The structure of three membrane protein complexes determined by electron diffraction methods. (a) Cytochrome b–c_1 complex of the respiratory chain, (b) cytochrome a–a_3 complex of the respiratory chain, (c) light-harvesting chlorophyll a/b–protein complex from chloroplast thylakoid membranes.

With knowledge of the complete primary structure of light-harvesting chlorophyll–protein complexes, it has been possible to prepare hydropathy plots based on the primary sequence. These, together with antibody binding data, are consistent with three (Karlin-Neumann *et al.*, 1985) or four (Anderson and Goodchild, 1987) membrane-spanning α-helical segments and in agreement with a value of 44% α-helical content derived from ultraviolet circular dichroism spectroscopy (Nabedryk et al., 1984). Chlorophyll proteins of the photochemical reaction centres of Photosystem I and Photosystem II have also been subjected to hydropathy plot analysis according to the primary amino acid sequence, and these proteins appear to possess a higher α-helical and lower β-structure content than found in light-harvesting chlorophyll protein II (Morris and Herrmann, 1984; Fish et al., 1985). Direct observation of the light-harvesting complexes of Photosystem I of chromophyte algae by ultraviolet CD spectroscopy support this view (Hiller et al., 1987).

The first photochemical reaction centre to be crystallised, and thus made suitable for X-ray diffraction analysis, was that from *Rhodopseudomonas viridis* (Michel, 1982). The photochemical activity of the crystalline material was subsequently confirmed (Zinth et al., 1983). An X-ray structure analysis of these well-ordered crystals was used to compute an electron density map at 0.3 nm resolution (Deisenhofer et al., 1984, 1985). From this map an atomic model of the arrangement of the prosthetic groups has been deduced and the central part of the complex is found to be divided into two subunits, each of which forms five membrane-spanning α-helical segments. The arrangement of the membrane-spanning α-helical segments in chloroplast membrane proteins has been confirmed by polarised infrared spectroscopy (Nabedryk et al., 1982; Breton and Nabedryk, 1984, 1987) as well as from hydropathy indices based on the primary amino acid sequence (Karlin-Neumann et al., 1985; Zuber, 1986). Other X-ray crystallographic studies of the photochemical reaction centre from *Rhodobacter sphaeroides* (Allen et al., 1986; Chang et al., 1986; Frank et al., 1987; Ducruix and Reiss-Husson, 1987) and *R. palustris* (Wacker et al., 1986) and cytochrome c from *Rhodopseudomonas capsulata* (Holden et al., 1987) have also been described.

4. Membrane ATPases

Another membrane protein that has been subjected to detailed X-ray crystallographic analysis is CF_1. This is a component of ATP synthase which consists of a membrane integrated component CF_0, which is believed to act as a membrane proton channel, and a hydrophilic part, CF_1, which contains the nucleotide binding sites. The structure of the mitochondrial, bacterial and chloroplast synthases appear to be similar (Amzel and Pedersen, 1983; Senior and Wise, 1983; Strotman and Bickel-Sandkotter, 1984). Gel filtration and electrophoresis experiments have shown that CF_1 consists of five different types of subunit, α (59 kDa), β (56 kDa), γ (37 kDa), δ (17.5 kDa) and ε (13 kDa). Each CF_1 molecule contains 3α, 3β, 1γ, and as yet unknown number of δ and ε subunits (McCarty and Nalin, 1986). The structure has been examined by X-ray diffraction (Amzel et al., 1982) and a low-resolution model has been proposed in which there are six distinct regions of electron density approximately equal in size. Electron microscopy of crystals of CF_1 and single oligomeric units suggest an image consisting of two layers of large subunits, α and β, in the form of a flattened trigonal antiprism with dimensions approximately 11.5 nm × 11.5 nm × 8 nm (Akey et al., 1983; Tiedge et al., 1983, 1985; Tsuprun et al., 1984; Lunsdorf et al., 1985; Boekema et al., 1986, 1988). These studies have indicated that the central mass of CF_1 is somewhat larger than observed in F_1 from mitochondria and a division of the central mass region into two parts is often observed. A scheme for the arrangement of subunits based on these methods is illustrated in Fig. 10.11.

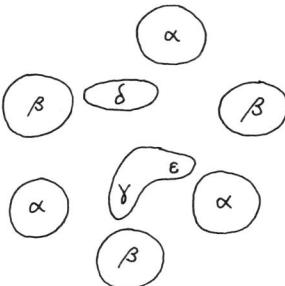

FIG. 10.11. Proposed arrangement of subunits in CF_1 based on image analysis of electron micrographs of negatively stained preparations of enzyme prepared from spinach chloroplast thylakoid membranes.

IV. MEMBRANE DYNAMICS

A. Membrane Fluidity

The concept of membrane fluidity and the idea that membrane lipids and proteins are normally able to diffuse relatively freely in the lateral plane of the membrane are central to current views of membrane organisation. The basic idea of membrane fluidity is a simple one. At temperatures below the transition temperature (T_c) characterising their gel-to-liquid crystal phase transition, membrane lipids are normally packed in an ordered crystalline form and as such show very restricted mobility. Above this temperature, the forces between the lipid molecules are insufficient to maintain an ordered lattice and the increased thermal motion of their fatty acid chains results in the relatively disordered state. Under these conditions, the lipid chains are free to flex and rotate. The lipid molecules are able to undergo rapid nearest neighbour exchange and the membrane is said to be in a fluid state.

1. General considerations

A number of difficulties arise, however, when attempts are made to quantify the extent of this fluidity. The most obvious is that there is no single parameter that can be used to describe membrane fluidity. The lipid bilayer is an anisotropic system. As such, the values of the physical parameters that describe its properties vary with position. The motional fluidity of lipid chains close to their headgroups, for example, is clearly very different from that at the centre of the membrane bilayer.

Two approaches have been adopted in an attempt to overcome this problem. One is to try to describe membrane fluidity in terms of an apparent viscosity, often referred to as microviscosity, based on a treatment of the motion of probe molecules in terms of standard hydrodynamic equations (Shinitzky and Barenholz, 1978). This approach allows an easy comparison between different systems but suffers from the grave disadvantage that it gives no information about the molecular basis of fluidity. It is thus possible for membranes to have the same apparent fluidity but very different molecular organisation.

10. STRUCTURE AND DYNAMICS OF PLANT MEMBRANES

The other approach which is now more commonly used is to define fluidity in terms of physical parameters that describe the motion of the fatty acyl chains of the membrane lipids or of probes attached to, or located adjacent, to such chains. This approach, whilst often very useful, has itself led to further problems. The main areas of controversy are: (1) the extent to which the various parameters in current use refer to different aspects of fatty acyl chain motion; and (2) the extent to which they report on an average fluidity in a system consisting of many different microenvironments, some of which may even be created by the presence of the probe itself.

The question of the different parameters available for the measurement of membrane fluidity and the extent to which they report on different aspects of chain motion has been elegantly covered by Stubbs (1983). In this treatment, a clear distinction is made between the range and the rate of such motion. This distinction is probably most easily explained in terms of the motion of a spin probe such as a nitroxide-labelled fatty acid. The probe, as illustrated in Fig. 10.12, tends to be oriented in the lipid bilayer with its fatty acid headgroup in the membrane surface. The motion of the probe is usually described in terms of a restricted random walk within a cone of half-angle θ. This motion can, under appropriate circumstances, be described in terms of an order parameter (S):

$$(S) = \frac{1}{2} \langle 3 \cos 2\theta - 1 \rangle \tag{10.2}$$

where the angular brackets represent a time average value of molecular motion. Alternatively, the motion can be described in terms of a correlation time (τ), which can be thought of as the average time for which the molecule moves in any given direction. The value of τ reflects the degree of immobilisation of the spin label and is normally defined in terms of the relative line heights of the nitroxide spectrum in the cases of weakly to moderately immobilised labels (Knowles *et al.*, 1976). Order parameters and correlation times are closely related, but clearly the former parameter emphasises the range of motion whilst the latter emphasises the rate of motion.

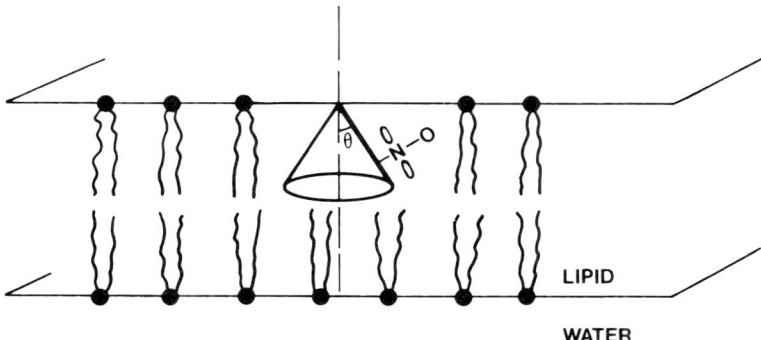

FIG. 10.12. Anisotropic motion of a nitroxide fatty acid spin label in the lipid region of a membrane. Probe motion takes place in a cone of half angle θ.

A similar situation exists in the case of the analysis of the motion of the fluorescent probe 1,6-diphenyl-2,3,5-hexatriene (DPH). DPH is a rod-shaped molecule with a high affinity for the hydrophobic regions of lipid bilayers. In a typical measurement, DPH is excited using plane polarised light that preferentially excites molecules with their transition dipoles oriented parallel to the plane of polarisation of the exciting light (photoselection). If these molecules maintain this orientation during the period between absorption and emission, the fluorescence they emit will show a similar polarisation. If, on the other hand, they reorient during this period, this polarisation will be lost. The extent of fluorescence depolarisation with time is reflected in changes in the value of the fluorescence anisotropy, $r(t)$ where:

$$r(t) = \frac{I_{\parallel}(t) - I_{\perp}(t)}{I_{\parallel}(t) + 2I_{\perp}(t)} \tag{10.3}$$

where $I(t)$ is the fluorescence measured through polarisers oriented parallel ($\parallel$) or perpendicular ($\perp$) to the plane of polarisation of the excitation beam.

In practice, most investigators measure the change in steady-state polarisation values given by:

$$p = \frac{I_{\parallel} - I_{\perp}}{I_{\parallel} + I_{\perp}} \tag{10.4}$$

rather than the time dependence of fluorescence decay, as the experimental equipment required is far simpler. Analyses of the motion of DPH in membranes suggest that it is best described in terms of a wobbling-in-cone model (Kinosita and Ikegami, 1984). This model shows many similarities to that described above for the fatty acid spin label leading to the derivation of order parameters based principally on the angle of the cone swept out by the probe and relaxation times based on the rate of motion within the cone.

TABLE 10.2. Physical parameters of membrane fluidity.[a]

Technique	Physical parameter[b]	
	Range	Rate
NMR	S_{C-D}	τ_c
ESR	S	τ_c
Optical anisotropy:		
Fluorescence anisotropy decay	r_∞, θ_c, S	φ, D_w
Differential phase fluorimetry	r_∞, S	R
Steady-state anisotropy[c]	— P, r^s, η	—
Fluorescence recovery after photobleaching	—	D_L
Fluorescence excimer formation		D_L
Raman spectroscopy	S_{trans}	—

[a] Data from Stubbs (1983).
[b] S_{C-D}, S, order parameter; r_∞, residual equilibrium anisotropy; θ_c, cone angle; P, polarisation value; r^s, steady-state anisotropy; η, microviscosity; τ_c, rotational correlation time; φ, apparent rotational relaxation time; R, rotational relaxation time; D_L, lateral diffusion constant.
[c] Parameters relate to both range and rate.

A list of physical parameters commonly used in the classification of membrane fluidity, taken from the work of Stubbs (1983), is set out in Table 10.2. A clear distinction is made in this list between parameters that predominantly reflect the range of fatty acid or probe motion and those that reflect its rate. It is extremely important that this distinction is appreciated when comparing the results of membrane fluidity measurements made using different physical techniques.

The problem of interpreting results of fluidity measurements made in highly heterogeneous systems such as biological membranes is one of distinguishing between different contributions to fluidity. Most of the methods currently used for measuring membrane fluidity were first developed using simple systems consisting of a single molecular species of lipid dispersed in water. The extrapolation of these techniques to membrane systems, or even to total membrane lipid extracts, is not necessarily straightforward. To cite an obvious example, it is clearly understood that the organisation of the chloroplast thylakoid membrane in the grana stacks is very different from that in the stroma lamellae. Abundant evidence is now available demonstrating that there is a lateral separation of the protein components of the light-harvesting apparatus of the two photosystems (see reviews of Andersson and Anderson, 1980; Anderson, 1981). There is also evidence to suggest that the membrane lipids are asymmetrically distributed both between the inner and outer leaflets of the membrane and within the plane of the membrane (Quinn and Williams, 1985). Clearly any measurements performed on such a system purporting to report on the fluidity of the membrane as a whole must be viewed with circumspection.

In the case of the chloroplast, it is relatively easy to prepare membrane fractions that are highly enriched in granal or stromal membranes. Steady-state fluorescence depolarisation studies on such fractions using DPH and spin probes (Ford *et al.*, 1982; Barber *et al.*, 1984) indicate that their membrane fluidities are appreciably different. More detailed analysis involving measurements of the decay of fluorescence polarisation anisotropy (Ford and Barber, 1983) suggest that these differences are probably reflections of the degree of restriction of the motion of the probe by membrane proteins rather than differences in the lipid component *per se*. This view is supported by studies of the effects of reconstituting chloroplast coupling factor into thylakoid lipid vesicles (Millner *et al.*, 1984a). Information of this type is extremely useful in building up a picture of the organisation of a given membrane. However, it also serves to underline the difficulties involved in the interpretation of the behaviour of probes in highly heterogeneous systems.

2. *Practical applications*

The majority of studies carried out to date have involved measurements on chloroplast thylakoid membranes or membrane lipid extracts of mitochondria, chloroplasts or whole leaf tissue. A number of studies have been carried out on other systems including purified plasma membrane preparations (Yoshida, 1984; Ishikawa and Yoshida, 1985; Yoshida *et al.*, 1986; Uemura and Yoshida, 1986), microsomal preparations (Paliyath *et al.*, 1984; Lynch *et al.*, 1987), protoplasts (Legge *et al.*, 1986) and unicellular algae (Thiery *et al.*, 1987). There have been two major spurs behind much of the work carried out to date. The first of these has been interest in the use of measurements of discontinuities in the temperature dependence of membrane fluidity as a possible indicator of lipid phase transitions in the context of chilling sensitivity. The second relates to measurements of the fluidity of the chloroplast thylakoid membrane in the context of the role of plastoquinone as a mobile electron carrier shuttling between the light-harvesting apparatus of Photosystem II and the cytochrome f/b_6 particle of the photosynthetic electron transport chain.

Although there are in principle a wide variety of techniques available for studying membrane fluidity, practical considerations effectively restrict measurements to those involving spin labels and fluorescence techniques. In practice spin label techniques have been most widely used for the study of putative phase transitions and fluorescence depolarisation techniques for the direct study of membrane fluidity *per se*.

There are numerous reports in the literature of the use of spin labels to measure the temperature dependence of membrane fluidity in plant membrane preparations and lipid extracts (see Quinn and Williams, 1978, 1983 for collected references). Although not technically difficult, the interpretation of such measurements is open to considerable question. Two main criticisms of such studies have been put forward. The first is that the formulae used to calculate the correlation times on which many of these analyses are based were originally derived for isotropic motion. The motion in membranes is clearly anisotropic. This makes the interpretation of changes in the temperature dependence of parameters derived from spectral changes, particularly of breaks in Arrhenius plots, extremely difficult. Such discontinuities, as pointed out by Cannon *et al.* (1975) and re-emphasised by Schreier *et al.* (1978), can readily arise from continuous rather than discontinuous changes in the rates and/or ordering of probe motion under conditions in which the assumptions underlying parameter calculation become increasingly inappropriate.

The second criticism is that the spin label spectra measured in membrane fractions normally consist of two components; a fluid component associated with the bulk-phase lipid and a motionally-restricted component associated with lipid in direct contact with membrane proteins and/or trapped between protein molecules (see Section IV.C). The relative effects of these two components on overall-spectral parameters varies markedly with temperature. In the absence of prior spectral subtraction to determine the spectrum of the fluid component, any analysis of such spectra, as pointed out by Gang (1988), is likely to be of dubious value. The same problem may occur to a lesser extent in studies performed on membrane lipid extracts. In this case, while there is no specific motionally-restricted component, the possibility of the occurrence of temperature-dependent rearrangements of bilayer-forming and non-bilayer-forming lipid components which lead to changes in relative ordering of lipid chains cannot be excluded.

The existence of these criticisms, it must be emphasised, does not preclude the use of spin label techniques in membrane studies. In many cases, such techniques provide results that are in good agreement with those obtained using other techniques. It is important, however, to appreciate that extreme care has to be taken in the interpretation of studies based solely on spin label measurements. In the case of measurements relating to lipid phase transitions in plant membranes this warning has, unfortunately, too often been ignored.

Measurements of the temperature dependence of fluorescence polarisation of DPH, like those involving spin labels, are technically simple. Changes in DPH fluorescence are, however, known to be relatively insensitive to phase transitions in biological membranes or membrane lipid extracts. This insensitivity is illustrated in the results obtained for a total lipid extract of the membrane lipids of the blue-green alga *Anacystis nidulans* presented in Fig. 11.13(a). As described below in Section IV.B.2, preparations of this type undergo a broad but clearly defined phase transition between about 0°C and 25°C that can readily be detected using differential scanning calorimetry and wide-angle X-ray diffraction. It is, however, extremely difficult to relate any particular feature of the fluorescence changes to such a transition.

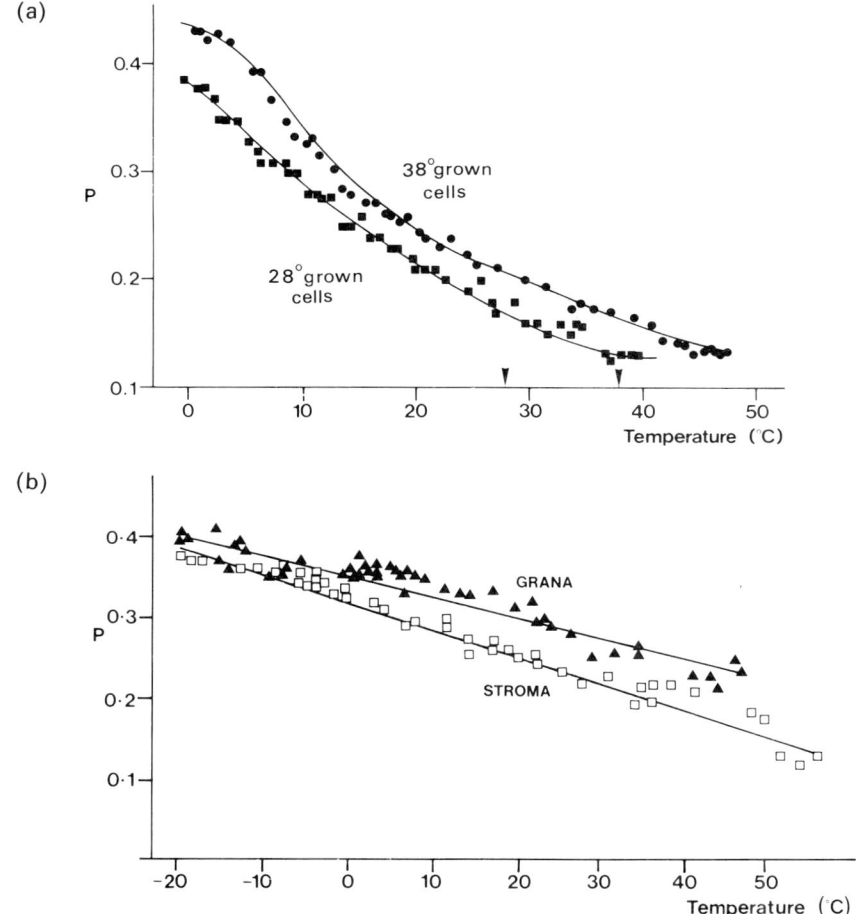

FIG. 10.13. Plots of fluorescence polarisation, P, as a function or temperature of DPH intercalated into (a) aqueous dispersions of membrane lipid extracts prepared from *Anacystis nidulans* grown at 28°C (■) and 38°C (●). (data from Mannock et al., 1985), (b) stromal and granal fractions of thylakoid membranes of pea chloroplasts (data from Barber et al., 1984).

Measurements of the steady-state polarisation of this type can be useful in distinguishing between the relative fluidities of closely related membrane fractions. Ford and Barber and their co-workers have used them widely in studies of the fluidity of the thylakoid membrane (Ford et al., 1982; Ford and Barber, 1983a,b; Barber et al., 1984). An example, taken from their work, is shown in Fig. 10.13(b). In this case measurements of DPH polarisation have been used to demonstrate differences in the relative fluidity of granal and stromal membrane preparations. The different fluidities of the fractions, as reflected by DPH measurements, are suggested to be a consequence of the increased protein levels of the appressed granal membranes as opposed to the non-appressed stromal membranes (Ford *et al.*, 1982). Barber *et al.* (1984) have made similar comparisons between the fluidity of thylakoid membranes isolated from peas (*Pisum sativum*) grown at different temperatures. In this case the fluidity of the membranes isolated from low-temperature-grown plants was shown to increase so that its value at the lower growth temperature was comparable to that found for plants grown at higher temperatures at their growth temperature. Raison *et al.* (1982) have reported similar results for spin label measurements made on polar lipid extracts of *Nerium oleander* grown at different temperatures.

Two other techniques worthy of brief mention in the context of measurements of the fluidity of plant membranes are the use of excimer formation and fluorescence recovery after photobleaching (FRAP). Under appropriate conditions, pyrene forms complexes between unexcited ground state molecules and excited state molecules. These excimers are characterised by an emission peak at about 470 nm. The rate of excimer formation in organic solvents is proportional to the collision frequency of excited and non-excited molecules. As such it is proportional to the concentration of the two molecular species and the local viscosity in the region of the excited molecules. Pyrene excimer formation in aqueous dispersions of lipids or detergents has been used to estimate lateral diffusion coefficients of pyrene and lipid–pyrene adducts in such systems. It has been generally assumed that excimer formation in membranes, as in organic solvents, is a diffusion controlled process. Recent studies reported by Blackwell *et al.* (1986) of pyrene fluorescence in model membrane systems cast doubt on this assumption and suggest that extreme caution be exercised in the interpretation of such data.

FRAP is based on the use of lipids or membrane proteins labelled with a marker such as fluorescein or rhodamine. These markers are introduced into cell membranes. Exposure of a small area of the cells to a high intensity laser beam leads to a photochemical bleaching of marker molecules within the area of exposure. Measurements of the subsequent increase in fluorescence intensity as unbleached markers from surrounding areas diffuse back into the bleached area allow an estimation of their lateral diffusion constant in the membrane phase. Metcalf and his co-workers have used FRAP to measure the mobility of lectin–receptor complexes (Metcalf *et al.*, 1983) and a range of fluorescent lipids (Metcalf *et al.*, 1986) in the plasma membranes of soybean protoplasts. In the case of phospholipid diffusion, they found evidence suggesting the existence of two populations of phospholipids which they tentatively ascribed to stable immiscible domains of fluid and gel-like lipids. Fragata *et al.* (1984) have also used FRAP to measure the lateral diffusion of plastocyanin in model membrane systems.

B. Phase Separations

There are two basic types of phase separations that might be expected to occur in biological membranes (Williams and Quinn, 1987). The first, and more widely studied, is associated with the lipid liquid-crystal bilayer to gel phase transition ($L_\alpha \rightarrow L_\beta$). The second is associated with the transition from the liquid crystal bilayer to a non-bilayer configuration ($L_\alpha \rightarrow H_{II}$).

1. General considerations

All biological membranes, with the possible exception of myelin, contain appreciable proportions of non-bilayer-forming lipids. Typically such lipids account for approximately 20% of the total membrane lipids. In the case of chloroplast and mitochondrial membranes such lipids may, however, account for 40–50% of the lipids (Quinn, 1985). Dispersions of total polar lipid extracts of chloroplast membranes in media of the type used in chloroplast isolation, as described in Section III.A, are characterised by the presence of extensive regions of phase-separated non-bilayer lipids. The absence of such structures in the native membranes has led to the conclusion that membrane proteins, possibly intrinsic membrane proteins, stabilise such lipids into bilayer configurations (Quinn et al., 1982; Williams et al., 1984). This in turn has led to the suggestion that the phase separation behaviour of such membranes might conveniently be discussed in the context of a diagram of the general form shown in Fig. 10.14 (Williams, 1988).

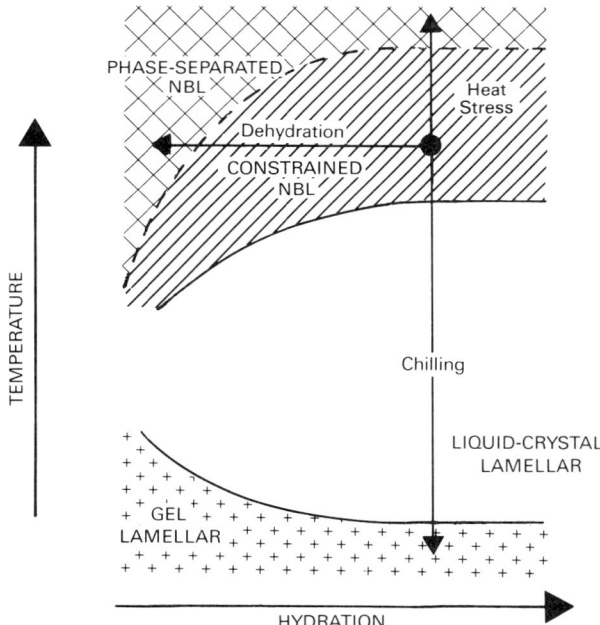

FIG. 10.14. Diagrammatic representation of the different phase separation regions of biological membranes in terms of temperature and hydration showing the relationship between liquid-crystal and gel lamellar phases and phase-separated and membrane-constrained non-bilayer-forming lipids (NBL).

In terms of this diagram, the normal physiological state of the membrane is thought to correspond to a region where both the bilayer-forming and non-bilayer-forming lipids are constrained into a liquid-crystal bilayer configuration. Lowering the temperature of the membrane, it is suggested, leads first to an increased tendency of the non-bilayer-forming lipids to revert to a bilayer configuration, and then at lower temperatures to phase-separate to form a gel-phase region while exposure to heat stress and/or dehydration of the membrane leads to a tendency for the non-bilayer lipids to phase-separate to form non-bilayer structures. Examples of all these types of phase-separation behaviour have been reported in plant membrane systems.

In terms of the physical methods available for the study of such phase separations, freeze-fracture electron microscopy has proved by far the most useful. The most convincing results, however, have been obtained in the few cases in which it has been possible to back up such studies using differential scanning calorimetry and X-ray diffraction. Phosphorus NMR has been used in an attempt to identify non-bilayer phase separations in dry plant membranes (Priestly and de Kruijff, 1982), but with no success to date.

A great deal of interest has been generated over the years by studies suggesting the occurrence of phase transitions in plant membranes based on the measurement of discontinuities in parameters associated with membrane fluidity. There are, however, good reasons (as discussed in Section IV.A.1) for treating such claims with some suspicion. Of the various fluorescence and spin probe techniques that have been used for the study of phase separations in plant systems, the use of *cis-* and *trans-*parinaric acid, which show different affinities for gel and liquid crystal regions of the lipid bilayer, is probably the most reliable. Its use, however, has been mainly if not entirely restricted to lipid extracts (Pike and Berry, 1980; Fork *et al.*, 1981; Pike, 1982; Murata and Yamaya, 1984).

2. Practical applications

Freeze-fracture electron microscopy is by far the most widely used technique for the study of phase separations in plant membranes. The great advantages of this technique are that the membranes can be examined under conditions approaching those of their native state and that small amounts of phase-separated lipid can be located. In practice, it must be admitted, most studies are performed on isolated organelles or protoplasts, but it is possible to use the technique to study seeds (Thomson and Platt-Alloia, 1978; Tovio-Kunnucan and Stushinoff, 1981; Bliss *et al.*, 1984; Vigil *et al.*, 1984), pollen (Kroh and Knuiman, 1985; Platt-Alloia *et al.*, 1986; Kerhoas *et al.*, 1987) and excised leaf tissue (Pearce and Willinson, 1985).

In the case of bilayer to non-bilayer phase transitions, the phase-separated lipid tends to reorganise itself into cylindrical inverted micelles. These, as illustrated in Fig. 10.15, have a very characteristic appearance which allows the detection of phase separation even when only a relatively small proportion of the total lipid is involved. Good examples of such phase separations have been reported for heat-stressed chloroplasts (Gounaris *et al.*, 1983b, 1984; Thomas *et al.*, 1986), chloroplasts exposed to low pH and phospholipase attack (Thomas *et al.*, 1985) and rye protoplasts isolated from non-hardened seedlings (Gordon-Kamm and Steponkus, 1984). The non-bilayer phase separations seen in heat-stressed chloroplasts can be readily explained in terms of the phase separation diagram in Fig. 10.14. The phase separations seen in rye protoplast were attributed to a dehydration of the membrane surfaces on freezing (Gordon-Kamm and Steponkus, 1984) and those seen in chloroplasts exposed to low pH and phospholipase A_2 to changes in the electrostatic balance of the membrane surface (Thomas *et al.*, 1985).

Simon (1978) explained the extreme leakiness of seeds on first imbibing water by suggesting that the lipids in dry membranes might phase separate to form hexagonal-II structures. Attempts to demonstrate the existence of non-bilayer lipid structures in seeds and pollen using freeze-fracture techniques have been largely unsuccessful. The only positive result to date has been that of Tovio-Kinnucio and Stushinoff (1981), who reported the occurrence of crystalline structures resembling hexagonal-II lipid in the lipid storage bodies of dried lettuce seeds.

A common feature of all the systems in which non-bilayer lipid phase-separation has been reported is that they contain regions of stacked or closely appressed membranes. It is possible that the presence of such features is an essential prerequisite for the observation of such phase separations and that it is their absence in seeds and pollen that accounts for the lack of positive results obtained in these systems.

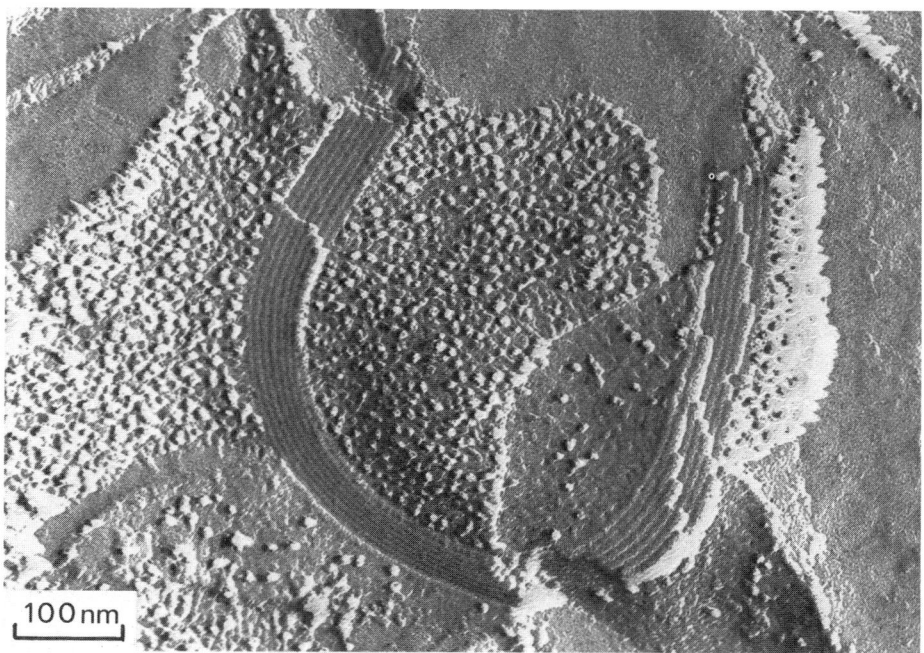

FIG. 10.15. Freeze-fracture electron micrographs showing phase-separated cylindrical inverted lipid micelles in the thylakoid membranes of isolated chloroplasts incubated at 42°C for 5 min prior to thermal quenching from 25°C.

In the case of liquid-crystal bilayer to gel phase transitions, identification of phase separations by freeze-fracture electron microscopy is more equivocal. The formation of gel-phase lipid in membranes is normally accompanied by an exclusion of freeze-fracture particles associated with intrinsic membrane proteins from the fracture face. As the liquid-crystal to gel phase transitions of most of the lipids found in plant membranes normally occur at temperatures well below 0°C (Quinn and Williams, 1983), the membranes of most plants show little or no tendency to undergo phase separations of this type at readily accessible temperatures.

The bulk of the work carried out on phase separations of this type in plant membranes has, therefore, been carried out using the thermophilic blue-green alga *Anacystis nidulans*. In this system there is clear evidence of the formation of particle-free patches of the type described above (Armond and Staehelin, 1979; Brand *et al.*, 1979; Furtado *et al.*, 1979; Verwer *et al.*, 1979; Ono and Murata, 1982). Typical examples are shown in Fig. 10.16. Other systems in which such patches have been identified are the plasma membranes of protoplasts isolated from hardened rye seedlings following freezing (Gordon-Kamm and Steponkus, 1984), the plasma membranes of leaf tissue from chilled (Pearce, 1985a; Pearce and Willinson, 1985) and drought-stressed (Pearce 1985b,c) wheat seedlings, and the thylakoid membranes of broad-bean chloroplasts subjected to catalytic hydrogenation (Thomas *et al.*, 1986). In the case of *Anacystis*, there is evidence from differential scanning calorimetry (Furtado *et al.*, 1979; Ono *et al.*, 1983; Mannock *et al.*, 1985) and X-ray diffraction studies (Tsukamoto *et al.*, 1980) confirming the existence of gel-phase lipid. In most cases this evidence is not available for the higher plant studies and the possibility that the particle-free patches seen in some of these systems arise for other reasons, as pointed out by Pearce (1985c), cannot be ignored.

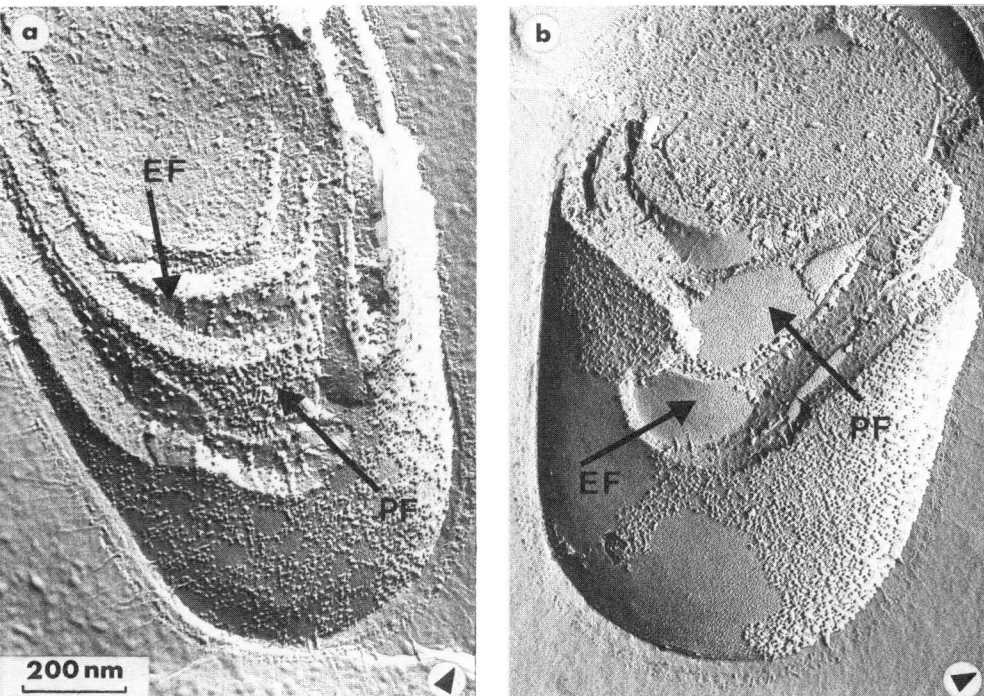

FIG. 10.16. Freeze-fracture electron micrographs of *Anacystis* cells grown at 38°C and pre-equilibrated at (a) 35°C and (b) 15°C just prior to thermal quenching. The formation of particle-free patches characteristic of the gel phase can be seen in the exoplasmic (EF) and protoplasmic (PF) faces of the photosynthetic membrane of the quenched cells from lower temperatures.

The two most unequivocable methods for demonstrating gel to liquid-crystal phase transitions in membranes are probably differential scanning calorimetry and wide-angle X-ray diffraction. Unfortunately, both methods, while comparatively simple in principle, are relatively insensitive and require highly concentrated membrane preparations. In the case of differential scanning calorimetric measurements, which are based on the measurement of the heat uptake (or release) associated with lipid phase transitions, most of the relevant studies have been performed on lipid extracts of plant mitochondria (Dalziel and Breidenbach, 1982; Raison and Orr, 1986), chloroplasts (Orr and Raison, 1987) or whole leaves (Raison and Wright, 1983). Whilst such studies have provided a great deal of potentially useful information, particularly in the context of the possible role of saturated species of phosphatidylglycerol in chilling sensitivity, some caution has to be taken in their intepretation. The occurrence of phase transitions, particularly those involving minor lipid components, in total lipid extracts is no guarantee that such transitions take place in the native membrane. Differential scanning calorimetric measurements have been used to search for evidence for lipid phase transitions in plant mitochondria (O'Neill and Leopold, 1982) and higher plant chloroplasts (Low et al., 1984), but these transitions, if they occur, are close to the detection limits of this technique. There is, however, increasing interest in the application of differential scanning calorimetric measurements to the study of thermal changes associated with protein reorganisation in chloroplasts (Cramer et al., 1981; Smith et al., 1986; Thomas et al., 1986; Thompson et al., 1986).

Wide-angle X-ray diffraction measurements, which are based on shifts of the diffraction maximum associated with the packing of the fatty acyl chains of the lipids from 0.46 nm in the liquid-crystal state to 0.42 nm in the gel state, are even more demanding in terms of sensitivity. These techniques have again been successfully applied to the measurement of phase transitions in *Anacystis* membranes (Tsukamoto et al., 1980). They have also been widely used in the study of the formation of gel-phase lipid domains as part of the sequence of events in stress-induced membrane disassembly in studies of senescence (McKersie et al., 1976; McKersie and Thompson, 1977, 1978, 1979; Pauls and Thompson, 1980; Legge et al., 1982; Faraqher et al., 1987), dehydration injury (Senaratna et al., 1984, 1985a,b, 1987) and freezing injury (Kendall et al., 1985).

C. Lipid–Protein Interactions

1. General considerations

In discussing lipid–protein interactions a clear distinction has to be made between the possible requirement of a particular protein for a small number of specific lipids bound at well defined sites that are required in order to maintain its activity, and general lipid–protein interactions arising from the fact that intrinsic membrane proteins are by their nature surrounded by a lipid matrix. Our knowledge of the former type of interaction is extremely fragmentary. Perhaps one of the best documented examples is the requirement of the mitochondrial enzyme β-hydroxybutyrate dehydrogenase for phosphatidylcholine (Grover et al., 1975).

The idea that lipids located at the lipid–protein interface may differ in some way from the bulk lipid of the membrane bilayer arose from the work of Jost et al. (1973a,b,c) on reconstituted lipid–protein preparations. Jost and her co-workers made two basic observations. First, that the spectra of nitroxide-labelled fatty acids intercalated into such preparations were characterised by two-component spectra. One of these components was shown to be associated with spin label situated in a fluid environment, and the other with a motionally restricted environment. Second, that restoration of the enzymatic activity of reconstituted preparations of this type appeared to require sufficient lipid to form a layer consisting of a single shell of phospholipids around each protein molecule. These studies led directly to the idea that intrinsic membrane proteins are surrounded by 'boundary layer lipid' or a 'lipid annulus' that might be able to influence protein activity.

It is now generally recognised that the motion of lipid molecules situated at lipid–protein interfaces, and probably also of pockets of lipid trapped between proteins, is of necessity more restricted than that of lipid in the bulk phase when viewed on the time-scale (10^{-10}–10^{-7} s) of electron spin resonance measurements. Studies using deuterated lipids suggest, however, that these restrictions are limited to this fast time-scale and that little or no restriction exists on the slower time-scale (10^{-6}–10^{-4} s) of NMR measurements. This does not preclude the possibility that certain classes of lipid tend to concentrate close to a given protein, either as the result of electrostatic interactions between the protein and the lipid headgroup or for reasons of structural compatibility that result in an increased thermodynamic stability of the overall system. The rates of exchange of such lipid with the bulk phase are, however, thought to be rapid, and there is little evidence to suggest that such selectivity has a pronounced effect on enzymatic activity. The possibility that such interactions reduce membrane leakage should not, however, be excluded.

2. Lipid–protein interactions in thylakoid membranes

The chloroplast thylakoid membrane is the only membrane that has been studied in terms of possible lipid–protein interactions. Four main areas of possible interaction have been suggested:

(1) the stabilisation of oligomeric forms of chlorophyll *a*/*b* light-harvesting protein (LHC-II) by phosphatidylglycerol;
(2) a general association of given phospholipids with the light-harvesting systems of the two photosystems;
(3) a non-specific association between monogalactosyldiacylglycerol and Photosystem II; and
(4) a tight association between sulphoquinovosyldiacylglycerol and the protein-translocating complex CF_0/CF_1.

The existence of these interactions has mainly been inferred by indirect measurements based on:

(a) the analysis of detergent-extracted membrane proteins;
(b) correlations between losses of specific photosynthetic activities associated with the hydrolysis of lipids by phospholipases and galactolipases;
(c) the requirement of specific lipids for the restoration of resonance energy transfer between pigment–protein complexes and the enzymatic/transport activities of the CF_0/CF_1, complex;
(d) correlations between the appearance of non-bilayer structures, the losses of electron transport activity and the physical dissociation of PS-II complexes as visualised by freeze-fracture electron microscopy.

Much of this work involves only indirect evidence of interactions and in any case has been recently reviewed elsewhere (Quinn and Williams, 1985; Gounaris *et al.*, 1986; Siegenthaler and Rawyler, 1986; Gounaris, 1988; Williams, 1988).

10. STRUCTURE AND DYNAMICS OF PLANT MEMBRANES

Little has appeared in the literature relating to the direct measurement of protein–lipid interactions using physical methods. Murphy and Knowles (1984) have shown that the motion of spin-labelled lipids is subject to similar motional constraints in chloroplast thylakoid membranes to those seen in other membrane preparations. The contribution of the motionally constrained lipids of spectra obtained for spin-labelled phospholipids intercalated into thylakoid membranes are clearly illustrated in Fig. 10.17.

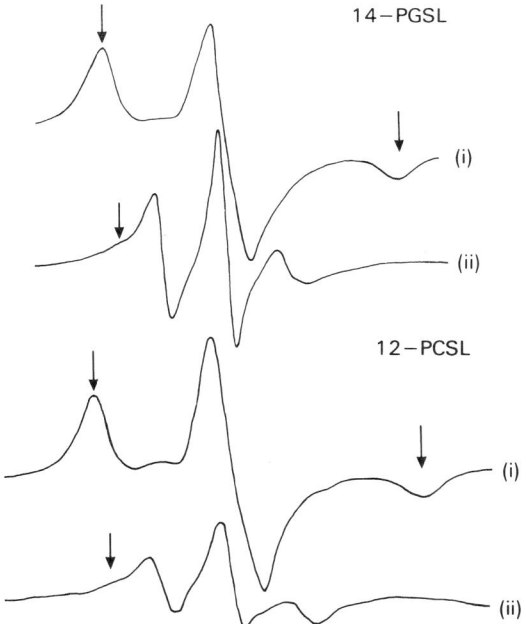

FIG. 10.17. Electron spin resonance spectra of phosphatidylglycerol and phosphatidylcholine, spin labelled on the 12-C and 14-C positions of the *sn*-2 acyl chain, incorporated into (i) appressed chloroplast thylakoid membranes and (ii) whole thylakoids. Arrows indicate the parts of the spectra diagnostic of motionally restricted lipid. Data from Murphy and Knowles (1984).

Gounaris *et al.* (1983b, 1984) showed that irreversible phase separations of non-bilayer-forming lipids can occur in chloroplast thylakoid membranes subjected to mild heat stress. Thomas *et al.* (1986) have since shown that the threshold temperature for such separations can be raised by the hydrogenation of membrane lipids in isolated chloroplasts. Parallel measurements of the thermal properties of such membranes using differential scanning calorimetry indicated that the threshold temperature of an irreversible endothermic transition, thought to be associated with the reorganisation of the Photosystem-II light-harvesting apparatus, shows a similar increase (Fig. 10.18). The precise origin of these changes is still a matter of debate. One obvious possibility, however, is that the reduced tendency of the non-bilayer-forming lipids to phase separate in such membranes leads to a preservation of lipid–protein interactions that stabilise the Photosystem II light-harvesting apparatus.

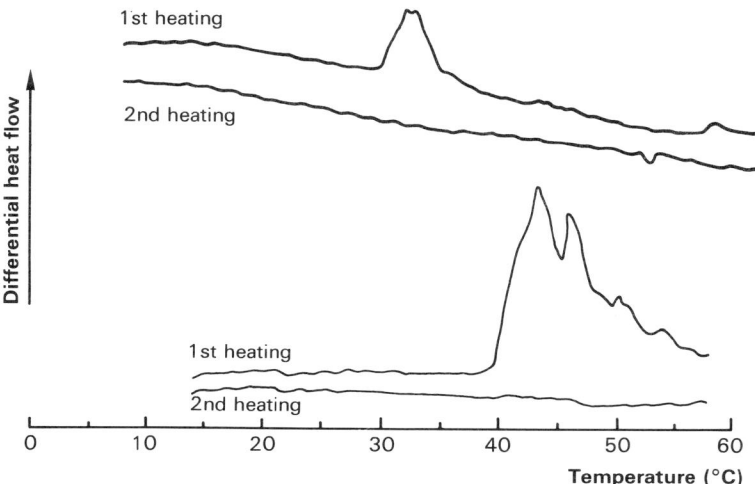

FIG. 10.18. Thermograms showing the first and second heating scans obtained from (a) control chloroplasts and (b) chloroplasts subjected to catalytic hydrogenation. Average number of double bonds per lipid molecule before and after hydrogenation was 5.29 and 1.17, respectively. Data from Thomas *et al.* (1986).

V. CONCLUSIONS

Techniques for the isolation of plant membranes and the separation and analysis of their constituents by chromatographic and electrophoretic methods have been developed over many decades of research. More recent applications of biophysical methods to study the structure of membrane components in isolation, in reconstituted systems and, in some cases, in intact membranes, has laid the foundation for a clearer understanding of the molecular basis of many biochemical functions performed by membranes.

Some of the outstanding questions that remain to be resolved are how oligomeric protein complexes are integrated into membranes and what regulates the interaction between these complexes in multicomponent systems like the electron transport chains of mitochondria and chloroplast membranes. The mechanisms that operate to enable plant membranes to function over widely varying environmental conditions is another problem that distinguishes membranes of plants from those of homeothermic animals.

In this chapter we have outlined some of the major experimental methods currently employed to address some of these questions. It is likely that future trends will involve a refinement of these techniques and more especially of methods employed to process the information derived from biophysical studies. Progress, however, must proceed together with improvements in the methods for preparing membrane fractions so as to ensure that they are in a condition as close as possible to their *in vivo* state.

REFERENCES

Akey, C. W., Crepeau, R. H., Dunn, S. D., McCarty, R. E. and Edelstel, S. J. (1983). *EMBO J.* **2**, 1409–1415.

Allen, J. P., Fecher, G., Yeates, T. O., Rees, D. C., Deisenhofer, J., Michel, H. and Huber, R. (1986). *Proc. Natl. Acad. Sci. USA*, **83**, 8589–8593.
Anderson, J. M. (1981). *FEBS Lett.* **124**, 1–10.
Anderson, J. M. and Goodchild, D. J. (1987). *FEBS Lett.* **213**, 29–33.
Andersson, B. and Anderson, J. M. (1980). *Biochim. Biophys. Acta* **593**, 427–440.
Amzel, L. M. and Pedersen, P. L. (1983). *Ann. Rev. Biochem.* **52**, 801–824.
Amzel, L. M., Narayana, P., Pedersen, P. L. and Sygusch, J. (1982). *Ann. N.Y. Acad. Sci.* **402**, 21–27.
Armond, P. A. and Staehelin, L. A. (1979). *Proc. Natl. Acad. Sci. USA* **76**, 1901–1905.
Arrondo, J. L. R., Goni, F. M. and Macarulla, J. M. (1984). *Biochim. Biophys. Acta* **794**, 165–168.
Baglioni, P., Gabrielli, G. and Guarini, B. G. T. (1980). *J. Colloid Interface Sci.* **78**, 347–353.
Barber, J., Ford, R. C., Mitchell, R. A. C. and Millner, P. (1984). *Planta* **161**, 375–380.
Bishop, D. G., Kenrick, H. R., Bayson, J. H., MacPherson, A. S. and Johns, S. R. (1980). *Biochim. Biophys. Acta* **602**, 248–259.
Blackwell, M., Gounaris, K. and Barber, J. (1986). *Biochim. Biophys. Acta* **858**, 221–235.
Blaurock, A. E. (1982). *Biochim. Biophys. Acta* **650**, 167–207.
Bliss, R. D., Platt-Alloia, K. A. and Thomson, W. W. (1984). *Plant Cell. Environ.* **7**, 601–606.
Boekema, E. J., Berden, J. A. and Van Heel, M. G. (1960). *Biochim. Biophys. Acta* **851**, 353–360.
Boekema, E. J., van Heel, M. and Graber, P. (1988). *Biochim. Biophys. Acta* **933**, 365–371.
Brand, J. J., Kirchanski, S. J. and Ramirez-Mitchell, R. (1979). *Planta* **145**, 63–68.
Breton, J. and Nabedryk, E. (1984). *FEBS Lett.* **176**, 355–359.
Breton, J. and Nabedryk, E. (1987). In "Topics in Photosynthesis" (J. Barber, ed.), Vol. 8, Ch. 4. Elsevier, Amsterdam.
Bull, T. and Lindman, B. (1974). *Mol. Cryst. Liq. Cryst.* **28**, 155–160.
Burnell, E. E., Cullis, P. R. and de Kruijff, B. (1980). *Biochim. Biophys. Acta* **603**, 63–69.
Caffrey, M. (1985). *Biochemistry* **24**, 4826–4844.
Cameron, D. G., Casal, H. L., Mantsch, H. H., Boulanger, Y. and Smith, I. C. P. (1982). *Biophys. J.* **35**, 1–16.
Cannon, B., Polnaszek, F. F., Butler, K. W. and Smith, I. C. P. (1975). *Arch. Biochem. Biophys.* **167**, 505–515.
Cary, P. R. (1982). "Biochemical Applications of Raman and Resonance Raman Spectroscopy", pp. 208–233. Academic Press, New York.
Casal, H. L. and Mantsch, H. H. (1984). *Biochim. Biophys. Acta* **779**, 381–401.
Casal, H. L., Smith, I. C. P., Cameron, D. G. and Mantsch, H. H. (1979). *Biochim. Biophys. Acta* **550**, 145–149.
Casal, H. L., Cameron, D. G., Smith, I. C. P. and Mantsch, H. H. (1980). *Biochemistry* **19**, 444–451.
Casal, H. L., Cameron, D. G., Jarrell, H. C., Smith, I. C. P. and Mantsch, H. H. (1982). *Chem. Phys. Lipids* **30**, 17–26.
Casal, H. L., Mantsch, H. H., Cameron, D. G. and Gaber, B. P. (1983). *Chem. Phys. Lipids* **33**, 109–112.
Chang, C.-H., Tiede, D., Tang, J., Smith, U., Norris, J. and Schiffer, M. (1986). *FEBS Lett.* **205**, 82–86.
Chapman, D., Williams, R. M. and Ladbrooke, B. D. (1967). *Chem. Phys. Lipids* **1**, 445–475.
Charvolin, J. and Rigny, P. (1971). *J. Magn. Reson.* **4**, 40–46.
Cline, K. (1981). *Proc. Natl. Acad. Sci. USA* **78**, 3595–3599.
Cline, K., Keegstra, K. and Staehelin, L. A. (1985). *Protoplasma* **125**, 111–123.
Costello, M. J. and Frey, T. G. (1982). *J. Mol. Biol.* **162**, 131–156.
Cramer, W. A., Whitmarsh, J. and Low, P. S. (1981). *Biochemistry* **20**, 157–162.
Cremers, F. F. M., Voorhout, W. F., van der Krift, Th. P., Lennissen-Bijvelt, J. J. M. and Verkleij, A. A. (1988). *Biochim. Biophys. Acta* **933**, 334–340.
Cullis, P. R. and de Kruijff, B. (1978a). *Biochim. Biophys. Acta* **507**, 207–218.
Cullis, P. R. and de Kruijff, B. (1978b). *Biochim. Biophys. Acta* **559**, 399–420.
Cullis, P. R., Hope, M. J., de Kruijff, B., Verkleij, A. J. and Tilcock, C. P. S. (1985). In "Phospholipids and Cellular Regulation" (J. F. Kuo, ed.), pp. 1–60. CRC Press, Boca Raton, FL.

Dalziel, A. W. and Breidenbach, R. W. (1982). *Plant Physiol.* **70**, 376–380.
Deisenhofer, J., Epp., O., Miki, K., Huber, H. and Michel, H. (1984). *J. Mol. Biol.* **180**, 385–398.
Deisenhofer, J., Epp, O., Miki, K., Huber, R. and Michel, H. (1985). *Nature (Lond.)* **318**, 618–624.
De Kruijff, B., Cullis, P. R., Verkleij, A. J., Hope, M. J., van Echfeld, C. J. A. and Tarashi, T. F. (1985). *In* "The Enzymes of Biological Membranes" (A. N. Martonosi, ed.), Vol. 1, pp. 131–204. Plenum Press, New York.
Demel, R. A., Geurts van Kessel, W. S. M. and van Deenen, L. L. M. (1972). *Biochim. Biophys. Acta* **266**, 26–40.
Douce, R. (1974). *Science* **183**, 852–853.
Douce, R., Holtz, R. B. and Benson, A. A. (1973). *J. Biol. Chem.* **248**, 7215–7222.
Ducriux, A. and Reiss-Husson, F. (1987). *J. Mol. Biol.* **193**, 419–421.
Edwards, G. E., Robinson, S. P., Tyler, N. J. C. and Walker, D. A. (1978). *Plant Physiol.* **62**, 313–319.
Eherenfest, P. (1933). *Proc. Akad. Sci., Amsterdam* **36**, 153–164.
Eriksson, P.-O., Khan, A. and Lindblom, G. (1982). *J. Phys. Chem.* **86**, 387–393.
Evans, P. J., Keates, A. G. and Cocking, E. C. (1972). *Planta* **104**, 178–181.
Faraqher, J. D., Wachtel, E. and Mayak, S. (1987). *Plant Physiol.* **83**, 1037–1042.
Fish, L. E., Kuck, U. and Bogorad, L. (1985). *J. Biol. Chem.* **260**, 1413–1421.
Ford, R. C. and Barber, J. (1983a). *Photobiochem. Photobiophys.* **1**, 263–270.
Ford, R. C. and Barber, J. (1983b). *Biochim. Biophys. Acta* **722**, 341–348.
Ford, R. C., Chapman, D. J., Barber, J., Pederesen, J. Z. and Cox, R. P. (1982). *Biochim. Biophys. Acta* **681**, 145–151.
Fork, D. C., Van Ginkel, G. and Harvey, G. (1981). *Plant Cell Physiol.* **22**, 1035–1042.
Fragata, M., Ohnishi, S., Asada, K., Ito, T. and Takahashi, M. (1984). *Biochemistry* **23**, 4044–4051.
Frank, H. A., Taremi, S. S. and Knox, J. R. (1987). *J. Mol. Biol.* **198**, 139–141.
Frey, T. G., Costello, M. J., Karlsson, B., Haselgrove, J. C. and Leigh, J. S. (1982). *J. Mol. Biol.* **162**, 113–130.
Fringeli, U. P. and Gunthard, Hs. H. (1981). *In* "Membrane Biology, Biochemistry and Biophysics" (E. Grell, ed.), Vol. 31, pp. 270–332. Springer-Verlag, Berlin.
Fuller, S. D., Capaldi, R. A. and Henderson, R. (1979). *J. Mol. Biol.* **134**, 305–327.
Furtado, D., Williams, W. P., Brain, A. P. R. and Quinn, P. J. (1979). *Biochim. Biophys. Acta* **555**, 352–357.
Gabrielli, G. and Guarini, G. G. T. (1978). *J. Colloid Interface Sci.* **64**, 185–187.
Gabrielli, G., Guarini, G. G. T. and Ferroni, E. (1976). *J. Colloid Interface Sci.* **54**, 424–429.
Gabrielli, G., Guarini, G. G. T. and Bastianini, F. (1979). *J. Colloid Interface Sci.* **69**, 352–353.
Gaines, G. L. (1966). "Insoluble Monolayers at Liquid–Gas Interfaces". Wiley-Interscience, New York.
Gally, H. U., Pluschke, G., Overath, P. and Seelig, J. (1980). *Biochemistry* **19**, 1638–1643.
Gang, L. (1988). PhD Thesis, University of Leeds, UK.
Gordon-Kamm, W. J. and Steponkus, P. L. (1984). *Proc. Natl. Acad. Sci. USA* **81**, 6373–6377.
Gounaris, K. (1988). *In* "Plant Membranes—Structure, Assembly and Function" (J. L. Harwood and T. J. Walton, eds), pp. 169–177. Biochemical Society, London.
Gounaris, K., Sunby, C., Andersson, B. and Barber, J. (1983a). *FEBS Lett.* **156**, 170–174.
Gounaris, K., Brain, A. P. R., Quinn, P. J. and Williams, W. P. (1983b). *FEBS Lett.* **153**, 47–52.
Gounaris, K., Brain, A. P. R., Quinn, P. J. and Williams, W. P. (1984). *Biochim. Biophys. Acta* **766**, 198–208.
Gounaris, K., Barber, J. and Harwood, J. L. (1986). *Biochem. J.* **237**, 313–326.
Gripsover, B., Morre, D. A. and Boss, W. F. (1984). *Protoplasma* **123**, 213–220.
Grover, A. K., Slotboom, A. J., DeHaas, G. M. and Hammes, G. G. (1975). *J. Biol. Chem.* **250**, 31–41.
Harlos, K. and Eibl, H. (1980). *Biochemistry* **19**, 896–899.
Hemminga, M. A. and Cullis, P. R. (1982). *J. Magn. Reson.* **47**, 307–323.
Hiller, R. G., Bardin, A.-M. and Nabedryk, E. (1987). *Biochim. Biophys. Acta* **894**, 365–369.
Holden, H. M., Meyer, T. E., Cusanovich, M. A., Daldal, F. and Rayment, I. (1987). *J. Mol. Biol.* **195**, 229–231.

Ishikawa, M. and Yoshida, S. (1985). *Plant Cell Physiol.* **26**, 1331–1344.
Janiak, M. J., Small, D. M. and Shipley, G. G. (1976). *Biochemistry* **25**, 4575–4580.
Jost, P. C., Capaldi, R. A., Vanderkooi, G. and Griffiths, O. H. (1973a). *J. Supramolec. Struct.* **1**, 269–280.
Jost, P. C., Griffith, O. H., Capaldi, R. A. and Vanderkooi, G. (1973b). *Proc. Natl. Acad. Sci. USA* **70**, 480–484.
Jost, P. C., Griffith, O. H., Capaldi, R. A. and Vanderkooi, G. (1973c). *Biochim. Biophys. Acta* **311**, 141–151.
Karlin-Neumann, G. A., Kohorn, B. D., Thornber, J. P. and Tobin, E. M. (1985). *J. Mol. Appl. Genet.* **3**, 45–61.
Karlsson, B., Hovmoller, S., Weiss, H. and Leonard, K. (1983). *J. Mol. Biol.* **165**, 287–302.
Kendall, E. J., McKersie, B. D. and Stenson, R. H. (1985). *Can. J. Bot.* **63**, 2274–2277.
Kerhoas, C., Gay, G. and Dumas, C. (1987). *Planta* **171**, 1–10.
Kinosita, K. and Ikegami, A. (1984). *Biochim. Biophys. Acta* **769**, 523–527.
Knoll, G. and Brdiczka, D. (1983). *Biochim. Biophys. Acta* **733**, 102–110.
Knowles, P. F., Marsh, D. and Rattle, H. W. E. (1976). "Magnetic Resonance of Biomolecules". Wiley, London.
Kreutz, W. (1970). *Adv. Bot. Res.* **3**, 54–165.
Kroh, M. and Knuiman, B. (1985). *Planta* **166**, 287–299.
Kuhlbrandt, W. (1984). *Nature (Lond.)* **307**, 478–480.
Kuhlbrandt, W. (1987). *J. Mol. Biol.* **194**, 757–762.
Kuhlbrandt, W., Thaler, T. and Wehrli, E. (1983). *J. Cell. Biol.* **96**, 1414–1424.
Kuo, A.-L. and Wade, C. G. (1979). *Biochemistry* **18**, 2300–2308.
Larsson, K., Fontell, K. and Krog, N. (1980). *Chem. Phys. Lipids* **27**, 321–328.
Legge, R., Thompson, J. E., Murr, D. P. and Tsijita (1982). *J. Exp. Bot.* **33**, 303–312.
Legge, R. L., Cheng, K. H., Lepock, R. J. and Thompson, J. E. (1986). *Plant Physiol.* **81**, 954–969.
Leonard, K., Haiker, H. and Weiss, H. (1987). *J. Mol. Biol.* **194**, 277–286.
Li, J. (1985). *Proc. Natl. Acad. Sci. USA* **82**, 386–390.
Lindblom, G. and Wennerstrom, H. (1977). *Biophys. Chem.* **6**, 167–171.
Lindblom, G., Larsson, K., Johansson, L., Fontell, K. and Frosen, S. (1979). *J. Am. Chem. Soc.* **101**, 5465–5470.
Lis, L. J. and Quinn, P. J. (1986). *Biochim. Biophys. Acta* **862**, 81–86.
Lis, L. J. and Quinn, P. J. (1987). *Mol. Cryst. Liq. Cryst.* **146**, 35–39.
Lord, R. C. and Mendelsohn, R. (1981). In "Membrane Spectroscopy" (E. Grell, ed.), pp. 377–436. Springer-Verlag, Berlin.
Lord, J. M., Kagawa, T., Moore, T. S. and Beevers, H. (1973). *J. Cell Biol.* **57**, 659–667.
Low, P. S., Ort, D. R., Cramer, W. A., Whitmarsh, J. and Marton, B. (1983). *Biochim. Biophys. Acta* **231**, 336–344.
Lunsdorf, H., Ehrig, K., Friedl, P. and Schairev, H. U. (1985). *J. Mol. Biol.* **173**, 131–136.
Luzzati, V. (1974). In "Perspectives in Membrane Biology" (O. S. Estrada and C. Gitler, eds), pp. 25–43. Academic Press, New York.
Luzzati, V. (1968). In "Biological Membranes" (D. Chapman, ed.), pp. 71–123. Academic Press, London.
Luzzati, V. and Tardieu, A. (1974). *Ann. Rev. Phys. Chem.* **25**, 79–94.
Luzzati, V., Gulik-Kryzwicki, T. and Tardien, A. (1968). *Nature (Lond.)* **218**, 1031–1034.
Lynch, D. V., Lepock, J. R. and Thompson, J. E. (1987). *Plant Cell Physiol.* **28**, 787–797.
Mannock, D. A., Brain, A. P. R. and Williams, W. P. (1985). *Biochim. Biophys. Acta* **821**, 153–164.
Mansourian, A. R. and Quinn, P. J. (1986). *Biochim. Biophys. Acta* **855**, 169–178.
Marsh, D. and Seddon, J. M. (1982). *Biochim. Biophys. Acta* **690**, 117–123.
McCarty, R. E. and Nalin, C. M. (1986). In "Photosynthesis III" (L. A. Staehelin and C. J. Arntzen, eds). Springer-Verlag, Berlin.
McElhaney, R. N. (1986). *Biochim. Biophys. Acta* **864**, 361–421.
McKersie, B. D. and Thompson, J. E. (1977). *Plant Physiol.* **59**, 803–807.
McKersie, B. D. and Thompson, J. E. (1978). *Plant Physiol.* **61**, 639–643.
McKersie, B. D. and Thompson, J. E. (1979). *Biochim. Biophys. Acta* **550**, 48–58.
McKersie, B. D., Thompson, J. E. and Brandon, J. K. (1976). *Can. J. Bot.* **54**, 1074–1078.

McLauchlin, A. C., Cullis, P. R., Hemminga, M. A., Hoult, D. I., Radda, G. K., Ritchie, G. A., Seeley, P. J. and Richards, R. E. (1975). *FEBS Lett.* **57**, 213–218.
Mendelsohn, R. and Koch, C. C. (1980). *Biochim. Biophys. Acta* **598**, 260–271.
Metcalf, T. N., Wang, T. J., Schubert, K. R. and Schindler, M. (1983). *Biochemistry* **22**, 3969–3975.
Metcalf, T. N., Wang, J. L. and Schindler, M. (1986). *Proc. Natl. Acad. Sci. USA* **83**, 95–99.
Michel, H. (1982). *J. Mol. Biol.* **158**, 567–572.
Millner, P. A., Chapman, D. J. and Barber, J. (1984a). *Biochim. Biophys. Acta,* **765**, 282–287.
Millner, P. A., Mitchell, R. A. C., Chapman, D. J. and Barber, J. (1984b). *Photosynth. Res.* **5**, 63–76.
Moor, H., Kistler, J. and Muller, M. (1976). *Experientia* **32**, 805–815.
Morre, D. J., Lem, N. W. and Sandelius, A. S. (1984). *In* "Structure, Function and Metabolism of Plant Lipids" (P. A. Siegenthaler and W. Eichenberger, eds), pp. 325–328. Elsevier, Amsterdam.
Morris, J. and Herrmann, R. G. (1984). *Nucleic Acids Res.* **12**, 2837–2850.
Murata, N. and Yamaya, J. (1984). *Plant Physiol.* **74**, 1016–1024.
Murphy, D. J. and Knowles, P. (1984). *In* "Structure, Function and Metabolism of Plant Lipids" (P. A. Siegenthaler and W. Eichenberger, eds), pp. 425–428. Elsevier, Amsterdam.
Murphy, D. J. and Woodrow, I. E. (1983). *Biochim. Biophys. Acta* **723**, 104–112.
Nabedryk, E., Tiede, D. M., Dutton, P. L. and Breton, J. (1982). *Biochim. Biophys. Acta* **682**, 273–280.
Nabedryk, E., Andrinambinintsoa, S., and Breton, J. (1984). *Biochim. Biophys. Acta* **765**, 380–387.
Neuman, R. D. (1976). *J. Colloid Interface Sci.* **56**, 505–510.
O'Neill, S. D. and Leopold, A. C. (1982). *Plant Physiol.* **70**, 1405–1409.
Ono, T.-A. and Murata, N. (1982). *Plant Physiol.* **69**, 125–129.
Ono, T.-A., Murata, N. and Fujita, T. (1983). *Plant Cell Physiol.* **24**, 635–639.
Orr, G. R. and Raison, J. K. (1987). *Plant Physiol.* **84**, 88–92.
Paliyath, G., Poovaiah, B. W., Munske, G. R. and Magnuson, J. A. (1984). *Plant Cell Physiol.* **25**, 1083–1087.
Pauls, K. P. and Thompson, J. E. (1980). *Nature* **283**, 504–506.
Pearce, R. S. (1985a). *J. Exp. Bot.* **36**, 369–381.
Pearce, R. S. (1985b). *J. Exp. Bot.* **36**, 1209–1221.
Pearce, R. S. (1985c). *Planta* **166**, 1–14.
Pearce, R. S. and Willinson, J. H. M. (1985). *Planta* **163**, 304–316.
Pike, C. S. (1982). *Plant Physiol.* **70**, 1764–1766.
Pike, C. S. and Berry, J. A. (1980). *Plant Physiol.* **66**, 238–241.
Platt-Alloia, K. A., Lord, E. M., DeMason, D. A. and Thomson, W. W. (1986). *Planta* **168**, 291–298.
Priestly, D. A. and de Kruijff, B. (1982). *Plant Physiol.* **70**, 1075–1078.
Quail, P. H. (1979). *Ann. Rev. Plant Physiol.* **30**, 425–484.
Quinn, P. J. (1985). *Cryobiology* **22**, 28–46.
Quinn, P. J. and Lis, L. J. (1987). *J. Colloid Interface Sci.* **115**, 220–224.
Quinn, P. J. and Williams, W. P. (1978). *Progr. Biophys. Mol. Biol.* **34**, 109–173.
Quinn, P. J. and Williams, W. P. (1983). *Biochim. Biophys. Acta* **737**, 223–266.
Quinn, P. J. and Williams, W. P. (1985). *In* "Photosynthetic Mechanisms and the Environment" (J. Barber and N. R. Baker, eds), pp. 1–47. Elsevier, Amsterdam.
Quinn, P. J., Gounaris, K., Sen, A. and Williams, W. P. (1982). *In* "Biochemistry and Metabolism of Plant Lipids" (J. F. G. M. Wintermans and P. J. C. Kuiper, eds), pp. 327–330. Elsevier, Amsterdam.
Raison, J. K. and Orr, G. R (1986). *Plant Physiol.* **81**, 807–811.
Raison, J. K. and Wright, L. C. (1983). *Biochim. Biophys. Acta* **731**, 69–78.
Raison, J. K., Roberts, J. K. M. and Berry, J. A. (1982). *Biochim. Biophys. Acta* **688**, 218–228.
Rand, R. P., Tinker, D. A. and Fast, P. G. (1971). *Chem. Phys. Lipids* **6**, 333–342.
Rathnam, C. K. M. and Edwards, G. E. (1976). *Plant Cell Physiol.* **17**, 177–186.
Reiss-Husson, F. (1967). *J. Mol. Biol.* **25**, 363–382.

Rilfors, L., Eriksson, P.-O., Arvidson, G. and Lindblom, G. (1986). *Biochemistry* **25**, 7702–7710.
Rivas, E. and Luzzati, V. C. (1969). *J. Mol. Biol.* **41**, 261–275.
Roeder, S. B. W., Burnell, E. E., Kuo, A.-L. and Wade, C. G. (1976). *J. Chem. Phys.* **64**, 1848–1849.
Schmidt, G. W. and Mishkind, M. L. (1986). *Ann. Rev. Biochem.* **55**, 879–912.
Schreier, S., Polnaszek, C. F. and Smith, I. C. P. (1978). *Biochim. Biophys. Acta* **515**, 375–436.
Scriven, L. E. (1976). *Nature (Lond.)* **263**, 123–125.
Seddon, J. M., Cevc, G., Kaya, R. D. and Marsh, D. (1984). *Biochemistry* **23**, 2634–2644.
Seelig, J. (1977). *Q. Rev. Biophys.* **10**, 353–418.
Seelig, J. (1978). *Biochim. Biophys. Acta* **515**, 105–140.
Seelig, J. and Niederberger, W. (1974). *J. Am. Chem. Soc.* **96**, 2069–2072.
Seelig, J. and Seelig, A. (1980). *Q. Rev. Biophys.* **13**, 19–61.
Sen, A., Williams, W. P., Brain, A. P. R., Dickens, M. J. and Quinn, P. J. (1981a). *Nature (Lond.)* **293**, 486–490.
Sen, A., Williams, W. P. and Quinn, P. J. (1981b). *Biochim. Biophys. Acta* **663**, 380–389.
Sen, A., Williams, W. P., Brain, A. P. R. and Quinn, P. J. (1982a). *Biochim. Biophys. Acta* **685**, 297–306.
Sen, A., Brain, A. P. R., Quinn, P. J. and Williams, W. P. (1982b). *Biochim. Biophys. Acta* **686**, 215–224.
Sen, A., Mannock, D. A., Collins, D. J., Quinn, P. J. and Williams, W. P. (1983). *Proc. Roy. Soc. Lond. B* **218**, 349–364.
Senaratna, T., McKersie, B. D. and Stinson, R. H. (1984). *Plant Physiol.* **76**, 759–761.
Senaratna, T., McKersie, B. D. and Stinson, R. H. (1985a). *Plant Physiol.* **77**, 472–474.
Senaratna, T., McKersie, B. D. and Stinson, R. H. (1985b). *Plant Physiol.* **78**, 168–171.
Senaratna, T., McKersie, B. D. and Borochov, A. (1987). *J. Exp. Bot.* **38**, 2005–2014.
Senior, A. E. and Wise, J. G. (1983). *J. Membrane Biol.* **73**, 105–124.
Shinitzky, M. and Barenholz, Y. (1978). *Biochim. Biophys. Acta* **515**, 367–394.
Shipley, G. G. (1973). In "Biological Membranes" (D. Chapman and D. F. M. Wallach, eds), Vol. 2, pp. 1–89. Academic Press, London.
Shipley, G. G., Green, J. P. and Nichols, B. W. (1973). *Biochim. Biophys. Acta* **311**, 531–544.
Siegenthaler, P. A. and Rawyler, A. (1986). In "Encyclopedia of Plant Physiology", New Series, Vol. 19, pp. 693–705. Springer-Verlag, Berlin and Hamburg.
Silvius, J. R. (1982). In "Lipid–Protein Interactions" (P. C. Jost and H. O. Griffith, eds), Vol. 2, pp. 239–281. Wiley, New York.
Simon, E. W. (1978). In "Dry Biological Systems" (J. H. Crowe and J. S. Clegg, eds), pp. 205–224. Academic Press, New York.
Sims, B. and Zografi, G. (1972). *J. Colloid Interface Sci.* **41**, 35–39.
Singer, S. J. and Nicholson, G. L. (1972). *Science* **175**, 720–731.
Small, D. M. (1986). In "Handbook of Lipid Research" (D. J. Hanaham, ed.), Vol. 4, pp. 1–672. Plenum Press, New York.
Smith, I. C. P. and Mantsch, H. H. (1979). *Trends Biochem. Sci.* **4**, 152–154.
Smith, K. A., Ardelt, B. K. and Low, P. S. (1986). *Biochemistry* **25**, 7099–7105.
Smith, R. D. and Berg, J. C. (1980). *J. Colloid Interface Sci.* **74**, 273–278.
Stilbs, P. (1986). *Progr. Nucl. Magn. Reson. Spectrosc.* **19**, 1–45.
Stilbs, P., Arvidson, G. and Lindblom, G. (1984). *Chem. Phys. Lipids* **35**, 309–314.
Stockton, G. W., Polnaszek, C. F., Leitch, L. C., Tulloch, A. P. and Smith, I. C. P. (1974). *Biochem. Biophys. Res. Commun.* **60**, 844–850.
Stokes, D. M. and Walker, D. A. (1971). *Plant Physiol.* **48**, 163–165.
Strotman, H. and Bickel-Sandkotter, S. (1984). *Ann. Rev. Plant Physiol.* **35**, 97–120.
Stubbs, C. D. (1983). In "Essays in Biochemistry" (P. N. Campbell and R. D. Marshall, eds), Vol. 19, pp. 1–39. Biochemical Society, London.
Sunder, S., Cameron, D. G., Mantsch, H. H. and Bernstein, H. T. (1978). *Can. J. Chem.* **56**, 2121–2126.
Tenchov, B. G., Lis, L. J. and Quinn, P. J. (1987). *Biochim. Biophys. Acta*, **897**, 143–151.
Thiery, R., Klein, R. and Tatischeff, I. (1987). *FEBS Lett.* **223**, 381–386.
Thomas, P. G., Brain, A. P. R., Quinn, P. J. and Williams, W. P. (1985). *FEBS Lett.* **183**, 161–166.

Thomas, P. G., Dominy, P. J., Vigh, L., Mansourian, A. R., Quinn, P. J. and Williams, W. P. (1986). *Biochim. Biophys. Acta* **849**, 131–140.
Thompson, L. K., Sturtevant, J. M. and Bradvig, G. W. (1986). *Biochemistry* **25**, 6161–6169.
Thomson, W. W. and Platt-Alloia, K. A. (1978). *Stain Technol.* **57**, 327–334.
Tiedge, H., Schafer, G. and Meyer, F. (1983). *Eur. J. Biochem.* **132**, 37–45.
Tiedge, H., Lunsdorf, H., Schafer, G. and Schairer, H. U. (1985). *Proc. Natl. Acad. Sci. USA* **82**, 7874–7878.
Tilcock, C. P. S., Cullis, P. R. and Gruner, S. M. (1986). *Chem. Phys. Lipids* **40**, 47–56.
Tomoaia-Cotisel, M., Sen, A. and Quinn, P. J. (1983a). *J. Colloid Interface Sci.* **94**, 390–398.
Tomoaia-Cotisel, M., Zsako, E., Chifu, E. and Quinn, P. J. (1983b). *Chem. Phys. Lipids* **34**, 55–64.
Tomoaia-Cotisel, M., Zsako, J., Chifu, E. and Quinn, P. J. (1989). *Chem. Phys. Lipids* **50**, 127–133.
Tovio-Kinnucan, M. A. and Stushinoff, C. (1981). *Cryobiology* **18**, 72–79.
Tristram-Nagle, S., Wiener, M. C., Yang, C.-P. and Nagle, J. F. (1987). *Biochemistry* **26**, 4288–4294.
Tsukamoto, Y., Yeki, T., Mitsui, T., Ono, T.-A. and Murata, N. (1980). *Biochim. Biophys. Acta* **602**, 673–675.
Tsuprun, V. L., Mesyanzhinova, I. V., Kozlov, I. A. and Orlova, E. V. (1984). *FEBS Lett.* **167**, 285–290.
Uemura, M. and Yoshida, S. (1986). *Plant Physiol.* **80**, 187–195.
Unwin, P. N. T. and Zampighi, G. (1980). *Nature (Lond.)* **283**, 545–549.
Verkleij, A. J. (1985). *Biochem. Biophys. Acta* **779**, 43–63.
Verkleij, A. R. and Ververgaert, P. H. J. Th. (1978). *Biochim. Biophys. Acta* **515**, 303–327.
Verwer, W., Ververgaert, P. J. J. T., Leunissen-Bijvelt, J. and Verkleij, A. J. (1979). *Biochim. Biophys. Acta* **504**, 231–234.
Vigil, E. L., Steere, R. L., Wergin, W. P. and Christiansen, M. N. (1984). *Am. J. Bot.* **71**, 601–606.
Wacker, T., Gad'on, N., Becker, A., Maentele, W., Kreutz, W., Drews, G. and Welte, H. (1986). *FEBS Lett.* **197**, 261–273.
Weisbeek, P., Hageman, J., Cremers, F., Keegstra, K., Banerle, C. and Smeekens, S. (1986). *Curr. Top. Plant Biochem. Physiol.* **5**, 88–104.
Wilkinson, D. A. and Nagle, J. F. (1978). *Anal. Biochem.* **84**, 263–271.
Williams, W. P. (1988). In "Plant Membranes—Structure, Assembly and Function" (J. L. Harwood and T. J. Walton, eds), pp. 47–64. Biochemical Society, London.
Williams, W. P. and Quinn, P. J. (1987). *J. Bioenerg. Biomembr.* **19**, 605–624.
Williams, W. P., Gounaris, K. and Quinn, P. J. (1984). In "Advances in Photosynthesis Research" (C. Cybesma, ed.), Vol. III, pp. 123–130. Nijhoff/Junk, The Hague.
Williamson, F. A., Morre, D. J. and Jaffe, M. J. (1975). *Plant Physiol.* **56**, 738–743.
Wong, P. T. T. (1984). *Ann. Rev. Biophys. Bioeng.* **13**, 1–24.
Yoshida, S. (1984). *Plant Physiol.* **75**, 38–42.
Yoshida, S., Kawata, T., Uemura, M. and Niki, J. (1986). *Plant Physiol.* **80**, 152–160.
Zingstein, H. P. (1972). *Biochim. Biophys. Acta* **265**, 339–366.
Zinth, W., Kaiser, W. and Michel, H. (1983). *Biochim. Biophys. Acta* **723**, 128–131.
Zuber, H. (1986). *Trends Biochem. Sci.* **11**, 414–419.

Index

A

Absorption detected magnetic resonance (ADMR), 249
Accumulated photon echo, 240
Absorption fluorescence, 232
[^{14}C]Acetate, 72, 94
Acetyl-CoA, 181
Acid rain, 108, 109
Actinic effect, 234, 235, 238, 243
Actinic flash, 269, 282
Actinic fluence rates, 208, 210
Actinometry, 210
Action spectroscopy, 187, 191–193, 219, 222, 224–225
Acyl-ACP(CoA):sn-glycerol-3-phosphate acyltransferase, 87, 94
Acyl carrier protein (ACP), 23
Acyl-CoA(S), 23, 24, 32, 36, 37, 38, 39, 40, 41, 42, 43, 44, 96
  preparation, 27–28
  purification, 33
  substrates, 28–29
  synthase, 95
Acylglycero-3-phosphate acyltransferase, 23, 24, 41
  assay, 42
Acyl pyrrolidides, 14
Acyltransferase, 94, 95
Alcohol dehydrogenase, 206
Alcoholysis, 124
Aldehydes
  in polyacetylenes, 160, 164, 175, 178
  in wax esters, 129
Aliphatic monoesters
  higher, 124–125
  of wax, 123–124
Aliphatic hydrocarbons
  wax components, 113, 114–119
Alkaline phosphatase, 205, 206, 208, 226
α,ω-Alkanediols, 124

2-Alkenyl-4,4-dimethyloxazonalines, 15
Alkyl esters, 112
9-Aminoacridine, 264, 287
  assay of fluorescence, 283, 287
Amyloplasts, 75, 84
  purification, 79–80
Analogue substitution
  photoreceptor identification, 222
8-Anilo-4-naphthosulphonic acid (ANSA), 57, 86, 87, 88, 91, 96
p-Anisoyl
  HPLC detection of phospholipids, 61
Anisoyldiacylglycerol, 90
Antenna complex, 260
Antenna pigments, 235, 245
Antenna quenching, 290
Anthracene, 113
Antibiotic marker genes, 226
Antimycin A, 264
Apoferritin, 207
Apomyoglobin, 248
Appressed membrane, 265, 267, 329
Arachidonic acid, 6, 28
  structure, 1
Arachidonic picolinyl ester, 15
Argentation thin layer chromatography (TLC), see Silver nitrate-TLC
Aromatic esters, 125
Aromatic hydrocarbons
  wax components, 113
Arrhenius plot, 324
ATPase, 319–320
ATP synthase, 288
Attenuated total reflection, 313
Autoradiography, 56, 91, 92, 93, 95, 96

B

Ba131, 218, 219
BamHI, 215, 216

342　INDEX

Bathochromic shift, 222
Beer–Lambert's law, 233
Benzoate
　phospholipid head group substitute, 60, 61
Benzotropylium cation, 174, 175
Benzoquinone, 271
Bilayer membrane, 63
Birefringence, 314–316
Bisalkylthiols, 13
Bithienylbutynene, 160, 166, 175, 177, 179
Bligh and Dyer extraction, 30, 37, 49, 50, 85, 98, 101
Borage, 26
　Δ6 desaturase activity, 39, 41
Boundary layer lipid, 334
Bovine serum albumin (BSA), 33
　in acyl-CoA utilising assays, 28, 29
Bragg reflection, 306, 307
tertButyl dimethylsilyl ethers, 13

C

Caffeine
　in centrifugal preparative chromatography, 166, 178
　in TLC of polyacetylenes, 163–165
Capric acid, 21
Calorimetry, 309–310, 325, 328, 331, 333
Carbazole, 112
Cardiolipin, see bisPhosphatidylglycerol, 302
iso-Cardopatine, 166, 175
β-Carotene, 236, 248
　absorption spectrum, 233
Carotenoids, 233, 240, 245, 247, 250, 253, 266
　electrochromic shift, 244, 246
Casparian band, 145
Catalase, 26
CDP-choline, 43, 44
　as precursor, 66
CDP-diacylglycerol, 66
CDP-choline-1,2-diacylglycerol choline-phosphotransferase—see Choline phosphotransferase
Cellulase, 75, 108, 299
Cellulose acetate, 110
Centrifugal preparative chromatography, 165–166
CF_1, 319, 332
　subunit arrangement, 320
CF_0, 319, 332
Chemical probes, 64
Chlorohydrins, 109
Chloroplast, 23, 93, 94, 234, 264, 265, 286, 299, 331
　CO_2 dependent O_2 evolution, 75

Chloroplast (cont.)
　determination of intactness, 81
　envelope membrane, 72, 74, 300
　isolation, 73
　preparation for chlorophyll fluorescence measurements, 271
　purification of envelope membrane, 83
　purity assessment, 81
　spinach, 82
　stroma, 74, 83
　thylakoids, 74, 83, 243, 245, 246, 267, 323, 324
Chlorophyll, 189, 203
　a, 240, 242, 243, 248, 260
　absorption spectrum, 233
　antenna, 243
　bacterio, 240, 247
　–chlorophyll interaction, 233, 234, 248
　determination, 75
　factors affecting fluorescence yield, 261–267
　fluorescence, 235, 243, 262
　measurement of fluorescence changes, 267–282
　measurement of fluorescence long term quenching, 282–289
　membrane organisation and fluorescence of, 266
　–protein interaction, 234, 248
　proteins of the thylakoid membrane (CP29, CP43, CP47), 260
　Q_y region, 253
　quantum yield, 289–293
　triplet, 243, 244
Choline
　head group precursor, 66–68
Cholinephosphotransferase, 23, 24
　assay, 43–44
Chromatotron, see Centrifugal preparative chromatography
Circular dichroism, 234, 248, 319
Coenzyme A (CoA), 23, 27, 28, 37, 39, 40, 44, 93, 96
Conidiation, 222
Colloidal gold conjugated antibodies, 300
Collodion, 110
Colorimetric assay
　of phospholipids, 51, 58
Column chromatography
　of cuticular waxes, 110–112, 113, 130, 133, 136
　of glycolipids, 88
　of polyacetylenes, 161–162, 163
　reverse phase, 130, 166, 167
Complex lipid
　acyl constituent analysis, 31–32
　extraction, 29–30

Complex lipid (*cont.*)
  purification, 31
  stereospecific analysis, 34–35
Cork layer, *see* Phelem
Conversion factors for radiometric units, 188
Crepenynic acid, 181
Cubic phases, 301, 307, 312
Cupric sulphate, 53
Cuticular wax, 108
  analysis, 109–135
Cutin
  analysis of cuticular waxes, 109–135
  analysis of cutin, 135–145
  cuticular membrane, 106–109
  monomer acid structures, 143
Cutinase, 144
Cuvette, 236
Cytochromes, 233, 244, 247, 253, 319
  a/a_3, 318
  b, 220, 221
  b/c_1 complex, 317
  b/f complex, 246
  c, 316
  f/b_6 complex, 324
  oxidase, 317

D

D-Amino acid oxidase, 220
D1 protein, 280
d-Spacing, 307
DC measuring light, 244–246
1-Deazariboflavin, 222
Deazaflavin, 224
Depolymerisation
  of cutin, 135, 136, 138, 139, 142, 144
  of suberin, 148
Dehydration injury, 331
Dehydromatricaria acid, 180
Dehydromatricaria ester, 172, 176, 179
  ^{1}H-NMR spectrum, 173
  structure, 177
Dehydromatricanol, 180
Delayed light emission, 250
Deoxyribonuclease I, 226
Deoxyribonucleate cyclobutane dipyrimidine photolyase, *see* DNA photolyase
Desaturases, 22, 25, 26
  assays, 39–41
  $\Delta 6$, 39
  $\Delta 12$, 39, 40, 41
  $\Delta 15$, 39, 41
Detecting flash absorption spectrometry, 246–247
Deuterolytic cleavage of cutin, 138, 144

Developing seeds
  in the study of TAG synthesis, 25–26
Diacylglycerol, 21, 23, 24, 31, 34, 35, 49, 52, 81, 83, 88, 90, 93, 97, 100
  interconversion with PC, 43–44
  precursor, 66
  purification, 98
Diacylglycerol acyltransferase, 23, 24
  assay, 44
1,2 Diacylglycerol-3-phosphate, *see* Phosphatidic acid
Diaphorase, 206
Diazomethane
  in methanolysis of fatty acids, 90, 130, 136
2,7-Dichlorofluorescein, 33, 112
3-(3,4-Dichlorophenyl)-1,1-dimethyl-urea (DCMU), 263, 264, 275–276, 277, 278, 280, 285, 287
Diels–Alder reaction, 177
Digalactosyldiacylglycerol (DGDG), 55, 87, 89, 95, 97, 99, 303
  structure, 73
Differential scanning calorimetry, 309, 310
Dihexosyldiacylglycerols, 302
9,16-Dihydroxyhexadecanoic acid, 142
β-Diketones, 110, 112, 119, 122
  substituted, 122–123
Dilatometry, 310
Dilinolenoylphosphatidylcholine, 302
Dillapiol, 178
p-Dimethylaminobenzaldehyde, 162
Dimethylethanolamine, 67
1,5-Dinitro-3-fluorobenzene (DNFB), 64
Dioleoyl-glycerol, 44
  purification, 98
Dioleoylphosphatidylcholine, 302
Dioxo-β-diketones, 122
Dipalmitoylphosphatidylcholine, 310
1,6-Diphenyl-2,3,5-hexatriene (DPH), 321, 322, 323, 324, 325, 326
Distearoyldiacylglycerol, 303, 304
Distearoylmonogalactosyldiacylglycerol, 306
Distearoylmonogalactosylglycerol, 303
Diterpenoids, 130
5,5 Dithiobis(2-nitrobenzoic acid) (DTNB), 38
cDNA
  of phytochrome, 213, 214
DNA photolyase, 186, 187, 219, 223–227
trans-2-docosenoic acid, 124
Dodecanoic acid, 125
Dot blot, 214

E

EcoRI, 215

EDTA
  inhibition of phosphatidate phosphohydrolase, 42
Eherenfest theory, 304
Electroblotting
  of fusion proteins, 218
  of phytochrome, 199, 201
Electron carriers, 233
Electron diffraction, 318
Electronic transition
  photosynthesis and, 233–249
Electron microscopy
  ATPase, 332
  of CF_1 crystals, 319
  epicuticular wax morphology, 107
  plant membranes, 303, 307–309
Electron spin resonance, (ESR), 232, 332
  of cuticular membranes, 108
Electron transfer kinetics, 250
Electron transport, 260
  photosynthetic, 324
  uncoupled thylakoid and, 286
Electrophoresis, 67
Endoplasmic reticulum, 22, 23, 25, 299
Endoproteolytic attack
  of phytochrome, 193, 195, 197, 198, 199
Energy fluence rate, 187–188
Enzyme alteration of phospholipid membranes, 63, 64
Enzyme linked immunosorbent assay (ELISA)
  for phytochrome, 203–208, 219
Epicuticular wax, 106, 107, 130, 147, 148
  analysis of 133–134
  extraction and isolation, 109–110
Epitope
  mapping, 213, 215–219
  phytochrome, 205, 207
Epoxyacids, 135
9,10-Epoxy-18-hydroxyoctadecanoate, 138, 139, 142
Epoxides, 175, 177–178
Erucic acid, 22
Estolides, 124, 125, 130, 135
  mass spectra, 126, 127
Ethanolamine
  head group precursor, 66–68
Exchange proteins, 63, 64–66
Excimer, formation, 326
Exciton theory, 233
Exonuclease, 218
Extrinsic membrane proteins, 299

F

F_1, 319

F_0, 319
Falcarinol, 180
Falcarinone, 180
Far red absorbing phytochrome (Pfr), 189, 194, 195, 201, 203
  as a function of wavelength, 212
Fatty acids, 1–16, 21, 22, 23, 24, 25
  ammonium salt preparation, 29
  analysis in glycolipids, 90–93
  hydroxy, 106, 122, 124, 125, 130, 135, 136
  identification, 6–16
  methyl esters, 4, 5, 31, 32, 33, 35, 36, 60, 90, 92
  nomenclature, 22
  quantitative analysis, 91–92
  separation procedures, 3–6
  stereospecific analysis, 34–35
  structures, 2
  TLC of molecular species, 92–93
  very long chain, 148
  wax components, 130
Fatty acid synthetase, 93
Fatty alcohols, 129, 135, 136, 138, 147, 148
Fatty aldehydes, 129
Femtosecond spectroscopy, 238
Ferulic acid, 149
Filicanone, 130
  structure 131
Filters
  bandpass, 191
  cutoff, 191
  interference, 191
  neutral density, 190–191
Flame ionization detector (FID), 60, 92, 137
  in HPLC, 59, 62, 63
Flash excitation fluorescence apparatus, 268–271, 273, 279–282
Flash photolysis, 235
Flavanoids, 131
Flavins, 187, 223
  receptor, 219
  analogues, 222
  semiquinone radical, 225
Flavones, 138
Flavoproteins, 219, 220, 221, 222
Flip flop of phospholipids, 63, 65
Fluid mosaic model, 297
Fluorescein, 326
Fluorescence, 192, 249, 251
  anisotropy, 322, 323
  chlorophyll transients, 259–293
  depolarisation, 324
  detected magnetic resonance (FDMR), 251
  in HPLC detection, 62
  lifetime measurements, 250
  measurement, 268–282

Fluorescence (*cont.*)
    measurement of long term quenching, 282–289
    modulated, 263
    nanosecond delayed, 250
    non-photochemical quenchers of, 263, 264, 265, 266, 283–285, 286, 287–288
    of phytochrome with Zn^{2+}, 201
    photochemistry and, 261
    photochemical quenchers of, 263, 283–285
    recovery after bleaching (FRAB), 326
    spectroscopy for photoreceptor identification, 220–221
    variable, 243
    yield, 261
Free flow electrophoretic separation, 300
Freeze injury, 331
Freeze fracture, 300, 307, 308, 310, 328, 329
Freeze substitution, 300
Furans, 135
Fusion peptides, *see* Fusion proteins
Fusion proteins, 213, 215, 216, 218

G

Galactolipases, 33
Galactolipids, 51, 71, 72, 88, 93
    biosynthesis, 82
    –galactolipid galactosyltransferase, 81, 93
    unnatural, 81
Garbus extraction, 49
Gas chromatography–mass spectroscopy (GC-MS), 2, 4, 11, 15, 130, 135, 136, 175
    of aromatic esters, 129
    of cutin monomers, 138–142
    of β-diketones, 119–122
    of fatty alcohols, 129
    of monoketones, 118, 119
    of suberin monomers, 148
    of substituted β-diketones, 122
    of triterpenoids, 130
Gas-liquid chromatography (GLC), 11, 15, 16, 31, 32, 33, 37, 41, 60, 90, 124, 129, 130
    capillary, 15, 31, 109, 112, 113, 123, 125, 136, 137, 138, 139
    of cuticular wax components, 112, 113, 119, 133, 134, 135, 136–138, 142
    of fatty acids, 4–6, 91–92
    of polyacetylenes, 166–167
    of suberin monomers, 148
Gel-liquid phase transition, 324, 326, 329, 330
    temperature, 320

Glucan-4-glucanohydrolase, *see* Cellulase
Gluconate-6-phosphate dehydrogenase, 84
Glycerol, 8, 20, 21, 28, 29, 37, 42, 43, 44, 88, 301
Glycerol galactose, 88, 89
Glycerol kinase
    in the preparation of [^{32}P]G3P, 29
Glycerol-3-phosphate, 21, 23, 24, 29, 36, 37, 38, 40, 43, 88, 89, 93, 94, 96, 97
    acylating enzymes, 41–42
Glycerophosphate acyltransferase, 23, 24, 41
    assay, 42
Glyceryl-quinovoside-6-sulphonate, 88, 89
Glycolipid
    biosynthesis, 93–101
    determination and characterisation, 85–93
    isolation of intact plastids, 72–82
    plastid envelope purification, 82–84
Granal stacks, 300, 318, 323
Grignard reagent, 34
Growth regulators
    polyacetylenes, 181

H

Hajra extraction, 85–86
Head group precursors of phospholipids, 66
    purification, 67–69
Hentriacontane-14-16-dione, 122
Heptadecanoic acid, 92
Herbicides, 109
    inhibitors of bound quinone, 277, 279, 281–282
Hexadecanedioic acid
    dihydroxy, 136
    hydroxy, 136
Hexadecane-1,16-dioic acid, 142
Hexadecanediol, 139
Hexadecanetriol, 139
Hexadecatrienoic acid
    16:3 plants, 72, 82
Hexagonal
    type I structure, 307
    type II structure, 301, 305, 306, 307, 317, 328, 329
High performance liquid chromatography (HPLC), 33
    Berthold radioactive monitor, 90
    mass detection, 62–63
    of cuticular waxes, 112
    of fatty acids, 3–4
    of glycolipids, 85, 88, 90
    of phospholipids, 51, 58–61, 62–63, 69
    of polyacetylenes, 166–167
    silver ion, 4

High performance thin layer chromatography (HPTLC), 52, 58
  of phospholipids, 54–55
pHindIII-5', 217
Hydrogenation, 12, 15, 16, 330, 333
  of cutin, 138
  of wax components, 113, 123, 130
Hydropathy indices, 319
Hydropathy plots, 318
p-Hydroxybenzaldehyde, 148
β-Hydroxybutyrate dehydrogenase, 331
Hydroxyhexadecanoic acid, 142
Hydroxymethylketone, 122
Hydroxyoxo-β-diketone, 122, 123
4-Hydroxy-25-oxo-hentriacontane-14,16-dione, 123
18-Hydroxy-7,16 hentriacontanedione, 123
Hypsochromic shift, 222, 225

I

Icosapentaenoic acid, 7
Immune immunoglobulins, 207
Immunoblotting
  of fusion proteins, 217, 219
  of phytochrome, 199, 200
Immunochemical quantitation
  of phytochrome, 203, 215
Immunostaining
  of fusion proteins, 218
  of phytochrome, 199, 201
Indenyl cation, 175
Infra red spectroscopy, 232
  fatty acids, 7
  in photosynthesis study, 252
  interferometric Fourier transform, 313
  plant membranes, 313–314
  polyacetylenes, 163, 171–172
  wax components, 113
Inositol
  head group precursor, 66
Intersystem crossing, 243, 282
Intracuticular wax, 108
  extraction and isolation, 110
Intrinsic membrane proteins, 298, 301, 316–320, 327, 329, 332
Iodine
  lipid detection, 57, 58
Iodoacetamide, 195, 197, 198
Iodohydrin, 109
Iron complexes, 248
Iron-sulphur proteins, 253, 313
Isobutylamine, 180
Isopropyl-β-D-thiogalactopyranoside (IPTG), 218

K

Kennedy pathway, 23, 24, 43
Ketols
  wax components, 118–119
β-Ketols, 112
Kinetic absorption spectroscopy, 232, 235–247
Kinetic flash absorption, *see* Kinetic absorption spectroscopy
Kornberg–Pricer pathway, 93, 94–97
Ketones
  polyacetylene, 160, 164, 175
  wax components, 118–123

L

Lactones
  polyacetylene, 160, 175, 179
Lasers
  argon pumped, 238, 242
  continuous wave (CW), 238, 242, 243, 250, 253
  diodes, 238, 242, 243
  dye, 242, 253
  excimer, 242
  ion, 253
  pulse, 267
  ruby, 242
  YAG, 239, 241, 242
Lauric acid, 21
Lecithin, 315
Lectin receptor complex, 326
Leukotriene, 6
Leupeptin, 195
Light absorption
  photosynthesis and, 233–249
Light detectors, 190
Light fluence rate, 186
Light harvesting antenna, 232
Light harvesting complex, 248, 260, 265, 267, 277, 319, 323, 332, 333
Light harvesting pigments, 267
Light harvesting chlorophyll–protein complex, 317–319
Light induced absorbance changes (LIAC), 219
  photoreceptor identification, 221–222
Light induced electron transfer, 246
Light sources
  arc lamps, 190
  fluorescent lamps, 189
  incandescent lamps, 189
Lignans, 178
Lignin, 106, 148, 149

Linear dichroism, 247–248, 249
Linoleate, 4, 8, 10, 21, 23, 24, 39, 42, 179
α-Linolenate, 4, 8, 10, 16, 21, 22, 23, 24, 39, 41, 92
   18:3 plants, 72, 82
γ-Linolenate, 4, 22, 39, 41
Linolenate isomers, 6
Linoleoyl-CoA, 42
Linoleoylphosphatidylcholine, 39
Linoleoylpyrrolidide, 14
Lipase, 90, 91, 299, 301
   pancreatic, 144
Lipid extraction, 29–30
   Bligh and Dyer, 30, 37, 49, 50, 85, 98, 101
   cuticular wax, 109–135
   garbus, 49
   glycolipids, 85–86
   lipid from suberin, 146–148
   lipid from whole tissue, 30
   lipid from microsomal membrane, 30
   phospholipase digest, 34
Lipid mutants, 55, 108, 152
Lipid–protein interaction, 331–334
Liquid crystal bilayer, 301, 312, 326
Liquid scintillation cocktails, 35–36
Low fluence, 214
Low temperature differential spectroscopy, 235
Luminescence 262
lyso-Monogalactosyldiacylglycerol, 91
lyso-Phosphatidic acid, 23, 24, 30, 34, 41, 42, 89, 93, 94, 95, 97
   extraction, 85
   preparation of, 96
lyso-Phosphatidylcholine, 37, 38, 39
lyso-Phosphatidylcholine acyltransferase (LPCAT), 24, 38, 39
   assay, 37–39

M

Magnetic circular dichroism (MCD), 245
Malonyl-CoA, 181
Mass spectrometry (MS)
   chemical ionization (CIMS), 11, 123, 138
   cuticular wax components, 112, 120–121, 122, 140–141
   electron impact (EIMS), 11, 113, 118, 119, 123, 125, 138
   estolides, 126, 127
   fatty acids, 10–15
   field desorption, 125, 134
   polyacetylenes, 163, 175
   tandem, 123, 124
Matricarialactones, 176, 177
Matricarianol, 180
Measuring light, 237
Mehler reaction, 292
Membranes, plant
   dynamics, 320–334
   isolation, 298–300
   structure, 300–320
Membrane fluidity, 320–326
Membrane proteins, 28
   structure, 316–319, 324
Membrane sidedness studies
   covalent bonding of probes, 63, 64
   enzymic alteration of phospholipids, 63, 64
   transfer with exchange protein, 63, 64–66
Methyl ketones, 122
Methyl-9,12,13(S)-trihydroxyoctadec-10-enoates, 3
Micelle, 28, 308, 328
   inverted, 311
Microsecond spectroscopy, 244
Microsomal membranes, 23, 25, 29, 32, 37, 38, 39, 40, 42, 299, 323
   lipid extraction, 30
   preparation, 26–27
Microspectrofluorometer, 221
Microspectrophotometry, 249
Millisecond spectroscopy, 24
Mitochondria, 299, 317, 323
Modulated fluorimeter, 282, 283–285
Molar extinction spectra
   of phytochrome, 212, 213
Molecular sieve chromatography, 113
Monoacylglycerol, 94
Monochromatic light, 187, 189
Monoclonal antibodies
   to phytochrome, 201, 203, 207, 213, 215, 218
Monogalactosyldiacylglycerol (MGDG), 55, 71, 73, 81, 82, 83, 87, 89, 91
   structure, 73
   synthesis, 93
   synthetase, 94
   synthetase assay, 97, 100
   synthetase partial purification, 98–101
Monohexosyldiacylglycerol, 302
Monohydroxyhexadecane-1,16-dioic acid, 142
Monoketones
   wax components, 118
Monolayer membrane, 302–304
Monomethylethanolamine, 67
Myelin, 301, 315, 327
Myristic acid, 21

N

NADH–ubiquinone reductase, 317

Nanosecond spectroscopy, 241–244
1-Naphthol, 86, 88
Naphthalene, 113
Neutron scattering, 300
Neutral solvent, *see* Solvent systems
Nigericin, 271, 287, 288
Nitrate reductase, 223
Nitroxide labelled fatty acids, 331
Non-appressed lamellae, 300, 326
Non-immune immunoglobulins, 207, 218
Northern blot, 214
Nuclear magnetic resonance (NMR) spectroscopy
   deuterium, 312, 314
   ^{31}P-NMR, 312, 313, 328
   pulsed field gradient, 311
   plant membranes, 310–313, 332
   two dimensional, 10
^{13}C-NMR:
   of cutin monomers, 144
   of fatty acids, 8–10, 11
   of suberin, 149
   of substituted β-diketones, 122, 123
   of triterpenoids, 130
^{1}H-NMR:
   of fatty acids, 7–8
   of triterpenoids, 130
Nuclease, 226

O

Octacosonol, 134
Octadecandiendiol, 139
Octadecanediol, 139
Octadecanetetraol, 139, 142
Octadecanetriol, 139
Octadecenetetraol, 139, 142
Oil body, 22
   preparation, 27
Oleate, 12, 23, 37, 38, 39, 40, 41, 181
Olefinic alkaloids, 180
Oleoyl-ACP, 96
Oleoyl-CoA, 24, 37, 38, 39, 40, 44, 96
Oleoyl-glycerol-3-phosphate, 96
Oleoylphosphatidyl choline, 39
Oleoylpyrrolidide, 14
Ontogeny, 144–145
Optical cryostats, 237
Optically detected magnetic resonance (ODMR), 251
Optical equipment
   for photoreceptor study, 187, 189–191
Optical rotation, 175
Ovalbumin, 207

Oxidation
   acid dichromate, 129
   cutin monomers, 144
   iodoform, 122
   permanganate-periodate, 113
   polyacetylene alcohols, 167, 171, 172
   suberin, 148, 149
Oxidative cleavage, 12, 15, 16
Oxime derivatisation, 135
Oxoalkanoic acid, 122
Oxo-β-diketones, 122
Oxohentriacontane-14,16-dione, 122
Oxohydroxyhexadecanoic acids, 145
Oxo-β-ketol, 122, 123
7-Oxopentadecan-2-ol, 124
Oxygen electrode, 75, 283
Oxygen evolving enzyme, 243, 245, 276
Oxygen evolution assay, 283
Ozone, 108
Ozonation, 16

P

Palmitate, 8, 21, 42
Palmitoyl-ACP, 96
Palmitoyl-CoA, 42
Palmitoyl-*lyso*-phosphatidylcholine, 38, 44
Paper chromatography, 67
Parinaric acid, 6
pAVA I, 218
Pectinase, 107, 108, 110, 299
Pectin glucosidase—*see* Pectinase
Pectolase, 79
Periderm, 106, 148
Percoll, 72, 74, 81, 82
   gradient, 76, 77
Petroselinate, 12
Phase behaviour of plant membranes, 301–316
Phase separation of membranes, 326–331
Phase transition
   of cuticular membranes, 108
Phenanthrene, 113
Phellem, 106, 145
Phenolic aldehydes, 148
Phenoxyacetic acid, 109
2-Phenylethyl alcohol, 129
Phenylethylamines, 180
1-Phenylhepta-1,3-diyn-5-ene, 166, 179, 180
   ^{1}H-NMR spectrum, 174
   structure, 163
1-Phenylhepta-3,5-diyn-1-ene, 173
   ^{1}H-NMR spectrum, 174
1-Phenylhepta-1,3,5-triyne (PHT), 166, 173, 180

1-Phenylhepta-1,3,5-triyne (PHT) (*cont.*)
   ^{1}H-NMR spectrum, 174
   structure, 163
Phenylmethylsulphonyl fluoride (PMSF), 195, 196, 299
Pheophytin, 235, 262
   bacterial, 240, 242, 247
   in PSII, 242
Phycobilins, 147
Phycobiliproteins, 233, 239, 249
Phycobilisomes, 239
Phylloquinone, 245
Phytochrome, 186, 187
   assay, 194–195
   DNA photolyase, 223–227
   isolation and purification, 193–199
   molecular biology, 213, 219
   molecular characterisation, 193–208
   phototransformation, 189, 193, 208–213, 214
   UV/blue light photoreceptors, 219–223
Phytosterols, 130
Phosphatidate phosphohydrolase (PAP), 23, 24, 29
   assay, 42–43
   EDTA inhibition, 42
Phosphatidic acid, 24, 29, 31, 41, 42, 44, 49, 50, 83, 88, 89, 93, 94, 95, 97
   as precursor, 66
   stereospecific analysis, 34
   structure, 21, 48
Phosphatidyl choline, 29, 30, 31, 35, 37, 38, 39, 40, 41, 43, 44, 54, 84, 87, 95, 99, 302, 315, 331
   ESR spectrum of spin labelled PC, 333
   HPLC separation, 60, 61
   in TAG synthesis, 23–24
   stereospecific analysis, 34
   structure, 21, 48
Phosphatidyl ethanolamine, 30, 43, 54, 302
   HPLC separation, 60, 61
   inner mitchondrial membrane marker, 81
   probe for, 64
   structure, 48
Phosphatidyl glycerol, 99, 100, 302, 331, 332
   ESR spectrum of spin labelled PG, 333
   HPLC separation, 60, 61
   structure, 48
Phosphatidyl inositol, 51, 54, 59, 87, 302
   structure, 48
bis-Phosphatidyl glycerol (cardiolipin), 95
   inner mitochondrial membrane marker, 81
   structure, 48
Phosphatidyl methanol, 30
Phosphatidyl serine, 59, 302, 317
   probe for, 64

Phosphatidyl serine (*cont.*)
   structure, 48
Phosphodiesterase, 226
Phosphogluconate dehydrogenase, 81
Phospholipase, 49
   A, C, +D, 49
   A_2, 34, 35, 41, 42, 64, 328, 332
   inhibition, 50
   as probes, 64, 65
Phospholipids, 21, 23, 34, 41, 47–69
   chromatography of, 51–63
   extraction, 49–50
   membrane sidedness, 63–66
   precursors, 66–69, 89
   separation, 50
   tissue selection, 49
   transfer proteins (PLTP), 64, 66
Phosphorescence, 192, 249, 250
Photoacoustic spectroscopy, 249
Photoactivation of nitrate reductase, 223
Photoactive reaction centres, 232, 235
Photochemical hole turning, 235
Photon counting, 250, 251
Photodegradation, 189
Photodetectors, 238, 242
Photoequilibrium of phytochrome, 208–212
Photoinhibition, 265
Photomorphogenesis, 186
Photonasty, 186
Photon fluence rate, 188, 191, 192, 193, 209, 210, 225
Photoprotection, polyacetylenes and, 178, 181
Photoreceptors, *see* Phytochrome
Photorepair, 186, 189
Photosynthesis, optical techniques in, 231, 232
   light absorption, 233–249
   light emission, 249–251
   vibrational spectroscopy, 251–254
Photosystem I (PSI), 234, 242, 244, 260, 265, 269, 274, 276, 292, 319
   fluorescence emission, 260
Photosystem II (PSII), 235, 244, 245, 250, 251, 261, 263, 264, 265, 266, 267, 268, 269, 273, 274, 275, 276, 279, 283, 288, 289, 292, 332
   connectivity between centres, 278
   electron transfer in, 236
   fluorescence emission, 260
   heterogeneity, 277–278
   manganese complex in, 234, 243
   membranes, 240
   secondary electron donation in, 243
Photosuppression, 222
Phototaxis, 186, 222
Phototropism, 186

Picolinyl esters, 14, 15
Picosecond spectroscopy, 238–239, 241, 246
Pigment-protein complexes, 240, 243, 247, 248, 251, 252, 260, 263
Piperideines, 180
Piperidines, 180
Planck's equation, 188
Plasmalemma, 106
Plasma membrane, 297
Plasmid, 218
Plastid, 23, 25, 72, 90
   isolation of, 72–82
   intactness, 81
   envelope membrane purification, 82, 84
Plastocyanin, 233, 244
Plastoglobules, 83
Plastoquinone, 232, 271, 272, 274, 275, 276, 280, 281, 324
   quenching by, 266
pNciI-5', 218
Pockels cell, 237, 242
Polyacetylenes, 160
   analysis, 161–178
   biosynthesis, 181
   biological activities, 181
   extraction, 161
   group differences and similarities, 179–181
Poly(A)$^+$-RNA, 215
Polar solvent, *see* Solvent systems
Polychromatic light, 187
Polyclonal antibodies
   to phytochrome, 207, 218, 219
Polyenic alkamides, 180, 181
Poly-α-1,4-galacturonide:glycanhydrolase, *see* Pectinase
Polyhydroxy esters, 3
Polymer structure of cutin, 142–144
Polymorphism of lipids, 301, 309, 312
Polyol, 14
Poly(trimethylsilyl)ether, 14
Poly(vinylpyrrolidine), 72
Pontica epoxide, 177
PPO-POPOP, 36
PpuMI, 218
Primary alcohol wax esters, 123
Primary electron donor of PSI (P_{700}), 234, 244, 283, 292
Primary electron donor of PSII (P_{680}), 235, 243, 251, 260, 262, 264, 266, 268, 272
Primary photochemical acts, 240
Probes, 63–66
Protein phosphatase, 265, 287
Proteolysis
   of phytochrome, 203, 218
Proteolytic enzymes, 81, 193, 197, 299
   inhibitors of, 84, 195, 201, 205, 207, 299

Proton gradient, 264, 266, 286, 288
Protoplasts, 79, 80, 264, 288, 299, 323, 326, 328, 330
Pterin, 187
Pyrene polycyclic hydrocarbons, 113
Pyridine dimers, 224, 226
Pyrrolideines, 180

Q

Quadrupolar splitting of deuterium, 311
Quantum efficiency, *see* Quantum yield
Quantum yield, 192, 235, 237, 238, 243, 262, 265, 266, 289–292
   far red phytochrome (Pfr), 208–211
   photosynthetic electron transport, 290
Quenching
   analysis, 283, 289–292
   by oxidised plastoquinone, 266
   light induced, 267
   non-photochemical, 263–264, 265, 266
   photochemical, 263
   proton gradient and, 264
   thermal, 309
   under light stress, 265–266
Quinone, 233, 244, 245
   bound (Q_B), 234, 245, 251
   equilibrium constant measurement, 280–281
   primary (Q_A), 235, 240, 243, 244, 245, 251, 262, 263, 268, 271–272, 274, 275, 276, 278, 279, 281–282, 291
   secondary, 263, 264, 268, 269, 271, 272–274, 276

R

Radiationless de-excitation, 192
Radioactivity determination, 35–37
   quantification, 57
   TLC plates, 55–57
Radio-GC, 32, 34, 36, 37, 40, 92
Radioimmunoassay, 207
Radiometric units, 187–188
Raman spectroscopy, 7, 232
   photosynthesis study, 252, 253–254
   plant membrane, 313
Reaction centre, 244, 245, 260
   bacterial, 248, 249, 253
   electron acceptors, 240
   primary biradical properties, 242–243
Reaction yield detected magnetic resonance, 249
Recording devices
   for kinetic absorption spectroscopy, 238

Rectangular phases, 307
Reflectance spectroscopy, 249
Restriction endonucleases, 215, 216, 217, 218, 219
Retinal rods, 301, 316
Rhodamine, 326
Rhodopsin, 187, 222, 252, 254
Rhombohedric phases, 307
mRNA, 214, 215
Rate constant of phototransformation, 192
Roseoflavin, 222

S

Sandwich immunoassay, see ELISA
Scanning densitometry, 58
Scintillation cocktails, 35
   biodegradable, 57
SDS-PAGE
   of fusion proteins, 218
   of phytochrome, 199–201
Secondary alcohols, 129
Secondary electron acceptor
   of PSI, 245
   of PSII, 245
Semicarbazone, 135
Semiquinone, 262
Senescence, 311
Serine
   head group precursor, 66–68
Sharp reflection, 307
Sieve effect, 234
Silver nitrate-TLC, 3, 15, 16, 35, 36, 40, 60, 63, 90, 92–93
   of polyacetylenes, 164
   of wax components, 113–130, 136
   preparation of plates, 32–33
Singlet oxygen, 244
Slot blot, 214
Sodium borodeuteride, 135
Solvent systems
   Bligh and Dyer extraction, 30, 49, 50, 85
   elution of phospholipids, 34
   extraction of epicuticular wax, 109–110
   extraction of phospholipase digest, 34
   Garbus, 49
   glycolipid fatty acid analysis, 90
   glycolipid separation, 90
   Hajra extraction, 85–86
   lipid extaction from whole tissue, 30
   neutral, 31
   phospholipid head group purification, 67
   phospholipid separation, 52–53
   phospholipid separation by HPLC, 59, 60
   polar, 31

Solvent systems (cont.)
   purification of acyl-CoA, 33
   purification of MGDG synthase, 99
   separation of DAG isomers, 35
   separation of FAMEs, 33
   TAG stereospecific analysis, 35
Spark imaging system, 57
Spectra characterisation of phytochrome, 201–203
Spectrophotometric quantitation of phytochrome, 203
Spheradene, 253
Spin probes, 321, 323, 324, 328, 333
Spin–spin relaxation, 249
Spiroenolethers, 175, 177
   structure, 176
Sporangiophore, 186, 222
Stark effect, 248
State transition, 264–265
Steady state absorption spectroscopy
   differential, 234–235
   optical perturbation, 234
Steady state fluorescence depolarisation studies, 323
Stereospecific analysis, 34–35
Sterols, 88
   of epicuticular wax, 132
Stress induced membrane disassembly, 331
Stromal lamellae, 300, 323
Suberin, 106, 136, 139
   analysis, 145–149
   components of, 150–153
Sulpholipid, 71, 72, 87, 88, 90, 91, 93, 94, 95, 302
   extraction, 85
   structure, 73
Sulphoquinovosyldiacylglycerol (SQDG), 332
Subcellular membranes, 298
Surface pressure-area isotherm, 302
Syringaldehyde, 148

T

Terpenoids, 130–131
α-Terthienyl, 160, 166, 179
trans-2-Tetracosenoic acid, 124
Tetragalactosyldiacylglycerol, (TETRA-GDG), 81, 87, 89, 98, 99
Tetramethyluric acid, 164
Thermoluminescence, 249, 251
Thermolysin, 81–82, 87
Thin layer chromatography (TLC), 3, 29, 31, 33, 34, 38, 64, 67, 129
   arsenites in, 3
   autoradiography, 56, 91, 92, 93, 95, 96

Thin layer chromatography (TLC) (*cont.*)
  biethylbutynene, 175
  borates in, 3
  boric acid in, 136
  caffeine in, 163–165
  cuticular waxes, 110, 112, 133
  cutin monomer derivatives, 136
  dehydromatricariaester isomers, 176
  fatty acids, 90–91, 92–93
  glycolipids, 85, 86–88, 95
  identification with hydrolysis, 58
  ketols, 118, 119
  modification of silica gel layer, 55
  one dimensional, 53, 54
  paraffin impregnation of plates, 93
  phospholipids, 51, 52–54
  plate activation, 52
  polyacetylenes, 162, 163–165
  radioactive detection, 55–57
  reverse phase, 112, 130
  spray reagents for lipid detection, 57–58
  tetramethyluric acid in, 164
  two dimensional, 53, 54, 85, 86
Thiohydrolases, 28
Thiophene, 163, 164, 166, 173, 175
  formation, 172
Thylakoid, *see under* Chloroplast
Thymine dimer, 224, 226, 227
TMS enol ethers
  of β-diols, 122, 123
  of triterpenoids, 130
Transcription
  nuclear run on, 215
  phytochrome control of, 214–215
Transfer protein, *see* Exchange protein
Trans/*gauche* isomerisation, 313
Transient electron acceptor, 262, 264, 266, 276
Translational diffusion coefficient, 310, 312
Transmethylation, 32, 37, 135
  mixture, 92
Triacontanol, 134
Triacylglycerol, 51
  assay of enzymic steps, 37–44
  experimental material, 25–27
  extraction, purification, and quantitative analysis, 29–33
  radioactivity determination, 35–37
  stereospecific analysis, 34–35
  structure and biosynthesis, 20–25
  substrates, 27–29
Trideca-1,11-dien-3,5,7,9-tetrayne, 178
Trideca-3,5,7,9,11-pentyn-1-ene, 167, 178
Trifluoroacetate esters, 130, 147
Trigalactosyldiacylglycerol (TRI-GDG), 81, 87, 89, 98, 99

9,10,18-Trihydroxyoctadecanoate, 138, 142
Trimethylsilylation, 112, 136
  of fatty alcohols, 129
  of ketols, 119
Trimethylsilyl ester, 11, 13
Trinitrobenzenesulphonic acid, (TNBS), 64
Triolein, 44
Triplet–triplet energy transfer, 243–244
Tris(triphenylphosphine) chloro-rhodium (I), 124
Triterpenoic acid, 130, 131, 132
Triterpenoids, 109, 130, 132–133
  esters, 112
  structures, 131
Triterpenols, 130, 131, 132
Triterpenones, 130, 133
Tyrozine$_z$, 235, 268

U

UDP-galactose, 93, 94, 96, 97, 98, 99, 101
1,2 diacylglycerol 3-β-D-galactosyl-transferase, *see* MGDG synthase
Ultraviolet (UV)
  /blue light receptors, 186, 187, 219–223
  endonuclease, 226
  irradiation of polyacetylenes, 175–177
  spectroscopy of fatty acids, 6–7
Unappressed stromal lamellae, 265
3-Undecyl-2-hydroxy-1,4-naphthoquinone (UHNQ), 273, 274
Urea clathration, 113

V

Valinomycin, 271
Vanallin, 148, 162
Very low fluence (VLF), 214, 215
Vibrational spectroscopy
  in photosynthesis study, 251–254
Violaxathin, 264, 266

W

Walz fluorimeter, 275, 282, 283
Water-oxidising enzyme, 233, 268, 277, 280
Water vapour loss, 106, 108
Wax esters, 112, 147
  identification, 123–129
Wax mutant, *see* Lipid mutants
Waxes
  analysis of cuticular wax, 109–135
  analysis of cutin, 135–145

Waxes (*cont.*)
   analysis of suberin, 145–153
   cuticular membrane, 106–109
Wettability of leaf surface, 109

X

Xanthophyl cycle, 264
Xenon flash ignition, 237, 242, 244
X-ray crystallography, 240, 248, 251, 319
X-ray diffraction
   of cutin, 144

X-ray diffraction (*cont.*)
   of membrane proteins, 316, 319
   of plant membranes, 300, 301, 304–307, 310, 325, 328, 330

Z

Z to E form conversion
   of polyacetylenes, 175–176
Zeaxanthin, 264, 266
Zn^{2+} visualisation of phytochrome, 201